Wildlife Conservation on Farmland

Wildlife Conservation on Farmland

Volume 1

Managing for Nature on Lowland Farms

EDITED BY

David W. Macdonald

and

Ruth E. Feber

OXFORD
UNIVERSITY PRESS

Wildlife Conservation on Farmland. Managing for Nature on Lowland Farms. Edited by David W. Macdonald and Ruth E. Feber. © Oxford University Press 2015. Published 2015 by Oxford University Press.

OXFORD
UNIVERSITY PRESS

Great Clarendon Street, Oxford, OX2 6DP,
United Kingdom

Oxford University Press is a department of the University of Oxford.
It furthers the University's objective of excellence in research, scholarship,
and education by publishing worldwide. Oxford is a registered trade mark of
Oxford University Press in the UK and in certain other countries

First Edition published in 2015

Impression: 1

Published in the United States of America by Oxford University Press
198 Madison Avenue, New York, NY 10016, United States of America

British Library Cataloguing in Publication Data
Data available

Library of Congress Control Number: 2015935287

ISBN 978–0–19–874548–8

Printed in Great Britain by
Clays Ltd, St Ives plc

Preface

Wildlife Conservation on Farmland. Volume 1. Managing for Nature on Lowland Farms

This book and her sister volume, *Wildlife Conservation on Farmland: Conflict in the Countryside*, offer perspectives on the same landscape, but viewed from different vantage points. Both books consider farmland, as the bulk, the matrix, and the engine of the countryside, and both address the detailed science that informs a big societal choice, namely what does society want from farmers and farmland and, considering the diversity of conflicting pressures on the farmscape, how do we make the best of it? Indeed, what do we, as citizens and as a nation, think is 'the best of it'? But the perspectives of these books, while from different hilltops, are through the same eyes. That is, while this volume, *Wildlife Conservation on Farmland: Managing for Nature on Lowland Farms*, focuses on understanding Nature (and all that affects her) with the purpose of fortifying the evidence base for managing outcomes, and while *Wildlife Conservation on Farmland: Conflict in the Countryside* focuses on understanding the very same Nature with the purpose of untangling and mitigating the tensions between people and wildlife in the countryside, both focus through a particular set of eyes, those of the broadly based team of Oxford University's Wildlife Conservation Research Unit—the WildCRU—and the colleagues and collaborators that comprise its extended family.

What journey, personal and professional, led us to pose the questions these books seek to answer? For one of us, the ten-year-old DWM, the roots of this enterprise may have lain in that first week's work on my Uncle Iain's sheep farm, Thornihill Farm, at Colvend in Dumfriesshire—a week spent learning to hand-milk Mary the house cow, and to rescue upturned ewes from the eye-pecking attentions of carrion crows (and that first ever wage packet containing, as I recall, a ten shilling note). Those roots later may have been nourished in the hours spent drinking welcome cups of tea while debating the love of, and conflict with, nature, in the kitchens of farms across which I followed the movements of some of the first foxes in Europe to be radio-tracked. But professionally, the roots of our work certainly lie in the conversations with many farmers in the early 1980s when DWM asked each of them to list, and explain, the issues with wildlife that most impacted their farming and which, if solved, would facilitate a better compromise between farming and wildlife—conversations that put the onus on farmers and conservationists to work together rather than in opposition, and led directly in 1986 to the Wytham field margins experiment, near Oxford. This was then the biggest farm-scale, controlled and replicated experiment in farmland conservation, and, subsequently, the framework of REF's doctoral thesis on butterflies of field margins, and the beginning of her captivation with farmland wildlife.

That work was a first step in much of the journey related here. Over the past 30 years, our work has broadened in scale as thinking about agro-ecology has developed: the field margins experiment manipulated patches of habitat around fields; later, for example, on the farms around the Royal Agricultural College (now University), we magnified the scale to whole-field comparisons; in our work on farming systems in the 1990s and 2000s, we contrasted both fields and whole farms (under organic or conventional management); and, most recently, we have focused our attention on the landscape scale—first on the Chichester coastal plain and latterly, with Wytham once again our hub, in the Upper Thames region, with projects on organisms as varied as bats, toads, hedgehogs, and dragonflies, and habitats from ditches and ponds to rivers, set in farmland, hedgerow, and woodland.

That this book is explicitly a WildCRU perspective is an important point, because, if readers do not grasp it, they may be puzzled by the selection and topics of the chapters that follow. Let us be clear what these books are not. Neither of these books is comprehensive, but both are compendious. Neither the books, nor the chapters within them, individually or in aggregate, pretend to be a complete, or systematically impartial review of a whole literature (although they certainly aspire to be fully informed of the salient realities that create the framework for understanding each topic). Rather, each chapter summarizes, synthesizes, and builds upon the research that the WildCRU team (and its antecedent the Oxford Foxlot) has undertaken on each topic between 1972 and 2015, with the hope that our findings shed light that illuminates the wider subject, advances understanding, and fortifies the evidence base for policy and action. However, while we have set out to write with authority about what we have done, and to link that with a wider literature as is necessary or fruitful, we have not set out to review that wider literature with an even hand. Of field experiments, we report on ours at Wytham, rather than on all similar experiments everywhere: of mustelids we report on badgers, rather than on pine martens, of raptors on buzzards rather than on red kites; not because we think or imply the one is more important than the other, but because it is what we have chosen (we believe, with good reason) to work on, and thus to report in these syntheses.

Mention of syntheses raises another point which may affect the reader's understanding, and thus enjoyment, of these two companion volumes. Not only is it plain to all that 'you can't do everything', but anybody with even a glancing acquaintance with the process of science will know that a hurly-burly of scholarship, expediency, urgency, practicality, funding, and luck dictates that sliver of 'everything' which one actually does at a given time, and indeed the way in which these slivers assemble over time. Thus, over the years, as the WildCRU team worked on water voles or bats or toads, or on field margins or ponds or hedgerows, some onlookers may have presumed that we were flitting, much like the gadflies we have also studied, between worthwhile but largely unrelated topics. Nothing could be further from the truth, and one purpose in writing these syntheses has been to demonstrate that. From the start, and growing organically and logically as the opportunity has allowed, our purpose has been to link projects, themes, and principles. Consider how, for example, the conservation of water voles can be effective only if it integrates a number of linked themes—local habitat management (riverbanks), landscape scale (reducing isolation of populations), control of invasive species (mink), and health and welfare of reintroduced water voles. So it is that you will find in these volumes that the factors which affect the movements of hedgehogs and of toads, in the contexts of linear habitats, fragmentation, and dispersal, are insightful to, and illustrative of, similar ecological principles, and have similar policy relevance. You will read of how the same principle—we call it perturbation—affects the transmission of infectious disease among both badgers and foxes, and makes relevant the study of the societies of both. Indeed, although only touched on in these books (which concentrate on the UK), these same linkages create the weave and waft of the entire, international, WildCRU research agenda, applying the same principles to the corridors through which moths move in the Oxfordshire countryside[1] to the movements of lions in Zimbabwe[2] or tigers in the Western Ghats[3], just as the principles governing the group-formation of badgers in Wytham (Chapter 4, Volume 2) seem to us to be aligned with those facilitating social groups of giant otters in Peru[4]. On this topic of common threads, readers will find cross-references herein to other shared principles, such as the importance of resource dispersion[5], or of economics and substitutability[6, 7], or payments for ecosystem services[8], or

[1] Merckx, T., Marini, L., Feber, R.E., and Macdonald, D.W. (2012). Hedgerow trees and extended field margins enhance macro-moth diversity: implications for management. *Journal of Applied Ecology*, **49**, 1396–1404.

[2] Elliot, N., Cushman, S., Macdonald, D.W., and Loveridge, A. (2014). The devil is in the dispersers: the metrics of landscape connectivity change with demography. *Journal of Applied Ecology*, **51**, 1169–1178.

[3] Gopalaswamy, A.M., Mohan, D., Karanth, K.U., Kumar, N.S., and Macdonald, D.W. (2015). An examination of index-calibration experiments: counting tigers at macroecological scales. *Methods in Ecology and Evolution*. doi: 10.1111/2041-210X.12351.

[4] Groenendijk, J., Hajek, F., Schenck, C., Staïb, E., Johnson, P.J., and Macdonald, D.W. (2015). Effects of territory size on the reproductive success and social system of the giant otter, south-eastern Peru. *Journal of Zoology*. doi: 10.1111/jzo.12231.

[5] Macdonald, D.W. and Johnson, D.D.P. (2015). Patchwork planet: the resource dispersion hypothesis, society and the ecology of life. *Journal of Zoology*, **295**, 75–107.

[6] De Barros, A.E., Macdonald, E.A., Matsumoto, M.H., Paula, R.C., Nijhawan, S., Malhi, Y., and Macdonald, D.W. (2013). Identification of areas in Brazil that optimize conservation of forest carbon, jaguars, and biodiversity. *Conservation Biology*, **28**, 580–593.

[7] Dutton, A.J., Hepburn, C., and Macdonald, D.W. (2011). A stated preference investigation into the Chinese demand for farmed vs. wild bear bile. *PLOS ONE* 6: e21243. doi:10.1371/journal.pone.0021243.

[8] Dickman, A.J., Macdonald, E.A., and Macdonald, D.W. (2011). A review of financial instruments to pay for predator conservation and encourage human-carnivore coexistence. *Proceedings of the National Academy of Sciences of the United States of America*, **108**, 13937–13944.

culture[9, 10], and indeed wider inter-disciplinarity[11, 12], all wrapped in the fabric of inconvenient truths and the elephants in the room that reveal conservation to be inherently about biology, but ultimately a branch of politics[13].

This mention of resonance, indeed synergy, between WildCRU's work on the farmland of lowland Britain, and our wider work, prompts the question of to whom this book might be useful. Much of what we have written is in the form of case studies and it goes without saying that we expect aficionados of water voles or damselflies to find interest in our chapters on these organisms, and it is a modest hope that enthusiasts of riparian or pond conservation would also find merit in these chapters. However, we hope for greater reach. Not only do we intend these cases to illustrate lessons that can be applied to different species, but also to different landscapes, and even different countries. Many of the studies have been designed to focus on broad ecological principles, with the aim that, whatever the direction of agricultural change and vagaries of agricultural policies, they will be a valuable source of information upon which policy makers and practitioners can draw. Our examples are drawn largely from the lowlands, and surely there are aspects of upland ecology even within Britain on which we do not touch at all. However, not only do we have some upland case histories—the affairs of grey and red squirrels in Perthshire and the rewilding activities of wild boar in Sutherland come to mind, alongside the beavers of Knapdale—but our lack of focus on peatbogs or heather moors does not disqualify most of our material from nationwide relevance. Indeed, it seems to us that our case studies apply directly and in detail and in principle to landscapes across much of continental Europe and indeed throughout the temperate world.

Further, we hope our chapters will be read from Patagonia to Parak by conservationists and policy makers seeking insight into the issues of ecosystem management and policy formulation that are common themes globally[14, 15].

Turning to the specifics, the chapters in this volume, *Farming and Wildlife: Managing for Nature on Lowland Farms*, describe WildCRU's studies of the impacts of different aspects of farmland habitat management on key species groups, conducted at a range of spatial scales, from within-field management to landscape-level management. Much of the information gained from these studies has been pivotal in the development of current day agri-environment schemes. The introductory chapter offers a perspective on the issues that we think are relevant, with reference to agriculture and biodiversity loss, ecological theory, and valuation. The first two case studies, Chapters 2 and 3, summarize results from our Wytham field margins experiment which is, as far as we know, unparalleled in its longevity, providing biological and policy-relevant insights into plant and animal community development on arable field margins—themselves among the sinews of the farmscape. Chapter 4 reveals the effects of different farming practices of both within-field and uncropped areas, on small mammals, from common species such as wood mice, to apparently declining species such as the harvest mouse; while Chapter 5 offers a viewpoint on the wider impacts of agri-environment schemes on farmland birds and the future for bird conservation in the agri-environment era. Much farming practice is carried out at a field level, for example, the type of crop that is sown in a field, whether or not a headland is sprayed, or how a hedgerow is managed. But farm practice at the level of the farming system, such as whether a farm is managed as a high or low intensity enterprise, will have other implications for biodiversity. This is the focus of Chapter 6, which synthesizes the results of our comparative studies of the biodiversity of organic and conventional farms in England; while Chapter 7 uses a modelling approach to reveal the complexities of the wider environmental impacts of different farming systems.

The spatial scale at which many organisms function is greater than that of both the field and the farm.

[9] Marchini, S. and Macdonald, D.W. (2012). Predicting ranchers' intention to kill jaguars: case studies in Amazonia and Pantanal. *Biological Conservation*, **147**, 213–221.

[10] Meena, V., Macdonald, D.W., and Montgomery, R. (2014). Managing success: Asiatic lion conservation, interface problems and peoples' perceptions in the Gir Protected Area. *Biological Conservation*, **174**, 120–126.

[11] Macdonald, D.W. (2013). From ethology to biodiversity: case studies of wildlife conservation. *Nova Acta Leopoldina*, **111**:380, 111–156.

[12] Macdonald, D.W., Loveridge, A.J., and Rabinowitz, A. (2010). Felid futures: crossing disciplines, borders and generations. In: *The Biology and Conservation of Wild Felids*, pp. 599–649. Oxford University Press, Oxford.

[13] Macdonald, D.W. and Willis, K.J. (2013). Elephants in the room: tough choices for a maturing discipline. In D.W. Macdonald and K. Willis, eds., *Key Topics in Conservation Biology 2*, pp. 469–494. Wiley-Blackwell, John Wiley & Sons Ltd., Oxford.

[14] Macdonald, D.W., Collins, N.M., and Wrangham, R. (2006). Principles, practice and priorities: the quest for alignment. In D. Macdonald and K. Service, eds., *Key Topics in Conservation Biology*, pp. 273–292. Blackwell Publishing, Oxford.

[15] Dickman, A.J., Hinks, A.E., Macdonald, E.A., Burnham, D., and Macdonald, D.W. (2015). Priorities for global felid conservation. *Conservation Biology*, **29**, 854–864.

While conservation measures taken at a field or farm scale will have impacts on some species, for others, impacts may be different, or only detectable, at the landscape scale. Chapters 8 to 10 describe the responses of key species groups (moths, bats, and damselflies and dragonflies) to different aspects of farmland management at both local and landscape scales in the Upper Thames region, and investigate the mechanisms underlying these responses. Chapters 11 and 12 focus on the management for biodiversity of key habitats on farmland—waterbodies (rivers and ditches) and woodlands—vital linear and stepping stone habitats that permeate and punctuate the farmscape, providing resources for a range of species.

Our landscape-scale work has provided an ideal framework for testing hypotheses relating to species restoration. When the species in question is a game bird, survival of reintroduced individuals is of particular economic importance: factors surrounding the behaviour and survival of reintroduced grey partridge are investigated in Chapter 13. Chapter 14 describes an ambitious restoration programme for the charismatic and dramatically declining water vole (covering a topic often neglected in reintroduction research—the health and welfare of the animals). In the penultimate chapter, Chapter 15, we use the water vole as a vivid example of the journey to conserve a species, from the discovery of its decline and the diagnosis of the causes, to the testing and implementation of strategies for its recovery; we argue that this journey holds many lessons for species far beyond the riverbank.

Sensing the mood of the moment, this first volume of the duo concludes by asking, in Chapter 16, 'Where next'? and exploring one of the possibilities for a radical future for the British countryside—rewilding. We, the editors, are children of the post-war agricultural miracle that has seen productivity burgeon, revolutionizing food security, but at the calamitous expense of the near death of the countryside. Then society began to change its mind, and to seek ways of nature cohabiting with the industry of agriculture—much of the last 30 years, including many of the studies in this book, was dedicated to holding the line, slowing and ideally halting further degradation and loss. Now may be the moment to reverse the retreat, and advance towards a new, imaginative, and enriched countryside and, as emblems of this aspiration, Chapter 16 considers whether the restoration of certain mammals could restore key processes and lead to a more self-sustaining ecosystem.

As we draw this project to an end we are awed by the number of people who have helped to bring it to fruition. This has been a long and arduous process and so we thank first and fervently our partners, Dawn and Guy, who have supported us throughout, keeping us afloat as we navigated steadfastly to our purpose through the choppiest months. It's a lofty aspiration, but we hope that the findings in these books will help lead to a more richly biodiverse, beautiful, but nonetheless functional, countryside for our children.

Each of the chapters is written by a small subset of the people who undertook the research, and we thank so gratefully the authorial teams, and the community of researchers who contributed to the underlying research. Equally, the books are a product of the friendships and tolerance of large numbers of farmers who have endured, indeed, encouraged, our antics on their land, and have been those most valuable of allies: the critical friends. Our editorial task was hugely aided by Dr Eva Raebel, and we thank also the many members and friends of WildCRU who have helped in myriad ways, from providing photographs to critically reviewing drafts. We thank also our editors at Oxford University Press: Ian Sherman and, successively, Helen Eaton then Lucy Nash, and, during the production process, Cherline Daniel of Integra. Our work on these books has been an immense task, and was only possible due to the support of the Holly Hill Trust and The Rivers Trust. We are grateful to the many sponsors of our projects, and they are mostly acknowledged in individual chapters, but several donors have kept the enterprise afloat more generally: we have in mind especially the Tubney Charitable Trust (and we remember fondly the particular enthusiasm of the late Terry Collins), Defra, the Esmée Fairbairn Foundation, the People's Trust for Endangered Species, Earthwatch, and the Environment Agency. Latterly, WildCRU has been based at The Recanati-Kaplan Centre at Tubney House—reverberating through these names is our gratitude to Tom and Daphne Kaplan for their friendship and for securing the future of WildCRU's headquarters.

It has been a privilege to bring together the work of so many people in these two books. We hope you will enjoy them.

David W. Macdonald and Ruth E. Feber

Wildlife Conservation Research Unit, Recanati-Kaplan Centre, Department of Zoology, University of Oxford and Lady Margaret Hall, University of Oxford.

Contents

List of Contributors

Richard B. Bradbury Royal Society for the Protection of Birds (RSPB), Sandy, UK

Christina D. Buesching Wildlife Conservation Research Unit, The Recanati-Kaplan Centre, Department of Zoology, University of Oxford, UK

Francis Buner Game & Wildlife Conservation Trust, Fordingbridge, UK

Dan E. Chamberlain University of Turin, Italy

Ruth E. Feber Wildlife Conservation Research Unit, The Recanati-Kaplan Centre, Department of Zoology, University of Oxford, UK

Leslie G. Firbank School of Biology, University of Leeds, UK

Robert J. Fuller British Trust for Ornithology, Thetford, UK

Merryl Gelling Wildlife Conservation Research Unit, The Recanati-Kaplan Centre, Department of Zoology, University of Oxford, UK

Lauren A. Harrington Wildlife Conservation Research Unit, The Recanati-Kaplan Centre, Department of Zoology, University of Oxford, UK

Barbara Hart Allergenic Mite Supplies Limited, Glasgow, UK

Ian D. Hodge Department of Land Economy, University of Cambridge, UK

Paul J. Johnson Wildlife Conservation Research Unit, The Recanati-Kaplan Centre, Department of Zoology, University of Oxford, UK

Xavier Lambin Institute of Biological and Environmental Sciences, University of Aberdeen, UK

Danielle Linton Wildlife Conservation Research Unit, The Recanati-Kaplan Centre, Department of Zoology, University of Oxford, UK

David W. Macdonald Wildlife Conservation Research Unit, The Recanati-Kaplan Centre, Department of Zoology, University of Oxford, UK

Will Manley Royal Agricultural University, Cirencester, UK

Fiona Mathews College of Life and Environmental Sciences, University of Exeter, UK

Thomas Merckx Behavioural Ecology and Conservation Group, Biodiversity Research Centre, Earth and Life Institute, Université catholique de Louvain (UCL), Belgium

Tom P. Moorhouse Wildlife Conservation Research Unit, The Recanati-Kaplan Centre, Department of Zoology, University of Oxford, UK

Lisa R. Norton Centre for Ecology and Hydrology, Lancaster Environment Centre, UK

Alison E. Poole Wildlife Conservation Research Unit, The Recanati-Kaplan Centre, Department of Zoology, University of Oxford, UK

Eva M. Raebel Wildlife Conservation Research Unit, The Recanati-Kaplan Centre, Department of Zoology, University of Oxford, UK

Elina Rantanen Zoological Society of London, London, UK

Philip Riordan Department of Zoology, University of Oxford, UK

Christopher J. Sandom Wildlife Conservation Research Unit, The Recanati-Kaplan Centre, Department of Zoology, University of Oxford, UK

Rosalind F. Shaw Environment and Sustainability Institute, University of Exeter, Penryn, UK

Eleanor M. Slade Department of Zoology, University of Oxford, UK

Helen Smith Consultant Ecologist, Diss, UK

Nick W. Sotherton Game & Wildlife Conservation Trust, Fordingbridge, UK

Rob Strachan (dec.) Wildlife Conservation Research Unit, The Recanati-Kaplan Centre, Department of Zoology, University of Oxford, UK

Fran H. Tattersall Wildlife Conservation Research
Unit, The Recanati-Kaplan Centre, Department of
Zoology, University of Oxford, UK

Tom Tew Environment Bank, UK

David J. Thompson Department of Evolution,
Ecology and Behaviour, Institute of Integrative
Biology, University of Liverpool, UK

Martin Townsend Consultant Ecologist, Oxford

Hanna L. Tuomisto Wildlife Conservation
Research Unit, The Recanati-Kaplan Centre,
Department of Zoology, University of
Oxford, UK

Jeremy D. Wilson Royal Society for the Protection of
Birds (RSPB), Edinburgh, UK

CHAPTER 1

Farming and wildlife: a perspective on a shared future

David W. Macdonald, Eva M. Raebel, and Ruth E. Feber

To accomplish great things, we must not only act, but also dream; not only plan, but also believe.

Anatole France, 1896.

An insightful idiom, when considering the current state of the planet is to 'consider the view from 40 000 feet'. From what height is it most useful to view the farmscape and the biological, economic, and regulatory processes that shape it? In this chapter, we take a journey in an imaginary hot-air balloon as we explore the views from 40 000 feet and those very much closer to the ground, from telescope to magnifying glass, from the vertiginous to the groundedly Lilliputian, which are complementary within agri-environment policy. Whatever scale we take, from quadrat, to plot, field, farm, landscape, ecosystem, through institutions from farm-business to globe: the reality is that from 12 km up, we're in a global extinction crisis, and agriculture is a big part of the story.

In the context of wildlife in the UK, drawing largely on examples from more than 30 years' research by our team, the Wildlife Conservation Research Unit (WildCRU), and with a focus on lowland farms in England, our purpose is to reveal insights and offer solutions. In the first of two sister books, *Wildlife Conservation on Farmland. Volume 1: Managing for Nature on Lowland Farms*, we use the WildCRU's work to reveal how—for plants, invertebrates, birds, and mammals—agricultural systems and wildlife interact, presenting examples from scales varying from landscape to microcosm, from populations to individuals. In the second book, *Wildlife Conservation on Farmland. Volume 2: Conflict in the Countryside* (Macdonald and Feber 2015), we continue through the WildCRU's case studies of the factors leading to war and peace for species ranging from badgers to buzzards, where research on the basic ecology of the problem species is used as a basis for understanding

and finding evidence-based solutions to human–wildlife conflict.

As his mother put it to the grieving Hamlet 'Thou know'st 'tis common: all that lives must die, Passing through Nature to eternity'; but Gertrude's consolation shouldn't soften our grief because there is nothing 'common' about current species loss: it is man-made. Although species have a general lifespan of one to ten million years, rates of extinction are now estimated at as much as 1000 times background rates (UNEP 2011). According to the IUCN Red List (2014) of threatened species, one-quarter of mammal species (5488 species) and over a third (41%) of amphibian species (6260) are globally threatened or extinct and many are declining. However, only a few of the likely total of, even conspicuous, species have been described by taxonomists and therefore some groups are not being fully assessed. For example, fewer than 800 of the world's estimated 3.5 million insect species appear in the IUCN Red List (Dickman et al. 2007); a Rumsfeldian perspective on the unknowns can catapult the percentage of threatened species, such as insects, from 0.06% to 73% (Stewart et al. 2007). A philistine might ask 'does this biodiversity loss matter?' The answer, of course, is 'yes': biodiversity is a multifaceted concept encompassing 'genes, species, and ecosystems' (Dickman et al. 2007), and intricate plant and animal communities, that are the engines for the ecosystem services that support us (Box 1.1). In terms of the 'planetary boundaries' proposed by Rockström et al. (2009), which determine the safe operating space for humanity with respect to the Earth system, the biodiversity loss's boundary has already been surpassed. Yet in spite of extensive conservation efforts, the EU

Box 1.1 Biodiversity and ecosystem services

Biodiversity is the material from which the 'building blocks' of ecosystems are made, providing goods and services vital for our well-being and economic prosperity (Defra 2011; UK National Ecosystem Assessment, NEA 2011). Ecosystem services comprise goods directly produced by the environment (e.g. food, fresh water, fuel wood, clean air, genetic resources), those resulting from environmental processes (e.g. climate regulation, carbon sequestration, watershed protection, nutrient cycle, soil formation), and cultural benefits (e.g. recreational, aesthetic, educational, spiritual/religious), all adding to human well-being (UN Millennium Ecosystem Assessment (ME) 2005). Over just the past 60 years, though, around 60% of ecosystem services have been degraded or used unsustainably (ME 2005)[1].

Agricultural land offers many of these ecosystem services (NE 2012), while agricultural production functions because of healthy ecosystems. A forceful illustration of this two-way link is provided by the keystone process of pollination. Pollination services directly link wild ecosystems with agricultural production (of fruits, nuts, oils, and animal feed) and hence human livelihoods; the global economic value of insect pollinator services to agriculture has been estimated to be 153 billion euros (in 2005: Gallai et al. 2009) with 35% of crop production coming from plants that rely on pollinators (Klein et al. 2007). In the UK, insect pollinated crops (e.g. oilseed rape, orchard fruit, soft fruit, and beans) have been estimated to account for c. 20% of cropland and c. 19% of crop value (Breeze et al. 2011). However, pollinators have declined globally due to a range of factors, including habitat loss, fragmentation and isolation, agrochemical use, and pathogens (Potts et al. 2010). Wild (unmanaged) pollinator species (e.g. wild bees, hoverflies, wasps, butterflies, moths) are very important: 600 wild solitary bees (200 species in the UK: POST 2010) can pollinate apple orchards as efficiently as 30 000 honey bees (two hives) (Delaplane and Mayer 2000).

Many ecosystem services are uncharismatic or inconspicuous, and may be overlooked (ME 2005). For example, dung beetles (Coleoptera: Scarabidae) are essential in temperate ecosystems for recycling nitrogen and incorporating organic matter into the soil, producing nutrient-rich soils while reducing pasture fouling, the spread of parasites, and the need for chemical fertilizers (Beynon et al. 2012), as well as lowering nitrogen emissions. In the USA, dung beetles have been estimated to save the cattle industry $380 million/year by recycling cattle waste and $58 million/year by reducing nitrogen loss (Losey and Vaughan 2006). However, dung beetles are susceptible to anthelmintic drugs, which are used to treat internal parasites of livestock (Beynon et al. 2012). As with pollinators, they suffer from other impacts due to agricultural intensification such as loss and fragmentation of habitat. To conserve biodiversity and ecosystem services, creating more sites for nature that are larger and better joined is crucial (Lawton et al. 2010).

Changes to ecosystems can have non-linear effects (e.g. irreversible alterations in water quality) and can also affect distant ecosystem services (ME 2005). A good example is provided by the difficulties in managing nitrogen [N] and phosphate [P] run-off from farms; these surplus nutrients encourage a build-up of algae in aquatic ecosystems, sometimes with distant effects which create marine 'dead zones'[2] (Conley et al. 2009)[3]. Toxic algal blooms (red tides) create a health hazard, with economic implications for livelihoods in tourist areas (for example, a cost of $11.4 million in 1989 to the Italian tourism industry: ME 2005), but also affect other ecosystem services provided by coastal wetlands. In the UK, coastal wetlands provide £1.5 billion annually in benefits by buffering the effects of storms and managing floods (NEA 2011) and, while many case studies have observed relationships between ecosystem structure and function in freshwaters, these relationships have never been fully quantified (Pacini et al. 2013).

In general, the value of many ecosystem services has been greatly overlooked (NEA 2011). The TEEB (Economics of Ecosystems and Biodiversity 2010) project seeks to help decision-makers to account properly for the value of nature by addressing 'market failure'. The failure is fundamental and almost universal: use of the environment is simply not being costed, and, being underpriced, nature is over-exploited. This 'invisibility of biodiversity values'[4] makes it difficult to provide convincing financial arguments for ecosystem conservation policies. This may especially be the case in poorer countries where the priority is resolving hunger, but where the link between hunger and biodiversity deterioration may not be conspicuous.

[1] Examination included fresh water, capture fisheries, air and water purification, regulation of regional/local climate, natural hazards, and pests (ME 2005).

[2] The greatest damage to biodiversity is caused by algal decomposition, which depletes oxygen in the water (hypoxia), creating 'dead zones' (Diaz and Rosenberg 2008).

[3] Drastic reductions in P loads in the Rhine River and surrounding rivers that drain into the North sea accomplished P limitations in the catchment, but generated greater N export, resulting in eutrophication of the Wadden Sea and part of the Norwegian coast (Skogen et al. 2004).

[4] Pavan Sukhdev, leading The Economics of Ecosystems and Biodiversity (TEEB) UNEP study stated: 'The invisibility of biodiversity values has often encouraged inefficient use or even destruction of the natural capital that is the foundation of our economies'.

did not achieve its target to halt biodiversity loss by 2010 (SCBD 2010; European Commission 2011). Nobody knows which species can be lost before—to use Jared Diamond's word—collapse. We are whipping up a blizzard of environmental perturbation in which hailstones of extinction increasingly batter us; are we already pushing our luck?

It would not be an exaggerated claim that agricultural development has had greater consequences for natural habitats and biodiversity than any other force (Tilman et al. 2011). The IUCN identifies agriculture as the main threat for 87% of already globally threatened bird species, and there's no reason to suppose it is a lesser threat to other taxa. Agriculture contributes, in complex and inter-related ways, to all five of the most important drivers of biodiversity loss (as identified by the Millennium Ecosystem Assessment 2005): habitat change, climate change, invasive species, overexploitation, and pollution. From the high altitude starting point of our hot-air balloon journey, the view reveals the full extent of agricultural impacts on the global terrestrial environment. The shift in lifestyle from nomadic hunting and gathering to settled agriculture prompted a handbrake turn in human history and sent the planetary 'Gaia' into a prolonged skid that continues today. One-quarter of the Earth's terrestrial surface is now under cultivation, undermining the capacity of ecosystems to maintain food production and to sustain services. This transformation probably took place independently in at least seven locations[5], about 10 000 years ago (Bogucki 1996). One of the agricultural movements was the human spread of domesticated plants and animals from the Fertile Crescent of Mesopotamia to Egypt, the Iranian Plateau, and finally, Europe (Bogucki 1996)[6]. Vere Gordon Childe, a renowned 1920s archaeologist of European pre-history, referred to this period as the 'Neolithic revolution', implying 'a revolution in technological and economic developments in the history of human society'. Unbeknown to our prehistoric ancestors, they also caused the first anthropogenic damage to the environment not long after settlement (e.g. soil erosion due to land use, van Andel

et al. 1990; depletion of vegetation by domesticated goats, Bellwood 2001; Macdonald and Feber 2015: Chapter 1). Resource over-exploitation permeates human history, with devastating effects from the outset, precipitating early extinctions from Easter Island to the Mayan Riviera. Exactly when humans stopped being part of nature is an unresolved philosophical debate (although Macdonald et al. (2010) defined naturalness as 'the point immediately before anthropogenic influence is detectable (either empirically or through deduction) for the first time in the history of the species'), but it is clear that something changed from the time when some four million hunter-gatherers (Cohen 1995) acquired food while living in ways that were broadly synergistic with nature, to the demands of the current seven billion people, of which one in seven people are still malnourished or lack food.

The topic of this book, then, can hardly be dismissed as of merely sectorial interest: what citizen is not affected by both food security and environmental services? Globally, the big question is: can seven billion people be fed, watered, and provided with energy equitably, healthily, and sustainably?[7] The ecosystem function, and thus biodiversity, of farmland will, for better or worse, be central to the answer. Humanity is living on tick—our ecological footprint overshooting the Earth's biocapacity by about 25%. Our perspective of this big picture is that the urgency of balancing current human welfare and future ecosystem services shouldn't distract us from safeguarding the detail likely to be lost from the canvas: the beauty and fascination of species and the tranquillity of wilderness—surely these are the business of conservation beyond the minimalist functionality of self-preservation.

As our hot-air balloon lowers, our new European perspective reveals, west of the marvellously pristine Carpathians, little of its original vegetation, which consisted of the Mediterranean evergreen forest and the deciduous forest of temperate Europe. Transition to agriculture in Europe around 9000 years ago occurred both by colonization of new areas and by cultural adaptation of resident foraging communities into farming (Bogucki 1996). The Neolithic spread of agriculture was closely related to the different times that it took crops brought from the Near East to adapt to the new climatic and seasonal conditions. So began, and in a world-changing way, the human habit of introducing non-native animal and plant species, and among the almost endless consequences of these first transportations was a change

---

[5] Near East (Fertile Crescent), South China (Yangtze River), North China (Yellow River), Sub-Saharan Africa, South-central Andes, Central Mexico, and Eastern USA.

[6] Bellwood (2001) summarizes the main source–expansion times: (1) Fertile Crescent (Levant and Zagros mountains of the Near East) to Britain (7000–4000 BC) over 3600 km; (2) Yangzi Basin to Island South East Asia (6500–2500 BC) over 5000 km, (3) Central Mesoamerica to the Southwest (3500–1500 BC) over 2500 km, (4) Pakistan into Peninsular India (7000–3000 BC) over 2000 km.

[7] As Sir John Beddington, then Government Chief Scientist, presciently asked in 2011.

in diet from a smorgasbord of animal resources (marine and terrestrial molluscs, crustaceans, fish, birds, and mammals), to a less diverse diet of domesticated plants (emmer, einkorn, wheat, barley, pea, chickpea, bitter vetch, and lentils) and domesticated animals (cattle, sheep, goats, and pigs) (Tresset and Vigne 2011). By 2000 BC nearly all of Europe was occupied by self-sufficient farmers and the 'natural' environment had been irreversibly transformed. The side-shows to this drama were scarcely less transformative, among them, wildcats domesticated themselves lured by rodents around the first grain stores (Driscoll et al. 2007), and wolves too, adapted themselves to man-made plenty (Driscoll and Macdonald 2010).

Descending further, we have a national-scale view of Britain, with its landscape history reminiscent of our global and European perspectives (Box 1.2). How one sees the landscape depends upon the hilltop from

which it is viewed, so between 1950 and 1975 there is an unhappy coincidence between what Oliver Rackham (2003) terms 'the locust years', when agriculture was at its most rapine, and the 'second agricultural revolution' or 'agricultural miracle', when, under the stress of the war, food imports were largely cut off and post-war pressure for home production and increased yields precipitated the development of technologies, agrochemicals, efficient machinery, and crop varieties. All this happened at breakneck speed: the countryside we regard as familiar is a very recent construction. Much semi-natural grassland was drained and tilled, while woods, hedges, and fens were destroyed to ensure that every acre did indeed count. Small holdings and peasant farms collapsed (Brassley et al. 2012) and, as Mabey laments, after three millennia of coexistence with agriculture, wildlife faced dire peril from changes wrought in scarcely 30 years (Box 1.3).

Box 1.2 Agriculture in Britain: from wildwood to 'butter mountains'

The wildwood that once covered postglacial Britain began to be cleared by Neolithic farmers around 4000 BC (Tresset and Vigne 2011). Agriculture took root later in Britain than on the Continent because a period of drier and warmer conditions was needed to allow key crops, such as barley, to thrive (Jones et al. 2012). By 500 BC, the area of wildwood had halved (Royal Forestry Society 2013) and, by 1086, William the Conqueror's great land survey (the Domesday Book) recorded that England had only 15% woodland and wood pasture remaining, and already 35% of land was arable (Rackham 1986).

Before 1750, open farms were small with three mix-farming components: an arable area, an area of grassland (some pasture and some cut for hay), and a third element of 'wasteland', including woodland marshes and bogs (Mabey 1980), from which people used resources such as timber, fish, and turf. Although agriculturalists of the sixteenth and seventeenth centuries viewed uncultivated wilderness areas—fens, moors, and mountains—as a 'standing reproach' (Thomas 1984), agriculture was still of low-intensity subsistence. This began to change with the Enclosure Acts of the eighteenth century (which removed the rights of local people to the land: Parliament 2014) and, hot on their heels, the Corn Laws of 1815, designed to ensure food and stable prices. Punitive import duties protected landowners from international grain trade, but limited the national food supply and thus inflated domestic food prices. In 1845, the

potato famine exacerbated the discontent of the emerging industrialized class, culminating in the repeal of the Corn Laws in 1846. In the absence of protective tariffs, Britain's agriculture see-sawed towards import dependency (lasting until 1931: Brassley et al. 2012) to the extent that, in the run up to World War I (1914), four-fifths of wheat flour was imported (Dewey 1989) and British agriculture was essentially based on livestock. However, with the start of World War I, the Agricultural Consultative Committee issued a series of press releases advising farmers to maximise food production: 'by increasing the area under wheat and cereals, [. . .], to break up grassland and to increase livestock' (Dewey 1989). By 1916, government interventionism secured minimum prices for farmers (e.g. wheat for four years) and deficiency payments thereafter, on the condition that they would increase their arable or grass land by at least a fifth.

The Scott report of 1942 into land use in rural areas shaped the British countryside, asserting that 'every acre counts' (Blunden and Curry 1988). During World War II, the rationing food system set up by the Ministry of Food allocated free basic foods (e.g. bread and potatoes) and farmers benefited during that time from 'guaranteed prices and assured markets'[8] (1947 Agricultural Act). However, post-war, European countries had been faced with food market instabilities, a disproportionate influence of food prices on inflation, and a need to maintain national food industries for political reasons. The Common Agricultural Policy (CAP) was

continued

---

[8] Rationing lasted until 1953.

Box 1.2 *Continued*

Western Europe's response, setting out its objectives in 1958 to 'encourage better productivity in the food chain, ensure fair standard of living to the agricultural community, market stabilization, and ensure the availability of food supplies to EU consumers at reasonable price' (European Commission 2013b). Britain joined CAP in 1973. Heavily subsidized food production distorted world markets bringing a high CAP budgetary cost: 87% of the total EU budget (compared to 43% in 2011: European Commission 2013a). The key problem was that stabilizing agricultural prices at high levels encouraged farmers to increase output, and in turn, supported large agri-businesses to the detriment of the small family farmers whose practices had traditionally been conducive to conservation (Martin 2011). Thus domestic production ran ahead of consumption, compelling the European Commission to purchase and store large surplus commodities, producing 'butter mountains' and 'wine lakes', which often had to be resold at a loss on world markets. Not only was this unsustainable for the European tax-payer, but escalating production let to intolerable environmental degradation. The agricultural intensification of the previous 40 years had resulted in profound losses of biodiversity (Box 1.3). By the late 1980s the reality had dawned that this model for agriculture was leading to both financial and environmental disaster—priorities had to be re-evaluated.

For the first time, rural conservation was emerging as a societal force with both popular and political support (Blun-

den and Curry 1988). The major impetus for change came via central regulation, and in England, farmers began to be paid for conservation by 1987 through agri-environment schemes (see Box 1.4). The CAP was gradually dismantled and the link between production and subsidy phased out (Rickard 2012). In 1988, CAP introduced budgetary stabilizers which set a CAP budget ceiling based on a maximum limit on quantities guaranteed to receive support payments (e.g. quotas for milk), and also introduced 'set-aside', paying famers for taking a percentage of land out of production, with the aim of reducing 'grain mountains' while delivering environmental benefits[9] (European Commission 2013b). By 1992, the MacSharry reform completely shifted financial support from product to producer and, as the twentieth century closed, it was acknowledged by the European Commission that future production would need to accommodate conservation (e.g. Krebs et al. 1999).

In 2008, the European Commission abolished set-aside in response to high cereal prices and low global production, notwithstanding its environmental and biodiversity benefits (Defra 2012a). The current shift from maximizing production to nurturing environmental co-benefits may or may not survive swings of the economic pendulum, and an interdisciplinary group of crystal ball gazers have concluded that food security will be a major issue for conservation planning in the first half of the twenty-first century (Kass et al. 2011).

Box 1.3 UK agriculture and biodiversity loss

The agricultural intensification of the post-war years has led to widespread and severe losses of much of the UK's farmland wildlife. Although quantitative trends can be assessed for only 5% of the 59 000 UK terrestrial and freshwater species (State of Nature 2013), an assessment of the population and distribution trends of 3148 species, undertaken by UK wildlife organizations, reveals that, over the past 50 years, 60% have declined and 31% have declined strongly (State of Nature 2013). Habitat changes are revealing; between

1998 and 2007, improved grassland has increased by 5.4%, the total length of woody linear features has decreased by 1.7%, and only 10% of managed hedges on arable land were in good structural condition and had appropriate margins (Carey et al. 2008). In Britain, 50% of ponds have been lost since the 1950s; although the number of ponds has recently increased by 12.5% (1998–2008), only 8% of lowland ponds are in good condition and the plant species richness of streamsides has decreased by 7.5% (Carey et al. 2008).

continued

---

[9] 'Set-aside' became compulsory in 1992, but was abolished in 2008 to increase EU cereal supply to the market and hence reduce prices, which had been high after low EU harvests.

Box 1.3 *Continued*

Of 333 UK farmland species (including broad-leaved plants, butterflies (Box Fig. 1.3.1), bumblebees, birds, and mammals) two-thirds were judged, using a trait-based approach to risk assessment, to have been threatened by agricultural intensification (Butler et al. 2009): 50% of plant species, 33% of insects, and 80% of bird species have declined and farmland habitats hold more threatened and scarce plant species than any other (Robinson and Sutherland 2002). In England alone, 492 species have become extinct since 1800[10] (NE 2010). For example, mitten's beardless moss *Weissia mittenii* (first described in 1851) was an English endemic species that thrived in fallow fields[11], but

current land use practices do not allow this slow grower to complete its life cycle; it was last recorded in 1920 and is now globally extinct (NE 2010). During the twentieth century, one species of plant was lost every two years (average from 23 English counties), with highest rates occurring since the 1960s (NE 2010).

One of the best monitored and most important indicator groups of farmland quality is farmland birds. From 19 indicator species of UK farmland birds, 12 have declined in populations from 1970 to 2010 (RSPB 2013). Birds have changed the political scene: the Birds Directive is the oldest piece of nature legislation in the EU, which created protection for all wild birds in the European Union and was adopted by all member states in 1979 (EU 2013). Many of the UK agri-environment scheme (AES) options are orientated towards bird conservation (see JNCC and Defra 2012).

The importance of monitoring these groups is to reveal population trends, ascertain the effectiveness of conservation policies and help to set targets. In 1992, the UK signed the Convention on Biological Diversity (CBD), leading, in 1994, to publication of the UK Biodiversity Action Plan (UK BAP). This identified priority habitats and species, together with plans to address their conservation (Defra 2012b). In 2008, the UK BAP concluded that 18% of priority habitats and 11% of priority species were increasing, but 42% of habitats and 24% of species were declining. However, the rate of decline was slowing down for 20% of BAP habitats and 8% of BAP species (UK Biodiversity Partnership 2010).

A key underlying cause of species decline is agricultural intensification linked to changes in management practices (NE 2010). Agri-environment schemes have therefore played a crucial part in UK BAP delivery. For example, delivery of the Hedgerow Habitat Action Plan was helped by the inclusion of hedgerow management options in Environmental Stewardship (see Box 1.4), and Environmental Stewardship has helped to facilitate the recovery of the large blue butterfly *Maculinea arion* through supporting changes in grassland management (UK Biodiversity Partnership 2010). In 2012, the UK Post-2010 Biodiversity Framework replaced the UK BAP, focusing on within-country-level biodiversity strategies (specified in the UK Strategic Plan for Biodiversity 2011–2020, Defra 2012b).

Box Figure 1.3.1 Butterflies are one of the many taxonomic groups affected by agricultural intensification. Photograph: marbled white *Melanargia galathea* © Roger Meerts/Shutterstock.

[10] 24% of butterflies, 22% of amphibians, 15% of dolphins and whales, 14% of stoneworts, 12% of terrestrial mammals, and 12% of stoneflies. This being an underestimate, for little is known about the existence and loss of many invertebrates, plants, and fungi (NE 2010).

[11] Also on woodland rides and roadside banks.

Yet, since even the pristine rainforest of Gabon is as it is only because it regrew following human agriculture a few hundred years ago[12], we might ask ourselves whether British farmland is a topic for conservation, since it is, by definition, unnatural in all its varied forms. How hard should we work to conserve iconic farmland specialists that have evolved dependence on bygone agriculture? For example, we now value farmland-dependent wildlife very highly but it has not always been the case: the cornflower *Centaurea cyanus* evolved seeds adapted to being sown with the crop and was so common in the UK until the mid-nineteenth century that it was considered 'a pernicious weed injurious to the corn and blunting the reapers' sickles'. Now, with seed cleaning, this arable plant is considered to be endangered, and largely relies for its survival upon subsidies that recreate arable habitats managed as they used to be. Similarly, the high brown fritillary butterfly *Argynnis adippe*'s life-support system is the artificial re-creation of agriculturally obsolete extensive, bracken-rich grazing systems; and the corncrake *Crex crex* of the Western Isles depends on corncrake-friendly mowing to avoid its chicks being killed (Chapter 5, this volume). Is this akin to protecting steam engines as interesting remnants of an obsolete transport system?

Even though only a philistine would abandon these species, this question raises two points: first, conservation is about consumer choices. In the same way that we treasure the once 'pernicious' cornflower now that it is endangered, society has to decide what Nature it wants, and work out how to deliver it. Second, the generality of whether, and how, to secure Nature in the countryside is relevant to a wider constituency than the stereotypical bearded, binocular-bound few, nostalgic for Constable's haywain. Government currently considers farmland biodiversity sufficiently important that over 6.5 million hectares of farmland in England are now covered by agri-environment schemes, seeking to modify human behaviour to conserve the landscape and its wildlife (Box 1.4).

Although in European landscapes the concept of multi-functionality (that is, that farmland must deliver

---

Box 1.4 Agri-environment schemes

Agri-environment schemes (AES) are voluntary agreements where farmers and other land managers are paid to manage their land in an environmentally friendly way. England's first agri-environment scheme (Box Table 1.4.1), the Environmentally Sensitive Areas (ESA) scheme, was launched in 1987, offering financial incentives to encourage farmers to 'adopt agricultural practices that would safeguard and enhance parts of the country of particularly high landscape, wildlife, or historic value' (NE 2012), and especially areas particularly vulnerable to agricultural practices. This was a targeted and tiered scheme, aimed at 'maintaining the conservation, landscape and historical value of the key environmental features of an area' (Defra 2000). Twenty-two ESAs were designated in England, covering 10% of geographically defined agricultural land (NE 2014). In 1991, the Countryside Stewardship Scheme (CSS) was launched, covering areas outside the delineated ESAs, but still within specific landscape types. While the emphasis of ESAs was on 'safeguarding' regions of national wildlife and landscape importance, CSS focused on 'improvements' to land management practices and hence the quality of the wider countryside (Defra 2000).

The ESA and CSS schemes were succeeded by Environmental Stewardship (ES) in 2005. ES gives farmers additional financial support when delivering effective environmental management on their land beyond cross-compliance and is delivered by Natural England on behalf of the Department for Environment, Food and Rural Affairs (Defra). There are three elements to ES (Box Table 1.4.1): Entry Level Stewardship (ELS, including Uplands ELS), Organic Entry Level Stewardship (OELS, and Uplands OELS), and Higher Level Stewardship (HLS). ESA and CSS only covered key areas and habitats designated for their unique environmental features, but ELS is a 'broad and shallow' scheme open to all farmers (an 'entry-for-all' voluntary approach). To qualify, each farm needs a certain amount of points, gained by choosing management options. While the focus of ESA's lower tiers was on payments to safeguard land from management changes that could harm key environmental features, ES pays to support beneficial management practices of the same features. OELS operates in a similar manner to ELS. Both ELS and OELS include options that may simultaneously target the conservation of various species: mammals, birds, and invertebrates

continued

---

[12] See Oslisly et al. (2013).

Box 1.4 *Continued*

Box Table 1.4.1 Agri-environment schemes (AES) recently and currently prescribed in England.

Agri-environment schemes	Start/end	
Environmentally Sensitive Areas **(ESA)**	1987–2004 Phase out 2014	Targeted and tiered. Payments made to farmers to maintain and enhance the conservation, landscape and historical value of key environmental features of designated areas.
Countryside Stewardship Scheme **(CSS)**	1991–2004 Phase out 2014	Operated outside ESAs: wider countryside but still within specific landscape types. Payments made to farmers to conserve and enhance typical landscapes, their wildlife, and history, and to help people to enjoy them.
Environmental Stewardship Scheme Entry Level Scheme **(ELS)**	2005–2013	Broad and shallow. Encourages a large number of farmers and land managers to undertake simple environmental management at a 'whole-farm' level to secure widespread environmental benefits.
Organic Level Scheme **(OELS)**	2005– ongoing	Similar to ELS but all or part of the land is managed organically at a 'whole-farm' level.
High Level Scheme **(HLS)**	2005– ongoing	Targeted. Combined with ELS and OLS options, it provides environmental benefits in high priority situations and areas. Concentrates on targeted and more complex management that needs advice and support; agreements are adapted locally to individual situations.
New scheme to be determined	2015–	New Rural Development Programme 2014–2020

will be encouraged by improving general boundary and linear features (stone walls, earth banks, woodland edges, beetle banks, and ditches) and landscape-scale issues such as diffuse pollution caused by fertilizers and pesticides adversely affecting riparian species and aquatic insects.

In contrast, HLS entrance is competitive, based on qualification from an initial Farm Environment Plan (FEP) assessment (NE 2013b). HLS is more demanding, with agreements tailored to local circumstances and farmers required to undertake higher levels of management to deliver focused environmental benefits in high priority situations and areas, and therefore attracting the highest payments. HLS only comprises about 10% of ES agreements, probably because of the levels of commitment required to deliver this more complex management (see review by Macdonald and Burnham 2011).

Across all schemes, farmers are, unsurprisingly, thought to select the options perceived to be easy (e.g. linear boundaries), which may not necessarily be those delivering the best environmental outcomes. In 2008, the 20 most popular options (out of 62) accounted for 90% of points scored within the total ELS scheme (Defra 2008). Nevertheless, by 2013, there were more than 45 000 agreements under this scheme, covering 70% of agricultural land in England (NE 2013a). However, as with other European countries, results of stud-

ies evaluating the impacts of AES on biodiversity have been mixed (Kleijn et al. 2006). An analysis by Baker et al. (2012) found strong evidence that AES management that provided winter food resources reduced the rate of population decline for several bird species, including Yellowhammer, Linnet, Reed Bunting, and Grey Partridge, while other measures had little effect. Better tailoring of options and wider uptake of AES is needed; the potential benefits of AES are inhibited by the implementation of ES at the level of single farms, and collaborative AES delivered by collectives of farmers may be more effective (Emery and Franks 2012). A landscape-scale approach would provide robust landscape-scale biodiversity able to sustain future challenges, such as climate change.

The new Rural Development Programme 2014–2020 (in line with the new CAP reform starting in 2015) potentially will provide an opportunity to improve scheme effectiveness. A Basic Payment Scheme (BPS), will include new payments for young farmers and for 'greening'. The new schemes were being developed in 2014 (transition year). To secure 30% of the greening payments, farmers will have three new obligations: crop diversification, maintaining permanent grassland, and the set-up of 5% of ecological focus areas on arable land (e.g. fallow land, buffer strips, use of nitrogen fixing crops; Defra 2014a,b).

a whole range of ecosystem services of which biodiversity is just one) has been an important driver behind the development of agri-environment schemes; depending on each culture's history and population pressures, different societies may expect their farmed landscapes to deliver different outcomes. If our hot-air balloon rises again to take the wider perspective, we see that, in the USA, for example, farmed land is generally viewed as being for maximum food production, with specific non-farmed wild areas set aside (National Parks/reserves) for wildlife. In Africa, the agricultural land of a small holder may only be expected to produce subsistence food for a single family. What of wildlife in the wake of intensification? Whether in developed agricultural systems in Europe or subsistence farming in developing countries, the answer is always the same: agricultural intensification coincides with biodiversity loss (Mattison and Norris 2005). The farming methods and the affected species may differ, but the underlying biology is inescapable; there is competition between species for nutrients, space, light, and water. So, too, similar problems recur: eutrophication and pollution of waters by agrochemicals, fertilizers, pest and weed control in Europe and, by extending crops into forests and wetlands, nutrient depletion and soil erosion in Africa.

What differs between farming systems is how we support and subsidize the value of the different components. Society sees farm-related wildlife as either desirable (e.g. beneficial: invertebrates, pollinators) or undesirable (e.g. pernicious weeds, other invertebrates (slugs), and some birds (pigeons)) because they reduce production. How do we get the balance between controlling those aspects we don't want, but with minimal damage to wildlife that we view as desirable or essential to ecosystem function? Food security and environmental protection are widely overlapping spheres, but not inevitably harmonious ones. Agriculture depends on the environment, but it can do so more, or less, rapaciously and, while some elements of biodiversity are essential to the functioning of agricultural ecosystems, others could be dispensable if food production were the only consideration. However, it isn't! Agricultural production is supported by complex communities of soil organisms and natural enemies of crop pests and animal diseases that, as a whole, ensure productivity and generate ecosystem services.

If society wants—needs—both food and a biodiverse, aesthetic, and culturally rich landscape, how is this to be delivered? The chapters in this book tackle this question at scales varying from within fields to between landscapes. Landing our balloon in the

seemingly homogeneous ocean of an industrial cereal field, for example, we might consider how farm practice impinges on the world of a wood mouse *Apodemus sylvaticus* (Fig. 1.1), one four thousandth human size (Chapter 4, this volume). In 1785, on ploughing up her nest, Burns addressed the famous 'wee sleekat, cow'rin, timorous beastie' in a poem that not only anticipates the essentials of agro-ecology, but offers a lament that touches the soul when Burns apologized to the stricken mouse: 'I'm truly sorry man's dominion has broken Nature's social union'. A modern day trauma faced by the wood mouse is the application of herbicides designed to kill the plants, and thus invertebrates, on which wood mice depend; but might some compromise mend Nature's broken social union? Gleaning fixes from radio-collared mice we found that they preferred unsprayed plots but spent almost as much time in plots sprayed with a reduced cocktail of herbicides, the so-called 'conservation headlands', offering a compromise to farming and wildlife. Further sleuthing revealed that, when the crops are high, the mice wander extensively, with male ranges three times the size of female ranges; following harvest, these collapse to tiny hedgerow-based enclaves.

Taking off again in our hot-air balloon and rising above the fields, we see that such uncropped farmland habitats emerge as variously joined up: hedgerows and the field margins associated with them ramify, like a

Figure 1.1 The wood mouse *Apodemus sylvaticus* is affected by farm operations at the field- scale, such as herbicide use and harvest. Photograph © Michael J. Amphlett.

vascular system, a *rete mirabile*, through the farmscape, where you might consider them either a source of crop pests or a refuge for nature. How might their management strike a balance? To find the answer, in 1987 we set up an experiment at the University of Oxford's farm at Wytham. We created 2 m-wide uncropped field margins which were either sown with a grass and wildflower seed mixture, or left to regenerate naturally, and which were managed by cutting at different times. Each resulted in different consequences for the plant communities that developed (Chapter 2, this volume), and the invertebrates they supported (Chapter 3, this volume). Increased plant species richness of margins (Fig. 1.2) catalysed a virtuous cascade: further up the food chain, we found that farmland weasels hunted almost exclusively along corridors and that hedgerow connectivity increased numbers of small mammals (Chapter 4, this volume), whereas hedgerow gappiness decreased bank voles (Chapter 4, this volume). Of course, a corridor, once created, is a toll-free passage for virtuous and villainous alike; released grey partridge that linger close to field margins, and therefore close to fox highways, are less likely to survive (Chapter 13, this volume), and the narrow swathes of tall riverine vegetation that are a crucial requirement for successfully reintroducing the endangered water vole *Arvicola amphibius* are rendered useless if patrolled by voracious American mink *Neovison vison* (Chapter 15, this volume). In our sister book, *Wildlife Conservation on Farmland: Conflict in the Countryside*, we delve deeper into the habits of invasive species, and their impacts on native farmland wildlife.

Rising higher in our balloon we see that the mosaic of fields comprises a farm, providing opportunities for agri-environment policies aimed at farm-, rather than field-scale management, such as support for organic farming. A reaction to the government's encouragement of chemical-based agriculture, the organic movement emerged in the UK with the formation of The Soil Association, in 1946. The movement had its intellectual roots in continental Europe, catalysed by publication of *The Living Soil* (1943) by Lady Eve Balfour, which described the first UK scientific study comparing organic and conventional farming. Seventy years on, we studied pairs of organic and non-organic farms across England and found that there were differences in their biodiversity, although the extent of the differences depended, among other things, on the taxonomic group we were looking at, whether butterfly, bat, or bird (Chapter 6, this volume). Some features of organic farms, such as the prohibition of mineral fertilizers and strict limits on pesticide use are defining, others, such as bigger and denser hedgerows could be adopted by non-organic farmers, but others, like the greater

Figure 1.2 Increasing the plant diversity of field margins across farmland benefits invertebrates and the species that depend on them. Photograph © Ruth Feber.

likelihood of joining AESs, suggest that the organic farmer's mindset is an asset to biodiversity. However, as our hot-air balloon gains height, the wider perspective reveals that organic cereal yields (in developed countries) are 60–70% those of conventional farms and so, all else being equal, farming organically needs more land, with knock-on effects for biodiversity.

So, moving up in our balloon, we see that a mosaic of farms comprises a landscape (Fig. 1.3). Would there be value in agri-environment policies that targeted farms to foster joined-upness? Considering the potential of corridors and the hazards of fragmentation, would it help if clusters of farmers were coordinated? Taking a landscape view, we measured the biodiversity benefits of targeting farms to foster joined-upness in the Upper Thames, by comparing two areas where neighbouring farmers had been persuaded to enrol in agri-environment schemes, versus two control areas where scheme uptake was *laissez faire*. Our studies showed that wide margins and hedgerow trees both increased moth diversity but, deconstructing this

finding to reveal its policy-relevant core, we found the effect of hedgerow trees on moth abundance and species diversity was significantly stronger where farmers had been targeted to join up (Chapter 8, this volume). By individually marking moths we discovered that the most mobile species responded to field margins only at a landscape scale. Similarly, highly mobile species of odonates only benefited from the provision of ponds when they were available at a landscape scale (Chapter 10, this volume), and most bats were more active in wetter, more connected landscapes than drier, unconnected ones (Chapter 9, this volume). So, implementing agri-environment schemes to increase landscape connectivity seems to provide the taxpayer with more biodiversity bangs per buck, but how to persuade the right farmers to join? We found that answers to questions revealing openness to new ideas and attitudes to agro-chemicals gave us a 79% success rate in predicting whether farmers would sign up to agri-environment schemes (Macdonald and Feber 2015: Chapter 14).

Figure 1.3 Measures to conserve farmland biodiversity can be more effective when applied at a landscape scale. Photograph © Matthew Dixon/Shutterstock.

The question 'how much is wildlife worth?' is a relatively modern one and it takes on considerable importance when we reflect on the damage inflicted on wildlife by farming. Twenty years ago, when we were pondering the prospects for integrating farming and conservation, we referred to a 'chasm of values' (Macdonald and Smith 1991). The chasm we alluded to was between two very different services provided to society by farming—the delivery of food security without exacting too severe a cost on the biodiversity that also depends on farmland. We lamented, in that 1991 perspective, that decisions about the management of farmland would depend on the *value* of these two services, which is problematic not only because people attribute very different values to them but, worse, often their values cannot be rendered in a common currency: the dilemma of incommensurables. Now in the twenty-first century, the balancing act between food and biodiversity is still much dependent on economic thinking: we question, how much to charge? One should be anxious about basing conservation priorities on monetary values since, in normal markets most actors on Nature's stage have none, and while we know that focus groups may cheerily pay $6000 per annum to keep a lion alive (Maclennan et al. 2009), nematodes might be a tougher sell. As the Joker said to the Thief: Businessmen they drink my wine, ploughmen dig my earth. None of them along the line know what any of it is worth[13].

Indeed, a monetary value promptly comes to mind if we think about Stonehenge and its tourist income, or the price of Van Gogh's sunflowers if comparable to other paintings, but suppose the buyers wanted to convert Stonehenge into a quirky fast-food restaurant, or to use Van Gogh's painting as kindling? What value would we put to stop them? The problem is that the value is greater than the price. Less hypothetically, what value would you put on stopping clouded leopards from being driven extinct by converting their habitat into an oilpalm plantation? Your horror is not traded and thus cannot readily be monetized by normal markets.

How do you find out how much people value things? In a local supermarket we saw shoppers voting between three good causes: youth football (largely for boys), Brownie groups (exclusively for girls), and bees (perhaps symbols of environmental health). Bees outstripped the football by 120%, and the Brownies

by 115%. Of course, creating this ranking doesn't answer the policy question of how to invest the money, but it is a helpful start to have a common currency (in this case votes). The difficulty is that most of nature on farmland is not traded. We have spent years conserving water voles (Chapters 14 and 15, this volume), but there is no market in water vole habitat. How might it be priced to reflect its value? The question remains: 'how should we value biodiversity'?

White et al. (1997) had already reported, using contingent valuation, that focus groups were willing (if you believe them) to pay an amount 'T' to conserve water voles at a site. Knowing the size of the site, we calculated the value of each metre of restored or preserved habitat (Dutton et al. 2010). The equation deals with the discount rate, r—the fall of value into the future—and the fact that people valued differently the preservation and restoration of habitat. Extrapolating to the whole UK revealed that citizens value each metre of water vole habitat at up to £12, as a one-off payment to ensure the well-being of water voles. Cheap at the price we might say, but how might we use this insight?

First, while the taxpayer will always want the best price, it tells us that they would judge as not worth it, a policy that cost more on average than £12 per metre of vole habitat. Second, by dealing simultaneously with the issues of incommensurables and marginal values, and by considering monetary value as a proxy for votes, it helps us with the apples and oranges problem of choosing between vole habitat and, for example, butterflies (or schools). Third, and most importantly, we were able to convert this average value for a metre of vole-inhabited waterway into a measure of the marginal benefit to the local vole population. An isolated metre of habitat is of use to neither man nor vole; it becomes worth £12 only if voles are likely to survive there. We calculated vole viability per extra metre using ecological models, then incorporated that into the valuation system, replacing the flat value with marginal values that guide policy to invest most where the marginal gains are greatest. Suppose you were considering investing in new water vole habitat, on a farm whose neighbour already had 900 metres of occupied water vole habitat. The marginal value of each new metre would be equal to its own contribution to housing voles, plus the value it confers on the nearby vole real estate. If you invested in 150 new metres, each would be worth £38 because of the extra viability they collectively add to the existing vole population. In contrast, if you created these 150 new metres in isolation, each would be worth only £3.50. In a well-designed AES these marginal values, set against such

---

[13] From 'All Along the Watchtower', written and originally recorded by Bob Dylan in 1967, but covered by Jimi Hendrix in 1968, which made the song famous.

costs as mink control (a topic explored in Macdonald and Feber 2015: Chapter 6), fencing, and the opportunity cost of the land, could maximize the efficiency of payments to farmers for vole production.

Perhaps better to value habitats than species, but value clearly hinges on education, itself a limited resource. Nevertheless, advocacy based on the non-market, even spiritual, value is so airy fairy in the face of hard cash, that we are pushed towards a single currency—money; so be it, but we must remember marginal values are what matter[14].

The bridge across the 'chasm of values' is currently constructed with the Ecosystem Services (ES) paradigm (Box 1.1). The idea is to attach values to different farming products by identifying the services they deliver—these services are then more tractable to valuation than any intrinsic aspect of the habitats themselves. There have been tensions and criticisms about commodifying and pricing ecosystem services and nature (see article by Dempsey and Robertson

2012), and Luck et al. (2012) point out that the 'reliance on economic metaphors' to value nature and services can exclude other reasons and intrinsic motivations for conserving ecosystems. Rackham (1994) speculated that economics, like aesthetics, is 'too brittle' to be the basis for conservation, although there can be little doubt that economic analysis will remain central for decision making at every level of the environment for the foreseeable future (Barrett et al. 2013). We may need to remind ourselves that the 'human well-being' dimension of ecosystem services does not only include utilitarian policies, but also cultural, societal, and individual factors (Polishchuck and Rauschmayer 2012; Box 1.5), and their inclusion will be the only way forward to allow us to preserve agricultural ecosystem services across developed and developing countries. Uncertainties in how to deal with the ethical imperative to conserve nature is one of the 'elephants in the room' highlighted by Macdonald and Willis (2013).

Box 1.5 Can biophilia help pay for conservation?

A conspicuous area for likely progress in both costing nature, and for exploiting its variation across society, concerns the value of nature for human health. Edward O. Wilson used the term 'biophilia' for the notion that contact with nature was a deep-seated requirement for human well-being; a definition hinting at an innate human affiliation with nature (Hughes et al. 2013), and how a 'dose' of nature might be prescribed as medicine to aid with health issues (Barton and Pretty 2010). As Leo Tolstoy anticipated, linkage with nature nurtures happiness ("One of the first conditions of happiness is that the link between Man and Nature shall not be broken"). One of the aspects of this which is as yet poorly understood is, if human health benefits can be established securely, then exactly what quantity and quality of experience is necessary to achieve the effect? (Hughes et al. 2013). Does a plant in the office equal a stroll in the countryside? Does a stroll in a monoculture equal a stroll in the rainforest? What is certain is that a small but detectable effect of nature on human health would have enormous implications for its value, and, if some degree of exposure to the rural environment is necessary, then, in the UK, this almost inevitably involves farmland.

So, can biophilia pay for conservation? Hughes et al. (2013) proposed the Biodiversity Leverage Hypothesis, in

which 'the health benefits of engagement with the natural environment are sufficient to lever, directly, conservation of health-giving aspects of the environment', and indirectly, propitiating co-benefits to biodiversity on a wider scale.

One of the direct health benefits from nature is its capacity for lowering levels of stress and promoting healthy activities. Work-related stress causes national financial losses, for example, in Britain alone, stress, depression, and anxiety cost 10.8 million workdays in 2010–2011 (HSE 2011). However, residents in housing with access to trees and grass coped better with life's major issues (Mitchell and Popham 2008). UK death rates, especially those from circulatory disease, are lower with proximity to green spaces (Mitchell and Popham 2008). Would society as a whole, and the economy, benefit from a reduction in sick-leave and health-care expenditure?

In the Biodiversity Leverage Hypothesis, we argue that even if the ecosystem service of health benefits provided by nature does not lever direct investment in wildlife conservation, indirect investment can be accrued through pleiotropic co-benefits. For example, if the aim is to halt forest habitat loss in an area, to provide health gains, the trees will be protected and the umbrella species effect will also help to conserve smaller organisms (e.g. beetles, arachnids) as a 'pleiotropic' co-benefit (Collins et al. 2011).

continued

---

[14] For example, see Owens (2008).

Box 1.5 *Continued*

We further reinforce the hypothesis by considering an 'engagement' dividend to biodiversity conservation, provided by the behavioural changes in individuals prompted by the natural environment to adopt healthier lifestyles, which in turn, encourage people to connect with nature. Motivation and enthusiasm for the environment, which would put pressure on environmental policy, can only be achieved with interaction and experiences—especially as children—with nature. Figures show that only a very small part of UK society visit protected areas (Booth et al. 2010), but since farmland in the UK is available to the public by

Rights of Way, the scope for engagement by the public is promising.

Ultimately, a full life-cycle economic analysis will be necessary to connect human health with well-being and nature (and the costs of accessing nature, e.g. car parks, maintenance, loss of delicate species due to trampling). In the meantime, The National Health Service might save money by investing in the Natural Health Service and, since the citizen largely pays for farmland, and sometimes lives within reach of it, what better service than health and well-being, to add to those being provided by agri-environment schemes.

The volatile nature of valuations is reflected in a survey of UK farmers that we undertook in 1981 (see Macdonald and Johnson 2000), showing that the strategy for 47% of farmers was one of hedgerow removal, mainly for 'efficiency', and also one of removing scrubland, open water, ponds, ditches, and trees (Macdonald and Feber 2015: Chapter 14). However, in 1998 we repeated the survey and by then, only 23% of farmers removed hedgerows, while scrubland, open water, pond, and ditch removal had decreased, and no farmer had removed trees. These farmers explained the motive for their changed behaviour in terms of a growing emphasis on 'wildlife' and the shift in support from production towards conservation, although, unsurprisingly, different sorts of people react differently when asked about their willingness to pay (WTP) for conservation: wealthier and more informed respondents were willing to pay more (White et al. 2001). Cultural values, as well as altruistic and egoistic attitudes, can also have a significant impact on WTP amounts (Ojea and Loureiro 2007), and we have found that people are willing to pay more for high profile charismatic species (e.g. tigers generate higher premium payments than duikers; Collins et al. 2011).

The understanding that everything is linked to everything else, in some quantifiable 'butterfly effect', has led to the necessity for Life Cycle Analysis (LCA). LCA assesses the overall environmental impact of products, processes, or services, and may help to quantify and value the environment. It follows the 'cradle-to-grave' approach on the life of a product: resource extraction (e.g. land use), material production, manufacture, consumption/use, and the end of life of the product, for example, is it recycled, how is it

collected and how is the waste disposed of (Rebitzer et al. 2004)? This allows for estimation of the environmental impacts, such as climate change, eutrophication, acidification, effects on human health and ecosystems, depletion of resources, water and land use, noise, etc. (Rebitzer et al. 2004). LCA has also been used to compare organic versus conventional farming in terms of greenhouse gas emissions from milk production (Flysjö et al. 2012), and between integrated, organic, extensive, and intensive farming production (e.g. Nemecek et al. 2011a,b; Chapter 7, this volume). The difficulty in accounting for non-obvious impacts is still a hindrance, for example, LCA of livestock usually omits impacts of water usage and the use of antibiotics (Flysjö et al. 2012; Reckmann et al. 2012), and there is a lack of development of environmental performance indicators that are adapted to tropical conditions (Eshun et al. 2011).

The vogue, globally, is to develop financial mechanisms to change human behaviour in order to benefit conservation (fully reviewed in Barrett et al. 2013). For example, those that treasure large carnivores (the international community) can directly pay those that detest them (local farmers), to tolerate them. The idea of compensation payments (e.g. for stock killed) is open to abuse, and, continuing with the extreme example of large carnivores, all too often the costs can outweigh the benefits (Dickman et al. 2011). An alternative, with greater market sophistication, is making conservation payments (rather than to tolerate damage) to, for example, Mexican ranchers, when camera traps record carnivores (e.g. jaguar, puma, ocelot) on their land (Dickman et al. 2011). This market-savvy approach is very familiar in the UK, in the guise of agri-environment schemes (AES; Box 1.4); the impacts of AES on farmland birds are explored in Chapter 5, this volume. The service in this case is the provision of nature and an environment in good heart; and financial rewards are provided to farmers and landowners to manage their land in an environmentally sensitive manner, enhancing landscapes, maintaining historical interest, and encouraging access to the countryside (Box 1.5). AESs are based on the concept of 'market failure' and the public goods that the environment produces. Since the public goods are socially desirable but unprofitable, to obtain a reward through the market place, the government intervenes to purchase them to improve social welfare (Barrett et al. 2013). AESs are ultimately a financial mechanism to change the behaviour of landowners, who will accept payment for the policy that offers most benefits at the time, whether production or conservation.

Many of the chapters in this book are about understanding the interface between agriculture in the British landscape, and how to conserve Nature. For most of our lifetimes, much of this conservation has been about damage limitation and, latterly, on farmland, the beginnings of repair. The WildCRU is proud to have played its part, along with many others, in that journey, as described in this book and its sister volume (Macdonald and Feber 2015). That said, to borrow from the title of Chris Patten's (2009) book about the future of nation states: 'What next?'. Many of the countryside's ailments remain the same, so certainly there is great value in prescribing more of the same: better evidence-based solutions, packaged within creative policies to hold the line for Nature. However, Patten's answer for the future of nation states (which, after all, are directly relevant, being the operational units that formulate and deliver the environmental policy that will shape both food production and nature conservation in the countryside) was to look for realistically radical leaps to a better future. As our balloon rises and extends our view from the English lowlands to the wider horizons of the UK's landscape, we think back with legitimate nostalgia to wonderful elements of the long-lost wild past. Large mammals have a special place in many human value systems (Macdonald et al. 2013), so think of the wolves *Canis lupus*, lynx *Lynx lynx*, and even bear *Ursus arctos* that once thrived here beside our recent ancestors, along with wild boar *Sus scrofa*, beaver *Castor fiber*, and elk *Alces alces*. There is nothing irrational or foolishly sentimental in seeking inspiration from these animals for a radical future, inspiration with a hard-nosed, science-based, and policy-relevant character. It is in this spirit that the last chapter of this book, Chapter 16, looks to the Scottish Highlands and the possibilities for restoring ecosystems with Caledonian pine, prey, and predators. We ask whether it is possible to restore natural processes in a landscape of human endeavour to balance biodiversity, ecosystem services, and self-sustainability, exploring a goal that is not to recreate the past, but rather to build a present informed by the past and fit for the future (Macdonald 2009).

In undertaking a journey it's well to keep in mind the destination. Taking the long view, where might we hope to reach? A good starting point is to remember that everything is connected to everything else, that when it comes to conservation, biology is necessary but not sufficient, and what is needed is extreme interdisciplinarity that weds the natural and social sciences and the environmental and human dimensions of the countryside. And just as no one academic discipline can provide all the answers, so too no one section of the community has the prerogative to do so. It takes two to tango, and while our perspective is as biologists, the landscape we have described from our hot-air balloon is the same one also overseen by farmers to whom it delivers a livelihood, and the same one ultimately delivering food and the countryside to the public. We are mindful that farmers, as delivery agents of the future countryside, may often be bemused by the efforts of policy wonks to create a framework for their efforts. Too often, the farmer, musing on the uncertainties of CAP reform that is stutteringly being articulated as we write, is reminded of *We, the unwilling, led by the unknowing, are doing the impossible for the ungrateful. We have done so much, for so long, with so little, we are now qualified to do anything with nothing*[15]. We are all in this together, and so one answer to Patten's question might be a human population enjoying a healthy, equitably high and sustainable standard of living, alongside functioning ecosystems populated with 'natural' levels of biodiversity. This vision for human well-being doubtless relies on advanced technology to sustain people in cleverly engineered 'green' towns supported by sophisticated communications networks. It relies on human ingenuity and rationality rising to Hamlet's challenge because, when it comes to saving the environment, and thus ourselves, there are indeed more things in heaven and earth to be considered, than have so far been dreamt of in our philosophy. It relies on valuing the spiritual in Nature. Optimizing

---

[15] Quote from Konstantin Josef Jireček.

the solutions will require the father-and-mother of all cost–benefit analyses—perhaps the environmental analogue of QALYs—the Quality Adjusted Life Years used to inform tough decisions in medicine (Macdonald 2013).

If each person is to enjoy a satisfyingly high standard of living, then the only way this can be achieved while leaving sufficient space to deliver the associated vision for the natural world (itself a prerequisite for the desired well-being of the people), will be for there to be only a fraction of the current number of people. Political compulsion and the biblical mechanisms of famine, pestilence, and strife are similarly horrific—let those apocalyptic horsemen keep their ghastly steeds stabled out of harm's way in Mordor—so the only hopeful road to this goal is one of phased population reduction over many generations. This would require an inter-generational pact for which we know of no precedent in human history.

So, while much of the research that populates these pages, and those of Macdonald and Feber (2015), concern the cogs of ecosystem function, and are earthy in their groundedness, they seamlessly connect not just to national policies, but to the grand issues of our age. It is the role of conservation biologists to hold a mirror to society of the environmental consequences of its choices. So, in 1968, Paul Ehrlich spoke prophetically about the Population Bomb and this is what he and Robert Pringle say now: 'we know where biodiversity will go from here in the absence of a rapid, transformative intervention: up in smoke; toward the poles and under water; into crops and livestock; onto the table and into yet more human biomass; into fuel tanks; into furniture, pet stores, and home remedies for impotence; out of the way of more cities and suburbs; into distant memory and history books'[16]. Averting this ghastly prediction may lie partly in choices impacting wild places and magnificent creatures in far-flung corners of the earth, but it lies also in our own backyard, indeed farmyard. The purpose of these two books is to provide some of the evidence for Ehrlich's 'transformative intervention', to secure a better future for farmland wildlife.

Acknowledgements

We are grateful to Paul Johnson for his input into this chapter.

---

[16] Ehrlich and Pringle (2008).

References

Baker, D.J., Freeman, S.N., Grice, P.V., and Siriwardena, G.M. (2012). Landscape-scale responses of birds to agri-environment management: a test of the English Environmental Stewardship scheme. *Journal of Applied Ecology*, **49**, 871–882.

Balfour, E.B. (1943). *The Living Soil*. Faber & Faber, London.

Barrett, C.B., Bulte, P., Ferraro, P., and Wunder, S. (2013). Economic instruments for nature conservation. In D.W. Macdonald and K.J. Willis, eds., *Key Topics in Conservation Biology 2*, pp. 59–73. Wiley-Blackwell, John Wiley & Sons Ltd., Oxford.

Barton, J. and Pretty J. (2010). What is the best dose of nature and green exercise for improving mental health? A multi-study analysis. *Environmental Science and Technology*, **44**, 3947–3955.

Bellwood, P. (2001). Early agriculturalist population diasporas? Farming, languages, and genes. *Annual Review of Anthropology*, **30**, 181–207.

Beynon, S.A., Mann, D.J., Slade, E.M., and Lewis, O.T. (2012). Species-rich dung beetle communities buffer ecosystem services in perturbed agro-ecosystems. *Journal of Applied Ecology*, **49**, 1365–1372.

Blunden, J. and Curry, N. (1988). *A Future for our Countryside*. Wiley-Blackwell, Oxford.

Bogucki, P. (1996). The spread of early farming in Europe. *American Scientist*, **84**, 242–253.

Booth, J.E., Gaston, K.J., and Armsworth, P.R. (2010). Who benefits from recreational use of protected areas? *Ecology and Society*, **15**, 19.

Brassley, P., Segers, Y., and Van Molle, L. eds. (2012). *War, Agriculture, and Food: Rural Europe from the 1930s to the 1950s*. Routledge.

Breeze, T.D., Bailey, A.P., Balcombe, K.G., and Potts, S.G. (2011). Pollination services in the UK: How important are honeybees? *Agriculture, Ecosystems and Environment*, **142**, 137–143.

Butler, S.J., Brooks, D., Feber, R.E., Storkey, J., Vickery, J.A., and Norris, K. (2009). A cross-taxonomic index for quantifying the health of farmland biodiversity. *Journal of Applied Ecology*, **46**, 1154–1162.

Carey, P.D., Wallis, S., Chamberlain, P.M., et al. (2008). *Countryside Survey: UK Results from 2007*. NERC/Centre for Ecology and Hydrology (CEH Project Number: C03259).

Cohen, J.E. (1995). *How Many People can the Earth Support?* Norton, New York.

Collins, M.B., Milner-Gulland, E.J., Macdonald, E.A., and Macdonald, D.W. (2011). Pleiotropy and charisma determine winners and losers in the REDD + game: all biodiversity is not equal. *Tropical Conservation Science*, **4**, 261–266.

Conley, D.J., Paerl, H.W., Howarth, R.W., et al. (2009). Controlling Eutrophication: Nitrogen and Phosphorus. *Science*, **323**, 1014–1015.

Defra (2000). *Economic Evaluation of the Countryside Stewardship Scheme*, MAFF, London.

Defra (2008). *Environmental Stewardship: Review of progress.* Defra—Natural England 2008.

Defra (2011). *Biodiversity 2020: a strategy for England's wildlife and ecosystem services.*

Defra (2012a). *2012 Review of Progress in Reducing Greenhouse Gas Emissions from English Agriculture.*

Defra (2012b). *UK BAP.* Available at: <https://www.gov.uk/>, accessed August 2014.

Defra (2014a). *Implementation of the new Common Agricultural Policy (CAP) regulations in England.* Available at: <https://www.gov.uk/>, accessed June 2014.

Defra (2014b). *Common Agricultural Policy (CAP) reform.* Available at: <https://www.gov.uk/>, accessed June 2014.

Delaplane, K.S. and Mayer, D.F., eds. (2000). *Crop Pollination by Bees.* CABI Publishing, Wallingford, pp. 360.

Dempsey, J. and Robertson, M.M. (2012). Ecosystem services: Tensions, impurities, and points of engagement within neoliberalism. *Progress in Human Geography*, **36**, 758–779.

Dewey, P.E. (1989). *British Agriculture in the First World War.* Routledge.

Diaz, R.J. and Rosenberg, R. (2008). Spreading dead zones and consequences for marine ecosystems. *Science*, **321**, 926–929.

Dickman, A.J., Macdonald, E.A., and Macdonald, D.W. (2011). A review of financial instruments to pay for predator conservation and encourage human–carnivore coexistence. *Proceedings of the National Academy of Sciences*, **108**, 13937–13944.

Dickman, C.R., Pimm, S.L., and Cardillo, M. (2007). The pathology of biodiversity loss: the practice of conservation. In D. Macdonald and K. Service, eds., *Key Topics in Conservation Biology*, pp. 1–16. Blackwell Publishing, Oxford.

Driscoll, C. and Macdonald, D. (2010). Top dogs: wolf domestication and wealth. *Journal of Biology*, **9**, 10.

Driscoll, C.A., Menotti-Raymond, M., Roca, A.L. et al. (2007). The near eastern origin of cat domestication. *Science*, **317**, 519–523.

Dutton, A., Edwards-Jones, G., and Macdonald, D.W. (2010). Estimating the Value of Non-Use Benefits from Small Changes in the Provision of Ecosystem Services. *Conservation Biology*, **24**, 1479–1487.

Ehrlich, P.R. and Pringle, R.M. (2008). Where does biodiversity go from here? A grim business-as-usual forecast and a hopeful portfolio of partial solutions. *PNAS*, **105**, 11579–11586.

Emery, S.B. and Franks, J.R. (2012). The potential for collaborative agri-environment schemes in England: Can a well-designed collaborative approach address farmers' concerns with current schemes? *Journal of Rural Studies*, **28**, 218–231.

Eshun, J.F., Potting, J., and Leemans, R. (2011). LCA of the timber sector in Ghana: preliminary life cycle impact assessment (LCIA). *International Journal of Life Cycle Assessment*, **16**, 625–638.

European Commission (2011). European Parliament resolution of 20 April 2012 on: *Our life insurance, our natural capital: an EU biodiversity strategy to 2020* (2011/2307(INI)). Available at: <http://ec.europa.eu/>, accessed January 2015.

European Commission (2013). *The Birds Directive.* Available at: <http://ec.europa.eu/environment/>, accessed June 2013.

European Commission (2013a). *CAP Post-2013: key graphs and figures.* European Commission: Agriculture and Rural Development. Available at: <http://ec.europa.eu/>, accessed 23 January 2013.

European Commission (2013b). *Agri-environment measures.* Available at: <http://ec.europa.eu/>, accessed November 2013.

Flysjö, A., Cederberg, C., Henriksson, M., and Ledgard, S. (2012). The interaction between milk and beef production and emissions from land use change—critical considerations in life cycle assessment and carbon footprint studies of milk. *Journal of Cleaner Production*, **28**, 134–142.

Gallai, N., Salles, J.M., Settele, J., and Vaissiere, B.E. (2009). Economic valuation of the vulnerability of world agriculture confronted with pollinator decline. *Ecological Economics*, **68**, 810–821.

HSE (2011). *Working Days Lost.* Health and Safety Executive, London.

Hughes, J., Pretty, J., and Macdonald, D.W. (2013). Nature as a source of health and well-being: is this an ecosystem service that could pay for conserving biodiversity? In D.W. Macdonald and K.J. Willis, eds., *Key Topics in Conservation Biology 2*, pp. 143–160. Wiley-Blackwell, John Wiley & Sons Ltd., Oxford.

IUCN (2014). The IUCN Red List of threatened species. Available at: <http://www.iucnredlist.org/>, accessed June 2014.

JNCC and Defra (2012). *UK Post-2010 Biodiversity Framework.* Published by JNCC and Defra on behalf of the Four Countries' Biodiversity Group.

Jones, G., Jones H., Charles, M.P., et al. (2012). Phylogeographic analysis of barley DNA as evidence for the spread of Neolithic agriculture through Europe. *Journal of Archaeological Science*, **39**, 3230–3238.

Kass, G.S., Shaw, R.F., Tew, T., and Macdonald, D.W. (2011). Securing the future of the natural environment: using scenarios to anticipate challenges to biodiversity, landscapes and public engagement with nature. *Journal of Applied Ecology*, **48**, 1518–1526.

Kleijn, D., Baquero, R.A., Clough, Y., et al. (2006). Mixed biodiversity benefits of agri environment schemes in five European countries. *Ecology Letters*, **9**, 243–254.

Klein, A.M., Vaissiere, B.E., Cane, J.H., Steffan-Dewenter, I., Cunningham, S.A., Kremen, C., and Tscharntke, T. (2007). Importance of pollinators in changing landscapes for world crops. *Proceedings of the Royal Society B*, 274 (1608), 303–313.

Krebs, J.R., Wilson, J.D., Bradbury, R.B., and Siriwardena, G.M. (1999). The second silent spring? *Nature*, **400**, 611–612.

Lawton, J.H., Brotherton, P.N.M., Brown, V.K., et al. (2010). *Making Space for Nature: a review of England's wildlife sites and ecological network.* Report to Defra.

Losey, J.E. and Vaughan, M. (2006). The economic value of eco‑
logical services provided by insects. *Bioscience*, **56**, 311–323.

Luck, G.W., Chan, K.M.A., Eser, U., et al. (2012). Ethical Con‑
siderations in On‑Ground Applications of the Ecosystem
Services Concept. *BioScience*, 62, 1020–1029.

Mabey, R. (1980). *The Common Ground : A Place for Nature in
Britain's Future?* Hutchinson in association with the Na‑
ture Conservancy Council, London.

Macdonald, D. (2009). Lessons Learnt and Plans Laid: Seven
Awkward Questions for the Future of Reintroductions. In
M. Hayward and M. Somers, eds., *Reintroduction of Top‑
Order Predators*. Wiley‑Blackwell, Hoboken, NJ, USA.

Macdonald, D.W. (2013). From ethology to biodiversity: case
studies of wildlife conservation. *Nova Acta Leopoldina*, **380**,
111–156.

Macdonald, D.W., Boitani, L., Dinerstein, E., Fritz, H., and
Wrangham, R. (2013). Conserving large mammals: are
they a special case? In D.W. Macdonald and K.J. Willis,
eds., *Key Topics in Conservation Biology 2*, pp. 277–312.
Wiley‑Blackwell, John Wiley & Sons Ltd., Oxford.

Macdonald, D.W. and Burnham, D. (2011). *The state of Brit‑
ain's mammals*. WildCRU, Oxford.

Macdonald, D.W. and Feber, R.E., eds. (2015). *Wildlife Con‑
servation on Farmland: Conflict in the Countryside*. Oxford
University Press, Oxford.

Macdonald, D.W. and Johnson, P.J. (2000). *Farmers and the
custody of the countryside: trends in loss and conservation of
non‑productive habitats 1981–1998. Biological Conservation*,
94, 221–234.

Macdonald, D.W., Loveridge, A., and Rabinowitz, A. (2010).
Felid futures: crossing disciplines, borders, and genera‑
tions. In D. Macdonald and A. Loveridge, eds., *The Biology
and Conservation of Wild Felids*, pp. 599–649. Oxford Uni‑
versity Press, Oxford.

Macdonald, D.W. and Smith, H.E. (1991). New perspectives
on agro‑ecology: between theory and practice in the agri‑
cultural ecosystem. In L.G. Firbank, N. Carter, J.F. Dar‑
byshire, and G.R. Potts, eds., *The ecology of temperate cer‑
eal fields* (*32nd Symposium of the British Ecological Society*),
pp. 413–448. Blackwell Scientific Publications, Oxford.

Macdonald, D.W. and Willis, K.J. (2013). Elephants in the
room: tough choices for a maturing discipline. In D.W.
Macdonald and K.J. Willis, eds., *Key Topics in Conserva‑
tion Biology 2*, pp. 469–494. Wiley‑Blackwell, John Wiley &
Sons Ltd., Oxford.

Maclennan, S.D., Groom, R.J., Macdonald, D.W., and Frank,
L.G. (2009). Evaluation of a compensation scheme to
bring about pastoralist tolerance of lions. *Biological Con‑
servation*, **142**, 2419–2427.

Martin, J. (2011). The transformation of lowland game shoot‑
ing in England and Wales since the Second World War:
the supply side revolution. *Rural History*, **22**, 207–226.

Mattison, E.H.A. and Norris, K. (2005). *Bridging the gaps be‑
tween agricultural policy, land‑use and biodiversity. Trends in
Ecology & Evolution*, **20**, 610–616.

Millennium Ecosystem Assessment (ME) (2005). *Ecosystems
and Human Well‑being: Synthesis*.

Mitchell, R. and Popham, F. (2008). Effect of exposure to nat‑
ural environment on health inequalities: an observational
population study. *Lancet*, **372**, 1655–1660.

NE (2010). *Lost life: England's lost and threatened species*. Avail‑
able at: <http://www.naturalengland.org.uk/>.

NE (2012). *Ecosystem services from Environmental Stewardship
that benefit agricultural production*. Natural England Com‑
missioned Report NECR102.

NE (2013a). *Entry Level Stewardship. Environmental Steward‑
ship handbook. Fourth edition‑January 2013*.

NE (2013b). *Higher Level Stewardship. Environmental Steward‑
ship handbook. Fourth edition‑January 2013*.

NE (2014). Environmentally Sensitive Areas scheme (ESA).
<http://www.naturalengland.org.uk/>, accessed July
2014.

Nemecek, T., Hugenin‑Elie, O., Dubois, D., and Gaillard, G.
(2011a). Life cycle assessment of Swiss farming systems: I.
Integrated and organic farming. *Agricultural Systems*, **104**,
217–232.

Nemecek, T., Hugenin‑Elie, O., Dubois, D., Gaillard, G.,
Schaller, B., and Chevert, A. (2011b). Life cycle assessment
of Swiss farming systems: II. Extensive and intensive pro‑
duction. *Agricultural Systems*, **104**, 233–245.

Ojea, E. and Loureiro, M.L. (2007). Altruistic, egoistic and
biospheric values in willingness to pay (WTP) for wild‑
life. *Ecological Economics*, **63**, 807–814.

Oslisly, R., White, L., Bentaleb, I., et al. (2013). Climatic and
cultural changes in the west Congo Basin forests over the
past 5000 years. *Philosophical Transactions of the Royal Soci‑
ety B*, **368**, 20120304.

Owens, S. (2008). Why conserve marine environments? *En‑
vironmental Conservation*, **35**, 1–4.

Pacini, N., Harper, D., Henderson, P., and Le Quesne, T.
(2013). Lost in muddy waters: freshwater biodiversity. In
D.W. Macdonald and K.J. Willis, eds., *Key Topics in Conser‑
vation Biology 2*, pp. 184–203. Wiley‑Blackwell, John Wiley
& Sons Ltd., Oxford.

Parliament (2014). Enclosure Act. Available at: <http://
www.parliament.uk/>, accessed June 2014.

Patten, C. (2009). *What Next? Surviving the Twenty‑first Cen‑
tury*. p. 544. Penguin, UK.

Polishchuck, Y. and Rauschmayer, F. (2012) Beyond 'bene‑
fits'? Looking at ecosystem services through the capabil‑
ity approach. *Ecological Economics*, **81**, 103–111.

POST (2010). *Insect Pollination POST Note 348*. Parliamentary
Office of Science and Technology, London.

Potts, S.G., Biesmeijer, J.C., Kremen, C., Neumann, P., Sch‑
weiger, O., and Kunin, W.E. (2010). Global pollinator de‑
clines: trends, impacts and drivers. *Trends in Ecology and
Evolution*, **25**, 345–353.

Rackham, Oliver (1986). *The History of the Countryside: The
full fascinating story of Britain's landscape*. J.M. Dent & Sons
Ltd., London.

Rackham, O. (1994). *The illustrated history of the countryside*.
Weidenfeld and Nicolson, London.

Rackham, O. (2003). *The illustrated history of the countryside*.
Weidenfeld & Nicolson, London.

Rebitzer, G., Ekvall, T., Frischknecht, R., et al. (2004). Life cycle assessment Part 1: Framework, goal and scope definition, inventory analysis, and applications. *Environmental International*, **30**, 701–720.

Reckmann, K., Traulsen, I., and Krieter, J. (2012). Environmental Impact Assessment—methodology with special emphasis on European pork production. *Journal of Environmental Management*, **107**, 102–109.

Rickard, S. (2012). *Liberating farming from the CAP: IEA Discussion Paper No. 37*. The Institute of Economic Affairs, London.

Robinson, R.A. and Sutherland, W.J. (2002). Post-war changes in arable farming and biodiversity in Great Britain. *Journal of Applied Ecology*, **39**, 157–176.

Rockström, J., Steffen, W., and Noone, K. et al. (2009). A safe operating space for humanity. *Nature*, **461**, 472–475.

Royal Forestry Society (2013). <http://www.rfs.org.uk/>, accessed on 24 January 2013.

RSPB (2013). <http://www.rspb.org.uk/>.

SCBD (2010). Secretariat of the Convention on Biological Diversity. COP-10 Decision X/2. Available at: <http://www.cbd.int/>, accessed January 2015.

Skogen, M.D., Soiland, H., and Svendsen, E. (2004). Effects of changing nutrient loads to the North Sea. *Journal of Marine Systems*, **46**, 23–38.

State of Nature (2013). *State of Nature Report*. The State of Nature partnership.

Stewart, A.J.A., New, T.R., and Lewis, O.T. (2007). *Insect Conservation Biology*. Proceedings of the Royal Entomological Society's 23rd Symposium. CABI.

TEEB (2010). *The Economics of Ecosystems and Biodiversity Ecological and Economic Foundations*. Edited by Pushpam Kumar. Earthscan, London and Washington.

Thomas, K. (1984). *Man and the natural world: changing attitudes in England, 1500–1800*. Penguin, London.

Tilman, D., Balzer, C., Hill, J., and Befort, B.L. (2011). Global food demand and the sustainable intensification of agriculture. *PNAS*, **108**, 20260–20264.

Tresset, A. and Vigne, J-D. (2011). Last hunter-gatherers and first farmers of Europe. *Comptes Rendus Biologies*, **334**, 182–189.

UK Biodiversity Partnership (2010). *The UK Biodiversity Action Plan: Highlights from the 2008 reporting round*, published by JNCC on behalf of the UK Biodiversity Partnership.

UK National Ecosystem Assessment, NEA (2011). The UK National Ecosystem Assessment: Synthesis of the Key Findings. UNEP-WCMC, Cambridge.

UNEP (2011). *UNEP Yearbook 2011. Emerging issues in our global environment*. United Nations Environment Programme.

Van Andel T, Zangger E, and Demitrack A. (1990). Land use and soil erosion in prehistoric and historical Greece. *Journal of Field Archaeology*, **17**, 379–396.

White, P.C.L., Bennett, A.C., and Hayes, E.J.V. (2001). The use of willingness-to-pay approaches in mammal conservation. *Mammal Review*, **31**, 151–167.

White, P.C.L., Gregory, K.W., Lindley, P.J., and Richards, G. (1997). Economic values of threatened mammals in Britain: a case study of the otter *Lutra lutra* and the water vole *Arvicola terrestris*. *Biological Conservation*, **82**, 345–354.

CHAPTER 2

From weed reservoir to wildlife resource—redefining arable field margins

Helen Smith, Ruth E. Feber, and David W. Macdonald

> . . . sweet flowers are slow and weeds make haste.
> **William Shakespeare, Richard III,**
> **Act 2, Scene 4.**

2.1 Introduction

From the late 1960s, the UK saw a progressive realization that effective biodiversity conservation would require more than a set of isolated, protected sites. The matrix in which these sites were embedded was very rapidly becoming impoverished by agricultural intensification, and action was urgently needed to retain biodiversity in the wider countryside (e.g. Barber 1970). But delivery of biodiversity conservation in intensively farmed arable areas was difficult to achieve and, for many, difficult to accept. Of many potential delivery mechanisms, the boundaries between arable fields soon emerged as areas where biodiversity could be promoted without significant loss of either cropped area or profit.

The traditional, post-enclosure role of field boundaries, delimiting fields and offering an impenetrable and sheltering barrier, were becoming obsolete. In the post-war era, field boundaries were seen increasingly as a waste of land, an impediment to the efficient use of farm machinery, and as sources of crop weeds. As a result, many were removed to amalgamate fields and most of those that remained were managed in ways that severely reduced their value to wildlife. Two-fifths of Britain's hedgerows were lost in the 40 years following the end of World War II (Rackham 1986), and hedgerow loss continued at an estimated 5378 miles per year between 1984 and 1990 (Barr et al. 1991, 1994). We conducted questionnaire surveys of farmers' attitudes in 1981 and 1989 which suggested that, over the ten-year periods preceding each survey, farmers' principal

motivations for the destruction of non-productive habitats, including hedgerows, were 'efficiency', 'tidiness', and 'weed control' (Macdonald and Johnson 2000; Macdonald and Feber 2015: Chapter 14).

Some of the greatest forces for reducing the impact of intensification on the countryside came initially from within the farming community. In 1983, leading farmer and conservationist Poul Christensen identified both the potential of field boundaries as key areas for biodiversity delivery on farms, and the 'woeful' lack of evidence-based advice available on how to achieve this (Christensen 1983). Published in a British Crop Protection Council monograph on vegetation management, this call to action was rapidly followed by a seminal series of monographs and conference proceedings which brought to the forefront new research on field boundary management (e.g. Bunce and Howard 1990; Firbank et al. 1990; Clarke 1992; Boatman 1994; Boatman et al. 1999).

Although field boundaries often comprise elements of several habitats—woodland in the form of hedges, wetland in ditches, tall herb grassland on adjacent margin strips, and ephemeral-dominated communities on field edges—they present unique management challenges that demand novel research. Their potential value lies not only in these individual elements, but also in the diversity created by their juxtaposition. However, realizing this potential is very difficult because their narrow linear nature exposes them disproportionately to outside influence and edge effects, particularly in the context of intensive arable systems.

One of the first major research programmes on the flora and invertebrate fauna of field edges was triggered by a dramatic collapse in grey partridge stocks in the 1950s (Potts 1980; Chapter 13, this volume). The Game Conservancy Trust (now the Game and Wildlife Conservation Trust) showed that up to 90% of Britain's partridge population depended on field margins for breeding. They instigated a wide ranging research programme to investigate the species' demise (e.g. Potts 1986; Potts and Aebischer 1995; Aebischer and Ewald 2004), an important part of which focused on the development of 'conservation headlands'—cropped edges of arable fields that received reduced inputs in order to encourage broad-leaved arable plants and associated invertebrates upon which adult grey partridge and their chicks depend. But there was no evidence that the value of uncropped field margins as game rearing habitat motivated even farmers with shooting interests to any extent (Macdonald and Johnson 2000).

While a growing minority of farmers shared Christensen's view that 'we must farm fields for food and the natural vegetation for posterity', the main incentives for changing management practice on field margins, and for the research needed to inform management prescriptions, were subsidy schemes that paid farmers to deliver biodiversity and other environmental benefits. Increasingly, these were generated by changes within the European Union's Common Agricultural Policy, and accompanied by cross-compliance regulations to ensure that support for biodiversity improvements was not undermined by other agricultural policy initiatives and regulations (e.g. Mitchell 1999; Chapter 1, this volume). Set-aside (Firbank 1998), Environmentally Sensitive Areas (see Dobbs and Pretty 2008), Countryside Stewardship (Carey et al. 2003), and, most recently, Environmental Stewardship (Natural England 2009), have all included options for improved management of field margins. Indeed, recognition of the importance of arable field margins for biodiversity conservation in the wider countryside resulted in their designation as a priority Biodiversity Action Plan (BAP) habitat in 1995. Field boundaries continue to be seen both as places where effective, low-cost delivery of biodiversity gains can be made in farmland and, increasingly, as a means of delivering wider landscape connectivity. By 2009, agri-environment schemes were supporting over 116 000 km of grass buffer strips on arable land and active management of over 163 000 km (41%) of all hedgerows in England (Natural England 2009). Overall, boundary options, especially hedgerow management, feature highly among the most popular Entry Level Scheme options taken up by farmers.

Support for field margin management is likely to continue; its incorporation into the England Biodiversity Strategy (Defra 2011) is intended to protect for the future the, effectively, rented gains made under the agri-environment schemes over the past 20 years.

A large body of applied research on field margin management has been generated by these initiatives both in the UK and continental Europe (see Vickery et al. 2009 for review). However, much of the focus of this has been short-term and goal-directed, specific to particular schemes—a reflection of the short-termism, not only of the schemes themselves, but also of the funding that accompanied them. Its wider applicability is often limited.

2.2 The University Farm field margin experiments

At an early stage in the development of field margin conservation initiatives, we established large-scale experiments that were designed to answer key questions about the management of arable field margins—the grassy strips between the field boundary structure (ditch or hedge) and the crop. Using newly created and existing arable field margins, we asked whether simple and practicable management techniques could be used to enhance their biodiversity in ways that were both sustainable in the medium to long term and agronomically acceptable. At the same time, our experiments provided the biological understanding needed to underpin a wide range of potential management options for enhancing the biodiversity of arable field margins. Although our experiments were all at a single site—the University of Oxford's farm at Wytham, west of Oxford—we used well replicated designs that took account of variation resulting from the wide range of soil types and cropping conditions across the farm. This made our results robust and widely applicable. Most of our monitoring was carried out in the first three years after establishment of the experiments, but we were able to continue one of the experiments for a further ten years, providing a unique understanding of the extent to which short-term successional change predicts longer-term outcomes for biodiversity in this habitat.

We evaluated the management options in terms of ease of management and a range of wildlife conservation outcomes, as well as their ability to deliver other ecosystem services, such as weed control and polyphagous predator abundance (Chapter 3, this volume). At the outset, we made some basic premises about field margin management for biodiversity. The first was that establishment of dense, grass-dominated, largely perennial swards that replicated semi-natural vegetation would provide the most appropriate vegetation for permanent field margins and provide an effective barrier to annual weeds, most of which require open gaps for germination (e.g. Tozer et al. 2008). This premise had two caveats: the first was that it largely precluded the conservation of rare arable (annual) plants, many of which are in severe decline. These species often occur on lighter soils and require specific management, such as regular cultivation, which limits competition from other plants (Walker et al. 2007). The second caveat was that seed mixtures used to establish these swards should be ecologically and visually appropriate for the situation into which they would be sown, whether for forage, green cover, amenity, or wildlife conservation. Particularly in the latter case, they should be of UK, and preferably local provenance (Sackville Hamilton 2001; Wilkinson 2001).

A second premise was that the erosion in width of field margins typical of modern arable fields needed to be reversed to buffer the boundary hedge or ditch from in-field operations and to allow enough space to develop, and manage effectively, grassy strips. Our third premise was that fertilizers and pesticides should be excluded from field margins. Not only are they wasted there but they are also detrimental to wildlife and exacerbate management problems. Their effects are well understood: direct effects, for example reducing populations of beneficial and benign invertebrates (e.g. Holland et al. 2000), and indirect effects, for example through indiscriminate fertilizer application perpetuating high soil nutrient status and an impoverished flora dominated by injurious agricultural weeds (De Cauwer et al. 2006; Hautier et al. 2009). The deliberate spraying-out of hedge bottoms with broad-spectrum herbicide as a means of combatting weeds remained common practice in the UK until it was banned in 2005 under cross-compliance regulations for the Single Payment Scheme (Defra 2004), but the problems of over-spraying and drift remain (Kleijn and Snoeijing 1997; Schmitz et al. 2013).

We carried out two experiments on field margins that were created by fallowing strips of land after ploughing, in autumn 1987. Our first experiment (referred to as the 'hay and silage margins') was used to examine how silage and hay management regimes influenced the development of two, contrasting sown leys on relatively wide, fallowed strips. One was a conventional, modern, timothy–ryegrass–clover ley and the other was a relatively diverse but still simple mixture of indigenous grasses and forbs, based on the

so-called herbal leys popular in the nineteenth century (Stapledon 1943; The Lawson Seed and Nursery Company (Limited) 1877; Foster 1988). We used diversity and species composition of both sown and naturally regenerating components of the swards to evaluate wildlife value and weed problems; their productivity was used to assess their agricultural potential and ease of management for amenity uses.

Our second experiment (referred to as the '2 m margins experiment') focused on the strips of land closest to the field boundary which, we hypothesized, could potentially provide the greatest wildlife gains on arable farmland, but which would also be likely to present the most significant challenges in terms of weed control and practicality of management. We extended the existing field margin width from around 0.5 m to 2 m, by fallowing an arable strip. Swards were created on these fallowed margin strips, either by allowing natural regeneration or by side-stepping a natural succession by sowing a grass and wildflower seed mixture. A range of mowing regimes was imposed on both sward types. We used the naturally regenerating swards to answer questions fundamental to the likely success or failure of natural regeneration as a means of establishing vegetation on restored field margins. We asked first whether the quantity and quality of colonizing species were likely to provide the basis for an acceptable or desirable flora and, second, whether the composition of the new flora could be reliably predicted from potential sources in the vicinity. We investigated whether sowing a grass and wildflower seed mixture, instead of allowing natural regeneration, could be used to deliver greater species richness, including species of more value for wider aspects of biodiversity. We then examined the extent to which the field margin flora could be manipulated by our management regimes to achieve more desirable outcomes for biodiversity and weed control in both the short and longer terms.

2.2.1 The hay and silage margins

A species-rich ley and a conventional ley, managed both with and without fertilizer, were established around the edges of three adjacent fields running down the slope of Wytham Hill. The species used in the leys are given in Box 2.1. Margins around the largest field were 9.6 m wide and were cut for silage. Those around the two smaller fields were 7.2 m wide and were cut for hay (the margin widths chosen to accommodate available harvesting machinery). The different timings of harvest and equipment required for hay and silage cropping regimes made it impracticable to use an experimental design that allowed direct statistical comparison between the results from the two experiments. However, because of our use of adjacent fields spanning the same soil gradient, and very similar initial results obtained for all fields before their management diverged, we were able to make qualitative comparisons between species richness and diversity under the two regimes.

The margins were divided into contiguous 50 m-long plots, half of which were sown with the conventional ley, and the remainder with the species-rich ley (Fig. 2.1). Both were drilled at a rate of 32 kg/ha. Among each ley type, half of the plots were treated with fertilizer levels appropriate for either hay or silage, and half received none (Smith et al. 1997). Hay was cut in the first half of July and silage in late May and late July.

Box 2.1 The seed mixtures used in the Wytham experiments

The hay and silage margins

Conventional ley		Seeds sown/m^2
Lolium perenne	perennial rye-grass (six varieties)	1250.0
Phleum pratense	timothy (early and late varieties)	560.0
Trifolium repens	white clover (cultivar)	700.0

Species-rich ley		
Cynosurus cristatus	crested dog's-tail	194.0
Dactylis glomerata	cock's -foot	508.0
Festuca pratensis	meadow fescue	750.0
Phleum pratense	timothy	1270.0
Poa pratensis	smooth meadow-grass	720.0
Trisetum flavescens	yellow oat-grass	363.0
Achillea millefolium	yarrow	30.0
Ranunculus acris	meadow buttercup	6.0
Rumex acetosa	common sorrel	6.0

The 2 m margins

Forbs		Seeds sown/m^2
Centaurea nigra	common knapweed	21.6
C. scabiosa	greater knapweed	7.5
Clinopodium vulgare	wild basil	90.0
Galium verum	lady's bedstraw	23.1
Hypericum hirsutum	hairy St. John's wort	172.8
H. perforatum	perforated St. John's wort	259.2
Knautia arvensis	field scabious	7.5
Leontodon hispidus	rough hawkbit	7.0
Leucanthemum vulgare	oxeye daisy	172.8
Primula veris	cowslip	39.6
Prunella vulgaris	selfheal	38.9

continued

Box 2.1 *Continued*

The 2 m margins

Forbs		Seeds sown/m²
Ranunculus acris	meadow buttercup	34.6
Ranunculus bulbosus	bulbous buttercup	12.1
Silene latifolia subsp. alba	white campion	23.1
Silene vulgaris	bladder campion	34.6
Torilis japonica	upright hedge parsley	14.4
Tragopogon pratensis	goat's beard	3.0

Grasses		
Cynosurus cristatus	crested dog's-tail	786.0
Festuca rubra ssp. commutata	red fescue	600.0
F. rubra ssp. litoralis	red fescue	360.0
Hordeum secalinum	meadow barley	60.0
Phleum bertolonii	smaller cat's tail	480.0
Poa pratensis	smooth-stalked meadow-grass	1080.0
Trisetum flavescens	golden oat-grass	720.0

Figure 2.1 Experimental silage margin at the University Farm at Wytham. ©Helen Smith.

We measured above-ground biomass (live shoot dry weight) and species composition of the plots three months after sowing and again three years later and used mean weights from each plot to estimate both individual species abundance and overall productivity. We found that our more species-rich grass ley had many advantages over the conventional ley. These included both direct benefits, measured in terms of plant species richness and composition, and indirect benefits in terms of ease of management.

2.2.1.1 Species richness and sward composition

Although numbers of species declined after the first year (Smith et al. 1997), the more species-rich ley not only remained richer in sown species than the conventional ley but also accommodated significantly more species regenerating naturally from the local flora over a three-year period (Table 2.1). These comprised up to 40% of the total in this ley but only 15% in the conventional ley. These attributes resulted from the species composition of the sown mixtures and were modified substantially by both fertilizer treatment and mowing regime.

Fertilizer application reduced the biodiversity of the swards (Table 2.1). Both species richness and diversity (Simpson's diversity index: Southwood 1978) were significantly reduced by the high fertilizer applications entailed in silage management, while only diversity (not species richness) was significantly reduced under the lower application rates entailed in hay

management (Smith et al. 1997). Under silage management, numbers of sown species were also significantly lower in fertilized than in unfertilized plots—this effect was apparent, though not significant, under the lower fertilizer applications of the hay management regime (Table 2.1).

These responses to fertilizer addition were largely driven by a small number of components of the seed mixtures. Two of the sown grasses, cock's-foot *Dactylis glomerata* in the more species-rich ley and perennial rye-grass *Lolium perenne* in the conventional ley, are highly competitive species, able to respond strongly to high soil nutrient status (Cowling and Lockyer 1968; Spedding and Diekmahns 1972). In the more species-rich ley, the poor retention of sown species at high fertility levels resulted from extreme dominance of *D. glomerata*, which comprised over 86% of the sward biomass under silage and 47% under hay management. Meadow fescue *Festuca pratensis*, which established well but then declined sharply, particularly in fertilized plots (e.g. Fig. 2.2), is known to suffer in competition with *D. glomerata* in established swards (Spedding and Diekmahns 1972). *L. perenne* showed a similar response to *D. glomerata*, becoming dominant under all treatments and responding strongly to fertilizer. This was particularly the case under the high application rates of the silage regime, where it eventually comprised over 90% of the biomass of the swards. This suppressed other species, including the sown Timothy *Phleum pratense*, and resulted in the development of extremely

Table 2.1 Mean numbers of all sown and unsown species (per 25 cm × 25 cm quadrat) in each treatment under hay and silage management in 1991. Treatment abbreviations: C − F = conventional ley minus fertilizer, C + F = conventional ley plus fertilizer, R − F = species-rich ley minus fertilizer, R + F = species-rich ley plus fertilizer. Significance levels ***P <0.001, ** P <0.01. * P <0.05, NS not significant.

Treatment	Hay						Silage					
	All		Sown		Unsown		All		Sown		Unsown	
	Mean	SE	Mean	SE	Mean	SE	Mean	SE	Mean	SE	Mean	SE
C − F	3.78	0.30	2.16	0.07	1.63	0.33	2.61	0.13	2.42	0.15	0.18	0.07
C + F	3.56	0.47	2.13	0.15	1.44	0.56	2.33	0.12	2.13	0.17	0.21	0.08
R − F	8.59	0.28	6.00	0.18	2.63	0.13	7.64	0.47	5.79	0.31	1.86	0.21
R + F	7.14	0.75	4.89	0.58	2.25	0.24	5.04	0.62	3.89	0.35	1.14	0.34
Significance of:												
Ley type	***		***		**		***		***		***	
Fertilizer	NS		NS		NS		**		***		NS	
Interaction	NS		NS		NS		*		*		NS	

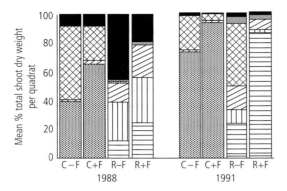

Figure 2.2 Productivity of individual species under silage management in 1988 and 1991. Mean percentage of the total shoot dry weight per quadrat, represented by the following: unsown species (excluding *Trifolium repens*) ■, *Trifolium repens* ▨, *Phleum pratense* ▧, *Lolium perenne* ▨, *Festuca pratensis* ⬚, *Dactylis glomerata* ⊟, Other sown species ⬚. Treatments are C, conventional ley; R, species-rich ley; −F, unfertilized; + F, fertilized. From: Smith et al. (1997).

species-poor swards (e.g. Fig. 2.2). These results show clearly that highly competitive species are unsuitable for inclusion in grass mixtures designed to encourage botanical diversity on soils with high residual fertility. Where the aim is to provide low-cost dense tussocky grass margins, for example, to provide habitat for overwintering polyphagous predators, the inclusion of competitive species such as *D. glomerata* is an option, although such margins will have limited value for other invertebrates, such as pollinators, which require nectar-providing plants (Thomas et al. 2002). Our results also underline the importance of excluding fertilizer drift from both existing field margin grass swards and naturally regenerating swards on expanded field margins.

A strong, but reciprocal response to fertilizer in a third species, the white clover *Trifolium repens* cultivar, included in the conventional ley, demonstrated that it too was unsuitable for inclusion in grass mixtures designed to maintain or enhance floristic species richness. The ability to fix nitrogen made this species highly competitive in unfertilized swards. The very high productivity of *T. repens* in the unfertilized plots, and of *L. perenne* in the fertilized plots, explains why the species richness of this ley remained both low and relatively unaffected by fertilizer addition; although *T. repens* declined over three years, it still remained significantly more abundant in the absence of fertilizer (Fig. 2.2; Smith et al. 1997). Largely as a result of this dominance by a small number of highly competitive

species, the conventional ley had around 50% less bare ground than the more species-rich ley in spring and summer of the second year of the experiment, and 75% less by the autumn (J.S. Rigby, unpublished data). This reduced the potential both for germination and establishment of less competitive colonists (Hautier et al. 2009). As a result, the more species-rich ley proved better able to accommodate colonists from the local flora than the simpler seed mixture. Although there is some evidence that more species-rich swards may be less invasible (Knops et al. 1999; Van der Putten et al. 2000), our experiment shows that, with this combination of species on agricultural soils, these swards were more prone to invasion. It supports the conclusion of Crawley et al. (1999) that the presence of competitive species is more important than species richness in determining the rate of species invasions on arable field margins, where the difficulties of enhancing species richness are often compounded by the low diversity of potential colonists (see '2 m margins experiments', below).

T. repens did not remain confined to the conventional ley where it was sown. It colonized neighbouring unfertilized plots of the more species-rich ley, where it became the single most abundant species after three years (e.g. Fig. 2.2). By contrast, it remained virtually absent from the fertilized plots where it is likely to have been suppressed by *D. glomerata* (Chestnutt and Lowe 1970; Spedding and Diekmahns 1972). Although *T. repens* added structural and visual diversity to our swards and was an important source of nectar for invertebrates, such as butterflies (Box 2.2), it was also likely to have been a factor in the poor establishment of less competitive sown species and a threat to floristic diversity in the longer term. Indeed, current agri-environment scheme advice is that where Italian ryegrass *Lolium multiflorum* and white clover *Trifolium repens* dominate the vegetation or soil seed bank, they should be controlled in advance of arable reversion to species-rich grassland, to reduce the risk of their out-competing the sown species (Natural England 2010).

Of the other sown grasses in this mixture, *P. pratensis* and golden oat-grass *Trisetum flavescens* increased slowly on the unfertilized plots after poor establishment and remained minor constituents of the swards. The forb species in the mixture were sown at low density (Smith et al. 1997) and, although they persisted, their frequencies were too low for analysis of the impacts of treatment. Such species might be included with more competitive grasses in leys, for example for nectar provision, visual amenity, or improving hay quality, but this is only likely to be cost-effective where competition from the latter is minimized, either by mowing or grazing, or by low residual soil fertility.

Although fertilizer addition had profound effects on the developing species richness and composition of the conventional and more species-rich leys, the different mowing regimes involved in hay and silage management were also influential. Mowing appeared to be a more important factor than fertilizer in determining the rate of loss of naturally regenerating species from the swards. This decline—mostly through loss of annual colonists—was conspicuously greater under silage than under hay management. Over a three-year period, unfertilized plots cut annually for hay lost 28% and 10% of species from the conventional and more species-rich leys respectively, but 49% and 25% of species when cut twice a year for silage (Smith et al. 1997). These two cuts during the main summer flowering period effectively prevented the majority of species from setting seed. In our 2 m margin experiment we showed that, while more frequent mowing can enhance species richness, its timing is critical, particularly in the first few years after establishment when swards are sufficiently open to provide opportunities for seedling establishment. Carefully timed mowing can thus be used to manipulate not only species richness but also the abundance and persistence of individual species.

2.2.1.2 Ease of management of the swards

For many landowners, the viability and acceptability of maintaining field margin swards, principally to benefit biodiversity, hinges critically on management issues. Management demands must be low, inexpensive, and practicable. They include the need to cut swards either to promote diversity or to manipulate species composition by increasing desirable species and controlling pernicious weeds. The ease with which this can be achieved is related to productivity. Once again, our more species-rich ley, particularly when left unfertilized, proved the best management option.

The greater numbers of naturally regenerating species in the more species-rich ley, particularly when it was left unfertilized, might suggest that it would present weed management problems. But this was not the case. After three years it harboured a lower mass of pernicious weeds than the conventional ley, even when unfertilized and colonized by *T. repens* (Smith

---

Box 2.2 Butterflies on the hay and silage margins

Butterflies are a well monitored group and widely regarded as useful indicators of the impacts of land use change on invertebrates, particularly other nectar feeders, such as bumblebees. Populations of many butterfly species have greatly reduced over the last 40 years, and some wider countryside butterfly species are declining at unprecedented rates in the UK (Fox et al. 2011). Appropriate management of arable field margins may help butterfly populations to survive on farmland.

We recorded butterfly abundance and species richness on the experimental hay and silage margins during June and July (Feber 1993). Only 11 species of butterfly were recorded during this period but it was nonetheless clear that the availability of nectar was a major factor in determining their species richness and abundance. On the hay and silage margins, *T. repens* was the dominant nectar source; the other two sown nectar sources, *R. acris* and *A. millefolium*, were much less frequent. Butterflies were most abundant under management regimes that favoured *T. repens*. Thus, for example, on the hay margins before cutting, significantly more butterflies were recorded on the unfertilized, more species-rich plots, which were invaded by *T. repens* (see text), than on any other plot type (P <0.001; Smith et al., 1993; Feber

1993). The highest numbers of butterfly species were also recorded on these plots, while numbers on the more species-rich, fertilized plots, where *D. glomerata* appeared to suppress *T. repens*, were significantly lower than on all other plot types (P <0.001). Overall, the attractiveness of these margins to butterflies was limited by their simple plant species composition, leading to a paucity of nectar sources and larval foodplants.

Although the design of the experiment did not allow direct statistical comparison of the results from the hay and silage margins, some qualitative comparisons can be made of the effects of the two mowing regimes. Butterfly abundance and species richness declined on both leys following cutting, but the management for hay was likely to have been more beneficial for maintaining plant species richness than the silage management. The timing of the two silage cuts was likely to remove, first, flowering shoots of most species and, second, any regrowth, and prevent species from producing seed. The timing of the hay cut was such that some species would flower and set seed before the cut, and flower on regrowth later in the season (Feber et al. 1996). Cutting field margins during the summer months is a practice known to be detrimental for butterflies (Feber et al. 1996; Chapter 3, this volume).

et al. 1997). Any tendency for its relatively open sward structure to allow greater persistence of pernicious weeds may have been countered by the relatively low soil fertility; the success of these species in modern agriculture is largely attributable to their ability to utilize high soil nutrient levels. However, none of our treatments harboured high weed populations. Annual grass weeds declined very rapidly as perennial species closed the swards. Three years after establishment, persistent and pernicious agricultural weeds were a very small component of the swards. Couch grass *Elytrigia repens*, the most abundant of these, occurred in only 7% of samples, even in the treatment in which it was most frequent.

Fertilizer substantially increased the biomass of both leys under hay and silage management, particularly amongst sown species. The sown species in the conventional ley were very significantly more productive than those in the more species-rich ley under both hay and silage management; fertilizer addition significantly increased their biomass under both regimes, but this increase was proportionately greater in the more species-rich ley because of the absence of *T. repens* (Table 2.2). In terms of attaining and managing for floristic diversity, the more species-rich ley had clear advantages, particularly when soil fertility was relatively low. The conventional ley was generally of limited value for wildlife (Box 2.2), even when unfertilized, and its high productivity presented management challenges in this context.

2.2.1.3 Forage potential

Although the species-rich ley was the least productive of our swards, it had good potential for forage, particularly for organic farms or situations where amenity and wildlife conservation interests are best served by extensified livestock management. Herbal leys, including grasses and forbs with good digestibility (Smith and Allcock 1985), may have considerable benefits in these situations. After three years, the productivity of our two leys differed by around 14% (Smith et al. 1997). Loss of productivity in the conventional ley when it was unfertilized was around 20%, and of the species-rich ley, around 30%, although these differences are likely to have been minimized by the greater biomass of clover in the unfertilized treatments.

2.2.1.4 Implications for conservation management

Our hay and silage margin experiments demonstrated that grass leys that are rather more species rich than conventional timothy/ryegrass/clover leys can confer significant advantages for wildlife and ease of management while still retaining forage potential. Although still extremely species poor when compared with semi-natural grassland, simple sown swards can make a considerable impact on visual amenity by encouraging common but attractive plant and animal species. For large-scale use they are more practicable than more expensive and complex wildflower seed mixtures, and usually more acceptable than allowing natural regeneration with its associated weed problems. But

Table 2.2 Mean shoot dry weight (g/m²) of all unsown and sown species in each treatment under hay and silage management in 1991. Treatment abbreviations: C – F = conventional ley minus fertilizer, C + F = conventional ley plus fertilizer, R – F = species-rich ley minus fertilizer, R + F = species-rich ley plus fertilizer. Significance levels ***P <0.001, ** P <0.01, * P <0.05, NS not significant.

Treatment	Hay						Silage					
	All		Sown		Unsown		All		Sown		Unsown	
	Mean	SE	Mean	SE	Mean	SE	Mean	SE	Mean	SE	Mean	SE
C – F	796.0	50.9	748.4	47.0	47.6	18.4	552.5	83.4	549.2	82.4	3.3	2.1
C + F	975.8	52.0	875.6	72.4	100.3	58.7	675.4	54.9	672.0	54.1	3.4	2.2
R – F	634.7	40.0	414.4	41.4	220.4	45.3	378.2	45.3	203.9	23.5	174.3	45.5
R + F	887.1	76.1	819.5	77.9	67.6	13.6	526.5	51.9	518.1	53.2	8.4	3.3
Significance of:												
Ley type	*		**		**		*		***		***	
Fertilizer	***		***		NS		*		***		NS	
Interaction	NS		**		NS		NS		**		NS	

although they can be advantageous in many situations they are only likely to be cost-effective when the component species are carefully chosen; they must be able to compete successfully under given levels of added or residual fertility and intended mowing regimes, and must be fit for purpose, whether amenity, wildlife, grazing, or a combination of functions. Our experiments on 2 m-wide field margins at Wytham strongly reinforced this message (see Section 2.2.2). Particularly, where more complex seed mixtures are used to emulate semi-natural grassland and prioritize wildlife outcomes, cost-effective choice of species for inclusion requires understanding of the interactions between sward management and composition, and of the phenologies and growth habits of individual species.

2.2.2 '2 m margin' experiments

In our second experiment at the University Farm, we contrasted ten different regimes for managing grassy field margins, using a rigorously designed experiment, replicated around the margins of fields at the University of Oxford's Farm at Wytham. The experimental design allowed us to distinguish the effects of the management regimes from those resulting from variability in soils and in cropping regimes around the farm.

The existing uncropped field margins on the farm were only about 0.5 m wide (the 'old' field margins). We expanded these by fallowing to create a total width of 2 m (the 'new' field margins), upon which our management regimes were imposed.

Most of our treatments involved simple cutting regimes imposed on two types of sward: those regenerating naturally following fallowing, and those sown

Figure 2.3 An experimental 2 m margin plot at Wytham, sown with grass and wildflower seed mixture and left uncut. ©Helen Smith.

with a wild grass and flower mixture in spring (Fig. 2.3). The seed mixture contained six 'non-aggressive' species of grass and 17 mostly perennial broad-leaved species (Box 2.1); we used this complex mixture to allow investigation of the responses to management of a range of species and to emulate a semi-natural flora. To ensure that the mixture was both visually appropriate and likely to succeed on the Wytham soils, we chose species typical of established grasslands in the area. We either left the margins uncut or cut them (with cuttings removed) in summer only, in spring and summer, and in spring and autumn. Two further treatments were imposed only on naturally regenerating plots: the first comprised cutting in spring and summer but with cut material *in situ* instead of removed, and the second comprised spraying annually in summer with a broad-spectrum herbicide (glyphosate) (Smith et al. 2010).

We monitored the plant species growing on both old and new zones of the field margins, and in the adjacent crop, throughout the first three years of the experiment. Ten years later, we monitored the new margins for a final time. Permanent quadrats were used to record species richness in each plot and the relative frequency of occurrence of each species. For the first three years of the experiment we also recorded all of the species growing in each plot and in the adjacent crop and field boundary (hedge or ditch). We quantified the seed bank in soil cores collected from each plot at the outset of the experiment (Smith et al. 1993). The results were used to help understand the processes involved in establishing new field margin floras and the extent to which these can be manipulated by management, asking whether (i) the process of colonization could provide the quality and quantity of elements needed for an acceptable flora, (ii) desirable species would persist during the subsequent succession, and (iii) outcomes could be significantly improved by our simple management regimes. Measuring the abundance of agricultural weeds was a key element in our work; only those prescriptions that deliver effective weed control are likely to be acceptable to farmers and land managers.

2.2.2.1 Colonization of field margins

Allowing natural regeneration on newly created field margins has the potential to deliver an ecologically and visually appropriate local flora for minimum outlay. But where the local flora is severely impoverished, as is often the case in intensively farmed areas, this may be difficult to achieve. We examined whether natural regeneration can supply the components of swards that are of wildlife conservation value, while at the same time being acceptable to farmers in terms of weed control. We looked at whether, first, the constituents of the colonizing flora, and, second, the relative abundances of the species comprising this flora, could be predicted from the composition of near-by source floras: the soil seed bank, the cropped area of the field, the old margin and boundary, and sources further afield on the Wytham Estate.

Most plant species that colonized the new field margins arrived very soon after fallowing (Fig. 2.4). Of 181 species recorded on the margins in the first three years of the experiment, 77% arrived in the first three months. These early colonists included not only annuals, but also 22 perennial grasses, 75 perennial forbs, 16 woody species, and three rush and sedge species, many of which might be considered desirable components of swards established for their wildlife value.

The species composition of the newly fallowed field margins initially had more in common with that of the adjacent crop edge and the soil seed bank than the original, narrow field margins (Table 2.3). Crop edge floras are constrained by annual ploughing and modified by cropping regimes and herbicide applications. Although half of the 146 species recorded in the Wytham crop edges were perennial, only those few which can reproduce from vegetative fragments, such as *E. repens* and field bindweed *Convolvulus arvensis*, established populations there. The crop edge was thus primarily a source of annual colonists that could seed into the new field margins (although many of the same species could also have colonized from the seed bank).

Table 2.3 Similarity between the species composition of the new margin flora and that of the old margins, crop, and 1987 seed bank in June, 1988–1990.

Year	Old margin		Crop edge		Seed bank	
	Cs[1]	SE	Cs	SE	Cs	SE
1988	0.418	0.021	0.502	0.030	0.471	0.017
1989	0.564	0.025	0.360	0.037	0.341	0.016
1990	0.599	0.014	0.419	0.028	0.341	0.015

[1] Similarity coefficients (Sorensen's coefficient of similarity) based on mean numbers of species per quadrat.

Although the seed bank on the fallowed field margins was rich (86 species) compared with many published estimates for arable soils, it was neither a major source of novelty nor of desirable perennial constituents of the new field margin flora. Ninety-five per cent of species recorded from the seed bank colonized the new field margins but 60% of these were annuals and most were also present in other very local sources; only three colonists were found exclusively in the seed bank (Smith et al. 1993).

On a very local scale though, the seed bank almost certainly contributed some new margin species that were not present in the immediate vicinity. For example, one 10 m stretch of margin developed a flora comprising a high frequency of rush species, none of which was present near-by, but all of which were found in the seed bank samples and are known to have long-lived seeds (Thompson and Grime 1979; Bakker et al. 1996).

The original narrow field margin and adjacent boundary feature were much the most important source of desirable perennials (that is, native perennial species that were not agricultural weeds) colonizing the new margin extensions. Twenty-four of the colonizing species were recorded only from these sources and all were forbs, grasses, and woody species typical of established meadow or woodland edge grasslands. In spite of this, the old margin and field boundary floras were less successful at colonizing than those of the crop and seed bank in the early years of the experiment. Grasses, and to a lesser extent forbs, were the most effective colonizers, with 80% and 73% respectively of those present on the old margins colonizing the new margins, compared with only 50% of woody species and 40% of sedge and rush species (Smith et al. 1993).

Although a very rich potential source of species on the Wytham Estate lay within a three mile radius of our experiment (623 species: Gibson and Brown 1991), most of the colonizing species were likely to have

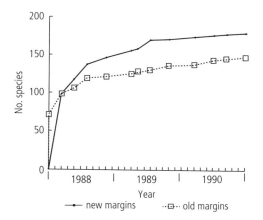

Figure 2.4 Cumulative total of plant species that colonized the field margins. From Smith et al. (1993).

come from very local sources. Ninety-nine per cent of the species recorded could have arrived from within the experimental fields and over 90% could have come from sources within 100 m.

These results on the species complement and likely origins of colonists on the new field margins suggest that, unless existing field boundaries remain rich in desirable perennial species, new floras that have the potential to meet our criteria for wildlife value and agronomic acceptability are very unlikely to establish. The likelihood of success is further limited by the quantity in which suitable species arrive, and then by their subsequent persistence as succession progresses.

Of the many annual and perennial species that colonized the new field margins at Wytham in the establishment years, most did so at very low frequencies, while a few were extremely common (Fig. 2.5). All of the biennials, woody species, and rush and sedge species colonized at low frequencies and remained very uncommon. Most of the dominant species were pernicious weeds. As with species complement, near-by source floras turned out to be good predictors of the relative abundances of species colonizing the new field

Table 2.4 The relationship between the rank abundance of species in the new margins and in the potential source floras in the first three years of the experiment. R_s is Spearman's rank correlation coefficient, with correlations based on rank abundance of species recorded from permanent quadrats in June each year and from seed bank samples collected at the outset of the experiment (Smith et al. 1993). The number of species in the samples varied from 41–61. Significance levels *** $P <0.001$, ** $P <0.01$, * $P <0.05$, NS not significant.

		1988		1989		1990	
Old margins	R_s	0.306	*	0.513	***	0.580	***
Crop edge	R_s	0.755	***	0.370	NS	0.369	*
Seed bank	R_s	0.660	***	0.273	NS	0.193	NS

margins. In the first year after fallowing, abundances in the new field margins were highly correlated with those in the seed bank and crop edge (Table 2.4), and the seed bank clearly made an important contribution to the abundance of some species. Stinging nettle *Urtica dioica*, for example, was the most abundant species in the seed bank and also the most frequent broad-leaved perennial on the field margins in the first year after fallowing. For many other species though, relative abundance in the seed bank was not a good predictor of their abundance in the colonizing flora, as has been found in other studies (e.g. Ball and Miller 1989; Stroh et al. 2012).

2.2.2.2 Succession

While the quota and abundances of species arriving defined the maximum potential for establishing desirable and acceptable perennial swards, successional processes rapidly modified this, with species richness declining over a thirteen-year period (Smith et al. 2010). Change was particularly rapid at a very early stage, largely as a result of decline in both numbers and relative abundances of annual colonists as perennials increased. Peaking at over 80% of species in the first summer after fallowing, annual numbers halved within a year (Fig. 2.6) as suitable gaps in which they could germinate declined in the rapidly closing perennial swards. The rapidity of this process is likely to have resulted from the high residual levels of soil fertility which promoted high productivity. This rapid loss of annuals, with positive implications for weed control, is consistent with other studies of colonization of former arable land (e.g. Gibson and Brown 1992; Steffan-Dewenter and Tscharntke 1997), although many factors, including soil type, nutrient levels, annual variations in climate (Morecroft et al. 2009), and the supply of propagules (Donath et al. 2007; Lepš et al. 2007; Leng et al. 2009) influence the composition

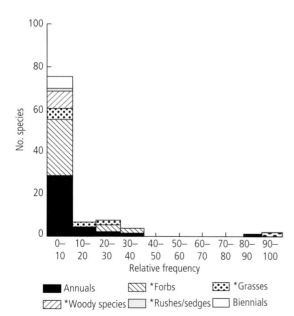

Figure 2.5 Relative frequencies (% occurrence in quadrat cells) of species colonizing the new margins three years after fallowing. *Grasses, forbs, rushes/sedges and woody species all refer only to perennial species. From Smith et al. (1993).

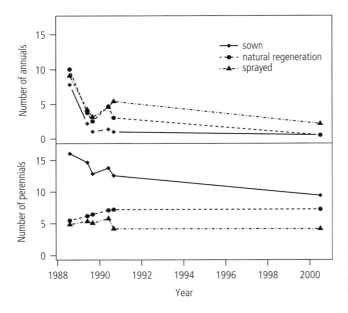

Figure 2.6 Comparison of mean numbers of annual and perennial species per quadrat in sown, natural regeneration, and sprayed plots. From: Smith et al. (2010). Reproduced with permission from Elsevier.

of the colonizing flora and the length of time taken to produce perennial-dominated swards.

Numbers of perennial species increased progressively, with the net result that species richness peaked in the first year after fallowing. Ten years later, however, perennial species numbers had changed very little in naturally regenerating swards (Fig. 2.6). Although Gibson and Brown (1992) suggested that progress towards species-rich swards is likely to continue for many years, our data did not support this in this arable edge situation.

These changes in species richness were mirrored in changes in the dominant species. Initially these were annual weeds, particularly barren brome *Anisantha sterilis*, wild oat *Avena* species, black grass *Alopecurus myosuriodes*, and knotgrass *Polygonum aviculare*, but these declined very rapidly (Fig. 2.7a). They were replaced within three years by the perennials, *U. dioica*, *C. arvensis*, *E. repens*, rough meadow grass *Poa trivialis*, and cut-leaved crane's-bill *Geranium dissectum*, only the latter two of which were not considered to be problem weeds in the context of the University Farm. However, after a further ten years, we found major changes in the abundance of many perennial species (Fig. 2.7b). Amongst the commonest, false oat-grass *Arrhenatherum elatius*, *D. glomerata*, *C. arvensis*, and Yorkshire-fog *Holcus lanatus* had continued to increase while some of the most pernicious species, including creeping thistle *Cirsium arvense*, *Urtica dioica*, and *E. repens*, had all declined over the experiment as a whole.

All of these changes meant that the developing field margin flora became progressively more similar to that of the adjacent old margin and field boundary, and less similar to those of the seed bank and adjacent crop. The numbers of species in common with the adjacent old margin and field boundary increased from 42% to 60% and their rank abundances also became significantly more similar (Tables 2.3 and 2.4). The existing field margin and boundary were thus the best available indicators of the potential for natural regeneration to result in desirable or acceptable new margin floras in the medium term, even though many species from this source failed to colonize.

2.2.2.3 Effects of management on sward composition

The study of colonization of our fallow field margins showed clearly that, where field boundary floras are severely impoverished, natural regeneration is not a viable option for establishing dense, grass-dominated, and largely perennial swards in the short or medium term. We tested the extent to which side-stepping secondary succession by using a seed mixture could produce more diverse and attractive swards and control weeds in this challenging situation. Even where the potential exists for natural colonization of new field margins by useful components of diverse perennial grass swards, we have shown that these are likely to be minority constituents of the swards. We used our contrasting mowing regimes to test the extent to which

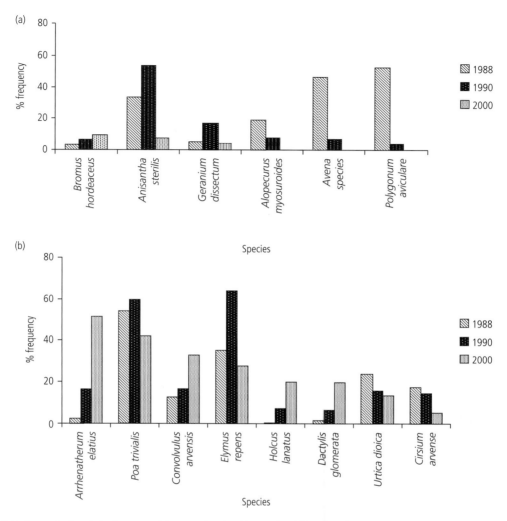

Figure 2.7 Changes in the relative frequency of the most abundant (a) annual and (b) perennial species on the 2 m margins. From: Smith et al. (2010). Reproduced with permission from Elsevier.

their abundance could be manipulated to produce richer and less weedy swards.

We found that both the rate of loss of annuals in the relatively short term, and the retention of perennial species over a longer period, could be considerably modified by the way in which swards were established and subsequently managed.

Sowing

Sowing with a perennial grass and wildflower seed mixture had major effects on the species richness and composition of the new field margin swards. Throughout the experiment these plots were dominated by

sown grasses and although, overall, the numbers of species declined over a thirteen-year period (Fig. 2.6), red fescue *Festuca rubra* was retained at a frequency of over 60% (see also Schippers and Joenje 2002) and smaller cat's tail *Phleum bertolonii* at just under 50%. Our experiment showed that careful seedbed preparation was essential for cost-effective return on the seed sown; the seed mixture was also sown into the old margin swards but establishment there was very poor, with sown species numbers only about 20% of those on the new margins after three years (Smith et al. 1999a). It is clearly unlikely to be cost-effective to sow wild flower seed mixtures into existing swards or weedy stubbles.

Establishing sown swards on field margins has been promoted in agri-environment schemes, originally because of their attractive and tidy appearance and likely benefits for weed control and, increasingly, to benefit farmland biodiversity (Critchley et al. 2006; Marshall et al. 2006). However, different studies, most of them short term, have reached differing conclusions about the maintenance of species richness in sown swards and their consequent cost-effectiveness for enhancing biodiversity. Some have found that species richness starts to decrease after the first year (e.g. Marshall and Nowakowski 1995; West and Marshall 1996; De Cauwer et al. 2005), while that of comparable naturally regenerating swards increases and stabilizes. De Cauwer et al. (2005) found significant convergence in species richness and vegetation composition between sown and unsown plots after only three years, and Warren et al. (2002) after six years. In an eight-year experiment, though, Lepš et al. (2007) found that, while sown species started to decline in numbers after three years, they still continued to increase their dominance of the swards. However, even with a high diversity seed mixture, naturally colonized plots tended to remain richer in species.

Our results, using a relatively species-rich mixture and longer term perspective, suggest that loss of species richness in sown swards is a protracted process. Species richness declined in sown plots during establishment (Fig. 2.6), but this resulted more from loss of unsown colonists than sown species, which increased to around 70% of the total (Fig. 2.8). Over the following ten years though, numbers of natural colonists stabilized, but an increase in abundance of rhizomatous perennial species with high potential growth rates probably accounted for a slow but progressive decline in numbers of sown species (see also Lepš et al. 2007). Nevertheless, even after 13 years, the sown plots on the new margins remained richer in perennials and accommodated fewer natural colonists than naturally regenerated plots (Fig. 2.6; Smith et al. 2010). At

Table 2.5 Relative abundance of the 12 most common species in the sown and naturally regenerating plots after 13 years, expressed as the percentage of the 576 25 cm × 25 cm quadrat cells in which they occurred in the paired treatments. * sown species. From: Smith et al. (2010). Reproduced with permission from Elsevier.

Sown plots	% of cells	Unsown plots	% of cells
Festuca rubra*	63.2	Arrhenatherum elatius	53.8
Arrhenatherum elatius	50.2	Poa trivialis	53.0
Phleum bertolonii*	47.9	Elymus repens	36.5
Convolvulus arvensis	42.4	Convolvulus arvensis	33.0
Elymus repens	25.5	Dactylis glomerata	28.3
Poa trivialis	24.8	Lolium perenne	24.3
Holcus lanatus	24.5	Holcus lanatus	21.2
Hordeum secalinum*	21.0	Phleum bertolonii	18.6
Centurea nigra*	14.6	Ranunculus repens	13.4
Trisetum flavescens*	8.2	Urtica dioica	13.0
Leucanthemum vulgare*	11.1	Festuca rubra	12.5
Lolium perenne	11.1	Agrostis stolonifera	10.4

this stage, up to 53% of the species in these plots were from the seed mixture (Fig. 2.8) and the sward composition remained very different from that in naturally regenerating plots (Table 2.5). The sown swards had become more similar in appearance, structure, and species composition to local semi-natural grasslands and the retention of sown forbs improved their quality for many invertebrates (Chapter 3, this volume; Smith et al. 1993; Feber et al. 1996; Baines et al. 1998).

These trends in numbers of species also tended to be mirrored by their abundances. The mat-forming habit, particularly of dominant sown grasses such as *F. rubra*, resulted in rapid sward closure and effective exclusion of annual species. Annual colonists declined significantly more rapidly in sown plots than

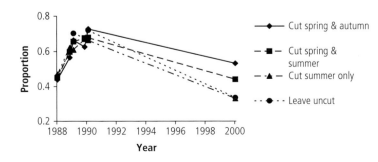

Figure 2.8 Mean numbers of sown species as a proportion of the total number of species in sown plots. From: Smith et al. (2010). Reproduced with permission from Elsevier.

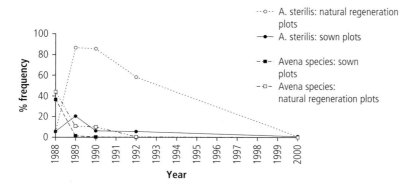

Figure 2.9 Changes in the mean frequency of *Anisantha sterilis* and *Avena* species in sown and natural regeneration plots averaged over cutting and vegetation removal treatments. From: Smith et al. (2010). Reproduced with permission from Elsevier.

in naturally regenerating swards, and common annuals, many of which were problem weeds, also peaked at lower frequencies in the sown plots (e.g. Fig. 2.9). Sowing thus delivered very rapid annual weed control. Similarly, work in the Netherlands found that all annuals in an annual/perennial mixture sown on an old arable field, where management ensured the development of closed perennial swards, were lost after two years (Schippers and Joenje 2002).

The rate of colonization by naturally regenerating perennial species was very effectively reduced by the wildflower seed mixture over the first three years of the experiment (Smith et al. 2010). Exclusion of rhizomatous perennial weeds by sown swards on this time-scale has been recorded from other short-term experiments. For example, Marshall (1990) showed that *E. repens* was excluded from perennial grass swards over a three-year period and De Cauwer et al. (2008) found that *E. repens* and *U. dioica* were significantly more frequent in unsown than in sown field margin swards three years after establishment. However, our experiment also showed that short-term results were not a good guide to the longer term. Some of the commonest naturally regenerating perennials that were initially suppressed increased progressively at the expense of less competitive sown grasses and, after thirteen years, had become as abundant as in naturally regenerated swards (Smith et al. 2010). Notable exceptions, which remained significantly less frequent, included *P. trivialis* and *D. glomerata* but, in terms of rhizomatous perennial weed control, the short-term benefits of sowing, particularly in the absence of any mowing, are unlikely to be sustained.

Mowing

In general, the numbers and relative abundance of species in our new field margins were modified less by mowing (i.e. mown versus unmown) than by sowing (i.e. sown versus natural regeneration). After 13 years we were unable to detect any significant effect of mowing on overall species richness although it had a significant influence in the establishment years. At this stage, any mowing increased the numbers of species but timing was also important; cutting in spring and autumn increased the numbers of perennial species, probably by increasing opportunities for seed return and for germination over the winter (Table 2.6).

When we left cut hay lying, rather than removing it from the margins of naturally regenerated plots, species richness was consistently, though not significantly, lower than in other treatments over thirteen years (Table 2.6). Rotting vegetation is likely to have caused both soil enrichment (Marrs 1993; Jacquemyn et al. 2003) and die-back of the smothered sward, affecting germination and seedling establishment opportunities and favouring competitive weedy species with high potential growth rates. The fact that this did not impact significantly on species richness on this time-scale is consistent with the suggestion that mid-successional species tend to be competitively equivalent and relatively unresponsive to nutrient change on fertile soils. They respond to changing nutrient levels through gradual changes in relative abundance that translate only slowly into changes in species richness (Huston 1994; Huberty et al. 1998).

Although the mowing regime had no significant effect on overall species richness after 13 years, it did influence both the establishment and persistence of sown species. Several studies have found that the rate of loss of sown species in the establishment phase of new field margins can be manipulated by mowing (Schippers and Joenje 2002; De Cauwer et al.

Table 2.6 Species richness based on mean numbers of species per quadrat. 2000 means with the same letter do not differ significantly. Except where indicated, vegetation was removed from all cutting treatments. From: Smith et al. (2010). Reproduced with permission from Elsevier.

Treatment	Date						
	08/88	06/89	09/89	06/90	09/90		07/00
Sow/cut spr + summer	23.1	16.7	14.3	15.0	14.4	a	11.5
Sow/cut summer	23.0	15.0	13.2	14.1	12.3	a	10.9
Sow/cut spr + autumn	24.1	20.3	15.2	18.0	14.9	ab	10.5
Sow/uncut	23.2	15.9	13.0	14.7	12.5	abc	8.8
Nat regen/cut spr + autumn	15.0	11.2	9.5	13.0	11.4	abc	8.5
Nat regen/cut spr + summer	13.7	10.1	9.7	12.4	10.2	abc	8.5
Nat regen/uncut	14.1	8.5	8.4	11.2	10.0	bc	8.0
Nat regen/cut summer	16.8	10.2	9.5	12.2	11.4	bc	8.2
Nat regen/cut spr + summer (hay left lying)	13.3	10.2	8.2	11.6	9.2	c	7.2
Nat regen/sprayed	12.9	9.3	8.6	10.4	9.7	c	7.0

2005). At Wytham, the timing of mowing was critical at this stage, with sown plots cut in spring and autumn being significantly the most species rich. After 13 years, plots cut twice—irrespective of timing—retained more sown species, forming a higher proportion of the total (Fig. 2.8), than those cut once or not at all. Plots left uncut retained fewest sown species (Smith et al. 2010). The better retention of sown species in mown plots may result from the selection of more stress tolerant species compared to competitors (sensu Grime et al. 1988).

Mowing also affected the abundance of many individual species although, again, these effects were usually less pronounced than those of sowing. Some annual species were significantly affected by both the frequency and timing of mowing, which influenced seed return and establishment opportunities. For example, *A. sterilis* decreased more rapidly as the frequency of mowing increased; by 2000, it was virtually eliminated when cut twice. This species lacks seed dormancy (Lintell Smith et al. 1999) and depends on seed production for population maintenance or increase. Black-grass *Alopecurus myosuroides* was found to be most persistent where hay cut in late June, containing ripe seed, was left lying rather than removed. Thus, where it is felt necessary to increase the rate of annual weed loss in establishing swards, the flowering phenology and seed dormancy characteristics of the dominant species

can be used to target the management regime (Watt et al. 1990).

Mowing had relatively little impact on the abundance of common, unsown perennials during the establishment years, but after 13 years it significantly influenced the abundance of many of these species. In these established swards the frequency of mowing was more important than its timing. The timing of mowing might be expected to have less influence, particularly on species that propagate by seed, once sward closure restricted germination opportunities, although it would continue to have a substantial influence on granivorous and nectar-feeding species (Shore et al. 2005; Marshall et al. 2006; Pywell et al. 2006; Vickery et al. 2009; Chapter 3, this volume).

Among the common species, *D. glomerata*, *C. arvensis*, and *H. lanatus* were all significantly more frequent in mown plots, with the latter two species also significantly more abundant in plots cut twice than in those cut once (Fig. 2.10). Other species were reduced by mowing. For example, *A. elatius* was 40% less frequent in plots mown twice than in those cut once or left uncut, while the pernicious species *C. arvense*, *U. dioica*, and *E. repens* responded negatively, often by degree, to the frequency of mowing. Because of their high potential growth rates these species are often regarded as intractable problems, assumed to require control by herbicide in high-fertility situations. However, our results show that consistent cutting, only twice per year,

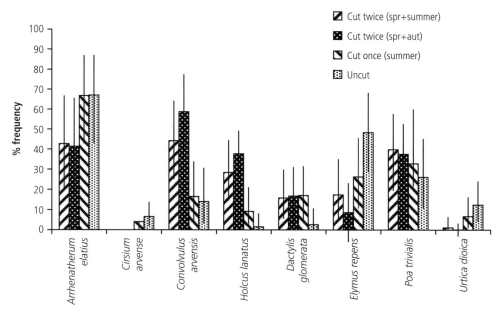

Figure 2.10 The effect of different mowing treatments on mean frequencies of common unsown perennial species in 2000. Spr: spring, aut: autumn. Bars represent 95% confidence intervals. From: Smith et al. (2010). Reproduced with permission from Elsevier.

can give good control of species such as *C. arvense* over this period.

Although frequencies of the pernicious weed *U. dioica* were very significantly reduced by more frequent mowing, when the mown vegetation was left lying rather than removed, it became significantly more abundant than under any other treatment (similar results were obtained by De Cauwer et al. 2008). By 2000, its frequency in these plots was 27.4%, compared with only 1.05% in equivalent plots from which the hay was removed (Smith et al. 2010). *U. dioica* illustrates well the potential impact on species richness of leaving cut material *in situ*, its high potential growth rate and ability to exploit high fertility soils (Pigott and Taylor 1964) eventually allowing it to form often monospecific stands. Although we did not test the impact of leaving cut hay on sown swards, the increase in *U. dioica* where hay was left lying suggests that nutrient demanding species (Marrs 1993; Hogg et al. 1995) would be likely to thrive at the expense of the slower growing species commonly used in seed mixtures, and result in more rapid loss of species richness than on naturally regenerated swards. Support for this comes from the work of De Cauwer et al. (2005), who found that leaving cut hay on sown

field margins had a significant negative impact on the retention of sown species richness after only three years.

Although all sown species declined over the thirteen-year experiment, the extent of their decline was often substantially modified by mowing. In contrast to some of the naturally regenerating perennial grasses, none of the sown perennial grasses was more abundant in the absence of mowing. In the establishment years, *T. flavescens*, *P. pratensis*, and crested dog's-tail *C. cristatus*, were all significantly more frequent when cut (Smith et al. 2010). After thirteen years, this difference remained significant for *T. flavescens* while *C. cristatus* had been lost from uncut plots. By contrast, *P. bertolonii* was most frequent in plots that were left uncut in the summer (either cut in spring and autumn or uncut). We were unable to detect significant effects of mowing on the commonest sown grasses because our monitoring method was insensitive to small changes in abundance of common and uniformly distributed species.

Most sown forbs were too uncommon for us to be able to detect any effects of mowing, but the most numerous also responded positively to mowing: *Leucanthemum vulgare* was significantly more abundant in cut than in uncut plots, field scabious *Knautia arvensis* was completely lost from uncut plots, and common knapweed *Centuarea nigra* responded by degree to the frequency of cutting (Smith et al. 2010).

Spraying

Spraying plots once a year with a broad spectrum herbicide consistently produced the most species-poor swards (although this effect was never significant) and, by opening up gaps, also perpetuated annual species as a conspicuous element in the sward (Smith et al. 2010). For example, after 13 years, the mean frequencies of *A. sterilis* and soft-brome *Bromus hordeaceus* were 32% and 55% in sprayed plots, compared with ranges between zero and 10% respectively, under other treatments. Conversely, numbers of perennial species on sprayed plots were significantly lower than under all other treatments from the second year of the experiment onwards (Fig. 2.6). After 13 years, the frequencies of some common perennial species were significantly reduced, to very low levels. For others, including some pernicious species such as *C. arvense*, our annual spraying regime had no significant effect on frequencies (Smith et al. 2010), although sub-lethal effects of herbicides on field margin perennials have been recorded in other studies (e.g. Schmitz et al. 2013).

In spite of the ban on the deliberate spraying-out of field margins (Section 2.2), the impact of herbicides on field margin vegetation remains a very relevant issue for two reasons. First of all, targeted use of herbicides has been discussed as a conservation management tool for field margins, for example, to encourage development of dense grass swards that exclude broad-leaved weeds (e.g. Marshall and Nowakowski 1994) or, conversely, to reduce competitive exclusion and promote diversity (Westbury and Dunnett 2008). However, there is little evidence to show that herbicide options are preferable to alternative and less expensive management regimes (e.g. Westbury et al. 2008); the Wytham experiment shows that swards that are species rich, attractive, and relatively weed-free, can be achieved by sowing and by simple mowing regimes. More pertinent to the future of field margin management than these suggestions, is the continuing threat to their biodiversity from both lethal and sub-lethal effects of overspraying and drift of herbicides. Our experiments help to illustrate the negative impacts of allowing this to continue and the need for both better risk assessment and consistent application of spray drift mitigation strategies (Schmitz et al. 2013).

2.3 Conclusions and recommendations

Our experiments at Wytham provide not only some specific and practical prescriptions for field margin management that can deliver both biodiversity and weed control, but also an understanding of ecological processes involved in sward development in this demanding situation. This understanding can be used to help tailor management options to achieve a wide range of outcomes that might be desirable or appropriate in different situations.

The demonstration that field margins created primarily to benefit biodiversity can also deliver effective weed control is essential to their acceptability to many farmers. In practice, only a tiny minority of pernicious weed species have the potential to spread effectively from field margins into crops (Marshall 1989) and, on the Wytham field margins, we found no relationship between the field margin management regimes and the abundance of common weeds in the adjacent crop (Smith et al. 1999b). Nevertheless, the commonly held perception that field margins present a weed risk is likely to influence willingness to create new field margins. Our results showed that the abundance of both annual and perennial weeds could be manipulated by simple and inexpensive management regimes.

We showed that the annual weeds that colonize newly fallowed field margins are likely to be a relatively short-lived problem and their decline can be expedited by simple management regimes that either increase the rate of perennial sward closure or use phenology and germination characteristics to target the problem species (Watt et al. 1990). Importantly, the results of the '2 m margins experiment' provided accurate insights not just into the short-term successional processes on field margins, but also into the longer-term outcomes for weed control and the development of biodiversity. For example, we found that swards established by sowing a wild flower seed mixture effectively excluded perennial as well as annual weeds in the establishment years but not in the longer term. However, the abundance of perennial weeds, a problem which initially appeared intractable, could be reduced by mowing in the longer term. Such long-term data can usefully inform management prescriptions for agri-environment scheme agreements, which usually last around ten years.

In terms of biodiversity, different mowing regimes enhanced plant species richness in the establishment years and in the longer term, with the frequency of mowing becoming more important than its timing. The flora of the adjacent old field margin and boundary was critical in determining the diversity of swards established by natural regeneration. Where this flora is impoverished and lacks desirable elements of semi-natural grassland, sowing a mixture of grasses and forbs appropriate to the area is recommended. At Wytham, despite losses of sown species from sown swards, these swards remained more species rich than naturally regenerated swards for at least thirteen years. Sown swards also contained a higher proportion of species typical of semi-natural grassland rather than of disturbed ground. Both of our experiments showed that the retention of many sown species typical of semi-natural grasslands could be improved by mowing, with removal of hay, to reduce competition from more competitive species that were better able to exploit these high fertility soils.

Although 2 m-wide margins are still included as an option in current agri-environment schemes, wider margins of 4 m or 6 m are commonly established. However, there was little evidence from our hay and silage margins experiment, or other studies that, over the short term at least, an increase in margin width results in greater species richness (e.g. Sheridan et al. 2008). One might predict the establishment phase to be more protracted as distance from the boundary and sources of many perennials increases, and a buffering effect of wider margins may become more apparent over the longer term. Our experiments highlighted the importance of keeping fertilizer and pesticide inputs away from field margins in order to increase species richness and control weeds most effectively.

Seed mixtures that include forbaceous species, whether just a few, as in our hay and silage margins experiment, or more complex mixtures such as that used in the 2 m margins experiment, will promote species richness in other taxa. Structural diversity—provided, for example, by tussock forming perennial grasses—will harbour overwintering polyphagous predators such as carabid beetles and spiders. Relatively small numbers of forb species can provide continuity of nectar sources for pollinators such as bumblebees, while specific invertebrate species or taxa can be targeted by providing their adult or larval foodplants. Management strategies will depend on desired outcomes—high floristic diversity may be less important than high beetle numbers or even a specific species. Our experiments provide the ground-rules for making informed decisions on reaching different outcomes—no one size fits all.

Acknowledgements

We are grateful to the many people who helped maintain the experiment and to the University of Oxford Farm for accommodating it. We particularly thank Katrina Rainey for assisting with the hay and silage margins, and Michael Morecroft and Michele Taylor for help with monitoring the 2 m margins in 2000. The experiment was funded by English Nature (now Natural England).

References

Aebischer, N.J. and Ewald, J.A. (2004). Managing the UK Grey Partridge *Perdix perdix* recovery: population change, reproduction, habitat and shooting. *Ibis*, **146**, 181–191.

Baines, M., Hambler, C., Johnson, P.J., Macdonald, D.W., and Smith, H. (1998). The effects of arable field margin management on the abundance and species richness of Araneae (spiders). *Ecography*, **21**, 74–86.

Bakker, J.P., Poschlod, P., Strykstra, R.J., Bekker, R.M., and Thompson, K. (1996). Seed banks and seed dispersal: important topics in restoration ecology. *Acta Botanica Neerlandica*, **45**, 461–490.

Ball, D.A. and Miller, S.D. (1989). A comparison of techniques for estimation of arable soil seed banks and their relationship to weed flora. *Weed Research*, **29**, 365–373.

Barber, D. (ed.) (1970). *Farming and Wildlife: a study in compromise.* Report of the Silsoe Conference, RSPB, Sandy, UK.

Barr C., Gillespie M., and Howard D. (1994). *Hedgerow Survey 1993.* Institute of Terrestrial Ecology.

Barr C., Howard B., Bunce B., Gillespie M., and Hallam C. (1991). *Changes in Hedgerows in Britain between 1984 and 1990.* Institute of Terrestrial Ecology.

Boatman, N.D. (ed.) (1994). *Field Margins—Integrating Agriculture and Conservation BCPC Monograph No. 58.* BCPC Publications, Farnham.

Boatman, N.D., Davies, D.H.K., Chaney, K., Feber, R.E., de Snoo. G.R., and Sparks, T.H. eds. (1999). *Field Margins and Buffer Zones: Ecology, Management and Policy.* Aspects of Applied Biology, **54**. Association of Applied Biologists, Wellesbourne, UK.

Bunce, R.G.H. and Howard, D.C. eds. (1990). *Species dispersal in agricultural habitats.* Bellhaven Press, London, New York.

Carey, P.D., Short, C., Morris, C., Hunt, J., Priscott, A., Davis, M., Finch, C., Curry, N., Little, W., Winter, M., Parkin, A., and Firbank, L.G. (2003). The multi-disciplinary evaluation of a national agri-environment scheme. *Journal of Environmental Management*, **69**, pp. 71–91.

Chestnutt, D.M.B. and Lowe, J. (1970). White clover/grass relationships: agronomy of white clover/grass swards. *White Clover Research* (ed. J. Lowe), pp. 191–213. Occasional Symposium of the British Grassland Society, No. 6. British Grassland Society, Hurley, Maidenhead.

Christensen, P.A. (1983). Management of natural vegetation on farms. In *Management of Vegetation*, (ed. J.M. Way), BCPC Monograph No. 26, pp. 169–173, BCPC, Croydon.

Clarke, J.(ed.) (1992). *Set-aside.* British Crop Protection Council, Farnham, Surrey, United Kingdom.

Cowling, D.W. and Lockyer, D.R. (1968). A comparison of the yield of three grass species at various levels of nitrogenous fertiliser grown alone or in a mixture. *Journal of Agricultural Science*, **71**, 127–136.

Crawley, M.J., Brown, S.L., Heard, M.S., and Edwards, G.R. (1999). Invasion resistance in experimental grassland communities: species richness or species identity? *Ecology Letters*, **2**, 140–148.

Critchley, C.N.R., Fowbert, J.A., Sherwood, A.J., and Pywell, R.F. (2006). Vegetation development of sown grass margins in arable fields under a countrywide agri-environment scheme. *Biological Conservation*, **132**, 1–11.

De Cauwer, B., Reheul, D., D'hooghe K., Nijs, I., and Milbau, A. (2005). Evolution of the vegetation of mown field margins over their first 3 years. *Agriculture, Ecosystems and Environment*, **109**, 87–96.

De Cauwer, B., Reheul, D., Nijs, I., and Milbau, A. (2006). Effect of margin strips on soil mineral nitrogen and plant diversity. *Agronomy of Sustainable Development*, **26**, 117–126.

De Cauwer, B., Reheul, D., Nijs, I., and Milbau, A. (2008). Management of newly established field margins on nutrient rich soil to reduce weed spread and seed rain into adjacent crops. *Weed Research*, **48**, 102–112.

Defra (2004). *Single Payment Scheme: Cross compliance handbook for England 2005.*

Defra (2011). *Biodiversity 2020: a strategy for England's wildlife and ecosystem services.* Crown Copyright.

Dobbs, T.L. and Pretty, J. (2008). Case study of agri-environmental payments: the United Kingdom. *Ecological Economics*, **65**, 765–775.

Donath, T.W., Bissels, S., Holzel, N., and Otte, A. (2007). Large scale application of diaspore transfer with plant material in restoration practice—impact of seed and microsite limitation. *Biological Conservation*, **138**, 224–234.

Feber, R.E. (1993). *The ecology and conservation of butterflies on lowland arable farmland.* D. Phil thesis, University of Oxford.

Feber, R.E., Smith, H., and Macdonald, D.W. (1996). The effects of management of uncropped edges of arable fields on butterfly abundance. *Journal of Applied Ecology*, **33**, 1191–1205.

Firbank, L.G. (ed.) (1998). *Agronomic and environmental evaluation of set-aside under the EC Arable Area Payments Scheme.* Report to the Ministry of Agriculture, Fisheries and Food. Institute of Terrestrial Ecology, Grange-over-Sands.

Firbank, L.G., Carter, N., Darbyshire, J.F., and Potts G.R., eds. (1990). *The Ecology of Temperate Cereal Fields.* BES Symposium 32, Blackwell, Oxford.

Foster, L. (1988). Herbs in pastures. Development and Research in Britain, 1850–1984. *Biological Agricultural and Horticulture*, **5**, 97–133.

Fox, R., Brereton, T.M., Roy, D.B., Asher, J., and Warren, M.S. (2011). *The State of the UK's Butterflies 2011.* Butterfly Conservation and the Centre for Ecology & Hydrology, Wareham, Dorset.

Gibson, C.W.D. and Brown, V.K. (1991). The effects of grazing on local colonisation and extinction during early succession. *Journal of Vegetation Science*, **2**, 291–300.

Gibson, C.W.D. and Brown, V.K. (1992). Grazing and vegetation change: deflected or modified succession? *Journal of Applied Ecology*, **29**, 120–131.

Grime, J.P., Hodgson, J.G., and Hunt, R. (1988). *Comparative Plant Ecology.* Unwin Hyman, London.

Hautier, Y., Niklaus, P.A., and Hector, A. (2009). Competition for light causes plant biodiversity loss after eutrophication. *Science*, **324**, 636–638.

Hogg, P., Squires, P., and Fitter, A.H. (1995). Acidification, nitrogen deposition and rapid vegetational change in a small valley mire in Yorkshire. *Biological Conservation*, **71**, 143–153.

Holland, J.M., Winder, L., and Perry, J.N. (2000). The impact of dimethoate on the spatial distribution of beneficial arthropods in winter wheat. *Annals of Applied Biology*, **136**, 93–105.

Huberty, L.E., Gross, K.L., and Miller, C.J. (1998). Effects of nitrogen addition on successional dynamics and species diversity in Michigan Old-Fields. *Journal of Ecology*, **86**, 794–803.

Huston, M.A. (1994). *Biological Diversity: The Coexistence of Species on Changing Landscapes.* Cambridge University Press, Cambridge.

Jacquemyn, H., Brys, R., and Hermy, M. (2003). Short-time effects of different management regimes on the response of calcareous grassland vegetation to increased nitrogen. *Biological Conservation*, **142**, 1340–1349.

Kleijn, D. and Snoeijing, G.I.J. (1997). Field boundary vegetation and the effects of drift of agrochemicals: botanical change caused by low levels of herbicide and fertilizer. *Journal of Applied Ecology*, **34**, 1413–1425.

Knops, J.M.H., Tilman, D., Haddad, N.M., Naeem, S., Mitchell, C.E., Haarstad, J., Ritchie, M.E., Howe, K.M., Reich, P.B., Siemann, E., and Groth, J. (1999). Effects of plant species richness on invasion dynamics, disease outbreaks, insect abundances and diversity. *Ecology Letters*, **2**, 286–293.

Leng, X., Musters, C.J.M., and de Snoo, G.R. (2009). Restoration of plant diversity on ditch banks: seed and site limitation in response to agri-environment schemes. *Biological Conservation*, **142**, 1340–1349.

Lepš, J., Doležal, J., Bezemer, T.M., et al. (2007). Long-term effectiveness of sowing high and low diversity seed mixtures to enhance plant community development on ex-arable fields. *Applied Vegetation Science*, **10**, 97–110.

Lintell Smith, G., Freckleton, R.P., Firbank, L.G., and Watkinson, A.R. (1999). The population dynamics of *Anisantha sterilis* in winter wheat: comparative demography and the role of management. *Journal of Applied Ecology*, **36**, 455–471.

Macdonald, D.W. and Feber, R.E., eds. (2015). *Wildlife Conservation on Farmland. Conflict in the Countryside.* Oxford University Press, Oxford.

Macdonald, D.W. and Johnson, P.J. (2000). Farmers and the custody of the countryside: trends in loss and conservation of non-productive habitats 1981–1998. *Biological Conservation*, **94**, 221–234.

Marrs, R.H. (1993). Soil fertility and nature conservation in Europe: theoretical considerations and practical management solutions. *Advances in Ecological Research*, **24**, 241–300.

Marshall, E.J.P. (1989). Distribution patterns of plants associated with arable field edges. *Journal of Applied Ecology*, **26**, 247–257.

Marshall, E.J.P. (1990). Interference between sown grasses and the growth of *Elymus repens* (couch grass). *Agriculture, Ecosystems and Environment*, **33**, 11–22.

Marshall, E.J.P. and Nowakowski, M. (1994). The effects of fluazifop-P-butyl and cutting treatments on the establishment of sown field margin strips. In: *Field Margins: Integrating Agriculture and Conservation.* (ed. N.D. Boatman). BCPC Monograph 58, pp. 307–312. The British Crop Protection Council, Farnham.

Marshall, E.J.P. and Nowakowski, M. (1995). Successional changes in the flora of a sown field margin strip managed by cutting and herbicide application. In Brighton Crop Protection Conference—Weeds, pp. 973–978. British Crop Protection Council, Farnham, Surrey.

Marshall, E.J.P., West, T.M., and Kleijn, D. (2006). Impacts of an agri-environment field margin prescription on the flora and fauna of arable farmland in different landscapes. *Agriculture, Ecosystems and Environment*, **113**, 36–44.

Mitchell, K. (1999). European policies for field margins and buffer zones. In: *Field Margins and Buffer Zones: Ecology, Management and Policy* (eds. N.D. Boatman, D.H.K. Davies, K. Chaney, R. Feber, G.R. de Snoo, and T.H. Sparks). *Aspects of Applied Biology*, **54**, 13–18.

Morecroft, M.D., Bealey, C.E., Beaumont, D.A., et al. (2009). The UK Environmental Change Network: emerging trends in the composition of plant and animal communities and the physical environment. *Biological Conservation*, **142**, 2814–2832.

Natural England (2009). *Agri-environment schemes in England 2009: a review of results and effectiveness*. Natural England, Sheffield.

Natural England (2010). *Arable reversion to species rich grassland: establishing a sown sward*. Natural England Technical Information Note TIN067. Natural England, Sheffield.

Pigott, C.D. and Taylor, K. (1964). The Distribution of Some Woodland Herbs in Relation to the Supply of Nitrogen and Phosphorus in the Soil. *Journal of Animal Ecology*, **33**, 175–185

Potts, G.R. (1980). The effects of modern agriculture, nest predation and game management on the population ecology of partridges (*Perdix perdix* and *Alectoris rufa*). *Advances in Ecological Research*, **11**, 1–79.

Potts, G.R. (1986). *The Partridge—Pesticides, Predation and Conservation*. Collins Professional and Technical Books, London.

Potts, G.R. and Aebischer, N.J. (1995). Population dynamics of the Grey Partridge *Perdix perdix* 1793–1993: monitoring, modelling and management. *Ibis*, **137**, 29–37.

Pywell R.F., Warman, E.A., Hulmes, L., et al. (2006). Effectiveness of new agri-environment schemes in providing foraging resources for bumblebees in intensively farmed landscapes. *Biological Conservation*, **129**, 192–206.

Rackham, O. (1986). *The History of the Countryside*. J.M. Dent & Sons, London/Melbourne.

Sackville Hamilton, N.R. (2001). Is local provenance important in habitat creation? A reply. *Journal of Applied Ecology*, **38**, 1374–1376.

Schippers, P. and Joenje, W. (2002). Modelling the effect of fertiliser, mowing, disturbance and width on the biodiversity of plant communities of field boundaries. *Agriculture, Ecosystems and Environment*, **93**, 351–365.

Schmitz, J., Schaefer, K., and Bruhl, C.A. (2013). Agrochemicals in field margins—assessing the impacts of herbicides, insecticides, and fertilizer on the common buttercup (*Ranunculus acris*). *Environmental Toxicology and Chemistry*, **32**, 1124–1131.

Sheridan, H, Finn, J.A., Culleton, N., and O'Donovan, G. (2008). Plant and invertebrate diversity in grassland field margins. *Agriculture, Ecosystems and Environment*, **123**, 225–232.

Shore, R.F., Meek, W.R., Sparks, T.H., Pywell, R.F., and Nowakowski, M. (2005). Will Environmental Stewardship enhance small mammal abundance on intensively managed farmland? *Mammal Review*, **35**, 277–284.

Smith, A. and Allcock, P.J. (1985). The influence of species diversity on sward yield and quality. *Journal of Applied Ecology*, **22**, 185–198.

Smith, H., Feber, R.E., Johnson, P.J., McCallum, K., Plesner Jensen, S., Younes, M., and Macdonald, D.W. (1993). *The Conservation Management of Arable Field Margins*. English Nature Science Series No. 18. English Nature, Peterborough.

Smith, H., Feber, R.E., and Macdonald, D.W. (1999a). Sown field margins; why stop at grass? In: Field Margins and Buffer Zones: Ecology, Management and Policy (eds. N.D. Boatman, D.H.K Davies, K. Chaney, R. Feber, G.R. de Snoo, and T.H. Sparks). *Aspects of Applied Biology*, 54, 275–282.

Smith, H., Feber, R.E., Morecroft, M.D., Taylor, M.E., and Macdonald, D.W. (2010). Short term successional change does not predict long term conservation value of managed arable field margins. *Biological Conservation*, **143**, 813–822.

Smith, H., Firbank, L.G., and Macdonald, D.W. (1999b). Uncropped edges of arable field margins managed for biodiversity do not increase weed occurrence in the adjacent crop. *Biological Conservation*, **89**, 107–111.

Smith, H., McCallum, K., and Macdonald, D.W. (1997). Experimental comparison of the nature conservation value, productivity and ease of management of a conventional and a more species-rich grass ley. *Journal of Applied Ecology*, **34**, 53–64.

Southwood, T.R.E. (1978). *Ecological Methods*. Chapman & Hall, London.

Spedding, C.R.W. and Diekmahns, E.W. (1972). *Grasses and Legumes in British Agriculture*. Bulletin of the Commonwealth Bureau of Pastures and Field Crops, no. 49. Commonwealth Agricultural Bureau, Farnham Royal.

Stapledon, R.G. (1943). Introduction. *The Clifton Park System of Farming* (ed. R.H. Elliot), pp. 5–16. Faber & Faber, London.

Steffan-Dewenter, I. and Tscharntke, T. (1997). Early succession of butterfly and plant communities on set-aside fields. *Oecologia*, **109**, 294–302.

Stroh, P.R., Hughes, F.R.M., Sparks, T.H., and Mountford, J.O. (2012). The Influence of Time on the Soil Seed Bank and Vegetation across a Landscape-Scale Wetland Restoration Project. *Restoration Ecology*, **20**, 103–112.

The Lawson Seed and Nursery Company (Limited) (1877). *Agrostographia: a Treatise on the Cultivated Grasses and Other Herbage and Forage Plants*. William Blackwood & Sons, Edinburgh and London.

Thomas, R., Noordhuis, J.M., Holland, J.M., and Goulson, D. (2002). Botanical diversity of beetle banks: Effects of age and comparison with conventional arable field margins in southern UK. *Agriculture, Ecosystems and Environment*, **93**, 403–412.

Thompson. K. and Grime, J.P. (1979). Seasonal Variation in the Seed Banks of Herbaceous Species in Ten Contrasting Habitats. *Journal of Ecology*, **67**, 893–921.

Tozer, K.N., Chapman, D.F., Quigley, P.E., Dowling, P.M., Cousens, R.D., Kearney, G.A., and Sedcole, J.R. (2008). Controlling invasive annual grasses in grazed pastures: population dynamics and critical gap sizes. *Journal of Applied Ecology*, **45**, 1152–1159.

Van der Putten, W.H., Mortimer, S.R., Hedlund, K., et al. (2000). Plant species diversity as a driver of early succession in abandoned fields: a multi-site approach. *Oecologia*, **124**, 91–99.

Vickery, J.A., Feber, R.E., and Fuller, R.J. (2009). Arable field margins managed for biodiversity conservation: a review of food resource provision for farmland birds. *Agriculture, Ecosystems and Environment*, **133**, 1–13.

Walker, K.J., Critchley, C.N.R., Sherwood, A.J., Large, R., Nuttall, P., Hulmes, S., Rose, R., and Mountford, J.O. (2007). The conservation of arable plants on cereal field margins: An assessment of new agri-environment scheme options in England, UK. *Biological Conservation*, **136**, 260–270.

Warren, J.M., Christal, A., and Wilson, F. (2002). Effects of sowing and management on vegetation succession during grassland habitat restoration. *Agriculture, Ecosystems and Environment*, **93**, 393–402.

Watt, T.A., Smith, H., and Macdonald, D.W. (1990). The control of annual grass weeds in fallowed field margins managed to encourage wildlife. *Proc. EWRS Symposium 1990, Integrated Weed Management in Cereals*. pp. 187–195.

West, T.M. and Marshall, E.J.P. (1996). Managing sown field margin strips on contrasting soil types in three environmentally sensitive areas. *Aspects of Applied Biology*, **44**, 269–276.

Westbury, D.B. and Dunnett, N.P. (2008). The promotion of grassland forb abundance: a chemical or biological solution? *Basic and Applied Ecology*, **9**, 653–662.

Westbury, D.B., Woodcock, B.A., Harris, S.J., Brown, V.K., and Potts S.G. (2008). The effect of seed mix and management on the abundance of desirable and pernicious unsown species in field margin communities. *Weed Research*, **48**, 113–123.

Wilkinson, D.M. (2001). Is local provenance important in habitat creation? *Journal of Applied Ecology*, **38**, 1371–1373.

CHAPTER 3

How can field margin management contribute to invertebrate biodiversity?

Ruth E. Feber, Paul J. Johnson, Fran H. Tattersall, Will Manley, Barbara Hart, Helen Smith, and David W. Macdonald

If all mankind were to disappear, the world would regenerate back to the rich state of equilibrium that existed ten thousand years ago. If insects were to vanish, the environment would collapse into chaos.

Edward O. Wilson

3.1 Introduction

Most invertebrates are small, inconspicuous, and less charismatic than their vertebrate cousins, yet invertebrates comprise the vast majority of recorded species on Earth (May 1988), and are an indispensable part of a healthy ecosystem. They are primary, secondary, and tertiary consumers, nutrient recyclers, and prey items for a wide range of other taxa. On farmland, invertebrates provide a range of ecosystem services such as pollination and pest control, essential for human welfare and economic prosperity. For example, production of at least one-third of the world's food, including 87 of the 113 leading food crops, depends on pollination by insects, bats, and birds (Klein et al. 2007); the value of insect pollination worldwide being estimated (in 2005) at €153 billion (Gallai et al. 2009).

Yet the outlook for many farmland invertebrates is bleak. How butterflies, among the best-monitored of terrestrial invertebrates, are faring provides an insight into the fortunes of the UK's invertebrates: an alarming 72% of butterfly species have decreased in abundance and 54% have decreased in distribution in the ten years to 2011 (Fox et al. 2011). These include wider countryside species such as the small tortoiseshell *Aglais urticae* and small skipper *Thymelicus sylvestris*. The statistics for moths make equally gloomy reading; records spanning nearly four decades show two-thirds of larger moth species in the UK are declining, amounting to around 220 species (Fox et al. 2013). More than a fifth of them plummeted by a third or more over a 10-year period (Conrad et al. 2006). Steep population declines have also been recorded for other groups, including bumblebees (Goulson et al. 2008) and carabid beetles (Brooks et al. 2012).

Many factors related to agricultural intensification are implicated in these and other invertebrate declines. For example, while some of the most notorious pesticides (e.g. DDT) are no longer used, non-target invertebrates continue to be vulnerable to modern insecticides. Neonicotinoid insecticides are currently the focus of considerable publicity. These are widely used and are highly toxic to most arthropods (Goulson 2013). Their use is strongly implicated in bumblebee declines: colonies of the bumble bee *Bombus terrestris* exposed in the laboratory to field-realistic levels of the widely used neonicotinoid imidacloprid have a significantly reduced growth rate and an 85% reduction in production of new queens compared with controls (Whitehorn et al. 2012); the European Commission is now legislating to restrict the use of neonicotinoids in the EU. Neonicotinoids have been called the 'new DDT', demonstrating the old truth that those not learning from history will repeat it (Monbiot 2013). Fungicides, widely used on farmland, can also affect invertebrates: the probability of infection by the gut parasite *Nosema* was increased where honeybees consumed pollen with a higher fungicide load (Pettis et al. 2013).

Farmland invertebrates face many other threats. Reduction in the abundance and diversity of non-crop plants on farmland through herbicide and fertiliser applications (Storkey et al. 2012) has consequences both for phytophagous species, such as Heteroptera, that feed directly on plants, and nectar and pollen feeders, such as bumblebees, that rely on the availability of foraging habitat and continuity of resource provision (e.g. Carvell et al. 2006). Loss, or erosion in quality, of non-cropped habitats, such as hedgerows, reduces the availability of breeding and overwintering habitat for invertebrates (Merckx and Berwaerts 2010) while habitat fragmentation at a landscape scale affects both mobile and more sedentary species (Warren et al., 2001; Chapter 8, this volume).

What refuge from these threats can be provided by changes in farmland management? An important option for mitigating impacts of farming intensification is the creation of uncropped field margins (Chapter 2, this volume). Field margins connect non-cropped habitats such as semi-natural grasslands (e.g. Sutcliffe et al. 2003). They are also part of wider landscape-scale matrix restoration (Donald and Evans 2006). Field margins increase resources for invertebrates through sward diversity and complex vegetation structure and can also reduce the exposure to pesticides of boundary-dwelling invertebrates, through a buffering effect. They may also provide invertebrates with a refuge from farming operations, offer undisturbed areas for breeding and overwintering, and act as sources from which overwintering invertebrates may colonize fields. In 1995, in recognition of the importance of field margins for wider biodiversity, cereal field margins were listed as a priority habitat by the UK Biodiversity Steering Group (UK Biodiversity Steering Group 1995a,b). Uncropped field margins are now one of the most widely adopted conservation measures on farmland and are an important component of agri-environment schemes in the UK (e.g. Grice et al. 2006) and Europe (e.g. Kleijn et al. 2006).

How should a field margin be created and managed to maximize benefits to wildlife, while at the same time accommodating the practical realities of farming? Finding practical ways to manage the uncropped edges of arable fields for biodiversity benefits was a key challenge in the mid-1980s. Two of our team, David Macdonald and Helen Smith, solicited opinions from an array of farmers and farm advisors. From this, it transpired that field margins posed many unanswered questions and that no reliable, evidence-based management prescriptions were available. In particular, the farmers asked us how could field margins be managed to foster desirable species of plants and

invertebrates while allowing for the control of weeds? This question, emerging as the priority from our 'market research' became the stimulus for developing a pioneering farm-scale experiment (see Chapter 2, this volume). Our first step was to bring together a team under the Manpower Services Commission[1]. This enabled us to transform the general conservation quality of the farm. Next, supported initially by English Nature (now Natural England) and the Co-Op Bank, we designed and established one of the largest farm-scale conservation ecology experiments in Europe at that time. In the meantime, we formed a liaison with the Farming and Wildlife Advisory Group (FWAG) who, throughout the project, were crucial in helping communicate the concerns and interests of farmers to us, and facilitating dissemination of our research results back to farmers and the wider practitioner community.

Our hypothesis was that simple methods for restoring and managing field margins could improve habitat quality for wildlife on farmland, without resulting in the proliferation of problem weeds. Successful management for both agriculture and conservation requires competing goals like these to be achieved. For example, many farmers will want to minimize weed problems and management costs, but many also want to promote game interests, or promote pollinators or aphid predators. Our experiments were primarily designed to gain an understanding of the critical ecological processes important in field margin management, so that the same principles could be applied in other situations. By using a blocked design and keeping our treatments constant, we minimized sources of variation and so maximized their predictive power, providing the robust information for extrapolating to different circumstances and sites.

In the previous chapter (Chapter 2, this volume), we examined the factors that affected vegetation on the experimental margins. Here, we look at how their botanical character affected the invertebrates that were found there. We briefly summarize the experimental design; full details are given in the previous chapter and in Smith et al. (1993).

3.2 Impacts of field margin management on invertebrates

We aimed to restore attractive and diverse communities on the field margins by establishing swards that

---

[1] This was a non-departmental public body, created in 1973 to coordinate employment and training services in the UK (such as training schemes to help to alleviate unemployment). It was abolished in 1988.

imitated semi-natural grassland (typically uncompetitive and perennial species). In autumn 1987, field margins around arable fields at Wytham were extended in width from 0.5 m to 2 m by fallowing arable strips, which were divided into 50 m-long plots. Since in many intensive arable situations the local flora is severely impoverished, we used two methods to establish our swards. Half of the plots were sown with a grass and wild flower seed mixture, which comprised six 'non-aggressive' grass species and 17 forbs in a 4:1 weight ratio. We selected a complex mixture to enable us to test the responses to management of a wide range of species, all of which were common components of semi-natural grassland in the Wytham area. The remaining plots were left to regenerate naturally: a cost-free option that would be likely to result in very different sward development, and which may have been especially appropriate where conservation of the local flora is a priority.

Both sown and naturally regenerated plots were managed in the following ways: either left uncut, or cut (with cuttings removed) in summer only, spring and summer, or spring and autumn. We predicted that mowing, and the timing of mowing, would have important consequences for plant communities, for example via differing impacts on flowering and seed production, or through competitive interactions between species (e.g. mowing, like grazing, can select for slow-growing species at the expense of more competitive ones), and on invertebrate communities, which were likely to be sensitive to vegetation structure and composition.

A ninth treatment, on unsown plots only, investigated the impact of leaving cut hay lying—advocacy of removal of cut hay has been a mantra of grassland conservation management because of its expected effect in reducing soil nutrient levels and increasing sward diversity. Finally, it was common practice in agricultural management at the time to spray field margins with glyphosate herbicide, annually in late June. Our tenth treatment was left uncut and managed in this way. Fertilizers and pesticides were not applied to the field margin swards, other than glyphosate on this tenth treatment.

Plots were monitored for plant and invertebrate diversity for up to twelve years. Here, we focus on the effects of field margin management on four important and ecologically contrasting groups of invertebrates: butterflies (Lepidoptera), spiders (Araneae), hoppers (Auchenorrhyncha), and heteropteran bugs (Heteroptera) (Feber et al. 1996; Baines et al. 1998; Haughton et al. 1999). We chose these groups because, between them, their ecological characteristics are such that they were considered broadly representative of the wider community of farmland invertebrates.

Butterflies and moths have complex life cycles, comprising egg, larval, pupal, and adult stages. Some species live as adults for only a few days or weeks, and spend the winter in the egg, larval, or pupal stage. The small skipper *Thymelicus sylvestris*, for example, hatches from its egg in late summer, spins a cocoon around itself, and overwinters as a tiny larva within the sheath of a grass stem. It feeds during the spring, pupates in June, and emerges in July, having just a few weeks as an adult in which to lay its own eggs. By contrast, larvae of the powerfully flying peacock *Aglais io* butterfly spend a short few weeks feeding voraciously on nettles in spring. They then pupate, emerging as adults in late summer, and surviving the winter as butterflies, hibernating until warm spring days when they seek nettle clumps on which to lay their eggs. With such different life histories, butterflies need a range of resources: sufficient nectar sources to supply them with food, suitable larval foodplants (which vary according to the species), and suitable habitat, often undisturbed, for the protection and development of the immature (immobile) stages. Furthermore, around 85% of British butterflies are relatively sedentary (Thomas 1984), so these functional habitats cannot be widely dispersed. We therefore predicted that butterflies would respond in contrasting ways to our different management treatments and would usefully indicate responses of other pollinator groups.

Spiders are highly successful predators and valuable pest control agents. They are often the first predators to enter a crop field after ploughing, and may become established before pests have an opportunity to colonize (Sunderland 1999), frequently reaching high numbers of individuals (e.g. up to 1000 individuals/m^2, Marc et al. 1999) and species (~50–60 species is not exceptional, e.g. Feber et al. 1998)). They have evolved the ability to withstand periods of starvation. Many spiders will actively (or passively via their webs) capture more prey than they consume (the reason for this behaviour is not understood); such 'superfluous killing' can further augment their effectiveness at controlling pests (Sunderland 1999). In laboratory experiments, Mansour and Heimbach (1993) found that three common spider species, *Erigone atra*, *Lepthyphantes tenuis*, and *Pardosa agrestis*, together could reduce numbers of the aphid *Rhopalosiphum padi* by as much as 58%. Variation in mobility is important. Some, such as linyphiids ('money' spiders), can travel for many miles using a dispersal method known as 'ballooning'. These spiders let out lines of silk which, if the air currents are favourable, act as parachutes and lift them to great heights (spiders have been captured an astonishing 5 km above sea level) from where they are carried to new locations. By contrast, other species, such as the larger, ground-dwelling lycosids (wolf spiders), move by walking, and may travel through only two or three fields during their entire lifetime. A range of feeding strategies is also exhibited: *Lepthyphantes tenuis* (Fig. 3.1) one of the commonest spiders on farmland, builds a sheet web, often anchored to vegetation and placed just above the soil, while wolf spiders will ambush or actively hunt down their prey. Field margin vegetation structure might therefore be predicted to be more important than the mix of plant species composing a sward in terms of potential impacts on spider communities (Gibson et al. 1992).

By contrast, the UK's 400 or so Auchenorrhyncha species (leafhoppers, planthoppers, and froghoppers) are exclusively herbivorous, sucking sap from their hostplants. The most well known are probably the immature stages of froghoppers, which produce the spittle-masses, often called cuckoo-spit, which appear in spring on many plants. Hoppers often have very specific feeding requirements and might be more sensitive to the species composition of the habitat than to its architecture.

Lastly, Heteroptera, the true bugs, are a useful indicator group for insect diversity in general, for various reasons. The Heteroptera (their name derived from the Greek 'hetero-' meaning different and 'ptera' meaning wings, referring to the contrasting texture of the front wings, leathery at the base and membranous at the apex) include a diverse assemblage of insects that have become adapted to a broad range of habitats—terrestrial, aquatic, and semi-aquatic. Like the hoppers, they are characteristic of temperate grasslands. The group includes phytophagous species (which feed on vascular tissues or on the nutrients stored within seeds), as well as predatory species and scavengers. The larval and adult stages are sensitive to changes in their environment, such as vegetation structure and flower abundance (Morris 2000; Zurbrügg and Frank 2006), and studies have shown that richness of the Heteropteran communities correlates strongly with total insect diversity (Duelli and Obrist 1998).

3.2.1 Margin management and butterflies

We monitored butterflies on the experimental margins, using transects, for three years and the experimental treatments had major, and very different, effects on them (Feber et al. 1996). Butterfly abundance was strikingly higher on sown compared to naturally regenerated treatments; as early as the second year of the experiment we recorded a mean of 90.9 butterflies per 50 m plot on sown, uncut treatments, compared to just 39.1 on their naturally regenerated (unsown) equivalents (Table 3.1).

Although sowing did not increase the total abundance of flowers, crucially, it increased the abundance of types and species of nectar sources that were preferred by butterflies. We discovered this by capturing, harmlessly marking with a permanent pen, and releasing, over 650 meadow brown *Maniola jurtina* (Fig. 3.2) butterfly individuals and 785 gatekeeper *Pyronia tithonus* individuals. For each of these 1400+ individuals, we recorded which experimental plots they were using and when, and what they were feeding on. These data, combined with over 40 hours of behavioural observations, and monthly counts of flowers on all of our margins, revealed that, of the 99 species of plant in flower on the margins in July and August, a mere 15 species were being used as nectar sources and of these, some were greatly preferred over others. Knapweeds *Centaurea nigra* and *C. scabiosa*, for example, despite together being ranked only as the 45th most numerous flowers in the sward, accounted for a remarkable

Figure 3.1 *Lepthyphantes tenuis* is a linyphiid spider commonly found on farmland. Photograph © Evan Jones.

Figure 3.2 We recorded flower visits by meadow brown butterflies to reveal their preferred nectar sources. Photograph © Neil Hardwick/Shutterstock.

12% of all butterfly visits. Oxeye daisy *Leucanthemum vulgare*, the fourth most abundant flower, accounted for a third of all flower visits by butterflies (Feber et al. 1994). All three of these species were components of the seed mixture, included with butterflies in mind, and they did not occur on the naturally regenerated swards (Smith et al. 1994). We had a rare glimpse of how these plant species had fared over the ensuing thirteen years by repeating our plant surveys in 2001 (Smith et al. 2010). Despite this length of time, two of the most favoured nectar sources, common knapweed *C. nigra* and oxeye daisy, were still among the most abundant species on sown plots, while naturally regenerated plots remained relatively poor in nectar sources (Smith et al. 2010). So, we had proven that wild flower seed mixtures could be very successful in supplying nectar resources for adult butterflies on farmland, even over the long term, if they are carefully designed. Such mixtures are likely to be most appropriate, and deliver the greatest biodiversity gain, when planted on fertile

soils with a history of intensive management, where other sources of suitable plant colonists are absent and the vegetation is impoverished or dominated by annuals (Smith et al. 1994). They will be most beneficial to adult butterflies and other pollinators if they include a mixture of early (e.g. cowslip *Primula veris*) and later flowering (e.g. field scabious *Knautia arvensis*) species to provide nectar throughout the season.

Mowing was one of the more important experimental interventions. In all years, mowing had substantial effects on mean butterfly abundance (Table 3.1). Butterfly numbers were highest on treatments which were uncut, or cut in spring and autumn. Numbers recovered slowly on mown treatments, particularly those that were sown, where oxeye daisy reflowered several weeks after mowing. However, neither butterfly abundance nor species richness regained levels comparable with those on treatments which had not been mown in the summer. Mowing in summer (at the end of June) was generally detrimental to butterfly populations since it removed nectar sources at a time which coincided with highest butterfly abundance. Although two months later there was little difference in flower abundance between cut and uncut treatments, in arable systems, where sources of nectar are patchy or scarce, large-scale mowing may remove all foraging areas except those accessible to the most mobile species, such as the vanessid butterflies (e.g. small tortoiseshell), during the period following the cut.

However, mowing is not always bad for butterflies. Summer mowing can generate late sources of nectar by delaying flowering or initiating reflowering. Currently, this approach is promoted by Entry Level Stewardship in England to extend forage availability for pollinators such as bumblebees (Natural England 2008). Where sward diversity is low, for example where simple seed mixtures have been sown with few forbaceous species, limited-scale mowing may thus be used to manipulate nectar availability throughout the season. Memmott et al. (2010) suggest that climate change could reduce the length of season in terms of nectar and pollen provision for bumblebees, but this could be ameliorated by adding extra forage species which flower early in the season (such as red or white deadnettle *Lamium purpureum*, *L. album*) and at the end of the flowering season (such as scabious).

In our experiment, the chief advantage of sown margins lay in their ability to provide abundant nectar resources for adults. Indeed, the criteria used for selecting wild flower species for seed mixture included their nectar source potential, but not their suitability as larval food plants. Nonetheless the sown margins

Table 3.1 Effects of management on the abundance of butterflies on uncropped field margins 1989–1991. Analyses were performed on log-transformed data. Standard errors are given in parentheses. From: Feber et al. (1996).

TREATMENT	Mean number of butterflies per 50 m length of margin		
	1989	**1990**	**1991**
Sown, cut spring and autumn	26.9 (5.47)	87.5 (13.56)	58.6 (9.18)
Sown, cut spring and summer	26.3 (3.44)	44.5 (7.44)	27.5 (5.22)
Sown, cut summer only	22.3 (3.60)	35.8 (4.62)	20.5 (3.84)
Sown, uncut	59.5 (13.32)	90.9 (16.97)	49.0 (6.96)
Unsown, cut spring and autumn	43.6 (13.07)	29.4 (8.80)	36.9 (5.25)
Unsown, cut spring and summer	27.0 (3.40)	14.1 (2.53)	18.3 (3.49)
Unsown, cut summer only	17.4 (5.39)	16.5 (3.53)	17.1 (3.78)
Unsown, uncut	40.1 (13.05)	39.1 (11.84)	28.4 (6.67)
Unsown, cut spring and summer, hay left	15.4 (4.02)	16.5 (3.05)	19.6 (4.89)
Unsown, sprayed	42.4 (15.43)	18.3 (4.88)	11.9 (2.42)

Main effects	d.f.	1989	1990	1991
Block	7,49	**	NS	**
Sowing	1,49	NS	***	***
Cutting	3,49	**	**	***
Sow × cut	3,49	NS	NS	NS
Planned comparisons between means				
Cut vs uncut	1,49	**	**	*
Cut in summer vs not cut in summer	1,49	**	***	***
Cut spring & autumn vs uncut	1,49	NS	NS	NS
Sprayed vs uncut in summer	1,49	NS	***	***

*** $P < 0.001$, ** $P < 0.01$, * $P < 0.05$, NS not significant

contained some grass species, such as red fescue *Festuca rubra* and small timothy *Phleum pratense bertolonii*, which were potentially important hostplants for the satyrid butterflies, including meadow brown, and gatekeeper. While sweep netting confirmed the presence of the larvae (in itself an important finding, as it showed that the butterflies were successfully breeding on the field margins), revealing which grass species they were actually feeding on was a challenging task. The larvae are extremely sensitive (falling from their plant at the slightest disturbance), unhelpfully green and hairy (making them difficult to see), and strictly nocturnal in their feeding habits. Undaunted, over spring nights, we crawled along the field margins and managed to locate by torchlight 22 meadow brown and 19 gatekeeper larvae, and tag the grass blades they were consuming, for identification in daylight. To our surprise, we found the larvae to be feeding on a wide range of grasses, including two common and weedy species of farmland, sterile brome *Anisantha sterilis* and blackgrass *Alopecurus myosuroides*, and also ryegrass *Lolium perenne*, a widespread species of intensive grassland, and widely believed to be of little value as a butterfly foodplant.

For these two satyrid species, the results suggested that, where permanent grassy swards existed, other factors such as nectar source abundance were more likely to limit their populations on farmland than the availability of the larval foodplant. However, many other farmland butterflies have more specific larval foodplant requirements. Common blue *Polyommatus icarus* larvae, for example, feed on bird's foot trefoil *Lotus corniculatus* or black medick *Medicago lupulina*, while small copper larvae rely on common sorrel *Rumex acetosa*, sheep's sorrel *Rumex acetosella*, or occasionally dock species *Rumex* spp.; all of these species were found in the naturally regenerated, rather than the sown, swards. While the sown margins had significantly more species in total than the naturally regenerated margins, at least in part because of the complex seed mixture used, the numbers of unsown species that they accommodated were substantially lower. This effect persisted even twelve years after the field margins were established (Smith et al. 2010). This implies that very widespread use of wild flower seed mixtures which do not include hostplants may not be advantageous to butterfly populations. In any large-scale restoration of uncropped field edges by wild flower seeding on intensively managed arable farmland, the inclusion of larval foodplants in the seed mixture, as well as adult nectar sources, is essential.

During the summer, adult females of many species were ovipositing and the larvae of some were feeding or completing their development on their hostplants. As well as affecting nectar source availability, mowing during any part of this period would have disruptive effects on these egg, larval, or pupal stages. For example, mowing during the spring or early summer can result in large-scale losses of larvae of species such as orange tip *Anthocharis cardamines*, whose larvae remain on a single plant of cuckoo-flower *Cardamine pratensis* or hedge garlic *Alliaria petiolata* to complete their development. The orange tip larva is so protective of its plant that the first one to hatch on a given plant will devour any other eggs that have been laid there—the conspicuous bright orange singly laid eggs are designed to deter other females from laying on the same plant. Other butterflies, such as small skipper *Thymelicus sylvestris* and large skipper *Ochlodes sylvanus*, lay eggs within the grass sheath, which must remain undisturbed if the larvae are to develop successfully.

Evidence in support of a patchy rather than widespread approach to mowing management on field margins is neatly provided by the egg-laying habits of two species of farmland butterfly, the small tortoiseshell and the peacock. Both species lay eggs in clumps on common nettle *Urtica dioica* plants, which hatch into clusters of larvae (Fig. 3.3). Data from the experimental margins on the location of these egg clusters revealed that ovipositing small tortoiseshell females almost exclusively preferred short nettle regrowth, in contrast to peacock butterflies which selected tall, mature nettle plants for egg laying (Fig. 3.4; Feber et al. 1999). Studies on small tortoiseshell larvae have shown that larval growth rates and pupal weights are significantly higher on nettle regrowth, the leaves of which are high in soluble nitrogen and water (Pullin 1987). The bigger clumps of peacock larvae may need the greater volume of plant material and physical support of larger plants to complete their development. Furthermore, the flight periods of both species, as with many butterflies, can vary between years by several weeks according to the weather: higher temperatures leading to earlier emergence. The different requirements of these two species alone, together with variation in their phenology, illustrate the importance of not managing all margins in the

Figure 3.3 Peacock *Aglais io* larvae on common nettle. Photograph © R. Feber.

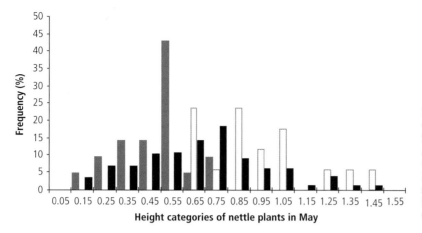

Figure 3.4 Height of nettle plants chosen for egg laying by small tortoiseshell *Aglais urticae* and peacock *Aglais io* butterflies on field margins. Grey bars: plants supporting *A. urticae* larval clusters; white bars: plants supporting *A. io* larval clusters; black bars: plants on field margins in May. Redrawn from Feber et al. (1999).

same way at the same time on any given farm. Mowing different margins on a farm, or different sections of the same margin, will result in a more heterogeneous sward providing breeding opportunities for a wider range of species. It may also have the benefit of extending flowering and thus forage availability for nectar and pollen feeders.

3.2.2 Margin management and Araneae, Auchenorrhyncha, and Heteroptera

During the first four years of the experiment, a total of 111 species of spider were recorded from a sample of

51 775 individuals. Distinct patterns emerged, the most obvious of which was the significantly higher abundance and species richness of Araneae on uncut compared to cut plots (e.g. Fig. 3.5). The timing of cutting was especially important. Both regimes involving cutting in summer were associated with a sharp reduction in the abundance of Araneae, which persisted throughout the year (Fig. 3.6), while cutting in spring and autumn, despite being the same frequency, had a much less severe effect. Immediately after cutting in spring, the abundance of Araneae was significantly lower compared to uncut plots, but numbers recovered relatively quickly (Fig. 3.6).

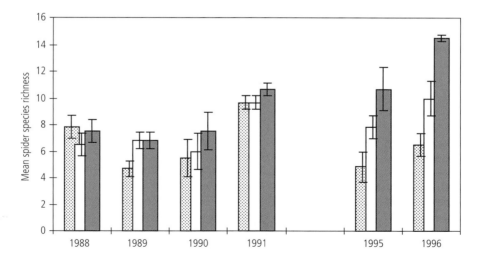

Figure 3.5 Effect of mowing (and collecting cuttings) on mean (± standard error) spider species richness in September on field margins under different management treatments. Dotted bars: margins cut in summer, open bars: margins cut in spring and autumn; solid grey bars: margins left uncut. From Macdonald et al. (1998).

(a) Cut in summer only

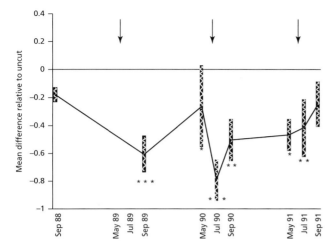

(b) Cut in spring and summer

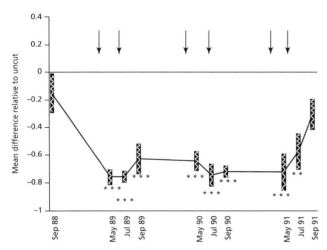

(c) Cut in spring and autumn

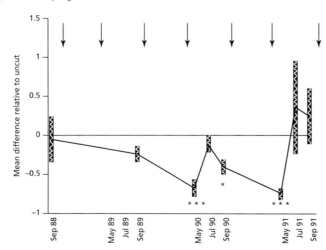

Figure 3.6 The effect of the cutting regimes on the abundance of spiders. Data are means, over six experimental blocks, of the proportional difference between plots under the relevant cutting regime and the uncut plots: (a) cut in summer only, (b) cut in spring and summer, and (c) cut in spring and autumn. Arrows denote the approximate timing of the seasonal cuts. Significance levels for the contrast between the cutting regime and uncut plots on different dates: $*P < 0.05$, $**P < 0.01$, $***P < 0.001$. From Baines et al. (1998).

Furthermore, the effects of summer mowing on spiders persisted into the following year (Baines et al. 1998; Smith et al. 1993). Although the effects on spiders of cutting in spring were less persistent than those of summer cutting, they may have particular agricultural significance. Cereal aphids, for example, which overwinter on grassland and hedgerow species, colonize crops in May and June (Hand 1989), and it is at this time that the predatory Araneae can potentially have most impact on crop pest control (Sunderland et al. 1986).

Management which increased the structural diversity of the field margin swards also increased the abundance and species richness of Araneae. This is likely to result from the requirement of many species of Araneae for specific web-building sites, and from higher prey densities in taller vegetation (Southwood et al. 1979). Thus, we found that the largest and most species-rich communities of spiders were fostered in the absence of regular cutting.

In common with butterflies, two of the other invertebrate groups we studied benefited from sowing. The abundance and species richness of both spiders and Auchenorrhyncha were higher on sown margins in the first four years of the experiment (Baines et al. 1998). It seems likely that this was mediated via habitat heterogeneity (Macdonald et al. 2000). Sown margins were likely to provide more heterogeneous habitat architecture, within which web-building niches were consequently more diverse. This hypothesis was supported by measurements of vegetation structure derived from combined measures of vegetation height and density. Vegetation structure was more complex on sown compared to naturally regenerated margins. The dominance of robust, branching species, such as oxeye daisy (Smith et al. 1994), in sown plots is likely to have been important. The greater plant species richness of these swards (Smith et al. 1999), again acting through structural changes, may also have increased the abundance of invertebrate prey.

In contrast to the Auchenorrhyncha, sowing had no significant or consistent effects on the Heteroptera. The dietary plasticity of many Heteropteran species (many predatory bugs, for example, are able to feed on plants in the absence of prey) is likely to make them less dependent on plant species composition. Among the herbivorous species, one of the most numerous in our samples was *Lygus rugulipennis*, the tarnished plant bug, which feeds largely on common nettle *Urtica dioica*; this plant was significantly less abundant in sown than naturally regenerated swards.

Auchenorrhyncha, while probably benefiting from enhanced physical heterogeneity of the habitat, are likely to be more dependent on the plant species composition of a habitat. While the group is considered to have rather generalized phytophagous habits (Morris 1971), the majority is known to prefer grasses to other vegetation and some feed exclusively on grasses (Morris 1990). Phytophagous groups may in general show a more direct link with plant species diversity rather than structural complexity. We know that the sown margins in our experiment contained more grass species compared with unsown (Smith et al. 1993).

Further evidence that spiders and Auchenorrhyncha were responding to different aspects of the changes in habitat heterogeneity resulting from the different management regimes was provided by their responses to herbicide spraying. Auchenorrhyncha showed a rapid but short-lived decline in the weeks immediately following spraying, while the effect on spiders was delayed by at least a month (Baines et al. 1998). This is what we would expect if spiders were more influenced by vegetation structure than were Auchenorrhyncha, as the sprayed, but dead, vegetation stems remained intact for some time after the herbicide application.

Eight years after the field margins were established they were once again sampled for Araneae, Auchenorrhyncha, and Heteroptera to investigate whether the pattern of effects of different management regimes on the invertebrate assemblages remained similar (Haughton et al. 1999). All of the mowing regimes continued to have a negative impact on the invertebrates, and the most severe impact was that of mowing in summer compared to leaving margins uncut (Table 3.2). This reduced the abundance of all of the invertebrates we identified. As in the first few years of the experiment, the timing, rather than the frequency of cutting, was more important. For example, cutting in summer only, or spring and summer, were more damaging to total invertebrate and spider abundance than cutting in spring and autumn (Haughton et al. 1999).

The impacts on invertebrates of sown compared to naturally regenerated margins were less marked eight years after the margins were established (Haughton et al. 1999). Of the three groups of invertebrates, only Auchenorryncha abundance was, by this stage, significantly increased by sowing. In our experiment, differences in plant composition between the sown and naturally regenerated sward types lessened over this time period (Smith et al. 2010; Chapter 2, this volume).

We found no evidence that removal of the cut material was of any benefit to invertebrates over the first four-year period (Baines et al. 1998); indeed, in the third year of the study, more spider species (mean = 15.8, SE = 0.8), were recorded where the cuttings were

Table 3.2 Mean number of individuals from treatments not cut and cut in summer only. Numbers in bold indicate significant differences (contrast analysis) at $P < 0.05$. Redrawn from Haughton et al. (1999).

Group	May		July		September	
	uncut	cut	uncut	cut	uncut	cut
Spiders	**116.2** (12.9)	**70.3** (9.0)	**173.3** (26.3)	**19.2** (2.6)	**432.2** (74.1)	**63.2** (20.5)
Heteroptera	**118.7** (15.4)	**49.5** (8.3)	**42.5** (5.9)	**13.3** (8.8)	**26.5** (5.2)	**8.2** (5.7)
Auchenorrhyncha	**201.8** (34.7)	**80.8** (15.8)	**152.0** (17.2)	**29.2** (6.2)	**245.8** (46.8)	**60.5** (9.8)

left, compared with 12.7 (SE = 1.13) species where they were collected. In a separate exercise, the rarely studied pseudoscorpions (also arachnids) were sampled from the field margins (Bell et al. 1999). The ancestral habitat of these invertebrates is leaf litter, with deep woodland leaf litter providing an ideal stable environment. After eight years, leaving hay lying appeared to ameliorate the effects of the cutting regimes on this group of invertebrates (Bell et al. 1999). Similarly, eight years after the start of the experiment, Heteroptera abundance was also significantly enhanced by leaving the hay *in situ*, perhaps through altering prey communities. Over the much longer term, one might predict a lowering of invertebrate diversity on these swards, mediated by nutrient addition, translating slowly into reduced plant species richness. However, even after 13 years, there was no evidence for reduced plant diversity on the naturally regenerated swards where hay had been left lying (Chapter 2, this volume; Smith et al. 2010).

Despite their contrasting ecologies, Araneae, Auchenorrhyncha, and Heteroptera all tended to be more abundant and species rich on uncut compared to cut margins (Smith et al. 1993; Baines et al. 1998; Bell et al. 2002). The removal of habitat structure, cover, and food by mowing make it likely that the majority of invertebrate groups would benefit, at least in the short term, from leaving margins uncut. However, in the short to medium term, some mowing is important if the plant species richness of the margins is to be maintained (Chapter 2, this volume; Smith et al. 1994); this also having knock-on effects for species richness of the invertebrate assemblages.

3.3 Does margin width matter for spiders?

We demonstrated that the ways in which uncropped field margins are established and managed have different consequences for the invertebrates living on them.

Other aspects of field margins might also be influential. For example, one question relates to their optimum width: are wider field margins better for biodiversity than narrow ones? Despite being relevant for the development of agri-environment schemes, we know rather little about this. Field margin widths under agri-environment schemes are determined primarily by economic and practical factors. The studies that have been undertaken suggest that there may be effects. For example, wider margins may have higher plant diversity (e.g. Schippers and Joenje 2002). Stoate and Boatman (2002) found that the width of perennial vegetation at the hedge base was associated with the presence of yellowhammer *Emberiza citrinella* and whitethroat *Sylvia communis* breeding territories (Stoate and Boatman 2002). Considering the intricacy of their communities, the effects of margin width on invertebrates are likely to be complex (Macdonald et al. 1998). Margin width may affect the movement of invertebrates into the crop during the summer and into overwintering habitats after harvest. A narrow margin might facilitate movement to and from a hedgerow, thus favouring species which overwinter in this habitat; but a wide margin might provide a physically greater area of overwintering habitat for species which require grassy habitats. Habitat area (patch size) has been shown to be important for both specialist and generalist species. Brückmann et al. (2010) showed that decreasing habitat connectivity dramatically decreased the abundance of specialists (up to 69%) in both plants and butterflies, while Osborne et al. (2008) found that linear features such as hedgerows held a greater proportion of bumblebee nests (20–37 nests per ha) than the equivalent area of non-linear habitats, such as woodlands and grassland (11–15 nests per ha). Conversely, species on narrow margins may be more susceptible to spray drift, inadvertent fertilizer application, and other management operations. Bundschuh et al. (2012) detected increasing grasshopper densities with increasing field margin width next to cereals and vineyards, and densities equivalent to those of grassland sites were observed only in field margins more than 9 m wide, except for field margins next to orchards. Considered together with their results from toxicological studies, they conclude that current insecticide risk assessments are insufficiently protective for grasshoppers in field margins. The positive relationship between habitat area and species richness derived from theories of island biogeography also suggests that larger areas may hold more wildlife; in their meta-analysis of set-aside and biodiversity, Van Buskirk and Willi (2004) found that larger parcels of set-aside increased species richness.

Working in collaboration with colleagues at the Royal Agricultural College (now the Royal Agricultural University) at Cirencester, we designed an experiment to tackle the question of whether field margin width affected spider abundance and richness in the margin and adjacent crop (Macdonald et al. 1998). We established, within a single field in autumn 1995, margins of 2 m, 8 m, and 20 m width (Fig. 3.7). We used a randomized block design, such that each of four blocks contained 80 m lengths of each of the three margin widths, in a random position relative to each other. Margins were sown with a simple grass and clover mixture comprising sheep's fescue *Festuca ovina* (15%), timothy *Phleum pratense* (15%), crested dogs-tail *Cynosurus cristatus* (10%), cocksfoot *Dactylis glomerata*

(15%), creeping red fescue *Festuca rubra* (24%), dwarf perennial ryegrass (15%), white clover *Trifolium repens* (3%), and red clover *Trifolium pratense* (3%) at a rate of 25 kg/ha. They were managed with a single cut in late July or early August, and cuttings were not removed.

Spiders were sampled using pitfall traps and suction sampling in each margin width at 1 m, 5 m, and 15 m from the field boundary; all samples in 20 m-wide margins were thus within the margins, while some samples in 2 m and 8 m margins were also in the crop (Fig. 3.7).

The results were not clear-cut or consistent. In May, there was a conspicuous tendency for more spiders to be caught in pitfall traps on the widest margins than on 8 m or 2 m margins (Fig. 3.8), regardless of distance from the boundary. There were more spiders on the 20 m margins at all three distances from the field boundary ($F_{2,9} = 16.9$, $P < 0.001$). At the 1 m distance, where all pitfalls were within the margins, this implied that, for spiders, the wider field margins constituted a distinct habitat type. In addition to the effect of margin width, there was also a significant tendency for fewer spiders to be caught further from the boundary ($F_{2,9} = 16.9$, $P < 0.001$). In spring, then, the evidence was that spiders were more abundant on wider margins, and more abundant closer to the field boundary.

In July, when spiders were everywhere more abundant than in May, the spring pattern was not repeated.

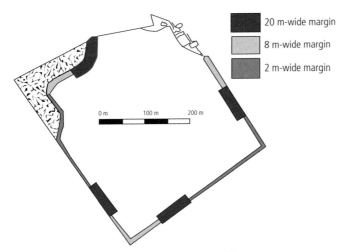

Location of invertebrate sampling points (•):

Figure 3.7 Experimental layout of variable width field margins. From Macdonald et al. (1998).

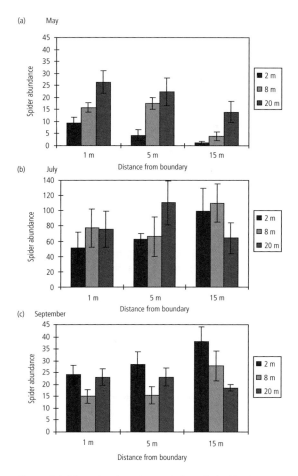

Figure 3.8 Mean abundance (± standard error) of spiders in pitfall samples at 1 m, 5 m, and 15 m from the boundary, in 2 m, 8 m, and 20 m-wide margins and adjacent crop. Samples were taken in (a) May, (b) July, and (c) September. From Macdonald et al. (1998): report to MAFF.

Similar numbers of spiders were captured in traps on the margins of different widths. Nor was there any evidence that the number of spiders was affected by distance from the field boundary.

By September, any effect of width was obscured by a tendency for more spiders to be caught in the cropped area. The pattern in September (after harvest) was complex. There were more spiders in the crop than the margin at distances of 8 m and 15 m from the margin[2]. Spiders will commonly disperse into crops from field margins; furthermore, in this experiment, the margins

were sown with a species-poor grass mixture rather than a complex seed mixture or being naturally regenerated, perhaps reducing their benefit to spiders compared to the crop habitat.

In summary, spiders were found to be more abundant on wider margins, and closer to the boundary in spring, suggesting that wider margins constituted a distinct habitat type. Wider margins may have higher plant diversity (Schippers and Joenje 2002), which might affect spider communities, but over the time-scale of this experiment this is an unlikely explanation. Aavik and Liira (2010) found that, for plant communities in agricultural habitats, more specialist species benefited from wider open boundaries, while narrow boundaries hosted more agro-tolerant species. They suggested that this may either be due to the buffering effect of wider boundaries (e.g. protecting species from pesticide or fertilizer drift), or to an increase in available microhabitats within wider boundaries, both of which may be true for spiders in this experiment.

3.4 Effects of boundary type on invertebrates

We have shown that field margin width affected the abundance and diversity of spiders. In this final section of the chapter we describe an experiment that explores the relationship between invertebrates in the boundary habitat and adjacent set-aside.

Set-aside was land removed from arable production for at least one year (rotational set-aside) and sometimes longer (long-term set-aside; Clarke 1992; Firbank et al. 1993). Since its introduction in 1992, the percentage of compulsory set-aside has varied (between 5 and 17.5%) according to market circumstances, reflecting its role as a production control. In 2007, the set-aside rate was reduced to 0% in response to the sharp rise in cereal prices as a consequence of lower world production, increased demand for food, feed, and fuel purposes, and low global stocks. Although its principal aim was to reduce agricultural surplus, set-aside was found to have a range of positive impacts on wildlife, if managed appropriately (Andrews 1992; Sotherton 1998; Henderson and Evans 2000; Firbank et al. 2003). The ways in which set-aside was established and managed also had implications for the development of agri-environment schemes in the UK.

Set-aside had enormous potential for enhancing the invertebrate and vertebrate populations of farmland, including birds (Henderson et al. 2000a,b),

---

[2] $F_{2,9} = 7.63$, $P = 0.011$ and $F_{2,9} = 4.1$, $P = 0.054$, respectively.

Box 3.1 Should set-aside be configured as margins or blocks?

We have demonstrated clearly that the ways in which field margins are managed affects their flora and fauna. What about the spatial configuration of fallowed land? Maybe configuring it as field margins is not the best option. For example, a different physical arrangement of fallowed land might affect how easily it can be colonized. There has been little experimental work to answer this, though much speculation. The process of colonization is fundamental to arable ecology, because of the constantly changing nature of the habitats, particularly through the annual harvest cycle, but little is known of factors which might promote or inhibit movement into arable habitats (Macdonald and Smith 1990).

The introduction of set-aside in 1992 (Chapter 1, this volume) provided an opportunity to investigate the impacts of configuration of fallowed areas of land on invertebrate communities. Set-aside could be arranged either as margins around fields or as whole fields (blocks). In the former case, relatively unproductive land would be lost and there might be benefits for wildlife. For example, field margin set-aside will be close to sources of colonists in the hedgerows, and strips of set-aside may provide a network of colonizing pathways through the farm. Margin set-aside may encompass a wider range of environments than whole-field set-aside, because to make up an equivalent area it must be present in more fields. Conversely, whole-field set-aside

might also benefit some wildlife. Many predators forage along hedgerows and margins, and therefore their prey species could benefit from large blocks of set-aside, distant from these routes. Blocks of set-aside will also have relatively less edge than similar areas arranged as margins, and therefore a smaller area will be susceptible to spray drift. We therefore asked the following question: Do invertebrate abundances differ on set-aside configured as blocks and margins?

We established four replicate margin networks on two farms, each of approximately 5 ha. A 5 ha block of set-aside was also associated with each margin network. Margins and blocks were sown in autumn 1995 with a grass and clover mixture at a rate of 25 kg/ha and managed with a single cut in late summer. Invertebrates were sampled using pitfall traps and suction sampling in May, July, and September 1997, at 88 locations within the margins and blocks. Adult Carabidae were identified to species. Staphylinids, other beetles, spiders and aphids were counted, but not identified to species.

Our results showed that there were no consistent differences in invertebrate numbers between set-aside arranged as blocks compared to set-aside arranged as margins. However, there were differences in some months for some groups. Numbers of beetles captured in pitfall traps in September were higher in set-aside blocks compared to margin

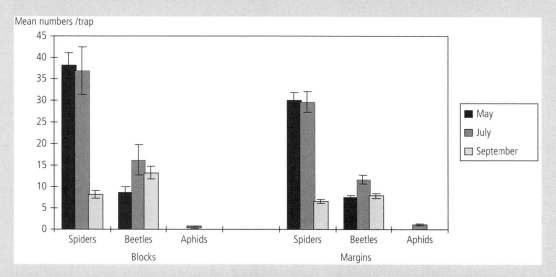

Box Figure 3.1.1 Mean (± standard deviation) abundance of key invertebrate groups in pitfall traps in blocks and margins in May, July, and September.

continued

Box 3.1 *Continued*

set-aside[3] (Box Figure 3.1.1). Aphids, which were sampled only in July, were more abundant in margin samples[4].

We found no evidence that the configuration of set-aside as blocks or margins had any effect on the abundance of spiders captured in pitfall traps. However, in the suction traps there were consistently more spiders in the set-aside blocks; this was statistically significant in both May and July[5].

Many aspects relating to the establishment and management of set-aside are relevant for efforts to reverse declines in farmland biodiversity. Maintaining areas of appropriately managed low-input grassland or fallowed (naturally regenerated) land on farms is one of the

strategies that is likely to continue to be part of agri-environment schemes, and this has also been the focus of voluntary measures in the UK to mitigate the loss of set-aside. Knowledge of how set-aside configuration influences biodiversity can help to focus resources for maximizing biodiversity benefit. The results suggested that block and margin set-aside may constitute qualitatively different habitats for some groups of invertebrates, and that the creation of field margins should not be regarded as a substitute for maintaining or establishing larger low-input sown or fallowed patches, which can offer other resources for wildlife in farmland landscapes.

---

invertebrates (Moreby and Aebischer 1992; Moreby and Sotherton 1995), and arable weed species (Firbank et al. 1993). The benefits of set-aside for biodiversity were influenced both by how it was established (Critchley and Fowbert 2000) and by its subsequent management (Hansson and Fogelfors 1998; Tattersall et al. 1999a; Bracken and Bolger 2006).

We showed that the configuration of set-aside, as margins or whole fields (blocks), had different impacts on the communities that developed within it (Box 3.1), and hypothesized that the type of boundary may also affect the development of biodiversity in adjacent set-aside. For example, Tattersall et al. (1999a) showed that margin set-aside that was situated next to hedgerows had a more abundant and diverse small mammal community (see Chapter 4, this volume), primarily because wood mice *Apodemus sylvaticus* were more numerous in margins, and bank voles *Myodes glareolus*, a hedgerow species, did not venture into the centre of set-aside fields. If some boundary features are rich sources of invertebrate colonists, we might hypothesize that greater abundances of invertebrates would be found in set-aside adjacent to those features. It follows that one would expect more individuals to disperse by chance into the set-aside habitat from such sites, and there may be an element of deliberate emigration from a high density population in which food or oviposition sites might be limited. The result would

be higher invertebrate abundance in the set-aside adjacent to high density populations, and a decline in abundance with increasing distance into the set-aside, at least in its first year of existence. The wider policy implication is that the creation of field margins or fallowed areas might be targeted next to certain features to maximize their biodiversity and pest control benefits. We therefore used a large-scale field experiment to ask the following questions: Are field boundaries a potential source of colonists into the set-aside? Are some field boundary types better sources of colonists than others? Does the invertebrate community in set-aside reflect that of potential colonists in adjacent field boundaries? Over what distance does the boundary have an impact on set-aside invertebrate communities?

To answer these questions, invertebrates were collected by D-vac suction sampling at 16 sites in four fields at two of the Royal Agricultural College farms, Coates and Harnhill Manor, in June 1995 (Gates et al. 1997). At each site, one sample was taken from the field boundary vegetation, and three samples from the adjacent whole-field set-aside land, at 2 m, 8 m, and 16 m from the edge of the field. Invertebrates were identified to order, except for abundant groups such as Diptera, which were identified to sub-order. At each site the characteristics of the boundary were recorded (presence or absence of hedge, wall, fence, ditch, and bank, and the heights of understorey and hedge vegetation). All set-aside land had been sown the previous autumn with a mixture of winter wheat and mustard, and vegetation on the set-aside was sparse.

---

[3] $F_{1,61} = 13.6$, $P = 0.0005$.

[4] $F_{1,61} = 4.8$, $P = 0.0362$.

[5] $F_{1,61} = 4.2$, $P = 0.045$ and $F_{1,61} = P = 0.0012$, respectively.

3.4.1 Field boundaries as potential sources of colonists

We found that invertebrates were much more abundant in the boundary samples compared to the adjacent set-aside (Gates et al. 1997, Fig. 3.9). Set-aside in this experiment consisted of a sparse covering of low, species-poor, vegetation, which plainly constituted an unattractive habitat compared to the adjacent boundary with its dense, species-rich vegetation. The set-aside appeared to be a particularly poor habitat for the plant-feeding Auchenorrhyncha. There was not strong support for declines in invertebrate abundance with increasing distance from field boundary across the set-aside (Gates et al. 1997, Fig. 3.9). While the field boundary is likely to be a potential source of colonists, invertebrates may also colonize set-aside from further afield. Furthermore, they may be sufficiently mobile that we could not detect differences on the scale of our sampling (up to 18 m from the boundary). Overall, sowing with a more species-rich seed mixture, or leaving the set-aside in place for more than a year, would have been likely to increase invertebrate abundance within the set-aside.

3.4.2 Are some field boundary types better sources of colonists than others?

Higher vegetation, the presence of a hedge, and presence of a wall all increased the abundance of invertebrates in the boundary (Gates et al. 1997). The presence of a hedgerow influenced the greatest number of invertebrate groups, while the presence of banks and ditches had no detectable effect on abundance of any invertebrate group in this study. Two groups, parasitoids and spiders, were not affected by any of the explanatory variables, and the remaining groups were affected by only one or two variables.

Although abundance of most groups in the field boundary vegetation was related to one or more of the environmental characteristics, no factor was a consistently good predictor. There may have been variation between sites in such factors as soil, microclimate, and humidity, and these may have played a part in determining a site's suitability for invertebrates. Asteraki et al. (1995) found that these factors played a more important role in determining the abundance of most species of carabid beetles than the structural attributes of field boundaries, and the same may be true for other groups of invertebrates.

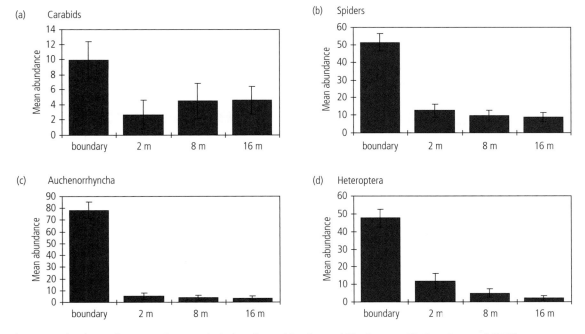

Figure 3.9 Abundances of representative groups in the boundary and 2 m, 8 m, and 16 m into set-aside. From Gates et al. (1997).

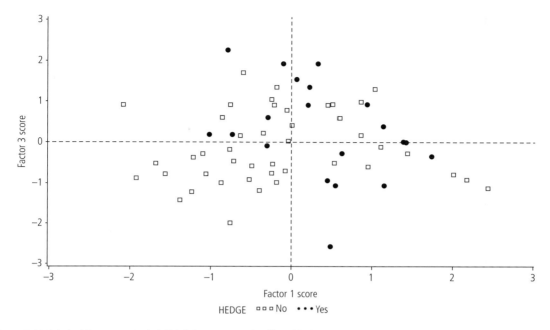

Figure 3.10 Principal Component Analysis (PCA) demonstrating the effect of hedgerow on the proportional composition of the invertebrate assemblage in the set-aside. From Macdonald et al. (1998).

Next, we looked at how a hedgerow affected the composition of the invertebrate assemblage in the set-aside, using Principal Component Analysis (Fig. 3.10), which enables us to visualize community structures in two dimensions—sample points that are close together on the plot have similar assemblages. Samples taken from set-aside with and without a boundary hedge tended to separate out into the lower left (an area of more harvestmen and fewer parasitoids and Brachycera) and upper right (an area corresponding to a higher proportion of Auchenorrhyncha and Heteroptera and fewer carabids) of the graph.

3.4.3 Are invertebrate assemblages in the boundary structured differently from those in the set-aside, and how do these assemblages change with distance from the boundary?

We carried out a PCA on proportions of different groups in the boundary (0 m) and at 2 m, 8 m, and 16 m into the set-aside (Fig. 3.11). Boundary samples were associated with an increasing proportion of Heteroptera, harvestmen, and Auchenorrhyncha, and a decreasing proportion of carabids. The composition of the samples in the set-aside (distances 2 m, 8 m, and 16 m) was similar. In other words, the invertebrate

communities in the boundary differed from those in the set-aside, but communities at different distances into the set aside did not differ from each other.

In summary, we found that boundary type influenced invertebrate abundance and species composition in set-aside. The presence of hedgerow was influential, increasing abundance in the set-aside and affecting community structure. Invertebrate abundance declined with distance into set-aside.

3.5 Conclusions and recommendations

We have shown that the ways in which arable field margins are established and managed can have profound effects on their invertebrate assemblages. Swards established by sowing with a grass and wild flower seed mixture attracted more adult butterflies and were also used as breeding habitat. In the short term, larger and more species-rich invertebrate assemblages were fostered on unmanaged margins than on those managed by cutting. Leaving field margins uncut for more than one or two years, however, leads to rapid succession to scrub (Smith et al. 1993) and is likely to be accompanied by loss of grassland invertebrate species (Usher and Jefferson 1987; Feber and Smith 1995). Some form of rotational cutting, ensuring some areas of tall sward are left uncut each year, is most likely to benefit

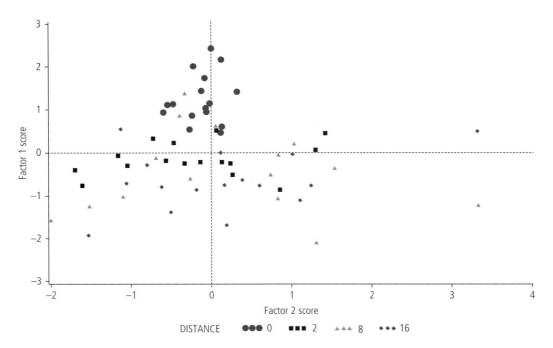

Figure 3.11 PCA on proportions of different groups in the boundary (0 m) and at 2 m, 8 m, and 16 m into the set-aside. Along the y-axis boundary samples separated from set-aside samples. From Macdonald et al. (1998).

butterflies and Araneae, as well as other invertebrate groups (Morris 1971; Smith et al. 1993; Feber and Smith 1995). On field margins this may be most practically achieved by mowing a swath next to the crop once a year, leaving taller vegetation adjacent to the boundary to be cut once every two or three years, or by mowing different sections of margin in different years.

The timing of cutting was critical, with mid-summer cutting having the most persistent, negative effects (Smith et al. 1993, Feber et al., 1996, Baines et al. 1998), so cutting at this time should be avoided. Cutting in spring and autumn is substantially less damaging to invertebrates in general, although the benefits to crop husbandry of Araneae as predators may be diminished by cutting in spring, and species of butterfly and other invertebrates that need undisturbed habitats in which to overwinter may be adversely affected by autumn cutting. However, cutting in autumn rather than summer increased the species richness of plants on the field margins over the shorter term (Smith et al. 2010) which may benefit invertebrates dependent on a diverse sward.

Field margin width had complex effects on invertebrate abundance and species richness. Field margins of different widths may have constituted different habitats for invertebrates in arable systems, and these might range from narrow margins to whole fields. The presence of a hedgerow influenced the invertebrates found in adjacent set-aside; sites with a boundary hedgerow were associated with increased numbers in most groups. Hedgerows are a valuable habitat in themselves for invertebrates (e.g. Pywell et al. 2005), and the presence of a hedgerow next to a field margin has been shown to increase numbers within the field margin of groups such as pseudoscorpions, probably by buffering the field margin from management (Bell et al. 1999). For other taxa, such as farmland birds, Vickery et al. (2009) conclude that the major influence on the value of a margin (at a field scale) is proximity to a good quality hedgerow (i.e. one that is well established, relatively species rich, and well managed). Margins near hedgerows may also be favoured by foraging birds (e.g. Henderson et al. 2004).

Other studies have shown that the numbers of plants (e.g. Marshall 1989), invertebrates (e.g. Holland et al. 1999), and small mammals (Tattersall et al. 2002) often decrease with distance from the field boundary and, where good quality field boundaries exist, field margins may be one of the most cost-effective ways of enhancing biodiversity (Vickery et al. 2009). Within current agri-environment schemes the greatest emphasis is on management of field margin and hedgerow habitat, although Butler et al. (2007) conclude that more effort needs to be made on improving resources within the cropped area, as well as on field margins and hedgerows, if the UK government's target for restoration of farmland birds is to be met. Our results suggested that fallowed land (in this case set-aside) configured in blocks constituted qualitatively different habitats for invertebrates compared to margin set-aside; this was also true for small mammals (Tattersall et al. 1999a). Farms with a mixture of configurations are likely to be the most wildlife friendly.

Our recommendation is that blanket management approaches for invertebrates at the farm scale are inappropriate. A diverse farmed landscape with margins of different sizes and different sward structures should be the aim. This will help invertebrate populations, with their differing ecological requirements, to survive and flourish in the characteristically unstable environment of arable systems.

Acknowledgements

This work was funded by English Nature (now Natural England) and MAFF (now Defra), with support from the Co-op Bank and the Ernest Cook Trust. We are grateful to the many staff, students, and volunteers who were involved with these projects. In particular, we thank Miriam Baines, Simon Gates, James Bell, Alison Haughton, and Martin Townsend. We are grateful to the University of Oxford and the Royal Agricultural College (now University) for their support in accommodating the experiments. Paul Johnson acknowledges the support of the Whitley Trust.

References

Aavik, T. and Liira, J. (2010). Quantifying the effect of organic farming, field boundary type and landscape structure on the vegetation of field boundaries. *Agriculture, Ecosystems and Environment*, **135**, 178–186.

Andrews, J. (1992). Some practical problems in set-aside management for wildlife. *British Wildlife*, **3**, 329–336.

Asteraki, E.J., Hanks, C.B., and Clements, R.O. (1995). The influence of different types of grassland field margin on Carabid beetle (Coleoptera, Carabidae) communities. *Agriculture, Ecosystems and Environment*, **54**, 195–202.

Baines, M., Hambler, C., Johnson, P.J., Macdonald, D.W., and Smith, H.E. (1998). The effects of arable field margin management on the abundance and species richness of Araneae (spiders). *Ecography*, **21**, 74–86.

Bell, J.R., Gates, S., Haughton, A.J., et al. (1999). Pseudoscorpions in field margins: effects of margin age, management and boundary habitats. *Journal of Arachnology*, **27**, 236–240.

Bell, J., Johnson, P.J., Hambler, C., et al. (2002). Manipulating the abundance of *Lepthyphantes tenuis* (Araneae: Linyphiidae) by field margin management. *Agriculture Ecosystems and Environment*, **93**, 295–304.

Bracken, F. and Bolger, T. (2006). Effects of set-aside management on birds breeding in lowland Ireland. *Agriculture, Ecosystems and Environment*, **117**, 178–184.

Brooks, D.R., Storkey, J.E. Clark, S.J., et al. (2012) Trophic links between functional groups of arable plants and beetles are stable at a national scale. *Journal of Animal Ecology*, **81**, 4–13.

Brückmann, S.V., Krauss, J., and Steffan-Dewenter, I. (2010). Butterfly and plant specialists suffer from reduced connectivity in fragmented landscapes. *Journal of Applied Ecology*, **47**, 799–809.

Bundschuh, R., Schmitz, J., Bundschuh, Albrecht Brühl, C. (2012). Does insecticide drift adversely affect grasshoppers (Orthoptera: Saltatoria) in field margins? A case study combining laboratory acute toxicity testing with field monitoring data. *Environmental Toxicology and Chemistry*, **31**, 1874–1879.

Butler, S.J., Vickery, J.A., and Norris, K. (2007). Farmland biodiversity and the footprint of agriculture. *Science*, **315**, 381–383.

Carvell C., Roy D.B., Smart S.M., Pywell R.F., Preston C.D., and Goulson D. (2006). Declines in forage availability for bumblebees at a national scale. *Biological Conservation*, **132**, 481–489.

Clarke, J. (1992). Set aside. BCPC Monograph No. 50, British Crop Protection Council, Farnham.

Conrad, K.F., Warren, M.S., Fox, R., Parsons, M.S., and Woiwod, I.P. (2006). Rapid declines of common, widespread British moths provide evidence of an insect biodiversity crisis. *Biological Conservation*, **132**, 279–291.

Critchley, C.N.R. and Fowbert, J.A. (2000). Development of vegetation on set-aside land for up to nine years from a national perspective. *Agriculture, Ecosystems and Environment*, **79**, 159–174.

Donald, P.F. and Evans, A.D. (2006). Habitat connectivity and matrix restoration: the wider implications of agri-environment schemes. *Journal of Applied Ecology*, **43**, 209–218.

Duelli, P. and Obrist, M.K. (1998). In search of the best correlates for local organismal biodiversity in cultivated areas. *Biodiversity and Conservation*, **7**, 297–309.

Feber, R.E. and Smith, H. (1995). Butterfly conservation on arable farmland. In A.S. Pullen, ed. *Ecology and Conservation of Butterflies*. Chapman and Hall, London.

Feber, R.E., Smith, H., and Macdonald, D.W. (1994). The effects of field margin restoration on the meadow brown butterfly (*Maniola jurtina*). In N.D. Boatman, ed. *Field Margins—Integrating Agriculture and Conservation*. BCPC Monograph No. 58. BCPC Publications, Farnham.

Feber, R.E., Smith, H., and Macdonald, D.W. (1996). The effects of management of uncropped edges of arable fields on butterfly abundance. *Journal of Applied Ecology*, **33**, 1191–1205.

Feber, R.E., Smith, H.E., and Macdonald, D.W. (1999). The Importance of spatially variable field margin management for two butterfly species. *Aspects of Applied Biology*, **54**, 155–162.

Feber, R.E., Bell, J., Johnson, P.J., Firbank, L.G., and Macdonald, D.W. (1998). The effects of organic farming on the surface-active spider (Araneae) assemblages in wheat in southern England, U.K. *Journal of Arachnology*, **26**, 190–202.

Firbank, L.G., Arnold, H.R., Eversham, B.C., et al. (1993). Managing set-aside land for wildlife. ITE Research publication 7, HMSO, London.

Firbank, L.G., Smart, S.M., Crabb, J., et al. (2003). Agronomic and ecological costs and benefits of set-aside in England. *Agriculture, Ecosystems and Environment*, **95**, 73–85.

Fox, R., Brereton, T.M., Asher, J., Botham, M.S., Middlebrook, I., Roy, D.B., and Warren, M.S. (2011). *The State of the UK's Butterflies 2011*. Butterfly Conservation and the Centre for Ecology & Hydrology, Wareham, Dorset.

Fox, R., Parsons, M.S., Chapman, J.W., Woiwod, I.P., Warren, M.S., and Brooks, D.R. (2013). *The State of Britain's Larger Moths 2013*. Butterfly Conservation and Rothamsted Research, Wareham, Dorset, UK.

Gallai, N., Salles, J.M., Settele, J., and Vaissiere, B.E. (2009). Economic valuation of the vulnerability of world agriculture confronted with pollinator decline. *Ecological Economics*, **68**, 810–821.

Gates, S., Feber, R.E., Macdonald, D.W., Hart, B.J., Tattersall, F.H., and Manley, W.J. (1997). Invertebrate populations of field boundaries and set-aside land. *Aspects of Applied Biology*, **50**, 313–322.

Gibson, C.W.D., Hambler, C., and Brown, ,V.K. (1992). Changes in spider (Araneae) assemblages in relation to succession and grazing management. *Journal of Applied Ecology*, **29**, 132–142.

Goulson, D. (2013). An overview of the environmental risks posed by neonicotinoid pesticides. *Journal of Applied Ecology*, **50**, 977–987.

Goulson D., Lye G.C., and Darvill B. (2008). Decline and conservation of bumble bees. *Annual Review of Entomology*, **53**, 191–208.

Grice, P.V., Radley, G.P., Smallshire, D., and Green, M.R. (2006). Conserving England's arable biodiversity through agri-environment schemes and other environmental policies: a brief history. *Aspects of Applied Biology*, **81**, 7–22.

Hand, S.C. (1989). The overwintering of cereal aphids on Gramineae in southern England, 1977–1980. *Annals of Applied Biology*, **115**, 17–29.

Hansson, M. and Fogelfors, H. (1998). Management of permanent set-aside on arable land in Sweden. *Journal of Applied Ecology*, **35**, 758–771.

Haughton, A.J., Bell, J.R., Gates, S., et al. (1999). Methods of increasing invertebrate abundance within field margins. *Aspects of Applied Biology*, **54**, 163–170.

Henderson, I.G., Cooper, J., Fuller, R.J., and Vickery, J. (2000a). The relative abundance of birds on set-aside and neighbouring fields in summer. *Journal of Applied Ecology*, **37**, 335–347.

Henderson, I.G. and Evans, A.D. (2000). Responses of farmland birds to set-aside and its management. In N.J. Aebischer, A.D. Evans, P.V. Grice, and J.A. Vickery, eds., *Ecology and Conservation of Lowland Farmland Birds*, pp. 339–422. British Ornithologists' Union, Tring.

Henderson, I.G., Vickery, J.A., and Carter, N. (2004). The use of winter bird crops on farmland by seed eating and insectivorous species in lowland England. *Biological Conservation*, **118**, 21–32.

Henderson, I.G., Vickery, J.A., and Fuller, R.J. (2000b). Summer bird abundance and distribution on set-aside fields on intensive arable farms in England. *Ecography*, **23**, 50–59.

Holland, J.M., Perry, J.N., and Winder, L. (1999). The within-field spatial and temporal distribution of arthropods within winter wheat. *Bulletin of Entomological Research*, **89**, 499–513.

Kleijn, D., Baquero, R.A., Clough, Y., et al. (2006). Mixed biodiversity benefits of agri-environment schemes in five European countries. *Ecology letters*, **9**, 243–254.

Klein, A.M., Vaissière, B., Cane J.H., Steffan-Dewenter I., Cunningham S.A., Kremen C., and Tscharntke T. (2007). Importance of crop pollinators in changing landscapes for world crops. *Proceedings of the Royal Society of London Series B-Biological Sciences*, **274**, 303–313.

Macdonald, D.W., Feber, R.E., Johnson, P.J., and Tattersall, F. (2000). Ecological experiments in farmland conservation. In M.J. Hutchings, E.A. John, A.J.A. Stewart, eds., *The Ecological Consequences of Environmental Heterogeneity*, British Ecological Society Symposium. Blackwell Scientific Publications, Oxford.

Macdonald, D.W., Hart, B.J., Tattersall, F.H., Johnson, P.J., Manley, W.J., and Feber, R.E. (1998). *The influence of configuration, juxtaposition, management and colonising sources on the agricultural and conservation consequences of set-aside*. Report to MAFF.

Macdonald, D.W. and Smith, H. (1990). Dispersal, dispersion and conservation in the agricultural ecosystem. In: Bunce, R.G.H. and Howard, D.C., eds., *Species Dispersal in Agricultural Habitats*, pp. 418–464, Belhaven Press. London.

Mansour, F and Heimbach, U. (1993). Evaluation of lycosid, micryphantid and linyphiid spiders as predators of Rhopalosiphum padi (Hom.: Aphididae) and their functional response to prey density—laboratory experiments. *Entomophaga*, **38**, 79–87.

Marc, P., Canard, A., and Ysnel, F. (1999). Spiders (Araneae) useful for pest limitation and bioindication. *Agriculture, Ecosystems and Environment*, **74**, 229–273.

Marshall, E.J.P. (1989). Distribution patterns of plants associated with arable field edges. *Journal of Applied Ecology*, **26**, 247–258.

May, R.M. (1988). How many species are there on Earth? *Science*, **241**, 1441–1449.

Memmott, J., Carvell, C., Pywell, R.F., and Craze, P.G. (2010). The potential impact of global warming on the efficacy of field margins sown for the conservation of bumblebees. *Philosophical Transactions of the Royal Society B*, **365**, 2071–2079.

Merckx, T. and Berwaerts, K. (2010). What type of hedgerows do Brown hairstreak (*Thecla betulae* L.) butterflies prefer? Implications for European agricultural landscape conservation. *Insect Conservation and Diversity*, **3**, 194–204.

Monbiot, G. (2013). *The Guardian*, 13 August 2013.

Moreby, S.J. and Aebischer, N.J. (1992). A comparison of the invertebrate fauna of cereal fields and set-aside land. In J. Clarke, ed., *Set-aside*, BCPC Monograph 50, pp. 181–187.

Moreby, S.J. and Sotherton, N.W. (1995). The management of set-aside land as brood-rearing habitats for gamebirds. In A. Colston and F. Perring, eds., *Insects, Plants and Set-aside*, pp. 41–44. Botanical Society of the British Isles, London.

Morris, M.G. (1971). The management of grassland for the conservation of invertebrate animals. In E. Duffey and A.S. Watt, eds., *The Scientific Management of Plant and Animal Communities for Conservation*, pp. 527–552. 11th Symposium of the British Ecological Society, Blackwell, Oxford.

Morris, M.G. (1990). The Hemiptera of two sown calcareous grasslands. I. Colonization and early succession. *Journal of Applied Ecology*, **27**, 367–378.

Morris, M.G. (2000). The effects of structure and its dynamics on the ecology and conservation of arthropods in British grasslands. *Biological Conservation*, **95**, 129–142.

Natural England (2008). *ELS handbook*, 2nd edn. October 2008.

Osborne, J.L., Martin, A.P., Shortall, C.R., et al. (2008). Quantifying and comparing bumblebee nest densities in gardens and countryside habitats. *Journal of Applied Ecology*, **45**, 784–792.

Pettis, J.S., Lichtenberg, E.M., Andree, M., Stitzinger J, Rose, R., and vanEngelsdorp, D. (2013). Crop pollination exposes honey bees to pesticides which alters their susceptibility to the gut pathogen *Nosema ceranae*. *PLOS ONE*, **8**(7): e70182.

Pullin, A. (1987). Changes in leaf quality following clipping and regrowth of *Urtica dioica* and consequences for a specialist insect herbivore, *Aglais urticae*. *Oikos*, **49**, 39–45.

Pywell, R.F., James, K.L., Herbert, I., et al. (2005). Determinants of overwintering habitat quality for beetles and spiders on arable farmland. *Biological Conservation*, **123**, 79–90.

Schippers, P. and Joenje, W. (2002). Modelling the effect of fertiliser, mowing, disturbance and width on the biodiversity of plant communities of field boundaries. *Agriculture, Ecosystems and Environment*, **93**, 351–365.

Smith, H., Feber, R.E., Johnson, P., et al. (1993). *The Conservation Management of Arable Field Margins*. English Nature Science No. 18. English Nature, Peterborough.

Smith, H., Feber, R.E., and Macdonald, D.W. (1994). The role of wild flower seed mixtures in field margin restoration. In N.D. Boatman, ed., *Field Margins—Integrating Agriculture and Conservation. BCPC Monograph No. 58*. BCPC Publications, Farnham.

Smith, H., Feber, R.E., and Macdonald, D.W. (1999). Sown field margins: why stop at grass? *Aspects of Applied Biology*, **54**, 275–282.

Smith, H., Feber, R.E., Morecroft, M.D., Taylor, M.E., and Macdonald, D.W. (2010). Short term successional change does not predict long term conservation value of managed arable field margins. *Biological Conservation*, **143**, 813–822.

Sotherton, N.W. (1998). Land use changes and the decline of farmland wildlife: an appraisal of the set-aside approach. *Biological Conservation*, **83**, 259–268.

Southwood, T.R.E., Brown, V.K., and Reader, P.M. (1979). The relationships of plant and insect diversities in succession. *Biological Journal of the Linnean Society*, **12**, 327–348.

Stoate, C., and Boatman, N.D. (2002). Ecological and Agricultural benefits of linear grassland features within arable systems. In D. Chamberlain, A. Wilson, Eds. *Avian Landscape Ecology: pure and applied issues in the large-scale ecology of birds*, pp. 191–194. Proceedings of 11th IALE Conference.

Storkey, J., Meyer, S., Still. K.S., and Leuschner, C. (2012). The impact of agricultural intensification and land-use change on the European arable flora. *Proceedings of the Royal Society B*, **279**, 1421–1429.

Sunderland, K.D. (1999). Mechanisms underlying the effects of spiders on pest populations. *Journal of Arachnology*, **27**, 308–316.

Sunderland, K.D., Frazer, A., and Dixon. A.F.G. (1986). Field and laboratory studies of money spiders (Linyphiidae) as predators of cereal aphids. *Journal of Applied Ecology*, **23**, 433–447.

Sutcliffe, O.L., Bakkestuen, V., Fry, G., and Stabbetorp, O.E. (2003). Modelling the benefits of farmland restoration: methodology and application to butterfly movement. *Landscape and Urban Planning*, **63**, 15–31.

Tattersall, F.H., Hart, B.J., Manley, W.J., Macdonald, D.W., and Feber, R.E. (1999a). Small mammals on set-aside blocks and margins. *Aspects of Applied Biology*, **54**, 131–138.

Tattersall, F.H., Macdonald, D.W., Hart, B.J., Johnson, P.J., Manley, W.J., and Feber, R. (2002). Is habitat linearity important for small mammal communities on farmland? *Journal of Applied Ecology*, **39**, 643–652.

Thomas, J.A. (1984). The conservation of butterflies in temperate countries: past efforts and lessons for the future. In R.I. Vane Wright and P. Ackery, eds., *Biology of Butterflies*, pp. 333–353. London: Academic Press.

Mora, C. Tittensor, D. P., Adl, S., Simpson, A.G.B., and Worm, B. (2011). How many species on earth and in the ocean? *PLOS Biology*, **9**: e1001127, 1–8.

UK Biodiversity Steering Group (1995a). *Biodiversity: the UK Steering Group Report. Vol. 2, Actions plans*. HMSO, London.

UK Biodiversity Steering Group (1995b). *Biodiversity: the UK Steering Group Report. Vol. 1: Meeting the Rio Challenge.* HMSO, London.

Usher, M.B, and Jefferson, R.G. (1987). Creating new and successional habitats for arthropods.In Collins, N.W. and Thomas, J. eds., *The Conservation of Insects and their Habitats.* Academic Press, pp. 261–281.

Van Buskirk, J. and Willi, Y. (2004). Enhancement of biodiversity within set-aside land. *Conservation Biology,* **18**, 987–994.

Vickery, J., Feber, R.E., and Fuller, R.J. (2009). Arable field margins managed for biodiversity conservation: a review of food resource provision for farmland birds. *Agriculture, Ecosystems and Environment,* **133**, 1–13.

Warren, M.S., Hill, J.K., Thomas, J.A., et al. (2001). Rapid responses of British butterflies to opposing forces of climate and habitat change. *Nature,* **414**, 65–69.

Whitehorn, P.R., O'Connor, S., Wackers, F.L., and Goulson, D. (2012). Neonicotinoid pesticide reduces bumble bee colony growth and queen production. *Science,* **336**, 351–352.

World Conservation Union, The (2010). *IUCN Red List of Threatened Species. Summary Statistics for Globally Threatened Species. Table 1: Numbers of threatened species by major groups of organisms (1996–2014).* Available at <http://www.iucnredlist.org>.

Zurbrügg, C. and Frank, T. (2006). Factors influencing bug diversity (Insecta: Heteroptera) in semi-natural habitats. *Biodiversity and Conservation,* **15**, 261–280.

CHAPTER 4

Small mammals on lowland farmland

David W. Macdonald, Lauren A. Harrington, Merryl Gelling,
Fran H. Tattersall, and Tom Tew

I'm truly sorry man's dominion,
Has broken nature's social union,
An' justifies that ill opinion,
Which makes thee startle
At me, thy poor, earth-born companion,
An' fellow-mortal!

Still thou art blest, compar'd wi' me
The present only toucheth thee:
But, Och! I backward cast my e'e.
On prospects drear!
An' forward, tho' I canna see,
I guess an' fear!

Robert Burns, *To a mouse*

4.1 Introduction

Lowland farmland in Britain comprises a diverse range of habitats, from semi-natural woodland to highly managed tilled land, pasture and unimproved grassland to hedgerows. This managed, man-made farmscape has for centuries provided habitat for at least 40 of the 54 species of terrestrial British mammals. Over the last two decades in particular, the potential impacts of post-war agricultural intensification on these species have caused increasing concern among conservationists (e.g. Macdonald and Smith 1991; Krebs et al. 1999; Stoate et al. 2001; Robinson and Sutherland 2002; Macdonald and Johnson 2003). The extent of such effects is difficult to assess because, in general, population status and change are poorly understood for most farmland mammals at a national level (e.g. Macdonald and Tattersall 2001; Battersby 2005). This is particularly true for the ten species of small mammal that inhabit the agricultural landscapes of mainland Britain.

In farmland habitats, communities of small mammals (by which we mean shrews, mice, and voles, under 50 g) are usually dominated by four species: the wood mouse *Apodemus sylvaticus* (Fig. 4.1), bank vole

Myodes glareolus (formerly *Clethrionomys glareolus*), field vole *Microtus agrestis*, and common shrew *Sorex araneus*; of these, the wood mouse is the most abundant and widespread (Tattersall and Macdonald 2003). The remaining six species present on farmland are the harvest mouse *Micromys minutus*, pygmy shrew *Sorex minutes*, yellow-necked mouse *Apodemus flavicollis*, house mouse *Mus domesticus*, water shrew *Neomys fodiens*, and dormouse *Muscardinus avellanarius*. Although there is a paucity of data on their national status and population trends, the autecology of each is relatively well known and has been widely described (Box 4.1; Harris and Yaldon 2008).

4.1.1 Why do small mammals matter?

The small mammal community is an important, but perhaps undervalued, component of farmland biodiversity, containing specialists and generalists across a range of habitats (Tattersall et al. 2002). The ecological and physiological characteristics of small mammals are such that the community as a whole may serve as a sentinel, whose richness, diversity, and abundance under different management regimes can reveal both

Wildlife Conservation on Farmland. Managing for Nature on Lowland Farms. Edited by David W. Macdonald and Ruth E. Feber.
© Oxford University Press 2015. Published 2015 by Oxford University Press.

Figure 4.1 Wood mouse *Apodemus sylvaticus.* Photograph ©Michael J. Amphlett.

ecosystem change and its causal factors. Additionally, some small mammals, notably dormice and harvest mice (both subject to Biodiversity Action Plans), are of conservation interest and concern in their own right.

Although the underlying data are poor, estimates suggest that the commonest farmland small mammals are very numerous indeed (Harris et al. 1995, 2008). Field voles are estimated to be the most abundant of all British mammals, with a pre-breeding population of 75 million, contributing the eighth largest source of wild mammalian biomass in Britain, at 2250 tonnes (Harris et al. 2008). However, their distribution is biased towards the uplands of Scotland and Wales, and abundances in the south have probably declined (Harris et al. 1995). Common shrews and wood mice are, respectively, the second and third most numerous British mammals, with estimated pre-breeding populations totalling 41.7 and 38 million, while the bank vole comes in sixth at 23 million. These three species also make a significant contribution to mammalian biomass, together weighing 1637 tonnes (Harris et al. 2008).

Given their abundance and biomass, and position midway in the food chain, small mammals play an important role in farmland food webs (Churchfield and Brown 1987; Smeding and de Snoo 2003). They impact on different trophic levels, feeding on plant material and invertebrates (Churchfield 1982; Hansson 1985; Plesner Jensen 1993), and are themselves preyed upon by a range of avian and mammalian predators

---

Box 4.1 A brief summary of the ecology of farmland small mammals

Up to ten species of small mammals (rodents and insectivores under 50 g) are to be found in farmland habitats. Below, we briefly summarize their ecology.

Bank voles *Myodes glareolus* (reviewed by Shore and Hare 2008) are largely herbivorous, taking seeds and forbs. Around half of their diet consists of green leaves, with woody plants favoured above herbs. They are found in woodland and hedgerows, where they prefer dense shrubby cover and have home ranges in woodland that range from 0.02–0.2 ha. They are active burrowers, creating runs and pathways through the ground vegetation in deciduous habitats.

Field voles *Microtus agrestis* (reviewed by Lambin 2008), are the heaviest and the largest of the farmland small mammals, averaging 30–40 g and are the chief prey of barn owls.

They feed predominantly on grass stems and green leaves. They are found mainly in rough, ungrazed grassland, although marginal woodland and hedgerows with long grass are also used. Young plantations and field margins can also be occupied by field voles in farmland, but hedgerows are often unsuitable, particularly if they lack a broad grassy margin. Their home ranges in England are poorly known, but in Sweden breeding males range up to 1.4 ha.

Wood mice *Apodemus sylvaticus* (reviewed by Flowerdew and Tattersall 2008; Fig. 4.1) are found in most lowland British habitats (including arable farmland), with highest densities in mixed deciduous woodland. They are the most widespread and abundant small mammal on farmland. Wood mice are opportunistic feeders, taking a variety of seeds and invertebrates. They are highly mobile, occupying

continued

Box 4.1 *Continued*

home ranges that are usually 0.02–3.6 ha, depending on habitat and time of year.

Common shrews *Sorex araneus* (reviewed by Churchfield and Searle 2008a) are found in most terrestrial habitats that offer low vegetation cover, including grass, woodland, and arable crops. They are insectivores and feed opportunistically on invertebrates, ranging from small mites, beetles, and snails, to large earthworms; they typically have small home ranges of 0.03–0.06 ha. Common shrews are active throughout the day as they need to eat 80–90% of their body weight daily.

Harvest mice *Micromys minutus* (Trout and Harris 2008) are one of the smallest farmland small mammals, at around 7 g. They are found in early successional grassland with long, dense stems, that enables them to build their above-ground breeding nests, and they eat fruits, seeds, and insects. They tend to be sedentary, with male ranges averaging 0.04 ha.

Pygmy shrews *Sorex minutes* (reviewed by Churchfield and Searle 2008b) weigh just 2–6 g. They are found almost everywhere where there is good cover and have larger home ranges than common shrews, up to 1.7 ha. They feed on surface-active invertebrates, taking smaller prey items than common shrews.

Yellow-necked mice *Apodemus flavicollis* (reviewed by Marsh and Montgomery 2008) are woodland specialists restricted to southern England and Wales. They can be common in hedgerow, but may be underrepresented by live trapping due to their arboreal nature. They have a similar diet to wood mice, but have larger home ranges where the species overlap.

House mice *Mus domesticus* (reviewed by Berry et al. 2008) are largely restricted to farm buildings in rural areas, where they can reach very high densities, but are occasionally found in hedgerows and fields. They are omnivorous, taking seeds and invertebrates when living away from human habitation.

Water shrews *Neomys fodiens* (reviewed by Churchfield 2008) are found near running or still water, but are occasionally recorded in hedgerows. They have small ranges (0.006–0.05 ha), which they shift frequently, and they feed on aquatic invertebrates, small fish, and amphibians.

Lastly, dormice *Muscardinus avellanarius* (reviewed by Bright and Morris 2008) are rare, arboreal, woodland specialists, limited to southern England and Wales. They depend on hedgerows to move between fragmented habitat patches, and eat flowers and pollen, nuts and berries, and some insects. They hibernate in winter.

(e.g. King 1985; Dyczkowski and Yalden 1998; McDonald and King 2008a,b; Meek et al. 2012). Small mammals may also play a part in tree regeneration in woodland and along hedgerows by consuming, caching, and dispersing seeds (Golley et al. 1975; Hayward and Phillipson 1979).

Their short lifespan, usually encompassing just one breeding season, and the potentially fast reproductive rate of many species, mean that small mammal populations can fluctuate rapidly in response to a change in their environment (Churchfield and Searle 2008a; Flowerdew and Tattersall 2008; Lambin 2008; Shore and Hare 2008). Their physical size and mobility means they can make active choices in response to changes in cover and food availability at the field scale; some, such as the wood mouse, are also mobile enough to respond rapidly to farm-scale changes (Tattersall et al. 2002; Tattersall and Macdonald 2003; Tattersall et al. 2004).

Their abundance, trophic position, and responsiveness make small mammals a useful community with which to study the impact of farm habitat management and agri-environment schemes. In addition, their presence in a highly managed and patchily structured landscape facilitates manipulations for pure and applied experiments in ecology. The Wildlife Conservation Research Unit's (WildCRU) small mammal research started in the early 1990s, with a series of studies of the ecology and behaviour of wood mice in arable fields at Oxford University's University Farm at Wytham, Oxfordshire (reviewed by Tew et al. 1994b; Tattersall and Macdonald 2003; Box 4.2). Subsequent work on lowland arable farms in Berkshire, Gloucestershire, and Wiltshire through to the early 2000s also focused on wood mice, but included studies of other species and the impacts of agri-environment schemes (reviewed by Tattersall and Macdonald 2003; Macdonald et al. 2007). More recently, WildCRU's work has covered pastoral landscapes (Gelling et al. 2007), and issues such as disease (Matthews et al. 2006) and mate choice (Brandt and Macdonald 2011; Box 4.3). The intention throughout our work has been to combine high calibre biological science with relevance to conservation. In this chapter we highlight the implications of our diverse findings for agri-environment policy, and summarize some of the broader science behind our

Box 4.2 Population dynamics and behavioural ecology of arable wood mice

In large woodlands (Wilson et al. 1993; Mallorie and Flowerdew 1994; Malo et al. 2012) and in other habitats as varied as sand dunes (Gorman and Akbar 1993) and reed beds (Canova et al. 1994), wood mouse populations are characterized by high winter and low summer abundances. Temporal patterns on farmland, however, are more variable. For example, wood mouse abundance in cereal fields and hedgerows adjacent to Oxford University's 400 ha Wytham Woods showed a steady increase over the summer, a sharp decline after harvest, and low abundance over winter (Tew and Macdonald 1993; Macdonald et al. 2000a).

Populations in small woodlots at Eysey Manor farm (Wiltshire, UK) showed a typical spring decline, followed by a period of relatively stable, low population density, increasing again in autumn and winter (Box Fig. 4.2.1, Tattersall and Macdonald 2003; Macdonald et al. 2000a), but there were no clear seasonal trends in wood mouse abundances in set-aside, boundary, or crop. Mice were present in cropped fields throughout the year; indeed, peak abundances occurred in November 1996, when fields were bare, tilled earth. Individuals started to come into breeding condition between January and March, and ceased breeding by November. Densities of mice in the woodlots ranged from 18–184/ha (averaging 75/ha), but densities in cropped fields ranged from 0–67/ha

(averaging 21/ha) and, in set-aside fields, we recorded average densities of 12/ha, ranging from 0–102/ha.

On arable land, wood mice can move considerable distances. Their home ranges tend to be larger than in woodland (Tew and Macdonald 1994; Macdonald and Tattersall 2003), and male ranges are larger than those of females, especially during the breeding season in March to October, when females defend small, discrete home ranges. On Wytham Farm, male home ranges, between March and August, were, on average, 2.36 ha (SE = 0.17, n = 110) whereas females' home ranges were much smaller, at 0.95 ha (SE = 0.22, n = 38) (Tew and Macdonald 1994). For both sexes, mean home-range size reached a maximum in the peak breeding period (May–July), at which point male home ranges were *c.* three times larger than those of females (Box Fig. 4.2.2). Tattersall et al. (2001) also reported a three-fold difference in home range size during the breeding season, and found that males moved further (in an average 10-minute period) (12.7 ± 0.47 m) than did females (5.8 ± 0.34 m). In general, males ranged widely and overlapped both with other males (mean % overlap = 18.6 ± 1.5, n = 290) and females (mean % overlap of females' range = 32.5 ± 4.3, n = 42), whilst female home ranges overlapped little or not at all with those of other females (mean % overlap = 4.1 ±1.3, n = 14)[1] (Tew and Macdonald 1994).

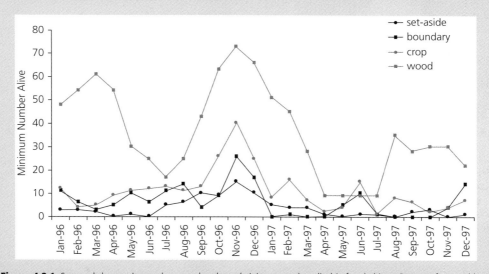

Box Figure 4.2.1 Seasonal changes in wood mouse abundance (minimum number alive) in four habitats. Data are for monthly trapping sessions at Eysey Manor Farm, Wiltshire, with four 0.36 ha trapping grids, one in each of set-aside, boundary, crop, and woodlot. From Tattersall and Macdonald 2003.

continued

---

[1] Mean values for all pairs where overlap was possible.

Box 4.2 *Continued*

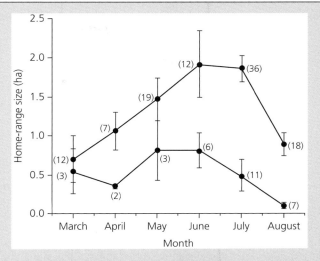

Box Figure 4.2.2 Male wood mice home ranges (upper line) are much larger than those of females (lower line), and the sizes of both are at a maximum in the peak breeding season (May–July). Data are mean ± SE home range size, calculated using restricted polygon areas, for wood mice on Wytham Farm; numbers of individuals in parentheses. From Tew and Macdonald (1994). With kind permission from Springer Science and Business Media.

Wood mice are polyandrous (females mate with several males, Bartmann and Gerlach 2001; Booth et al. 2007; Box 4.3) and polygynous (males mate with several females; Booth et al. 2007), and males with larger ranges potentially have access to more females than are available to males with smaller ranges. However, there was no correlation between male body weight and home range size, minimum nightly distance travelled, mean speed of movement, or proportion of time active (Tew and Macdonald 1994). Having a large range may be an advantage, but being big appears not to be the way to get it. Alternatively, a large range may be a requirement driven by low habitat quality (Godsall et al. 2013). Nevertheless, although potentially of benefit to mating males, extensive movements are not always a sign of health and vigour: in some mice, long movements were associated with infection with the intestinal nematode *Heligmosomoides polygyrus*, although the separation of cause and effect is difficult (Brown et al. 1994a).

In woodland, as in our Wiltshire farmland sites, the main breeding season was from spring to early autumn, but density dependent mechanisms relating to female spatial and reproductive behaviour postponed population increases (Montgomery 1989a,b; Wilson et al. 1993; Mallorie and Flowerdew 1994). Density dependence in woodland is sufficient to limit peak densities in early winter and promote the recovery of low populations in the summer, although in years with a good tree seed crop, density dependence may be overridden. In cropped fields, the availability of spilt or sown seed might have the potential to play a similar role. Fluctuations of wood mice are synchronous in widely separated woodlands (Mallorie and Flowerdew 1994), possibly because of synchronized seed crops. The same mechanism could synchronize woodland and hedgerow populations, where hedgerow management allows fruiting.

In a highly mobile species such as the wood mouse, landscape context is important, although largely neglected, when interpreting abundances in different habitats. We tested two models of how wood mouse populations in different farmland habitat patches might interact, using population parameters estimated from capture-mark-recapture data in set-aside, crop, boundary, and woodlot at two sites in arable farmland (Tattersall et al. 2004). In the source-sink model, populations in 'source' patches had fitness > 1, while in 'sink' patches fitness was < 1 and there was net dispersal from sources to sinks. In the balanced dispersal model, patches may have varied in quality and carrying capacity, but fitness was equal across patches and there were no constraints on dispersal. Our results broadly supported balanced dispersal: individuals were more likely to move to nearby patches but there was no directionality in their dispersal,

continued

Box 4.2 *Continued*

and no evidence of density dependence (Tattersall et al. 2004). Population growth rates were similar in all patches and occasional extinctions were stochastic. One crop patch, however, may have been an 'attractive sink' (high quality habitat where some external factor causes mortality): extinction followed high abundance, and coincided with poor re-cruitment. Recruitment rates were highest in boundary and crop, and survival rates were lowest in crop. Males on set-aside had a smaller December bodyweight than elsewhere. Our results demonstrate that more than one model for dispersal dynamics between populations may apply within the same landscape, for the same species.

Box 4.3 Reproductive behaviour in wood mice, harvest mice, and common shrews

Wood mice, harvest mice, and common shrews have a long mating season, beginning in March (May for harvest mice) and extending to September/October (dependent on species, region, and weather conditions), during which female wood mice and harvest mice produce up to seven litters; common shrews one or two.

One of the big questions in behavioural ecology is: who mates with whom, and consequently, how do these species choose their mates? In the laboratory, we were able to show that exposing female wood mice to short (24 h) olfactory and visual contacts with several unfamiliar males (from which they were physically separated) induced vaginal receptivity[2] more strongly than did the presence of the same male over the 24 h period (Stopka and Macdonald 1998), suggesting preference for unfamiliar males. Preference for unfamiliar males is common, and generally attributed to inbreeding avoidance (Cheetham et al. 2008).

Nevertheless, females do sometimes choose familiar males—the benefits being prior knowledge of individual quality, health, or social status (Cheetham et al. 2008)—and this seems to be the case for harvest mice. Female harvest mice in oestrus (n = 15) given access to one familiar and one unfamiliar male in separate chambers within an experimental laboratory arena spent proportionally more time, over a four-hour trial, in the chamber of the familiar male than in that of the unfamiliar male, and this effect was enhanced the heavier ('better quality') the familiar male was relative to the unfamiliar male (Brandt and Macdonald 2011). Dioestrous females (n = 14) appeared to be more interested in investigating novel (unfamiliar) individuals.

In the absence of prior knowledge, one mechanism by which females and males can obtain information about potential mates is reciprocal grooming, and in small mammals this is a common pre-mating behaviour (e.g. Stopka and Graciasová 2001). In large enclosures, we saw that males assess a female's phase of oestrus by smelling the anogenital region but that females seemed to permit males to attain anogenital contact only when they have been groomed for a certain amount of time (Stopka and Macdonald 1998). Males appear to groom reluctantly and we speculate whether the females are influenced by evaluation of any of the males' other qualities, or whether they are influenced simply by receipt of grooming.

In the wild, female wood mice and common shrews (as well as American mink *Neovison vison*, Yamaguchi et al. 2004; Macdonald and Feber 2015, Chapter 6; and probably many other mammals) mate with several males (i.e. they are polyandrous), and give birth to multiple paternity litters (Stockley et al. 1993; Booth et al. 2007). In common shrews, for example, the number of fathers per litter ranged from one to six (mean = 3.33 ± SE 0.47) in eight of nine wild litters (Stockley et al. 1993). Polyandry is probably due to the inability of males to guard females effectively in natural populations. The benefit to females of multiple paternity is not clear but possibilities include increased genetic diversity of litters (Booth et al. 2007) and

continued

---

[2] Female wood mice come into oestrus for variable periods (one to two days) and at unpredictable intervals over the duration of the mating season (usually every 4–6 days, Jonsson and Silverin 1997), in response to social conditions (Cinquetti and Rinaldi 1989); the presence of males tends to prolong oestrus, the presence of females suppresses it (Stopka and Macdonald 1998).

Box 4.3 *Continued*

reduction of inbreeding risk (Stockley et al. 1993). For males, this leads to competition for fertilizations, which might include sperm competition[3], as well as direct competition between individuals, and competitive tactics likely differ according to individual circumstances.

We found that male shrews that had matured early and established large home ranges in areas of high female density (Type B—familiar, advantaged—males, n = 9) fathered more offspring than did males that had matured later, occupied relatively small territories during the mating season, and made repeated long-distance excursions to locate females (Type A—presumably disadvantaged—males, n = 9), despite the latter having significantly higher epididymal sperm counts (Stockley et al. 1994). Further, the number of offspring fathered by Type B males was positively correlated

with both the number of female ranges overlapped and testis mass, whereas the number of offspring fathered by Type A males was positively correlated only with their epididymal sperm counts, suggesting that Type B males compete directly to maximize the number of females inseminated, whereas Type A males compete via sperm competition (Stockley et al. 1996).

Litter size for all three species is highly variable, and may be up to ten young, which for common shrews at least is far more than can be weaned[4]. Given the variation in genetic quality of offspring within litters and the significant reduction in fitness associated with inbreeding[5] (Stockley et al. 1993), this seemingly wasteful reproductive strategy might be of value if it allows competition between siblings to select for the fittest (Stockley and Macdonald 1998).

conclusions. We review first, what is known of the current status of small mammals on farmland, and then look at ways in which our research demonstrates this can be influenced by habitat management.

4.2 The status of farmland small mammals

In the absence of statistically sound, long-term monitoring programmes (Macdonald et al. 1998; Battersby 2005), we can use only indirect evidence to examine the likelihood of a wide-scale change in small mammal populations on lowland farmland, such as changes in quality and availability of their preferred habitats. Below, we review what little is known of population trends at a national level, and then assess changes in availability of habitat and food for small mammals.

4.2.1 Population trends

Population trends of small mammals in the UK are generally poorly known (Macdonald and Tattersall 2001; Battersby 2005). Among small mammals found on farmland, only the hazel dormouse, an ancient woodland specialist that uses hedgerows as corridors, has a national monitoring programme, the National Dormouse Monitoring Programme. This has revealed

a long-term decline in the last 25 years, with a 24% decline in the decade up to 2007 (TMP 2009), largely due to habitat loss. Being almost entirely arboreal, and therefore not captured in standard ground-level live-trapping studies, dormice are not usually recorded as part of the farmland small mammal community, and we will not consider them further here. There is a large body of literature relating to their monitoring, ecology, and restoration (Bright and Morris 2008).

Harvest mice are also relatively difficult to record through live trapping, partly due to their very small size (*c.* 7 g), but also because they are less active at ground level, where traps are set, than above ground, among the stalks of rough grassland plants (Trout and Harris 2008). Nevertheless, their populations have almost certainly declined significantly since the early twentieth century (Perrow and Jowitt 1995; Macdonald and Tattersall 2001; Battersby 2005). Although potentially able to occupy a wide variety of habitats (inferred from historical distribution records throughout the UK, Harris 1979), there is a paucity of more recent

---

[3] Competition between sperm of different males for fertilization of the ova of a female (Parker 1970).

[4] Offspring of shrews of the genus *Sorex* are probably among the most altricial of any eutherian mammal in terms of development at birth. Even among the Soricidae, *Sorex* gestation periods are short, and their neonates small (20 days and 0.4 g, for common shrews; Innes 1994).

[5] Juveniles from matings between full or half siblings are small at weaning and less likely to survive to maturity than are more outbred individuals (Stockley et al. 1993).

published studies but, such as they are, they show a patchy occurrence at low densities in most lowland rural landscapes. In the mid 1990s, a national Mammal Society survey found that harvest mice were no longer present at 29% of 800 sites where they had been recorded in the 1970s (Sargent 1999). Similarly, a BTO/Mammal Society survey of 240–455 transects of 2 km each, located in randomly selected 1 km squares, and undertaken each winter from 2001 to 2003, recorded harvest mouse presence in just 2.85% of squares (Battersby 2005). Love et al. (2000) surveyed small mammals' remains in barn owl pellets from across Britain, and were able to compare abundances of several species in pellets collected between 1956 and 1974 with pellets collected between 1993 and 1997. In contrast to Sargent (1999) and Battersby (2005), they found that harvest mice became more prevalent in owl pellets (Love et al. 2000).

Our unpublished surveys of hedgerows and fence-lines with long grass margins in the Upper Thames Tributaries region in 2004–2006, covering a total of 8574 ha, revealed that harvest mouse breeding nests occurred at an overall density of 0.36 nests per km^2 (decreasing to 0.07 nests per km^2 over the two years of the study) and were highly clustered within only a few areas (Riordan et al. 2009). Trapping in spring, autumn, and winter on the ground and at 0.5–1 m above the ground within the hedgerow, at six sites in areas where breeding nests had previously been found, resulted in the capture of only eight harvest mice at two sites. This equated to a density of 0.02 animals per hectare (based on estimates of minimum number alive, n = 810 traps deployed over a total of 67 nights), compared with densities of wood mice and bank voles at the same sites of 27.5 and 16.8 per hectare, respectively.

Field voles also tend to be patchily distributed on lowland farmland and there is some concern that, due to habitat loss, they have declined substantially since the early twentieth century when numbers sometimes reached plague proportions; further declines since the 1970s are suspected due to increasing competition with rabbits Oryctolagus cuniculus (Harris et al. 1995; Battersby 2005), with potential consequences for both terrestrial (e.g. weasels Mustela nivalis, Box 4.8) and avian (e.g. barn owl Tyto alba) predators. Assessing their status is made more difficult by their cyclic population changes (Lambin et al. 2000). There are no long-term national or regional studies to provide any evidence of population change, but a comparison of field voles in the diet of barn owls showed no difference in their prevalence in owl pellets collected from 1956–1974 and 1993–1997 (Love et al. 2000).

In spite of (or perhaps because of) their abundance and prevalence, nothing is known of national or regional population trends for wood mice, bank voles, and common shrews. Love et al.'s (2000) barn owl pellet study, described above, reported a large increase in Apodemus spp. and a smaller increase in bank voles, but a decline in common shrews. However, there are many forms of bias in using skeletal remains in owl pellets as a proxy for population abundance (Meek et al. 2012), as, indeed, is the case with other survey methods such as live trapping (e.g. Tew et al. 1994a; Plesner Jensen and Honess 1995) and surveying for field signs.

4.2.2 Habitat use and long-term trends in habitat availability

Although the basic habitat requirements of small mammals are well known (Box 4.1) there are few controlled comparisons of small mammal community composition in different farmland habitats with which to assess their relative value. In one such study (Tattersall et al. 2002), we live-trapped small mammals (n = 20, 384 trap nights) in rough grassland (long-term set-aside), field boundaries (defined as a hedgerow, with a field margin and crop edge), arable crop and small (< 1 ha) farm woodlots, at four sites in Wiltshire and Gloucestershire. The crop had the least diverse small mammal community (Simpsons Index 0.76), and was the habitat with the least individuals captured (229, of which 96% were wood mice). Set-aside had the most diverse small mammal community (Simpsons Index 0.563), though numbers captured were low (275 individuals); here again, wood mice dominated numerically, at 54.5% of individuals, followed by field voles (36.7%). Woodland and field boundaries had the highest populations of wood mice, common shrews, and bank voles (Fig. 4.2). Field voles were predominantly (75% of 135 individuals) caught in grassland and boundary. Yellow-necked mice (n = 4) were only caught in woodland, a habitat in which house mice (n = 4) and harvest mice (n = 5) were never found; pygmy shrews (n = 7) were never caught in the crop.

There are clear seasonal changes in habitat use by wood mice that do not appear to be shown by other species. Our extensive radio-tracking studies at Eysey Manor Farm, Wiltshire (Tattersall et al. 2000), and Wytham Farm, Oxfordshire (Todd et al. 2000), showed that wood mice ranged widely in crops and other habitats during the summer breeding season, before harvest (coinciding with a heavy parasite burden—Box 4.4). At this time, male wood mice had very large home ranges, averaging < 1.5 ha and

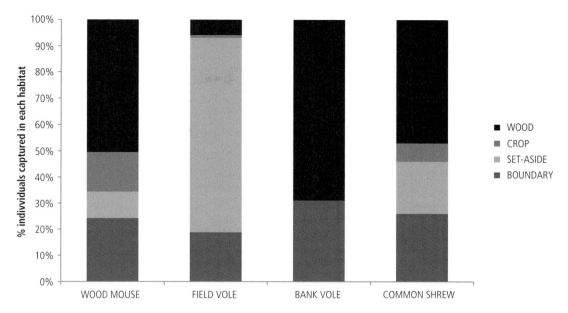

Figure 4.2 Habitat use by small mammals on farmland. Total number of individuals captured: wood mouse 1439; field vole 135; bank vole 186; common shrew 87.

overlapping other males and females, while females had smaller, exclusive ranges averaging around 0.5 ha (Box 4.2). After harvest and over winter, however, both males and females had small (about 0.2 ha) overlapping ranges which, with respect to occurrence in the landscape, contained more hedgerow than any other habitat (Tattersall et al. 2000; Fig. 4.3). Interactions between populations of wood mice in different farmland habitats are best described by a 'balanced dispersal' model (Box 4.2), but some crop fields might

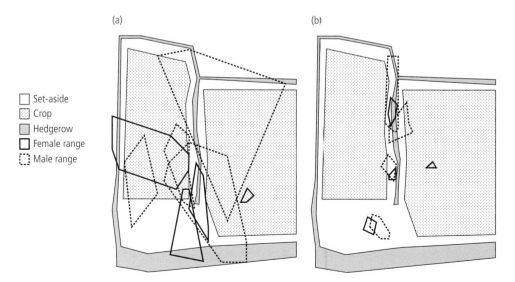

Figure 4.3 Location of male and female wood mice home ranges (measured as minimum convex polygons) before (a) and after (b) harvest, showing a significant (P = 0.001) reduction in size after harvest and restriction to hedgerows and set-aside margins. Range size before harvest averaged 1.29 ±0.29 ha, and after harvest 0.16 ± 0.04 ha. From Tattersall et al. (2001). Reproduced with permission from John Wiley & Sons.

Box 4.4 *Heligmosomoides polygyrus* infection affects wood mouse movements

Parasitic infections can alter the behaviour of hosts in the laboratory (reviewed in Barnard and Behnke 1990), possibly as a result of the parasite manipulating its host to increase transmission rate (Holmes and Zohar 1990; Berdoy et al. 2000; Macdonald and Feber 2015: Chapter 11), but few studies have attempted to investigate whether such effects are expressed under natural conditions. To test the hypothesis that infection with *Heligmosomoides polygyrus* (an intestinal nematode, ubiquitous in free-living populations of wood mice, Lewis 1987) affects the movements of male mice in the breeding season in ways that might affect their survival or reproductive success, we radio-tracked 25 male mice, infected (n = 16) and uninfected (n = 9) with *H. polygyrus*, for 1–13 nights (providing 41–620 fixes per mouse, Brown et al. 1994a). Infected mice moved significantly further and faster than did uninfected mice (average of 1255 m per night at a mean moving speed of 3.3 m per minute versus 888 m per night at 2.5 m per minute), and showed a (non-significant) tendency to spend more time moving (average of 84% versus 77%) and to occupy a larger home range than did uninfected mice (average of 2.5 ha versus 1.6 ha). Two possible reasons for this are: (1) the infected mice increase their food intake and thus have to move further given the patchy distribution of weeds and inverte-brates in the farmland habitat (although gastro-intestinal nematodes usually cause a reduction in food intake, Macdonald and Feber 2015: Chapter 11), and (2) competition between males for females may force infected (subordinate) male mice to move further than uninfected (dominant) ones for mating opportunities. Both could be the result either of *H. polygyrus* specifically manipulating the host to increase the distribution of its larvae and/or increase its food supply (and thus supporting a larger worm population), or the behaviour of the host influencing its susceptibility to infection.

Estimates of the eggs per gram of faeces (EPG) (using the mean of up to three daily samples, because faecal egg production varies over a 24 h cycle, Brown et al. 1994b) revealed seasonality in prevalence and intensity of infection (Box Fig. 4.4.1), coinciding with the annual population dynamics of mice. Peaks (80% prevalence, with a mean EPG value > 2500, in June) may be a reflection of higher average mouse age (and increased exposure) and decreased immunocompetence in mice that are in reproductive condition, but may also be affected by agricultural practices, such as harvest, that might alter transmission rate (by destroying burrows in which contaminated faeces may accumulate, or as an indirect effect of changes in the movements of mice— Tew and Macdonald 1993).

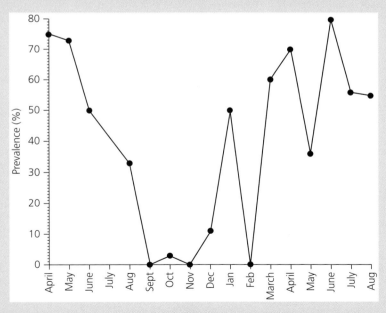

Box Figure 4.4.1 Seasonal variation in prevalence of infection with the gut nematode worm, *H. polygyrus*, in a wood mouse population on a farm in Oxfordshire (n = 3–42 mice, depending on month). Intensity of infection (based on estimated eggs per gram of faeces) followed an almost identical seasonal trend. Redrawn from Brown et al. (1994a). Reproduced with permission from John Wiley & Sons.

be 'attractive sinks', that is, highly suitable habitat, in which animals nevertheless experience high mortality or breeding failure (Tattersall et al. 2004).

4.2.2.1 Increases in intensively managed habitats

Through the twentieth century there were profound changes in agricultural systems and the wider countryside in the UK, which accelerated with a push for self-sufficiency after the 1947 Agriculture Act, and have continued to the present day (Stoate et al. 2001; Robinson and Sutherland 2002). These changes included an increase in intensively managed grassland and land cultivated for crops, with concomitant loss of semi-natural grassland, woodland, and linear features, and the polarization of agriculture into pastoral west and arable east (Hindmarch and Pienkowski 2000; Mead 2000). There has also been a loss of non-cropped habitats, such as hedgerows, ponds, and woodlots (Macdonald and Johnson 2000). As one example, close to our study sites in Oxfordshire, Barreto et al. (1998) examined ten 2 km^2 quadrats in the Thames floodplain, for which there were aerial photographs in 1947, 1961, and 1991. Semi-natural grassland covered 49% of quadrats in 1947, but just 12% in 1991; tilled acreage rose from 11% to 53% over the same period. On a national scale, these types of change mean that as much as 51% of Britain's 23 million hectares are now under intensive agriculture according to the 2007 Land Cover Map (Morton et al. 2011); in England the figure is 67%. The Countryside Survey 2007 (Carey et al. 2008) classified 4.6 million hectares as 'Arable and Horticultural' and 4.5 million hectares as 'Improved Grassland', making these the two most widespread of the 19 Broad Habitats measured.

Cropped fields have an impoverished small mammal community (Fitzgibbon 1997; Tattersall et al. 2002; Moore et al. 2003). Wood mice are the only common species here, but even their densities are much lower than in their preferred woodland habitat (Chapter 12, this volume) where >40 wood mice can be found per hectare (Flowerdew 1991; Tattersall et al. 2004). On three arable fields at Wytham Farm, Tew (1991) reported a mean annual density of just 2.57 wood mice/ha. Survival rates on cropped fields are lower than in other farmland habitats (Tattersall et al. 2004). An increase in cropped areas at the expense of natural and semi-natural habitats will certainly have lowered habitat availability to small mammals over the course of the past half century and more.

Improved grassland (fields managed intensively for pasture or mown annually for silage, where natural species are replaced with agricultural grasses and there are regular chemical inputs and other interventions) is, perhaps, for small mammals, the most species-poor of all farmland habitats. Although there are only a handful of published studies, these have reported exceptionally low capture rates in traps set in this habitat (Montgomery and Dowie 1993; Gelling et al. 2007). The pre-eminence of improved grassland in the British countryside means that a significant amount of land may be effectively uninhabitable for small mammals.

4.2.2.2 Losses of semi-natural grassland and linear features

Britain's 1.69 million hectares of semi-natural grassland are characterized by low-intensity management and a lack of recent cultivation, re-sowing, or fertilization; however there are only 111 150 ha of lowland semi-natural grassland, mainly within agri-environment schemes or protected areas (Bullock et al. 2011). Some grassland habitats within agri-environment schemes are a potentially important refuge for small mammals on farmland. Beetle banks, for example, can provide harvest mice with the uncut stems of species like cocksfoot and thistles *Cirsium* spp., that they need to support their nests (Bence et al. 2003). Minimally managed grasslands in permanently fallowed fields or field margins can provide a useful habitat for field voles and other species (Tattersall et al. 2000, 2002), and grasslands in various stages of succession may host up to eight species of small mammal (Churchfield et al. 1997). However, habitats within agri-environment schemes are a moveable feast, subject to the whim of policy. Furthermore, they have in no way compensated for what has been lost—the UK's lowlands have been depleted by around 90% of their semi-natural grassland since 1945 (Bullock et al. 2011), and this represents a major loss of habitat for species that require rough grassland, such as shrews, field voles, and harvest mice.

Hedgerows and associated features like ditches and grassy margins host a diverse and species-rich small mammal community (Tattersall et al. 2002; Gelling et al. 2007). They are especially valuable for bank voles, which require thick ground cover (Kotzageorgis and Mason 1997). Wood mice can also be trapped in very high numbers in field boundaries and hedgerows, which they are able to exploit in three dimensions (Buesching et al. 2008). Tattersall et al. (2004), for example, reported 22 mice/ha in field boundaries (1–2 m hedgerow bordered on either side by a 20 m uncropped margin) in the Upper Thames region, but only 8/ha in crops.

Hedgerows have an additional importance in linking patches of woodland and grassland, allowing small mammals to migrate through the landscape (Fitzgibbon 1997). In modern pastoral systems, hedgerows are often the only semi-natural habitat available (Pollard et al. 1974; Gelling et al. 2007). However, the availability of these vital features has drastically declined since World War II (e.g. Barr et al. 1993; Rackham 1997). From 1947 to 1963 an estimated 8000 km of hedgerow were lost annually, with a further 4800 km every year in the 1970s (Hooper 1974). Our research has demonstrated that a substantial minority of farmers continued to remove hedgerows in the 1990s (Macdonald and Johnson 2003). Since 1997, hedgerows have had limited legal protection and farmers are encouraged to adopt enhanced hedgerow management strategies (Defra 2006), and to regenerate lost hedgerows (Defra 2006), but the decline, now largely through lack of management, has continued, albeit at a reduced rate. The total length of 'managed' hedges decreased by 6% between 1998 and 2007 in Great Britain and slightly less than half (48%) were classified as being in good structural condition in 2007 (Carey et al. 2008).

4.2.3 Changes in habitat quality

In addition to widespread loss of favoured habitats for small mammals on lowland farmland, there have been changes in habitat quality, largely resulting from modern industrial farming practices. Notable among these is the huge increase in the number and extent of chemical applications on all types of arable, horticultural, and improved grassland fields (Stoate et al. 2001; Robinson and Sutherland 2002). The mechanization of farming has also transformed the speed and efficiency of the cropping cycle, and very large machinery has necessitated removal of some unproductive habitats (Macdonald and Johnson 2000), with consequences for small mammals.

4.2.3.1 Declines in invertebrate and plant foods associated with chemical inputs

Farmland small mammals feed on a wide range of plants, seeds, and invertebrates (Box 4.1). At Wytham Farm, our research showed that wood mice favoured wheat, barley, and oilseed rape over several other non-crop seeds and fruits, but food selection was not entirely opportunistic: individuals chose different foods, with characteristic nutrient and energy contents, according to a circadian cycle (Plesner Jensen 1993; Box 4.5). Thus, even an adaptable species such as the wood mouse needs to be able to access a complex set

of different foods with different nutritional qualities, to fulfil their daily nutritional requirements. Nutritional constraints of a different kind operate on another generalist species—common shrews need to catch over 570 invertebrate prey items per day (Churchfield 2002).

Over 200 varieties of agrochemicals (excluding fertilizers, but including herbicides, fungicides, and insecticides) are applied to arable land in Britain, often repeatedly within a year (Shore et al. 2003). The many herbicide applications on which modern farming relies, reduce botanical abundance, biomass, and species richness (Gardner and Brown 1998). High inputs of fertilizer further reduce botanical diversity—many species of dicotyledonous herb are unable to compete at high soil nutrient levels. The resultant impoverished flora has knock-on effects on invertebrates (Campbell et al. 1997) which, in addition, are directly affected by insecticides and molluscides. Spray drift is likely to result in a similar effect at the field boundary. This type of change is not restricted to arable habitats: the latest Countryside Survey (2007) found that species richness of plants growing in fields, woods, heaths, and moors decreased by 8% in Great Britain between 1978 and 2007. However, plant species richness in the Arable and Horticulture Broad Habitat increased by 30% between 1998 and 2007, largely due to set-aside and other agri-environment schemes (Carey et al. 2008).

That there have been widespread declines in arable invertebrates and plants is well documented (Potts 1980; Andreasen et al. 1996; Donald 1998; Robinson and Sutherland 2002). There is no evidence to suggest whether or not this has yet impacted on small mammals, but it does seem probable that, at the very least, the diversity and availability of their foods have declined.

4.2.3.2 Other changes in farming practice

The ways in which crops are sown and harvested, and the ground tilled, have changed beyond all recognition since 1947. Then, crops were harvested by hand and the grain stored in fields in corn ricks prior to threshing, which provided overwintering habitat for large numbers of house mice (Southern and Laurie 1946) and harvest mice (Rowe and Taylor 1964). Stubbles were left over the winter, and the ground tilled and crop sown in the spring.

Modern combine harvesting has long rendered corn ricks unnecessary, and the brutal efficiency of ever larger machines working the fields at the height of their breeding season may be one reason for the decline in harvest mice (Perrow and Jordan 1992). For wood mice, harvest is the most marked and traumatic period of disruption in the farming year. We tracked 16

Box 4.5 Wood mouse diet selection through the night

Asking how food preferences might affect wood mouse behaviour, we found that wheat, barley, and oilseed rape were among the five most favoured foods out of 26 types (of seeds and fruits) presented in diet choice experiments to wild wood mice, with wheat ranking higher than the other crops (the other two favoured items were maize *Zea mays*, and blackberry *Rubus* spp.) (Plesner Jensen 1993). This experiment, involving 25 trials, and each involving eight wild mice temporarily confined to cages in the field, revealed that the number of food types taken (but not the number of items, or mass of food eaten—which declined from the beginning to the end of the night) followed a bimodal pattern that corresponded with peaks in activity (at the beginning

and the end of the night) (Box Fig. 4.5.1). We presume that wood mice are selecting dietary items on the basis of energy content and digestibility, and certainly they maintained a remarkably constant protein:carbohydrate ratio through the night (as found in rats under self-selection conditions; Theall et al. 1984). During the first active period the percentage content of sugar in the diet selected by the mice was significantly higher than during the second active period (perhaps to provide a quickly digestable source of energy), and during both active periods, the foods chosen were higher in carbohydrates and proteins than those chosen during the inactive period in the middle of the night (which were higher in sugars and water).

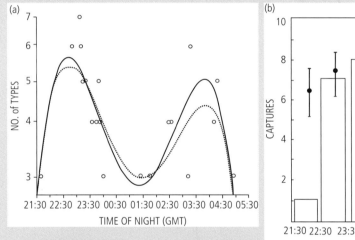
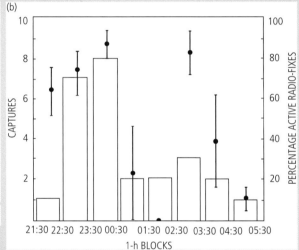

Box Figure 4.5.1 Bimodal distribution in number of food types taken by wood mice (a) (n = 25 trials, eight individual mice), corresponding with variation in the number of wood mice captures and mean percentage active radio-locations for four wild mice (b), throughout the night. From Plesner Jensen (1993). With kind permission from Springer Science and Business Media.

wood mice whilst the combine harvester passed over or near their underground nest sites. Of these, one was killed, six left the burrow and bolted, and ten stayed underground (Tew and Macdonald 1993). Following harvest, there was an 80% decline in number of wood mice captured in the fields. Tew and Macdonald (1993) concluded that harvest itself has little direct effect on mouse populations resident in arable fields, but the dramatic reduction in cover greatly increases their exposure to predators.

There is no information on the impact of deep ploughing on small mammals in Britain, but in Continental Europe, ploughing destroys the burrow network of common voles *Microtus arvalis* (Brügger et al. 2010). The increasing use of direct drilling and minimum tillage in Britain in more recent years may benefit wood mice and common shrews, as well as reducing damage to soil macrofauna such as earthworms (Stoate et al. 2001).

Other less visible farm practices, such as the use of rodenticides and molluscicides, are also potentially directly lethal to small mammals (Shore et al. 2003). The molluscicide methiocarb is widely used in a pelleted form, which is broadcast or drilled in the growing crop. Wood mice find the pellets palatable (Tarrant and Westlake 1998), but need only ingest less than five to receive a lethal dose (Shore et al. 1997). Shore et al. (1997) hypothesized that spring or summer applications would have the greatest impact on mouse populations because their numbers are naturally low at this time of year. They set up an experiment in which fields of winter wheat were broadcast with methiocarb in both autumn (n = 3 fields, 4.5–10 ha) and spring (n = 2, 4.5, and 5.2 ha) at the commercially recommended rate of 5.5 kg per hectare, resulting in an approximate density of 30 pellets/m^2. Total wood mouse numbers on treated fields declined by 78% in autumn, and 33% in spring. However, no dead mice were found, and 18 mice tested for a common biomarker of exposure to methiocarb (Thompson 1991) had levels within the normal range reported for non-exposed wood mice. It may be that there is an 'all or nothing' effect insofar as mice that find and consume a lethal amount of pellets die quickly, undetected, while others fail to locate or ingest pellets at all. We also found that secondary poisoning of wood mice by slug pellets could be reduced by incorporating Bitrex™ (denatonium benzoate) as a repellent (Kleinkauf et al. 1999).

4.3 Managing lowland farmland for small mammals

Given that small mammals have experienced serious declines in their habitat availability and quality, and considering their importance in the food web, it is reasonable to ask how we could enhance their populations. Wood mice are abundant and easy to trap and radio-track (and see Box 4.6 for way-marking), so have proved a useful model species with which to address this question in lowland arable farmland. Background information about their behaviour and ecology in arable habitats is given in Box 4.2. Where possible we have widened our research to encompass other small mammals as well.

4.3.1 Enhancing cropped areas for wood mice

Weed patches are an important resource for wood mice in crop fields. We radio-tracked the movements of 48 mice (n = 9000 fixes) at Wytham Farm during the summer breeding season, and found that they were greatly influenced by patches of foodplants within the superficially homogenous crop (Tew et al. 2000). Five key plant

Box 4.6 Way-marking in wood mice—the use of portable signposts

The relatively huge home ranges occupied by wood mice and the seemingly homogeneous environments (e.g. expanses of ploughed fields or uniform cereal crops) over which they roam, suggests that navigation would be particularly challenging for this species, yet the fine-grained spatial complexity of, and temporal shifts in, the distribution of their foodplants requires a reliable mechanism of orientation.

Wood mice are macrosmatic (have highly developed organs for smell) and nocturnal, but have good vision. Following the mice in the fields at Wytham Farm, we saw accumulations of small objects (leaves and twigs) where mice had been active—leading us to suspect that wood mice were distributing these objects as visual reference points. Using a series of experiments in which wild mice were provided with small portable plastic discs (5 cm diameter, 1.5 g) in laboratory enclosures, Stopka and Macdonald (2003) showed that mice moved the discs and positioned them within the enclosures in such a way that they appeared to function as visual points of reference (or way-marks) during exploration. Mice moved discs to areas where they were temporarily active, and then orientated their own exploratory movements to and from these discs. Detailed sequential behaviour analysis revealed that mice transported discs only during movements identified as 'exploratory'.

Way-marking may enable wood mice to orientate quickly back to a place of temporary interest (e.g. a rich foraging site) following disturbance by a predator, or on subsequent foraging trips and, being readily portable, may be a less confusing method for marking ephemeral sites than scent marks. Way-marks may also be a safer option for local navigation than scent marks, which are easily detectable by a predator (Stopka and Macdonald 2003).

species, all considered to be agricultural pests (black grass *Alopecurus myosuroides*, common chickweed *Stellaria media*, wild oat *Avena fatua*, goosegrass *Galium aparine*, and sterile brome *Anisantha sterilis*) were statistically significantly more abundant in 25 cm^2 quadrats (n = 1980) within which radio-tracked mice were stationary for more than ten minutes, than in quadrats through which mice travelled without pause. Stationary mice were feeding, a behaviour directly observed during radio-tracking, or subsequently inferred from discarded seed husks found on the quadrats. Quadrats in which the radio-tracked mice were stationary included less bare ground than those in which they travelled, or random ones within the home range. Mice may have been avoiding bare ground not only because of a perceived lack of weed seeds (they did not respond to 'hidden' seeds in the soil seed bank), but also to minimize predation risk from avian predators (Tew et al. 2000).

One way to enhance cropped areas (regardless of crop type) as habitats for wildlife might be to create weedy resource-rich areas through management. Reduced agrochemical application increases weed and invertebrate diversity (Fuller et al. 2005), and selectively reducing the application of broad-leaved herbicides and insecticides to the outer 4–6 m margins of cereal fields (headlands) benefits game birds through increasing invertebrate food supplies for chicks (Potts 1986; Sotherton 1988).

We asked whether wood mice could also detect, exploit, and benefit from such localized food-rich patches (Tew et al. 1992). We radio-tracked 15 wood mice (n = 7298 fixes) on 10 m x 20 m experimental headland plots at Wytham Farm, subject to three herbicide spraying regimes: unsprayed, reduced sprayed (where the headland missed out on a fiuroxyp treatment), and fully sprayed; applications of insecticides, fungicides, growth regulators, and fertilizers were identical across the three treatments. Mice preferred reduced-sprayed and unsprayed headlands over sprayed headlands and mid-field areas (Fig. 4.4); Tew

et al. (1992). Reduced-sprayed plots contained more field forget-me-not *Myosotis arvensis*, than sprayed or unsprayed headlands. Unsprayed plots contained more of the three most abundant grass weed species than did the sprayed plots (black grass, wild oat, and sterile brome), and also more of four of the fifteen invertebrate orders present (springtails (Collembola), true bugs (Hemiptera), flies (Diptera), and parasitoid wasps (Parasitica)). Direct observations suggested that mice in the food-rich patches were generally feeding: mice observed by torchlight in the unsprayed plots were typically seen eating black grass seeds and their feeding locations could be identified later by the characteristic patches of debris (ejected seed husks, and fallen seeds) on the ground.

The percentage cover of understorey vegetation (forbs and grasses) tends to be greater within organic crops than conventional crops (Macdonald et al. 2007). Organic crops are also less dense and invertebrate communities not only more abundant, but also more diverse (Feber et al. 1997, 1998; Macdonald et al. 2000b; Chapter 6, this volume). For small mammals, the composition and structure of organic crops may enhance the availability of certain limiting food items (weeds and invertebrates), which in turn would have positive effects at individual and population levels (Macdonald et al. 2007). We undertook a small-scale study comparing one conventional and one organic farm in Gloucestershire over three years, and more wood mice were caught, on average, in organic than conventional crops (Fig. 4.5a; Macdonald et al. 2007). The number of females in breeding condition was higher in most months on the organic farm, and there was evidence of a trend towards an earlier peak in numbers of breeding females on organic farms compared with conventional farms (Fig. 4.5b). Females were heavier on the organic farm in two out of three years, and more juveniles were present than on the conventional farm (Macdonald et al. 2007).

One implication of the lack of agro-chemical inputs in organic farming is that farmers cannot compensate for excess soil nitrogen losses over winter, and therefore cannot advance their growing season with winter-sown crops. As a result, summer sward height is significantly higher in conventional than organic systems. Wood mouse abundances in cereal fields in May and June correlated positively with sward height, and at this time densities were lower in organic (1.3/ha) than conventional (3.4/ha) fields (Macdonald et al. 2000a). Abundances are also low in conventionally managed spring-sown crops (Loman 1991).

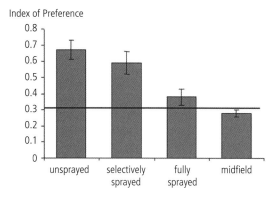

Index of Preference

Figure 4.4 Wood mouse prefer unsprayed and selectively (reduced) sprayed headlands compared to fully sprayed headlands and (sprayed) mid-field areas. Data are mean ± SE preference indices (PI) for 12 mice on Wytham Farm radio-tracked to a 5 m accuracy. PI = log (1 + U/A), where U = percentage usage of a habitat, A = percentage availability of that habitat; values > 0.3 indicate preference, values < 0.3 indicate avoidance. Adapted from Tew et al. (1992). Reproduced with permission from John Wiley & Sons.

(a)

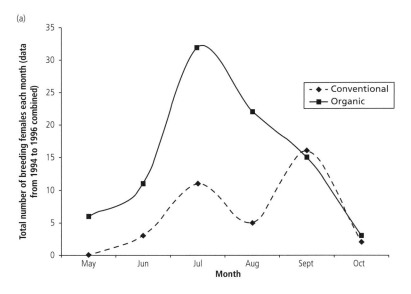

(b)

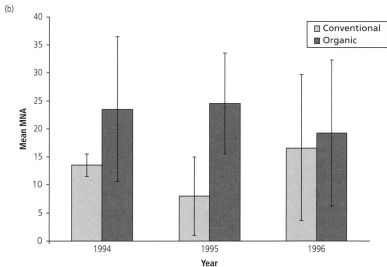

Figure 4.5 Total number of wood mice over three years (a) (mean MNA ± SD averaged by month, May to October) and (b) number of breeding females (three years combined) on an organic farm in Gloucestershire compared with a conventional farm. MNA = Minimum Number Alive. From Macdonald et al. (2007). Reproduced with permission from John Wiley & Sons.

4.3.2 Enhancing grassland for small mammals

Set-aside (land removed from agricultural production: Clarke 1992) was a widespread feature of the British countryside between 1992 and 2008. The set-aside scheme was at least partly responsible for an overall increase in neutral grassland (differentiated from improved grasslands by a less lush sward, greater range, and higher cover of herbs, and < 25% cover of perennial ryegrass *Lolium perenne*) and plant species richness on farmland in Great Britain between 1998 and 2007 (Carey et al. 2008). Although set-aside was recognized

as potentially beneficial for biodiversity (Macdonald and Smith 1991; Henderson et al. 2000), there was a great variety of management options. At the time of our research, the basic requirements were that a green cover be established over winter to prevent leaching (MAFF 1995, 1997). Cover could be established by sowing or by natural regeneration, and had to be controlled in August by cutting or spraying with a non-residual herbicide; insecticides and fungicides were banned. Farmers could relocate their set-aside each year (rotational set-aside) or leave it in one place for up to ten years (permanent set-aside), and it could be established as strips, fields, or blocks within fields. Although the set-aside scheme is no longer operating, farmers are encouraged to create or retain semi-natural grassland under the current agri-environment scheme, Environmental Stewardship, and our conclusions regarding management of this habitat remain useful and transferable.

We used the set-aside scheme as an opportunity to answer questions about the impacts of different configurations and management methods of grassland on wood mice and other small mammals (Tattersall et al. 1999a,b, 2000, 2001, 2002, 2004). Our main experimental sites at Eysey Manor Farm, Wiltshire, and Harnhill Manor Farm, Gloucestershire, were sown at 24.7 kg per hectare with a mix of six agricultural grasses and two clovers, including dwarf perennial ryegrass *Lolium perenne* (for speedy establishment of cover), cocksfoot *Dactylis glomerata* (for a tussocky structure), and timothy *Phleum pratense* (for height). The experiments were designed in such a way that they would not only inform the set-aside policy of the time, but would also provide information more widely relevant to the management of unused areas of farm grassland for biodiversity. Over three years, the team trapped and radio-tracked >2000 mice, voles, and shrews, on four farms in Wiltshire, Berkshire, and Gloucestershire to assess the suitability of set-aside for small mammals. We asked specifically (i) whether set-aside becomes a more suitable habitat with age; (ii) whether it should be sown (and if so, using what seed mixture) or naturally regenerated; (iii) whether it should be mown or left uncut; and (iv) whether it should be configured as blocks or strips.

4.3.2.1 Does set-aside improve with age as small mammal habitat?

We live-trapped small mammals (n = 3000 trap nights) on newly established rotational set-aside on four sites at two farms in Gloucestershire (Coates and Harnhill Manor farms) for five nights, monthly in June, July, and August 1995. Our results indicated that, during the first year after establishment, set-aside was not a suitable habitat for small mammals (Tattersall et al. 1997). This was true for set-aside configured as whole field blocks and as 20 m-wide field margin strips with associated hedgerow; and for set-aside naturally regenerated or sown with a mix of wheat and oilseed rape. Of the eight species captured, wood mice (n = 67 individuals) were the only species trapped on set-aside, and trapping success was considerably lower on the set-aside (0.6%) than in the adjoining hedgerow (30%) or

crop (13%), probably due to the lack of cover on newly established set-aside. Mice only occurred on set-aside after harvest, when the crops no longer provided cover for predators.

Further live-trapping in 1996 and 1997 (n = 7350 trap nights), quarterly, at two different sites on Harnhill Manor Farm, and two sites at Eysey Manor Farm in Wiltshire, found five species of small mammal on 5 ha blocks of set-aside, and an additional three species on 5 ha sites covering two 20 m strips of set-aside either side of a hedgerow, plus associated crop edges. In the first year of the study, the permanent set-aside was in its first year of establishment (with our standard mix of grasses and clovers) and cover was sparse, but by the middle and end of the second year vegetation was well established, with dense cover and a thick litter layer. Both species richness and diversity increased significantly between 1996 (mean richness = 1.91 ± 1.0; mean diversity = 0.18 ± 0.16) and 1997 (mean richness = 2.8 ± 1.24; mean diversity = 0.36 ± 0.24). Field voles and common shrews were both more abundant in the second year after establishment than in the first (Tattersall et al. 1999a). No field voles were recorded until nine months after the set-aside was sown; thereafter captures were sporadic until populations started to increase 20 months after sowing (Tattersall et al. 2000).

Colonization of set-aside is, of course, dependent on proximity of existing populations and, in farmland, field voles are generally patchily distributed (Richards 1985), so voles may simply require time to reach set-aside. Older set-aside might, therefore, have a higher density of field voles for this reason. We tested this hypothesis by trapping field voles in 16 set-aside fields aged 2–9 years old, distributed over five farms in southern England (Tattersall et al. 2000). We found that in set-aside older than two years, vole abundance was not related to age (Pearsons correlation coefficient r = 0.13), but increased with the proportion of grasses (r = 0.56) and litter (r = 0.62) in the sward (Tattersall et al. 2000). Once plant cover was established, on set-aside of two years or older, age per se provided no additional benefits, and vegetative characteristics were a more important predictor of set-aside suitability for field voles.

4.3.2.2 How should set-aside grassland be established?

Our data suggest that field voles will benefit most from management that results in speedy establishment of vegetation, maintains a thick layer of living and dead plants comprising > 50% grasses, and retains established set-aside in the same location from year to year. To this end, a seed mixture containing a high proportion of grasses, including some tall and tussocky species, such as timothy and cocksfoot, and rapidly growing species such as ryegrass, will probably be most useful for field voles (Tattersall et al. 2000). The requirements of other species might be different, however.

Tattersall et al. (1999b) compared the suitability for wood mice of set-aside areas established using two different seed mixtures. Eighteen male wood mice were each radio-tracked for at least three nights and 92 fixes on set-aside at Eysey Manor Farm, Wiltshire, and at ZENECA Agrochemical's Jealotts Hill Farm, in Berkshire. Set-aside at Eysey Manor was established using our standard mixture of grasses and clovers, while set-aside at Jealotts Hill was established using one of two species-rich mixes of grasses and native forbs (6% grass *Poa annua* plus 15 broadleaf species sown at 4 kg/ha; and 80% grass (8 species) plus 15 broadleaf species sown at 10 kg/ha). The richer seed mixtures at Jealotts Hill established a more species-rich sward than at Eysey Manor (species richness: 17 ± 3.0 versus 5 ± 0.9), but lower sowing rates resulted in lower cover at Jealotts Hill (91 ± 11% versus 75 ± 11%). Set-aside at Eysey Manor was taller (1.0 ± 0.1 m versus 0.6 ± 0.1 m, n = 34, ten 1 m² quadrats, respectively). Surprisingly, at Eysey Manor the mice avoided the taller, denser set-aside relative to crop and other habitats, but did not avoid the shorter, sparser, species-rich set-aside at Jealotts Hill. It seemed, therefore, that mice were selecting uncropped habitats on the basis of plant diversity rather than protection from predators, although the relative importance of cover probably depends on the availability of alternative sources of protection from predators, such as nearby hedgerows. More diverse vegetation might provide more food resources for wood mice, by providing seeds over a longer fruiting period, and by harbouring more invertebrates (Feber et al. 1995) and, in any case, even the short vegetation was, on average, 0.6 m tall and thus probably provided sufficient cover.

4.3.2.3 Does mowing affect small mammals?

Mowing is a major determinant of the amount of cover available for small mammals on grasslands. Plesner Jensen and Honess (1995) found that fewer small mammals were captured in plots of short (5 cm) grass than in tall (60 cm) grass, which provided them with cover from predators. We tested the hypothesis that wood mice would favour areas of longer vegetation by creating alternate cut and uncut patches of set-aside; cut patches were mowed annually in August, and cuttings

were left *in situ*. Patches were 50 m long and 6 m wide (mean height cut sections: 28 ± 8.7 cm, n = 55; uncut sections: 106 ± 22 cm, n = 60) and located in set-aside between crop and hedgerow (Tattersall et al. 2001). We radio-tracked 34 wood mice that ranged over our experimental patches, before (June–July) and after harvest (September–early November). Mice used both hedgerow and uncut sections of set-aside strips more than expected based on its availability, but used bare crop fields and cut sections of set-aside less than expected (Tattersall et al. 2001; Fig. 4.6).

Food availability as well as cover from predators were likely to be greatest in uncut set-aside. Aerial predators are less successful in habitats with long vegetation than in open habitats (Simonetti 1989; Longland and Price 1991), although this may be less important for wood mice than for other species of British small mammals (Plesner Jensen and Honess 1995). Invertebrate abundance on field margins is much higher in uncut plots than in plots cut in summer (Feber et al. 1995), and fresh grass seeds remain available in uncut but not cut vegetation.

There was no evidence to suggest that cutting was detrimental for field voles; indeed it provided a dense, flat, litter cover under which voles tunnelled (see also Box 4.7). Tattersall et al. (2000) suggested that field voles would benefit most from speedy establishment of vegetation in uncropped areas, and from the

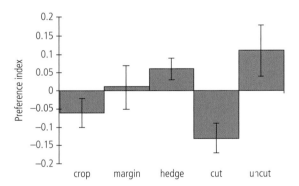

Figure 4.6 After harvest, radio-tracked wood mice (n = 34) at Eysey Manor Farm used hedgerow and uncut sections of set-aside more than expected based on their availability (indicated by a positive preference index), but avoided (used less than expected, indicated by a negative preference index) bare crop and cut sections of set-aside. Preference indices were calculated using compositional analysis, which compares habitat use (based on the number of radiolocations within a habitat) with habitat availability (as the proportion of each habitat type present within a home range). From Tattersall et al. (2001). Reproduced with permission from John Wiley & Sons.

Box 4.7 The effects of grassland management and grazing on field voles

Field voles *M. agrestis* depend on good supplies of young grass shoots for feeding, and therefore are heavily affected by grassland management regimes. To investigate the effects of different grassland management strategies on field vole population dynamics, Christina Buesching and Chris Newman, together with volunteers from the Earthwatch Institute (see Macdonald and Feber 2015: Chapter 15), studied field vole densities over the summer months of 2002 to 2004 in Wytham Woods. They compared grasslands under three different management regimes: (1) where bracken was controlled by regular spraying of Asulox (a commercial product containing the herbicide asulam); (2) in areas which were subjected to heavy sheep grazing pressure; and (3) in areas grazed by deer but otherwise unmanaged.

The results revealed some clear differences. Firstly, grasslands sprayed with Asulox were consistently the least suitable habitat for field voles, as, under this management regime, bracken was allowed to grow up to full size, shading the ground and largely preventing the formation of grass tussocks necessary for field vole nests, as well as limiting the fresh growth of new grass shoots, the field voles' main

food source. Secondly, on sheep grazed pastures, population densities varied considerably among years, depending on climatic conditions. In wetter years, sheep pastures supported the highest vole population densities (approximately 208 individuals/ha), as grass growth was good, resulting in good ground coverage and producing plenty of fresh shoots. In dry and hot conditions, however, grass growth was poor, resulting in patches of exposed dry soil with no ground cover and few fresh shoots, and thus very low densities of voles (c. 10 individuals per hectare). Thirdly, naturally grazed grasslands supported consistently high vole densities, as the comparatively stable conditions in these areas provided many tussocks and consistent growth of new grass shoots (Box Fig. 4.7.1; Buesching et al., unpublished data). Nevertheless, whilst bracken growth in these areas was largely prevented by the thick cover of old grass, and by the effects of trampling by deer, hawthorn encroachment on Wytham's calcareous grasslands posed a threat to these habitats (now exacerbated by a significant decrease in deer numbers following stringent deer control measures; see Chapter 12, this volume).

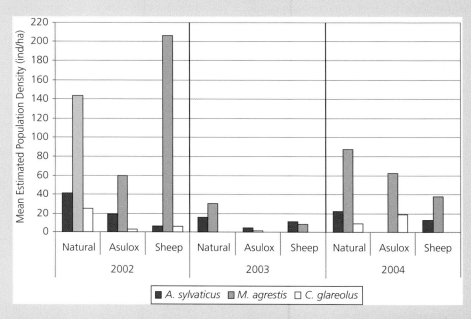

Box Figure 4.7.1 Estimated small mammal abundances across different grassland management regimes in Wytham Woods, 2002–2004. Buesching et al. (unpublished data).

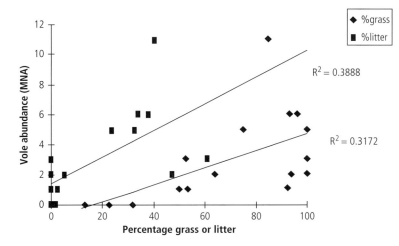

Figure 4.7 Field vole abundance (Minimum Number Alive) increases with increasing percentage of grass and leaf litter in whole field set-aside > 2 years old (n = 16 sites). MNA based on a five-night trapping period in June, all sites censused once. From Tattersall et al. (2000). Reproduced with permission from Elsevier.

establishment of permanent uncropped areas with a thick layer of living and dead plants, comprising 50% grasses (Fig. 4.7).

4.3.2.4 Is set-aside configuration important?

We hypothesized that set-aside configured as field margin strips would be more beneficial to small mammals than blocks or field set-aside (Tattersall et al. 1999b, 2001). Set-aside around field margins is close to sources of colonists in the hedgerows, and might, therefore, be expected to contain a greater density of small mammals than blocks or fields of set-aside. Indeed, MAFF (1995) specifically recommended siting set-aside next to, or linking, existing areas valuable to wildlife, such as hedgerows and rivers.

On naturally regenerated, poorly established, year-old set-aside, trap success was exactly equal (0.6%) on blocks and field margin strips of set-aside, and only wood mice were caught in either configuration (Tattersall et al. 1999b). There was some evidence that the (relatively barren) set-aside strip acted as a barrier to dispersal out of the hedgerow: there were significant differences between trap rows 24 m out from the hedge in crop directly adjacent to hedgerow (17.2% success) and in crop separated from the hedge by a set-aside strip (0.5% success). This was probably not an effect of higher densities towards the crop centre, because there was no significant difference between rows 12 m into the crop (15.2% success) and rows 24 m into the crop (17.2% success).

A somewhat different picture emerged on one-year-old and two-year-old sown set-aside where species other than wood mice were caught (Tattersall et al.

2002). Bank voles were present in 20 m-wide set-aside field boundary strips (mean minimum number alive MNA = 2.1 ± 0.57 in 1996 and 3.5 ± 1.4 in 1997) but never in the centre of blocks of set-aside, implying that they moved short distances out of their preferred hedgerow habitat into the adjacent edge habitats. Overall, total abundance of small mammals was greater on set-aside strips (total MNA = 21.1 ± 10.8) than on blocks (13.1 ± 13). Species richness, but not diversity, was also greater on strips (mean richness = 2.63 ± 1.07; mean diversity = 0.3 ± 0.19) than on blocks (mean richness = 2.06 ± 1.27; mean diversity = 0.23 ± 0.24).

In addition to live-trapping, we radio-tracked 34 mice on 2–3-year-old set-aside between June and November in 1996 and 1997, for a minimum 92 fixes over three nights, to monitor their habitat preferences. There was no evidence that strips of set-aside were more attractive to wood mice than blocks; indeed, after harvest, wood mice avoided margin set-aside significantly more than block set-aside (P = 0.021; Tattersall et al. 2000). One reason for this rather surprising result might be that wood mice are avoiding foraging close to the hedgerow, where their mammalian predators, such as stoats *Mustela erminea* and weasels *Mustela nivalis* (see Box 4.8), tend to patrol (King 1991a,b).

4.3.3 Enhancing linear features for small mammals

Lowland farmland is typically dissected by a network of hedgerows and other linear features (Macdonald et al. 2007). Any field boundary that incorporates shrubby vegetation (hedgerows, field margins, ditches,

Box 4.8 Weasels on farmland

Small mammals are a major part of weasel diet (e.g. King 1991b), and prey availability is an important determinant of weasel density and habitat use. Knowing how small mammal density and distribution varies through the year and across habitat types, in response to farmland management, Macdonald et al. (2004) asked whether their predators are similarly affected by the artificially managed farmscape and, specifically, how they might use different components of the cereal ecosystem.

Ten weasels (three females, three subadult males, and four adult males; body weight 48–69 g, 86.5–96 g, and 100–128 g, respectively) were radio-tracked at the University farm at Wytham for 2–18 days each (mean = 10.2 days), between January and November over two years. As might be expected for mustelids (see Macdonald and Feber 2015: Chapter 7), females had smaller ranges than did adult males (as did subadult males), but all ranges were unusually large[6], probably as a result of low prey densities on farmland. Weasels rarely travelled more than 5 m from linear habitats (mainly woodland edge and hedge with ditch), where small mammal abundance was locally high. Male home ranges overlapped but radio-tracked individuals were never seen together (although aggressive interactions with unknown males were observed on three occasions).

On average, weasels spent almost half their time (48.6% ± 20) apparently sleeping or resting; least time was spent travelling (on average 8.9% ± 2.9), and movements were faster when in the open (weasels under cover, such as along a hedge bottom, travelled at 3–5 km per hour, but up to 10 km per hour in the open). Weasels were observed hunting along hedge bottoms, field margins, and tracks, as well as gardens, farm tips, and straw bales in barns, and were occasionally seen digging for sleeping wood mice. On arable land, but not woodland, weasels frequently climbed in hedges and mature trees, and one was observed eating blackbird *Turdus merula* eggs. In contrast to the more usual arrhythmic patterns of regularly spaced short activity periods throughout 24 hours (King 1991b), weasels in this study were predominantly diurnal, perhaps to capitalize on the availability of vulnerable mice (that spend daylight hours in burrows where they are largely defenceless from predators able to access the burrows), or perhaps to avoid nocturnal predators (tawny owls, foxes, and cats) that were common on farmland. Similar temporal shifts in activity patterns, in response to predators or dominant competitors, have been observed in other species (such as rats: Fenn and Macdonald 1995, Macdonald and Feber in press: Chapter 11; and American mink: Harrington et al. 2009, Macdonald and Feber in press: Chapter 6).

---

tracks, streams, and dry-stone walls) has the potential to be a biological corridor for small mammals moving between larger areas of preferred habitat (Soule and Terbough 1999), and linear features have long been a focus of farmland conservation policy (Boatman et al. 1999). However, the usefulness of this approach is relatively poorly understood.

We tested the non-exclusive hypotheses that edge effects render narrow linear habitats inhospitable to specialists (bank voles) but not to generalists (wood mice); and that edge effects result in specialists being present in atypical habitat through excursions at the interface between the two habitats (Tattersall et al. 2002). Contrary to expectations, live-trapping small mammals in three non-linear farmland habitats (woodlots,

set-aside, and arable fields), and one linear habitat (field boundaries, consisting of 1–2 m-wide hedgerow bordered on either side by 20 m-wide set-aside strips) at Harnhill Manor and Eysey Manor farms, revealed no clear evidence that linear habitats are inhospitable to specialist small mammals, or (for field margins) that edge effects are detrimental (Tattersall et al. 2002). Bank voles were actually more abundant in trap lines in hedgerow than in equivalent trap lines in woodlots (mean MNA ± SE = 1.81 ± 0.58 versus 0.125 ± 0.085), but there was no evidence of linearity having an effect on field voles, wood mice, or common shrews.

Given the importance of hedgerows to small mammals (demonstrated at the field scale, above), and with an eye to how they could be improved as habitats, the influence of hedgerow characteristics on small mammal abundance was explored further in a larger-scale study across twelve farms in four different rural lowland areas (Central England, SW Wales, the Wales–England border, and SE England).

---

[6] Minimum convex polygons: female mean ± SD = 28.6 ± 0.9 ha, subadult male = 18.3 ± 5.4 ha, adult male = 113.3 ± 57.9 ha; mean length of linear habitat used: female mean ± SD = 1532 ± 478 m, subadult male = 1277 ± 577 m, adult male = 5867 ± 3095 m.

4.3.3.1 What features of hedgerows are important?

Hedgerow characteristics are known to affect bats (Wickramasinghe et al. 2003; Chapter 9, this volume), birds (Green et al. 1994; Macdonald and Johnson 1995; Hinsley and Bellamy 2000), and moths (Merckx et al. 2012; Chapter 8, this volume), but relatively little is known about which particular hedgerow features are important to the small mammal community.

Along 180 hedgerows on pastoral farmland, Gelling et al. (2007) explored the relationship between the relative abundance of small mammals and hedgerow connectivity (including proximity to woodland), hedgerow width, height, and length, the presence of an adjacent conservation buffer (a strip of unimproved grass verge ≥ 2 m wide), and the presence of a ditch, hedgebank, and standard trees. Population age structures of five species of small mammal (yellow-necked mouse, wood mouse, bank vole, field vole, and common shrew) were entirely consistent with those reported for each species in woodland habitats (or rough grassland for voles) (Harris and Yaldon 2008), evidence that these hedgerow animals were resident and breeding, rather than only using the hedgerows as migration routes (Gelling et al. 2007).

The greater the width of the hedgerow, and the greater its connectivity to other hedges and woodland, the greater was the number of small mammals occupying it (Gelling et al. 2007). For each one metre increase in hedgerow width per 100 m hedgerow length there was an estimated increase of 56.2 g (± 29.3 SE) in mammal biomass, a mean increase of 1.2 (± 0.8) bank voles, and 1.1 (± 0.7) wood mice. Conversely, the more gaps (defined as a break in the living canopy of the hedge; Clements and Toft 1992) there were in the hedgerow, the fewer bank voles were found there. However, this was not true for the other species, and proximity to woodland had no impact on the abundance of any small mammal. The presence of a conservation buffer strip, which provided the field voles' preferred habitat of rough, ungrazed grassland, was positively associated with the abundance of field voles. The presence of a hedgebank (an earth bank or mound associated with hedgerow) was associated positively with all species combined. Bank voles and yellow-necked mice were more prevalent where a ditch was present; ditches may increase the habitat available for a burrow system for bank voles, but the association with yellow-necked mice is less clear. Wood mice were associated with the presence of standard trees, possibly reflecting a greater availability of their seed-mast food (Mallorie and Flowerdew 1994; Montgomery and Dowie 1993). Responses to management are similarly species-specific:

for example, mechanically flailing hedgerows significantly reduces bank vole populations but increases field vole populations (Gelling et al. 2007).

4.3.3.2 Linear features for harvest mice

To discover which features of linear habitats were important for harvest mice, we surveyed hedgerows and fence-lines with long grass margins in the 8574 ha Upper Thames Tributaries region in 2004–2006 (Riordan *et. al.* 2009). Harvest mouse nest site locations (n = 41) were positively associated with diverse, large hedgerows along ditches, and were more prevalent in hedges along roadsides and unimproved pasture. Harvest mice are widely believed to have been affected by unsympathetic hedgerow management or hedgerow removal, as well as early harvest of winter cereals (Harris and Trout 1996; Perrow and Jowitt 1995; Macdonald and Tattersall 2001), and have recently been the subject of a number of reintroduction projects. Current recommendations are for harvest mice to be released into margins, ditches, or long grass associated with arable fields, where agricultural practices are kept to a minimum (Forder 2006).

4.4 Conclusions

Post-war arable intensification has profoundly changed the British countryside, with widespread losses of biodiversity that have continued into the twenty-first century (Carey et al. 2008). Simplification of plant and animal communities on farmland has disrupted food webs and resulted in declines across many taxa (Stoate et al. 2001; Robinson and Sutherland 2002), including invertebrates and plants taken by small mammals, and in their mammalian and avian predators. With perhaps two exceptions, the dormouse and harvest mouse, small mammals have been largely overlooked in farmland conservation policies, despite their significance in the food chain. Apart from dormice, there are no long-term monitoring schemes with which we could detect a downward population trend in any species of small mammal. With their largely nocturnal behaviour and / or small size, small mammals are inconspicuous, and considering their relative abundance, anything other than catastrophic decline is likely to go (or indeed, may already have gone) unnoticed. These are good reasons why we need to know how best to manage farmland habitats to benefit their populations.

In considering our recommendations, it is important to be mindful that there is no single prescription for managing habitat for small mammals; as is the case with birds (Vickery et al. 2009), different management regimes are appropriate for different species. For example, Gelling et al. (2007) recommend flailing different aspects of a single hedge in rotation to benefit both bank voles and field voles. Habitat management for small mammals should remain mindful also of the home ranges of different species, and hence of the scale at which habitat modification can impact them.

Almost all small mammals will benefit from provision of uncropped, grassy habitat, as long as a thick sward has been established and there is minimal disturbance from farming operations. Field voles will particularly favour a thick litter layer, and wood mice and harvest mice will favour (for different reasons) a tall sward: harvest mice need strong upright stems on which to build their nests, while highly mobile wood mice need cover from predators. Botanically diverse swards may provide a greater variety of seed, plant and invertebrate food resources for small mammals. Structurally complex hedgerows will also benefit a wide range of small mammals, particularly hedgerows associated with ditches (for bank voles) and grassy strips (for field voles). Linear configuration appears to provide no additional benefits over a non-linear configuration for common, mobile species like wood mice, field voles, and shrews, but linking habitat patches is widely considered important for woodland specialists like dormice and yellow-necked mice. In addition to excluding pesticides and fertilizers, organic farming is associated with many of the habitat features favoured by small mammals, such as more and larger hedgerows and more uncropped land (Fuller et al. 2005), and may deliver some of these benefits, particularly in simplified landscapes (Fischer et al. 2011).

There are many areas of research still to be explored. There is relatively little work on the impact of insecticides or other agro-chemicals on shrews and other small mammals at the population level. We do not fully understand the dynamics of field vole and harvest mouse populations in the patchy farmscape, nor do we understand how they colonize new areas. Of the different farm habitats, improved grasslands managed for silage or pasture are especially lacking in studies of small mammals. There is some evidence for inter-specific competition between yellow-necked mice and wood mice (e.g. Marsh 1999), and given the extensive vertical habitat use by wood mice in hedgerows, it is possible that, in some habitats (where their niches overlap) wood mice also compete with harvest mice. However, the biggest gap in our knowledge remains long-term population trends, arising from the lack of a robust, nationwide monitoring programme.

Acknowledgements

We are extremely grateful to the team of people who contributed to this work over more than two decades—Pavel Stopka, Ruth Brandt, Paula Stockley, Philip Riordan, Ian Todd, Liz Brown, Suzanne Plesner Jensen, Richard Shore, Ruth Feber, Christina Buesching, Chris Newman, and countless volunteers and field assistants. We thank Barbara Hart, Will Manley, and The Royal Agricultural University for their help with the set-aside work. We are grateful to the funders of the projects described in this chapter, particularly NERC's Joint Agricultural and Environmental Programme, MAFF (now Defra), and the People's Trust for Endangered Species. Carolyn King, Tim Coulson, and Aurelio Malo made helpful comments on earlier drafts of the chapter. Ruth Feber and Eva Raebel provided valuable editorial input.

References

Andreasen, C., Stryhn, H., and Striebig, J.C. (1996). Decline in the flora in Danish arable fields. *Journal of Applied Ecology*, **33**, 619–626.

Barnard, C.J. and Behnke, J.M. (1990). *Parasitism and host behavior*. Taylor and Francis Ltd., London.

Barr, C.J., Bunce, R.G.H., Clarke, R.T., et al. (1993). *Countryside Survey 1990: Main Report*. Department of Environment, Eastcote, UK.

Barreto G.R., Macdonald, D.W., and Srachan, R. (1998). The tightrope hypothesis: an explanation for plummeting water vole numbers in the Thames catchment. In Bailey, R.G., Jose, P.V., and Sherwood, B.R., eds., *United Kingdom floodplains*, pp. 311–327. Westbury Academic and Scientific Publishing, Otley.

Bartmann, S. and Gerlach, G. (2001). Multiple paternity and similar variance in reproductive success of male and female wood mice (*Apodemus sylvaticus*). *Ethology*, **107**, 889–899.

Battersby, J. (2005). *UK Mammals: Species Status and Population Trends*. JNCC/Tracking Mammals Partnership. JNCC, Peterborough.

Bence, S.L., Stander, K., and Griffiths, M. (2003). Habitat characteristics of harvest mouse nests on arable farmland. *Agriculture Ecosystems & Environment*, **99**, 179–186.

Berdoy, M.L., Webster, J.P., and Macdonald, D.W. (2000). Fatal attraction in rats infected with *Toxoplasma gondii*. *Proceedings of the Royal Society B*, **267**, 1591–1594.

Berry, R.J., Tattersall, F.H., and Hurst, J. (2008). House mouse. In S. Harris and D.W. Yalden, eds., *Mammals of the British Isles: Handbook, 4th Edition*, pp. 141–149, The Mammal Society.

Boatman, N.D., Daview, D.H.K., Chaney, K., Feber, R., de Snoo, G.R., and Sparks, T.H. (1999). *Field margins and buffer zones: Ecology, Management and Policy*. Aspects of

Applied Biology 54. Association of Applied Biologists, Warwick, UK.

Booth, W., Montgomery, W.I., and Prodöhl, P.A. (2007). Polyandry by wood mice in natural populations. *Journal of Zoology*, **273**, 176–182.

Brandt, R. and Macdonald, D.W. (2011). To know him is to love him? Familiarity and female preference in the harvest mouse, *Micromys minutus*. *Animal Behaviour*, **82**, 353–358.

Bright, P.W. and Morris, P.A. (2008). Hazel dormouse. In S. Harris and D.W. Yalden, eds., *Mammals of the British Isles: Handbook, 4th Edition*, pp. 76–81, The Mammal Society.

Brown, E.D., Macdonald, D.W., Tew, T.E., and Todd, I.A. (1994a). *Apodemus sylvaticus* infected with *Heligmosomoides polygyrus* (Nematoda) in an arable ecosystem: epidemiology and effects of infection on the movements of male-mice. *Journal of Zoology*, **234**, 623–640.

Brown, E.D., Macdonald, D.W., Tew, T.E., and Todd, I.A. (1994b). Rhythmicity of egg production by *Heligmosomoides polygyrus* in wild wood mice, *Apodemus sylvaticus*. *Journal of Helminthology*, **68**, 105–108.

Brügger, A., Nentwig, W., and Airoldi, J-P. (2010). The burrow system of the common vole (*M. arvalis*, Rodentia) in Switzerland. *Mammalia*, **74**, 311–315.

Buesching, C.D., Newman, C., Twell, R., and Macdonald, D.W. (2008). Reasons for arboreality in wood mice *Apodemus sylvaticus* and bank voles *Clethrionomys glareolus*. *Mammalian Biology*, **73**, 318–324.

Bullock, J.M., Jefferson, R.G., Blackstock, T.H., et al. (2011). *Semi-natural grasslands*. Cambridge, UK, UNEP–WCMC. In: Technical Report: The UK National Ecosystem Assessment, pp. 162–195.

Campbell, L.H., Avery, M.I., Donald, P., Evans, A.D., Green, R.E., and Wilson, J.D. (1997). *A review of the indirect effects of pesticides on birds*. JNCC Report No. 227. JNCC, Peterborough.

Canova, L., Maistrello, L., and Emiliani, D. (1994). Comparative ecology of the Wood mouse Apodemus sylvaticus in two differing habitats. *Zeitschrift Fuer Saeugetierkunde*, **59**, 193–198.

Carey, P.D., Wallis, S., Chamberlain, P.M., et al. (2008). *Countryside Survey: UK Results from 2007*. NERC/Centre for Ecology & Hydrology, pp. 105. (CEH Project Number: C03259).

Cheetham, S.A., Thom, M.D., Beynon, R.J., and Hurst, J.L. (2008). The effect of familiarity on mate choice. In J.L. Hurst, R.J. Beynon, S.C. Roberts, and T.D. Wyatt, eds., *Chemical Signals in Vertebrates 11*, pp. 271–280. Springer, New York.

Churchfield, S. (1982). Food availability and the diet of the common shrew Sorex araneus, in Britain. *Journal of Animal Ecology*, **51**, 15–28.

Churchfield, S. (2002). Why are shrews so small? The costs and benefits of small size in northern temperate Sorex species in the context of foraging habitats and food supply. *Acta Theriologica*, **47**, suppl. 1: 169–184.

Churchfield, S. (2008). Water shrew. In S. Harris and D.W. Yalden, eds., *Mammals of the British Isles: Handbook, 4th Edition*, pp. 271–275. The Mammal Society.

Churchfield, S. and Brown, V.K. (1987). The trophic impact of small mammals in successional grasslands. *Biological Journal of the Linnean Society*, **31**, 273–290.

Churchfield, S., Hollier, J., and Brown, V.K. (1997). Community structure and habitat use of small mammals in grasslands of different successional age. *Journal of Zoology*, **242**, 519–530.

Churchfield, S. and Searle, J.B. (2008a). Common shrew. In S. Harris and D.W. Yalden, eds., Common shrew. In S. *Mammals of the British Isles: Handbook, 4th Edition*, pp. 257–265. The Mammal Society.

Churchfield, S. and Searle, J.B. (2008b). Pygmy shrew. In S. Harris and D.W. Yalden, eds., *Mammals of the British Isles: Handbook, 4th Edition*, pp. 267–271. The Mammal Society.

Cinquetti R. and Rinaldi L. (1989). Effects of changed social conditions on modulation of the estrous cycle in mice. *Bollettino Di Zoologia*, **56**, 137–142.

Clarke, J. (ed.) (1992). *Set-aside*. BCPC Monographs No. 50. Farnham: British Crop Protection Council.

Clements, D.K. and Toft, R.J. (1992). *Hedgerow evaluation and grading system (HEGS). A methodology for the ecological survey, evaluation and grading of hedgerows*. Countryside Planning and Management, UK.

Defra (2006). *Farming and Landscape features: Hedgerows*. Available at: <http://www.defra.gov.uk/>, accessed June 2014.

Donald, P.F. (1998). Changes in the abundance of invertebrates and plants on British farmland. *British Wildlife*, **9**, 279–289.

Dyczkowski, J. and Yalden, D.W. (1998). An estimate of the impact of predators on the British field vole *Microtus agrestis* population. *Mammal Review*, **28**, 165–184.

Feber, R.E., Bell, J.J., Firbank, L.G., Johnson, P.J., and Macdonald, D.W. (1998). The effects of organic farming on surface-active spider assemblages in wheat in southern England, UK. *Journal of Arachnology*, **26**, 190–202.

Feber, R.E., Firbank, L.G., Johnson, P.J., and Macdonald, D.W. (1997). The effects of organic farming on pest and non-pest butterfly abundance. *Agriculture, Ecosystems & Environment*, **64**, 133–139.

Feber, R.E., Johnson, P.J., Smith, H., Baines, M., and Macdonald, D.W. (1995). The effects of arable field margin management on the abundance of beneficial arthropods. In R.G.McKinlay and D. Atkinson, eds., BCPC Monographs 63. *Integrated crop protection: towards sustainability?*, pp. 163–170. British Crop Protection Council, Farnham.

Fenn, M.G.P. and Macdonald, D.W. (1995). Use of middens by red foxes: risk reverses rhythms of rats. *Journal of Mammalogy*, **76**, 130–136.

Fischer, C., Thies, C., and Tscharntke, T. (2011). Small mammals in agricultural landscapes: Opposing responses to farming practices and landscape complexity. *Biological Conservation*, **144**, 1130.

Fitzgibbon, C.D. (1997). Small mammals in farm woodlands: the effects of habitat, isolation and surrounding land-use patterns. *Journal of Applied Ecology*, **34**, 530–539.

Flowerdew, J.R. (1991). Wood mouse. In G B Corbett and S Harris, eds., *The Handbook of British Mammals*, pp. 220–229. Blackwell Scientific Publications, Oxford.

Flowerdew, J. R. and Tattersall, F.H. (2008). Wood mouse. In S. Harris and D.W. Yalden, eds., *Mammals of the British Isles: Handbook, 4th Edition*, pp. 125–137, The Mammal Society.

Forder, V. (2006). *Captive breeding and reintroduction. The harvest mouse Micromys minutus*. The Wildwood Trust, Kent.

Fuller, R.J., Norton, L.R., Feber, R.E., et al. (2005). Benefits of organic farming to biodiversity vary among taxa. *Biology Letters*, **1**, 431–434.

Gardner, S.M. and Brown, R.W. (1998). *Review of the comparative effects of organic farming on biodiversity*. Report to MAFF for project number OF0149.

Gelling, M., Macdonald, D.W., and Mathews, F. (2007). Are Hedgerows the Route to Increased Biodiversity? Small Mammals' Use of Hedgerows in Pastoral Farmland in Britain. *Landscape Ecology*, **22**, 1019–1032.

Godsall, B., Coulson, T., and Malo, A.F. (2013). From physiology to space use: energy reserves and androgenization explain home-range size variation in a woodland rodent. *Journal of Animal Ecology*, **83**, 126–135.

Golley, F.B., Ryszkowski, L., and Sokur, I. T. (1975). The role of small mammals in temperate forests, grasslands and cultivated fields. In F.B. Golley, K. Petrusewicz, and L. Ryszkowski, eds., *Small mammals: their productivity and population dynamics. International Biological Programme 5*, pp. 223–241. Cambridge University Press, Cambridge.

Gorman M.L. and Akbar, Z. (1993). A comparative study of the ecology of woodmice Apodcmus sylvaticus in two contrasting habitats: deciduous woodland and maritime sand-dunes. *Journal of Zoology*, **229**, 385–396.

Green, R.E., Osbourne, P.E., and Sears, E.J. (1994). The distribution of passerine birds in hedgerows during the breeding season in relation to characteristics of the hedgerow and adjacent farmland. *Journal of Applied Ecology*, **31**, 677–692.

Hansson, L. (1985). The food of bank voles, wood mice and yellow-necked mice. *Symposia of the Zoological Society London*, **55**, 141–168.

Harrington, L.A., Harrington, A.L., Yamaguchi, N., Thom, M., Ferreras, P., Windham, T.R., and Macdonald, D.W. (2009). The impact of native competitors on an alien invasive: temporal niche shifts to avoid inter-specific aggression? *Ecology*, **90**, 1207–1216.

Harris, S. (1979). History, distribution and habitat requirements of the Harvest Mouse in Britain. *Mammal Review*, **9**, 159–171.

Harris, S., Morris, P., Wray, S., and Yalden, D. (1995). *A review of British mammals*. JNCC, Peterborough.

Harris, S., Morris, P.A., Wray, S., and Yalden, D. (2008). The mammal fauna of the British Isles in perspective. In S. Harris and D.W. Yalden, eds., *Mammals of the British Isles: Handbook, 4th Edition*, pp. 6–16. The Mammal Society.

Harris, S, and Trout, E. (1996). Rodents: order Rodentia. In G.B.Corbett and S.Harris, eds., *The Handbook of British Mammals*, 3rd Edition. Blackwell Scientific Publications, Oxford.

Harris, S. and Yaldon, D.W. (2008). *Mammals of the British Isles: Handbook, 4th Edition*. The Mammal Society, Southampton.

Hayward, G.F. and Phillipson, J. (1979). Community structure and functional role of small mammals in ecosystems. In D.M. Stoddard (ed). *Ecology of small mammals*, pp. 136–211. Chapman and Hall, London.

Henderson, I.G., Cooper, J., Fuller, R.J., and Vickery, J.A. (2000). The relative abundance of birds on set-aside and neighbouring fields in summer. *Journal of Applied Ecology*, **37**, 335–347.

Hindmarch, C. and Pienkowski, M.W. (2000). *Land management: the hidden costs*. Blackwell Science for the British Ecological Society, London.

Hinsley, S.A. and Bellamy, P.E. (2000). The influence of hedge structure, management and landscape context on the value of hedgerows to birds: a review. *Journal of Environmental Management*, **60**, 33–49.

Holmes, J.C. and Zohar, S. (1990). Pathology and host behaviour. In C.J. Barnard and J.M. Behnke, eds., *Parasitism and host behaviour*, pp. 34–63. Taylor and Francis Ltd., London.

Hooper, M.D. (1974). Hedgerow removal. *The Biologist*, **21**, 81–86.

Innes, D.G.L. (1994). Life histories of the Soricidae: a review. In J.F. Merritt, G.L. Kirkland Jr., and R.K. Rose eds., *Advances in the biology of shrews*, 111–136. Special Publication of the Carnegie Museum of Natural History, Pittsburgh, No. 18.

Jonsson, P. and Silverin, B. (1997). The estrous cycle in female wood mice (*Apodemus sylvaticus*) and the influence of male. *Annales Zoologici Fennici*, **34**, 197–204.

King, C.M. (1985). Interactions between woodland rodents and their predators. *Symposia of the Zoological Society of London*, **55**, 219–247.

King, C.M. (1991a). Stoat. In G.B. Corbet and S. Harris, eds., *The Handbook of British Mammals*, pp. 377–387. Blackwell Scientific Publications, London.

King, C.M. (1991b). Weasel. In G.B. Corbet and S. Harris, eds., *The Handbook of British Mammals*, pp. 387–396. Blackwell Scientific Publications, London.

Kleinkauf, A., Macdonald, D.W., and Tattersall, F.H. (1999). A bitter attempt to prevent non-target poisoning of small mammals. *Mammal Review*, **29**, 201–204.

Kotzageorgis, G.C. and Mason, C.F. (1997). Small mammal populations in relation to hedgerow structure in an arable landscape. *Journal of Zoology*, **242**, 425–434.

Krebs, J.R., Wilson, J.D., Bradbury, R.B., and Siriwardena, G.M. (1999). The second Silent Spring? *Nature*, **400**, 611–612.

Lambin, X. (2008). Field vole. In S. Harris and D.W. Yalden, eds., *Mammals of the British Isles: Handbook, 4th Edition*, pp. 100–107, The Mammal Society.

Lambin X., Petty, S.J., and Mackinnon, J.L. (2000). Cyclic dynamics in field vole populations and generalist predation. *Journal of Animal Ecology*, **69**, 106–118.

Lewis, J.W. (1987). Helminth parasites of British rodents and insectivores. *Mammal Review*, **17**, 81–93.

Loman, J. (1991). The small mammal fauna in an agricultural landscape in Southern Sweden, with special reference to the wood mouse *Apodemus sylvaticus. Mammalia*, **55**, 91–96.

Longland, W.S. and Price, M.V. (1991). Direct Observations of Owls and Heteromyid Rodents: Can Predation Risk Explain Microhabitat Use? *Ecology*, **72**, 2261–2273.

Love, R.A., Webon, C., Glue, D.E., Harris, S., and Harris, S. (2000). Changes in the food of British Barn Owls (*Tyto alba*) between 1974 and 1997. *Mammal Review*, **30**, pp. 107–129.

Macdonald, D.W. and Feber, R.E., eds. (2015). *Wildlife Conservation on Farmland. Conflict in the Countryside*. Oxford University Press, Oxford.

Macdonald, D.W., Feber, R.E., Tattersall, F.H., and Johnson, P.J. (2000b). Ecological Experiments in Farmland Conservation. In M.J. Hutchings, E.A. John, and A.J.A. Stewart eds., *The Ecological Consequences of Environmental Heterogeneity*, pp. 357–378, Oxford.

Macdonald, D.W. and Johnson, P.J. (1995). The relationship between bird distribution and the botanical and structural characteristics of hedges. *Journal of Applied Ecology*, **32**, 492–505.

Macdonald, D.W. and Johnson, P. (2000). Farmers and the custody of the countryside: trends in loss and conservation of non-productive habitats 1981–1998. *Biological Conservation*, **94**, 221–234.

Macdonald, D.W. and Johnson, P.J. (2003). Farmers as conservation custodians: links between perception and practice. In F.H. Tattersall and W.J. Manley, eds., *Conservation and Conflict: Mammals and Farming in Britain*, pp. 2–16. Linnean Society Occasional Publication, Westbury Publishing, Yorkshire.

Macdonald, D.W., Mace, G.M., and Rushton, S. (1998). *Proposals for future monitoring of British mammals*. A report produced for DETR and JNCC by the Wildlife Conservation Research Unit, University of Oxford.

Macdonald, D.W. and Smith, H. (1991). New perspectives on agro-ecology: between theory and practice in the agricultural ecosystem. In L.G. Firbank, N. Carter, J.E. Darbyshire, and G.R. Potts, eds., *The ecology of temperate cereal fields*, pp. 413–448. Blackwell Scientific, Oxford.

Macdonald, D.W. and Tattersall, F.H. (2001). *Britain's Mammals: The challenge for conservation*. People's Trust for Endangered Species, London.

Macdonald, D.W. and Tattersall, F. (2003). *The State of British Mammals 2003*. WildCRU for People's Trust for Endangered Species.

Macdonald, D.W., Tattersall, F.H., Service, K.M., Firbank, L.G., and Feber, R.E. (2007). Mammals, agri-environment schemes and set-aside—what are the putative benefits? *Mammal Review*, **37**, 259–277.

Macdonald, D.W., Tew, T.E., and Todd, I.A. (2004). The ecology of weasels (*Mustela nivalis*) on mixed farmland in southern England. *Biologia*, **59**, 235–241.

Macdonald, D.W., Tew, T.E., Todd, I.A., Garner, J.P., and Johnson, P.J. (2000a). Arable habitat use by wood mice (*Apodemus sylvaticus*). 3. A farm-scale experiment on the effects of crop rotation. *Journal of Zoology*, **250**, 313–320.

MAFF (1995). *Arable Area Payment 1995–1996. Explanatory Guide: Parts 1 and 2*. Ministry of Agriculture, Fisheries and Food, London.

MAFF (1997). *Arable Areas Payments 1997/1998, Explanatory Guide: Parts 1 & 2*. Ministry of Agriculture, Fisheries and Food, London.

Mallorie, H.C. and Flowerdew, J.R. (1994). Woodland small mammal population ecology in Britain—a preliminary review of the mammal society survey of wood mice *Apodemus sylvaticus* and bank voles *Clethrionomys glareolus*, 1982–1987. *Mammal Review*, **24**, 1–15.

Malo, A.F., Godsall, B., Prebble, C., et al. (2012). Positive effects of an invasive shrub on aggregation and abundance of a native small rodent. *Behavioral Ecology*, **24**, 759–767.

Marsh, A. (1999). The national yellow-necked mouse survey. The Mammal Society Research Report No.2. The Mammal Society, London. 16 pp.

Marsh, A.C.W. and Montgomery, W.I. (2008). Yellow-necked-mouse *Apodemus flavicollis*. In S. Harris and D.W. Yalden, eds., *Mammals of the British Isles: Handbook. 4th Edition*, pp. 137–141. The Mammal Society.

Mathews, F., Macdonald, D.W., Taylor, G.M., Gelling, M., Norman, R.A., Honess, P.E., Foster, R., Gower, C.M., Varley, S., Harris, A., Palmer, S., Hewinson, G., and Webster, J.P. (2006). Bovine Tuberculosis (Mycobacterium bovis) in British farmland wildlife: the importance to agriculture. *Proceedings of the Zoological Society of London B*, **273**, 357–365.

McDonald R.A. and King, C.M. (2008a). Stoat. In S. Harris and D.W. Yalden, eds., *Mammals of the British Isles: Handbook, 4th Edition*, pp. 456–467, The Mammal Society.

McDonald R.A. and King, C.M. (2008b). Weasel. In S. Harris and D.W. Yalden, eds., *Mammals of the British Isles: Handbook, 4th Edition*, pp. 467–476, The Mammal Society.

Mead, C. (2000). *The state of the nation's birds*. Whittet Books, Suffolk.

Meek, W.R., Burman, P.J., Nowakowski, M., Sparks, T.H., and Burman, N.J. (2012). Barn owl release in lowland southern England—a twenty-one year study. *Biological Conservation*, **109**, 271–282.

Merckx, T., Marini, L., Feber, R.E., and Macdonald, D.W. (2012). Hedgerow trees and extended-width field margins enhance macro-moth diversity: implications for management. *Journal of Applied Ecology*, **49**, 1396–1404.

Montgomery, W.I. (1989a). Population regulation in the wood mouse, *Apodemus sylvaticus*. I. Density dependence in the annual cycle of abundance. *Journal of Animal Ecology*, **58**, 465–475.

Montgomery, W.I. (1989b). Population regulation in the wood mouse, Apodemus sylvaticus. II. Density dependence in spatial distribution and reproduction. *Journal of Animal Ecology*, **58**, 477–494.

Montgomery, W.I. and Dowie, M. (1993). The distribution and population regulation of the wood mouse Apodemus sylvaticus on field boundaries of pastoral farmland. *Journal of Applied Ecology*, **30**, 783–791.

Moore, N.P., Askew, N. and Bishop, J.D. (2003). Small mammals in new farm woodlands. *Mammal Review*, **33**, 101–104.

Morton, R.D., Rowland, C., Wood, C. Meek, L., Marston, C., Smith, G., Wadsworth, R., Simpson, I.C. (2011). *Final Report for LCM2007—the new UK land cover map*. Countryside Survey Technical Report No 11/07 NERC/Centre for Ecology & Hydrology 112pp.(CEH Project Number: C03259).

Parker, G.A. (1970). Sperm competition and its evolutionary consequences in insects. *Biological Reviews*, **45**, 525–567.

Perrow, M.R. and Jordan, A.J.D. (1992). *The influence of agricultural land use upon populations of harvest mouse (Micromys minutes (Pallas))*. Report to TERF, Hoechst, UK, East Winch, Norfolk.

Perrow, M. and Jowitt, A. (1995). What future for the harvest mouse? *British Wildlife*, **6**, 356–365.

Plesner Jensen, S.P. (1993). Temporal changes in food preferences of wood mice (Apodemus sylvaticus L.). *Oecologia*, **94**, 76–82.

Plesner Jensen, S.P. and Honess, P. (1995). The influence of moonlight on vegetation height preference and trappability of small mammals. *Mammalia*, **59**, 35–42.

Pollard, E., Hooper, M.D., and Moore, N.W. (1974). *Hedges*. Collins, London.

Potts, G.R. (1980) The effects of modern agriculture, nest predation and game management on the population ecology of partridges. *Advances in Ecological Research*, **11**, 1–79.

Potts, G.R. (1986). The partridge: pesticides, predation and conservation. Collins, London.

Rackham, O. (1997). *The Illustrated History of the Countryside*. Weidenfeld & Nicolson, London.

Richards, C.G.J. (1985). The population dynamics of Microtus agrestis in Wytham, 1949–1978. *Acta Zoologica Fennica*, **173**, 35–38.

Riordan, P., Lloyd, A., and Macdonald, D.W. (2009). *Do harvest mouse nest survey results predict population size?* Report to People's Trust for Endangered Species.

Robinson, R.A. and Sutherland, R.J. (2002). Post-war changes in arable farming and biodiversity in Great Britain. *Journal of Applied Ecology*, **39**, 157–176.

Rowe, F.P. and Taylor, E.J. (1964). The numbers of harvest-mice (Micromys minutus) in corn-ricks. *Proceedings of the Zoological Society of London*, **142**, 181–185.

Sargent, G. (1999). Harvest mouse in trouble. *Mammal News*, **111.1**.

Shore, R.F., Feber, R.E., Firbank, L.G., Fishwick, S.K., Macdonald, D.W., and Nørum, U. (1997). The impacts of molluscicide pellets on spring and autumn populations of wood mice Apodemus sylvaticus. *Agriculture Ecosystems and Environment*, **64**, 211–217.

Shore, R.F., Fletcher, M.R., and Walker, L.A. (2003). Agricultural pesticides and mammals in Britain. In: Tattersall,

F.H. and Manley, W.J. eds., *Conservation and conflict: mammals and farming in Britain*. Linnean Society Occasional Publications, Westbury Publishing, pp. 37–50.

Shore, R.F. and Hare, E.J. (2008). Bank Vole. In S. Harris and D.W. Yalden, eds., *Mammals of the British Isles: Handbook, 4th Edition*, pp. 88–99. The Mammal Society.

Simonetti, J. A. (1989). Microhabitat use by small mammals in central Chile. *Oikos*, **56**, 309–318.

Smeding, F.W. and de Snoo, G.R. (2003). A concept of foodweb structure in organic arable farming systems. *Landscape and Urban Planning*, **65**, 219–236.

Sotherton, N.W. (1988). The cereals and gamebirds research project: overcoming the indirect effects of pesticides. In Harding, D.J.L., ed. *Britain since silent spring*, pp. 64–72. Proceedings of a symposium of the Institute of Biology, London.

Soule, M.E. and Terbough, J. (1999). *Continental conservation: scientific foundations of regional reserve networks*. Washington DC, Island Press.

Southern, H.N. and Laurie, E.M.O. (1946). The house mouse (Mus musculus) in corn ricks. *Journal of Animal Ecology*, **15**, 134–149.

Stoate, C., Boatman, N.D., Borralho, R.J., Carvalho, C.R., de Snoo, G.R., and Eden, P. (2001). Ecological impacts of arable intensification in Europe. *Journal of Environmental Management*, **63**, 337–365.

Stockley, P. and Macdonald, D.W. (1998). Why do female common shrews produce so many offspring? *Oikos*, **83**, 560–566.

Stockley, P., Searle, J.B., Macdonald, D.W., and Jones, C.S. (1993). Female multiple mating behaviour in the common shrew as a strategy to reduce inbreeding. *Proceedings of the Royal Society of London B*, **254**, 173–179.

Stockley, P., Searle, J.B., Macdonald, D.W., and Jones, C.S. (1994). Alternative reproductive tactics in male common shrews: relationships between mate-searching behaviour, sperm production, and reproductive success as revealed by DNA fingerprinting. *Behavioural Ecology and Sociobiology*, **34**, 71–78.

Stockley, P., Searle, J.B., Macdonald, D.W., and Jones, C.S. (1996). Correlates of reproductive success within alternative mating tactics of the common shrew. *Behavioral Ecology*, **7**, 334–340.

Stopka, P. and Graciasová, R. (2001). Conditional allogrooming in the herb-field mouse. *Behavioural Ecology*, **12**, 584–589.

Stopka, P. and Macdonald, D.W. (1998). Signal interchange during mating in the wood mouse (Apodemus sylvaticus): the concept of active and passive signalling. *Behaviour*, **135**, 231–249.

Stopka, P. and Macdonald, D.W. (2003). Way-marking behaviour: an aid to spatial navigation in the wood mouse (Apodemus sylvaticus). *BMC Ecology*, **3**.

Tarrant, K.A. and Westlake, G.E. (1998). Laboratory evaluation of the hazard to wood mice, Apodemus sylvaticus, from the agricultural use of methiocarb molluscicide pellets. *Bulletin of Environmental Contamination and Toxicology*, **40**, 147–152.

Tattersall, F.H., Avundo, A.E., Manley, W., Hart, B., and Macdonald, D.W. (2000). Managing set-aside for field voles *Microtus agrestis. Biological Conservation*, **96**, 123–128.

Tattersall, F.H., Fagiano, A.L., Bembridge, J.D., Edwards, P., Macdonald, D.W., and Hart, B.J. (1999a). Does the method of set-aside establishment affect its use by wood mice? *Journal of Zoology*, **249**, 472–476.

Tattersall, F.H., Hart, B.J., and Manley, W.J. (1999b). Small mammals on set-aside blocks and margins. *Aspects of Applied Biology*, **54**, 131–138.

Tattersall, F.H. and Macdonald, D.W. (2003). The arable wood mouse. In: Tattersall, F.H. and Manley, W.J. eds., *Conservation and conflict: mammals and farming in Britain*, pp. 82–96. Linnean Society Occasional Publication, Westbury Publishing.

Tattersall, F.H., Macdonald, D.W., Hart, B.J., Johnson, P., Manley, W., and Feber, R. (2002). Is habitat linearity important for small mammal communities on farmland? *Journal of Applied Ecology*, **39**, 643–652.

Tattersall, F.H., Macdonald, D.W., Hart, B.J., and Manley, W. (2004). Balanced dispersal or source-sink—do both models describe wood mice in farmed landscapes? *Oikos*, **106**, 536–550.

Tattersall, F.H., Macdonald, D.W., Hart, B.J., Manley, W.J., and Feber, R.E. (2001). Habitat use by wood mice (Apodemus sylvaticus) in a changeable arable landscape. *Journal of Zoology*, **255**, 487–494.

Tattersall, F., Macdonald, D., Manley, W., Gates, S., Feber, R., and Hart, B. (1997). Small mammals on one year set-aside. *Acta Theriologica*, **42**, 329–334.

Tew, T.E. (1991). Radio-tracking arable dwelling wood mice. In I G Priede and S M Swift, eds., *Wildlife telemetry—remote monitoring and tracking of animals*, pp. 561–569. Ellis Horwood, Chichester, UK.

Tew, T.E. and Macdonald, D.W. (1993). The effects of harvest on arable wood mice *Apodemus sylvaticus. Biological Conservation*, **65**, 279–283.

Tew, T.E. and Macdonald, D.W. (1994). Dynamics of space use and male vigor amongst wood mice, *Apodemus sylvaticus*, in the cereal ecosystem. *Behavioral Ecology and Sociobiology*, **34**, 337–345.

Tew, T.E., Macdonald, D.W., and Rands, M.R.W. (1992). Herbicide application affects microhabitat use by arable wood mice (*Apodemus sylvaticus*). *Journal of Applied Ecology*, **29**, 532–539.

Tew, T.E., Todd, I.A., and Macdonald, D.W. (1994a). The effects of trap spacing on population estimates of small mammals. *Journal of Zoology*, **233**, 340–344.

Tew, T.E., Todd, I.A., and Macdonald, D.W. (1994b). Field margins and small mammals. *British Crop Protection Council Monographs*, **58**, 85–94.

Tew, T.E., Todd, I.A., and Macdonald, D.W. (2000). Arable habitat use by wood mice (*Apodemus sylvaticus*). 2. Microhabitat. *Journal of Zoology*, **250**, 305–311.

Theall, C.L., Wurtman, J.J., and Wurtman, R.J. (1984). Self-selection and regulation of protein:carbohydrate ratio in foods adult rats eat. *Journal of Nutrition*, **114**, 711–718.

Thompson, H.M. (1991). Serum 'B' esterases as indicators of exposure to pesticides. In P. Mineau, ed. *Cholinesterase-inhibiting insecticides, and their impact on wildlife and the environment*, pp. 109–125. Elsevier, Amsterdam.

TMP (2009). *Tracking Mammals Partnership Update 2009*, JNCC

Todd, I.A., Tew, T.E., and Macdonald, D.W. (2000). Arable habitat use by wood mice (*Apodemus sylvaticus*). 1. Macrohabitat. *Journal of Zoology*, **250**, 299–303.

Trout, R.C. and Harris, S. (2008). Harvest mouse. In S. Harris and D.W. Yalden, eds., *Mammals of the British Isles: Handbook, 4th Edition*, pp. 117–125, The Mammal Society.

Vickery, J., Feber, R.E., and Fuller, R. (2009). Arable field margins managed for biodiversity conservation: a review of food resource provision for farmland birds. *Agriculture, Ecosystems & Environment*, **133**, 1–13.

Wickramasinghe, L.P., Harris, S., Jones, G., and Jennings, N.V. (2003). Bat activity and species richness on organic and conventional farms: impact of agricultural intensification. *Journal of Applied Ecology*, **40**, 984–993.

Wilson, W.L., Montgomery, W.I., and Ellwood, R.W. (1993). Population regulation in the wood mouse *Apodemus sylvaticus* (L.). *Mammal Review*, **23**, 65–92.

Yamaguchi, N., Sarno, R.J., Johnson, W.E., O'Brien, S.J., and Macdonald, D.W. (2004). Multiple paternity and reproductive tactics of free-ranging American minks, *Mustela vison. Journal of Mammalogy*, **85**, 432–439.

CHAPTER 5

Agri-environment schemes and the future of farmland bird conservation

Jeremy D. Wilson and Richard B. Bradbury

We are losing half the subject-matter cf English poetry.
Aldous Huxley, after reading *Silent Spring*.

5.1 Introduction

Over 50 years ago, Rachel Carson (1962) published *Silent Spring* as a clarion call, alerting society to the catastrophic environmental impacts of early generations of agricultural pesticides. The possibility that the soundtrack of nature may fade in the face of human attempts to control the environment has remained in the public consciousness ever since.

The pesticide impacts that prompted *Silent Spring* were the consequence of a technological drive to increase agricultural productivity, backed by legislation and Government subsidy, and born of post-war determination to increase food security. In the UK, there was growing evidence that toxic impacts of persistent organochlorine residues were having severe impacts on survival and breeding success of top predators. These chemicals killed sparrowhawks *Accipiter nisus*, and induced a fatal 10–20% thinning of the shells of newly laid eggs (Newton 1986). The resulting catastrophic population decline, with the species becoming effectively extinct over large areas of arable England, became a *cause celebre* for nature conservation. By the 1980s, organochlorines were banned from UK agricultural use, in an early and successful example of evidence-based conservation policy. Since then, however, agricultural intensification has continued apace and involves many changes in land use and husbandry with less direct impacts on farmland birds (Newton 2004). Loss of field boundary habitats, simplification of crop rotations, development of autumn-sown crop varieties leading to elimination of winter fallow periods, drainage and re-seeding of grasslands, heavy fertilizer usage, increasing livestock densities, and effects of pesticides

in reducing invertebrate weed populations—all were interdependent changes that contributed to a continuing erosion of biodiversity on agricultural land, including populations of many bird species, that had become clearly apparent in the UK by the 1990s (e.g. Chamberlain et al. 2000; Robinson and Sutherland 2002). As an example, England and Wales had lost 97% of their unimproved, flower-rich, lowland grasslands by the mid 1980s (Fuller 1987). Krebs et al. (1999) noted that these agricultural changes were concerned with ensuring that as great a proportion of primary production as possible is diverted to the human food chain and that, to the extent this is achieved, the rest of nature might be expected to suffer; a 'second Silent Spring'. Nature's soundtrack was indeed fading.

Partly because statutory nature conservation in post-War Britain was so focused on sparing land for wildlife through designation, legal protection, and management of nature reserves, a belief that agricultural and nature conservation objectives should be shared on actively farmed land was slow to gain ground (Marren 2002). In the end, the accumulation of Common Agricultural Policy (CAP) food surpluses across Europe during the 1980s provided the greatest impetus to reform an agricultural system in which, as Oliver Rackham (1986) put it 'we contrive to subsidize agriculture much more than any other industry and to have . . . a ravaged countryside'. Reforms came through three routes. First, set-aside policies implemented to reduce production surpluses were harnessed to deliver some wildlife benefits through modification of management requirements (Sotherton 1998). Second, the growth of organic farming brought

Wildlife Conservation on Farmland. Managing for Nature on Lowland Farms. Edited by David W. Macdonald and Ruth E. Feber.
© Oxford University Press 2015. Published 2015 by Oxford University Press.

some associated biodiversity benefits (e.g. Hole et al. 2005). Third, and most importantly, agri-environment schemes (AES) were introduced in which a proportion of subsidy funding to farmers was tied to the delivery of environmental public goods, including biodiversity conservation. In the UK, agri-environment policy making began in 1987 and successive generations of AES have followed with bespoke approaches developing in the increasingly devolved four countries of the UK. They remain a cornerstone of biodiversity conservation management on farmland to this day. This growth of the agri-environment approach prompted Buckwell and Armstrong-Brown's (2004) view that, following pre-industrial and industrial agricultural eras, the turn of the century might herald a 'post-industrial' agricultural era. In this era, recognition of the environmental costs of production-driven agriculture would lead to a balance between food production and the delivery of public goods, with public subsidy directed much more at the latter and much less at the former.

In the remainder of this chapter, we review briefly the successes and failures of the agri-environment era for the conservation of birds on farmland, focusing on the growth of the evidence base for AES design and conservation impact, and the role that research at Oxford has played. We conclude by considering what we have learned from measuring the responses of bird populations to these interventions in production-focused agricultural landscapes, and what the prospects might be for farmland bird conservation in the changing policy contexts of the early twenty-first century. Would Buckwell and Armstrong-Brown reach the same conclusions today?

5.2 The evidence base for agri-environment management for bird conservation

One of the successes of the publication of the UK Biodiversity Action Plan (UKBAP) in 1994 was that it identified clearly the species of highest conservation concern based on threat and population decline. Amongst bird species, many were associated with farmland (Table 5.1), prompting a concerted research programme in the UK designed to understand the impacts of agricultural change on these species and to seek potential management solutions.

This research programme was led both by NGOs with research capacity—notably the British Trust for Ornithology (BTO), Royal Society for the Protection of Birds (RSPB), and Game Conservancy Trust (GCT—now Game & Wildlife Conservation Trust)—and

Table 5.1 UK Biodiversity Action Plan priority bird species with important breeding and/or wintering populations in agricultural landscapes, indicating those species that were the subject of study by Oxford's Ecology and Behaviour Group (in bold).

Species	Date species listed as UKBAP priority (1995 or 2007 review)
Skylark *Alauda arvensis*	1995
White-fronted goose *Anser albifrons*	2007
Brent goose *Branta bernicla*	2007
Stone-curlew *Burhinus oedicnemus*	1995
Linnet *Carduelis cannabina*	1995
Twite *Carduelis flavirostris*	2007
Corncrake *Crex crex*	1995
Cuckoo *Cuculus canorus*	2007
Corn bunting *Emberiza calandra*	1995
Cirl bunting *Emberiza cirlus*	1995
Yellowhammer *Emberiza citrinella*	2007
Reed bunting *Emberiza schoeniclus*	1995
Black-tailed godwit *Limosa limosa*	2007
Grasshopper warbler *Locustella naevia*	2007
Yellow wagtail *Motacilla flava*	2007
Spotted flycatcher *Muscicapa striata*	1995
Curlew *Numenius arquata*	2007
House sparrow *Passer domesticus*	2007
Tree sparrow *Passer montanus*	1995
Grey partridge *Perdix perdix*	1995
Dunnock *Prunella modularis*	2007
Bullfinch *Pyrrhula pyrrhula*	1995
Turtle dove *Streptopelia turtur*	1995
Starling *Sturnus vulgaris*	2007
Black grouse *Tetrao tetrix*	1995
Song thrush *Turdus philomelos*	1995
Ring ouzel *Turdus torquatus*	2007
Lapwing *Vanellus vanellus*	2007

also by some University groups, including Oxford's Ecology and Behaviour Group. This group, affiliated to the Edward Grey Institute of Field Ornithology, worked alongside WildCRU over a decade from the mid 1990s. Funded both by Research Councils (BBSRC—Biotechnology and Biological Sciences Research Council—and NERC—Natural Environment Research Council) and central Government (mainly Defra—Department of Environment, Food and Rural

Box 5.1 An Oxford connection

(i) *Single-species ecology*

In the early 1990s, understanding of species-specific re-sponses to agricultural change remained relatively poor. In Oxford, work focused on passerines of varying ecologies and population trends (skylark *Alauda arvensis*, barn swallow *Hirundo rustica*, linnet *Carduelis cannabina*, bullfinch *Pyrrhula pyrrhula*, chaffinch *Fringilla coelebs*, house sparrow *Passer domesticus*, and yellowhammer *Emberiza citrinella* (Box Fig. 5.1.1)) across contrasts from organic to intensive agricultural management. This work contributed to under-standing the dependence of habitat quality on crop struc-ture for open-ground nesting species such as skylark (e.g. Wilson et al. 1997), and on hedgerow structure and field boundary management for shrub-nesting species, includ-ing yellowhammer and bullfinch (Macdonald and Johnson 1995; Bradbury et al. 2000; Proffitt et al. 2004). Foraging and diet studies demonstrated strong associations of yellow-hammers with sparse crop structures (Morris et al. 2002), further contributing to a growing understanding of the critical importance of vegetation density and heterogeneity on cropped farmland for nesting and feeding birds (Wilson et al. 2005). Chaffinches were found to be less susceptible to agricultural change due to their exploitation of arboreal invertebrate foods (Whittingham et al. 2001), and exploit-ation by linnets of ripening oilseed rape as a nestling food source (up to 80% of nestling gullet content in July) buffered them against historical losses of arable and grassland weed seed foods (Moorcroft et al. 2006).

(ii) *Bird associations with agricultural systems*

Early work in Oxford showed the value of over-winter stub-bles (Box Fig. 5.1.2) as foraging habitats for many farmland birds, especially finches, buntings, and sparrows for which Wilson et al. (1996) found 88% of individuals on stubble fields, which occupied only 11–14% of the surveyed field area. This contributed to recognition of both the conser-vation value of set-aside and the value of herbicide-free winter stubbles as an arable AES measure. Similarly, Ox-ford studies were amongst those to consider how winter use of grasslands by farmland birds varied with manage-ment. These studies showed the preference of many soil invertebrate-feeding groups, such as corvids, thrushes, plov-ers, and starlings, for pastures over leys or cropped fields (Perkins et al. 2000), especially where these grasslands are agriculturally improved (Barnett et al. 2004). These findings were consistent with the higher densities of earthworms and soil-dwelling invertebrate larvae in established, improved pastures (Tucker 1992), and may be an indication of a role for grassland intensification in recent population increases of corvids. Although short-grazed swards and patches of bare ground were favoured by most foraging birds in winter,

Box Figure 5.1.1 Yellowhammer *Emberiza citrinella*. ©Chris Knights (rspb-images.com).

Box Figure 5.1.2 Weedy stubble is an important food source for farmland birds. ©Andy Hay (rspb-images.com).

continued

Box 5.1 *Continued*

exceptions included preference by common snipe *Gallinago gallinago* for the cover, and of seed-eating species such as skylark, linnet, goldfinch *Carduelis carduelis*, yellowhammer, and corn bunting *Emberiza calandra* for the seed-bearing grasses and forbs, that were provided by ungrazed fields.

(iii) *Behavioural and demographic mechanisms*
The Oxford group worked in close collaboration with BTO over several years, developing improved models of long-term national trends in farmland bird populations (Siriwardena et al. 1998; Fewster et al. 2000). These modelled trends were then used in conjunction with long-term demographic data sets, such as national ring-recovery data and the nest record scheme, to provide strong evidence that trends of some granivorous species were being driven by change in mortality rates rather than by change in individual nest success (Siriwardena et al. 2000). Field studies of the effect of winter food supplementation on survival rates of house sparrows supported these findings by showing that seed supplementation could increase November–March survival rates from 39 to 65%, though only on a farm where a severe prior population decline had been experienced, and not on farms where populations had remained stable (Hole et al. 2002). These studies further contributed to acceptance of the importance of over-winter stubble retention or provision of other seed-rich habitat patches as AES measures, and led on to more detailed behavioural studies which sought to understand how vegetation structure interacted with seed availability and predation risk to determine habitat quality for foraging birds. For many flocking, seed-eating species such as finches and buntings, which rely on flight as an escape mechanism, these studies demonstrated the importance of stubble structures that combined lower stubble height with areas of bare ground, thus reducing visual obstruction (Butler et al. 2005; Whittingham et al. 2006). For example, the likelihood of occupation of stubble fields by linnets, reed buntings *Emberiza schoeniclus*, and corn buntings increased from less than 10% to over 50% as the unvegetated area increased from zero to 70% of the ground surface (Moorcroft et al. 2002). Other species such as skylark and grey partridge *Perdix perdix* are more reliant on crypsis and favoured more densely vegetated, longer stubbles (Box Fig. 5.1.2).

(iv) *Prediction*
One of the main challenges in translating increasing understanding of the causes of declines of farmland bird populations into robust management recommendations has been the trade-off between studies over large geographical areas which show patterns of correlation but tell us little about mechanisms, and smaller-scale field studies which give us insight into ecological mechanism but whose generality is uncertain (Wilson et al. 2009). Studies in Oxford sought to bridge this gap in several ways. For example, by drawing together data across multiple study areas including those centred round Oxford, Morris et al. (2001) and Whittingham et al. (2003) found, respectively, that foraging habitat associations of yellowhammers and breeding territory associations of skylarks and yellowhammers were robust across large areas of southern England. Bradbury et al. (2005) showed how (then) novel technologies such as LiDAR could be used to measure fine-grained heterogeneity in vegetation structure over large geographical areas, and thus parameterize habitat-association models such as those developed for skylarks in ways that would not be possible using field survey, thus bridging the gap between grain and extent in organism-habitat modelling (Whittingham et al. 2007).

Affairs, and its predecessor, MAFF—Ministry of Agriculture, Fisheries & Food), the group also worked in close collaboration with the BTO and RSPB, and delivered much of its work through PhD studentships. Its contributions can be summarized across four main areas (Box 5.1).

The brief synopsis outlined in Box 5.1 of some of the work carried out by the Ecology and Behaviour Group and WildCRU in Oxford gives a flavour of the evidence base that has developed in response to the recognition in the late 1980s and early 1990s of widespread farmland bird declines, and has informed the development of conservation measures delivered through AES. Indeed, a decade after the publication of the UK Biodiversity Action Plan, Grice et al. (2004) were able to conclude:

The results of scientific research have played a central role in convincing decision-makers of the plight of farmland birds and in the development of land management measures and policies designed to reverse the population declines, (and) have contributed to . . . a considerable knowledge base on the ecological requirements of key species, the reasons for their declines and the remedial measures necessary to bring about population recovery.

5.3 Bird conservation in the agri-environment era

5.3.1 Characteristics of successful AES management for birds

The conclusion reached by Grice et al. (2004) seems a fair one in a UK context given that most of the species in Table 5.1 now benefit either from bespoke AES measures or measures likely to benefit a species assemblage of similar ecology. The former include, for example, unsown patches in cereal fields to provide nesting opportunities for skylarks and stone curlews *Burhinus oedicnemus*, or late cutting of grass meadows to reduce nest loss rates of corncrakes *Crex crex* or corn buntings. The latter include field margin management designed to increase invertebrate and seed resources for a wide range of passerine species (Wilson et al. 2009). Indeed, in some cases, the use of research studies to inform the design of AE measures, monitor effectiveness of those measures when deployed in schemes, and to alter those measures based on monitoring outcomes, has permitted adaptive improvement of AES over time. In one example, Perkins et al. (2011) monitored breeding corn bunting populations on AES and control farms in north-east Scotland, both to identify and to refine AES options (notably late-cutting of silage fields) to benefit this species. They showed that the combination of both targeted AE measures and specialist advice to farmers was necessary to convert rapid declines in the wider landscape (14.5% per annum) to local increases on farms benefiting from these measures (5.6% per annum). Even so, the same study calculated that current AES provision was targeting only 24% of the population, whereas it would need to target 72% to halt declines at the national scale. Examples of reversals of national or regional population declines as a consequence of AES implementation remain regrettably few. In the UK, reversals of long-term declines of corncrakes, stone curlews, and cirl buntings *Emberiza cirlus* have all been achieved (reviewed by Wilson et al. 2009), as has a modest recovery from near-extinction of the population of little bustards *Tetrax tetrax* in western France (Bretagnolle et al. 2011). However, in all these cases, the species had become rare and localized, with remnant populations reduced to no more than a few hundred breeding pairs before targeted AES implementation halted and reversed the decline. These species' populations and breeding ranges remain very small in long-term historical context and wholly dependent on the continuation of those AES and other

conservation management interventions designed to support them.

Where species are more widespread, success has only been measurable at smaller scales, suggesting that the scale and density of AES implementation are still typically far too limited. For example, AES management has been shown to be capable of reversing declines of grassland-nesting waders such as lapwing *Vanellus vanellus* (Fig. 5.1), redshank *Tringa totanus*, and black-tailed godwit *Limosa limosa* (Ausden and Hirons 2002; Verhulst et al. 2007; Wilson et al. 2007; O'Brien and Wilson 2011), but the latter study estimated that appropriate management was only benefiting 2% of the lapwing population on Scottish farmland. Assuming that a population growth rate of 2–3% per annum were achievable where appropriate AES management is implemented, then it would need to have benefited 44–54% of the population to offset observed declines of 2.4% per annum in the wider farmed countryside and stabilize the national population decline. Similarly, Ewald et al. (2010) found positive effects of conservation headlands, beetle banks, and grass and cover crops on grey partridge numbers at the farm scale in England, but also found that, outside farms enrolled in the Partridge Count Scheme, only 1–2% of AES agreements included conservation headlands and beetle banks, far too few for AES to deliver any large-scale population benefit for that species.

Overall, the evidence shows that when there has been thorough diagnosis of the causes of population decline, when remedial management to be delivered through AES has been tested, and when deployment of these measures is well targeted and supported by high quality advice on implementation, then AES are capable of reversing national bird population declines (Evans and Green 2007). Examples of them doing so are rare simply because the required AES management is not often being deployed at anything like a sufficient scale, except for species whose declines have been so severe that they have been reduced to small geographical areas and remnant populations.

5.3.2 Monitoring and evaluation studies are important

If we are to know whether AES interventions are effective and to improve them through adaptive management, then well-designed monitoring and evaluation studies are crucial, focusing on biological effectiveness, economic cost and practicality, and effectiveness of implementation. In a review of the success of AES

Figure 5.1 Lapwing *Vanellus vanellus* on grassland. ©Andy Hay (rspb-images.com).

expenditure for biodiversity conservation, Kleijn and Sutherland (2003) found that evaluation studies were limited in number and often not designed robustly, despite the fact that the EU had, by then, spent €24.3 billion on AES. Perhaps spurred on by Kleijn and Sutherland's critique, the subsequent increase in the rate of publication of studies testing the effectiveness of AES implementation for biodiversity has been marked (Whittingham 2007; Pywell et al. 2012). Wilson et al. (2009) found that these studies showed good evidence of beneficial effects of AES management—at least at field and farm scales—for a wide range of taxonomic groups, including arable and grassland flora, arthropods (including pollinators), birds, and small mammals, although there are exceptions. For example, Fuentes-Montemayor et al. (2011) found that levels of bat activity and abundance of their insect prey were lower on AES than non-AES farms in Scotland. At larger (e.g. regional and national) scales, however, studies of biodiversity benefits of AES remain few (Kleijn et al. 2011).

There have been well-designed studies of bird population responses to AES in some parts of the UK,

although many of these have reported very quickly after scheme implementation, and perhaps without sufficient time for scheme impact to accrue (e.g. Davey et al. 2010). Most recently a large-scale study of breeding population changes of farmland birds in England over eight years and across over 2000 one-km-square plots, in relation to the type and extent of Environmental Stewardship management in the surrounding landscape, did find clear evidence of strong positive responses of a wide range of seed-eating species to greater areas of winter seed-providing management options (Baker et al. 2012). These included linnet, reed bunting, yellowhammer, and tree sparrow *Passer montanus* (Table 5.2).

As we go to press, another study, this one focused on the more targeted 'Higher Level' options within Environmental Stewardship, has similarly found more positive population trends of several species of high conservation concern, including lapwing, on farms with these measures than those without them (Bright et al. 2015).

Baker et al.'s (2012) results accord with previous research which had suggested that limited seed supplies

Table 5.2 Significant effects (bold = positive; regular type = negative) of stubble management (S) and wild bird seed management (W) provided under the English Entry Level Stewardship agri-environment scheme on nine-year population change of seed-eating farmland birds at three spatial scales around focal Breeding Bird Survey one-km-squares in three different landscape types: arable (A), mixed (M), and pasture-dominated (P). Summarized from Baker et al. (2012).

Species	1 km²	9 km²	25 km²
Corn bunting *Emberiza calandra*	**S(P); W(P)**	W(M)	
Chaffinch *Fringilla coelebs*	**W(P)**		**S(P); W(M)**
Goldfinch *Carduelis carduelis*	**S(A)**, S(P)	S(A); W(P)	
Greenfinch *C. chloris*			ST(P); W(M)
House sparrow *Passer domesticus*[1]			
Linnet *C. cannabina*	**S(A,M,P)**	**S(M)**; W(A,M)	**S(P)**; W(A,M)
Grey partridge *Perdix perdix*	**S(P)**		
Reed bunting *E. schoeniclus*	**S(A,P); W(P)**		**S(M)**
Skylark *Alauda arvensis*	**S(M); W(M)**		W(P)
Stock dove *Columba oenas*		**S(M)**	W(P)
Tree sparrow *P. montanus*	**W(A)**; W(M)	**S(A); W(A)**	**S(A); W(A)**
Yellowhammer *E. citrinella*	**S(A,P); W(A)**	**S(M)**	**S(A)**; W(M)

[1] House sparrows were studied but no significant effects were found.

had driven declines of many granivorous species (e.g. Siriwardena et al. 2000) and are encouraging evidence that, if implemented widely enough, seed-providing measures such as over-winter fallows and unharvested crops could reverse declines of these birds. At present, however, Baker et al. (2012) conclude that limited uptake of management options that provide seed resources in-field, and a tendency for seed supplies to become exhausted by late winter, may limit effectiveness. This illustrates how well-designed monitoring studies can inform adaptive improvements in AES implementation. In contrast, in some UK countries, there has been very little investment in rigorous monitoring of the effectiveness of AES for biodiversity conservation objectives. In Scotland, for example, government has not funded any comprehensive, peer-reviewed assessment of bird population responses to AES. Given that the study by Baker et al. (2012) was impressively cost-effective, in making use of data collected almost entirely by volunteers under the UK's Breeding Bird Survey (Risely et al. 2012), and that the same approach could be adopted in Scotland, this is surprising. It is also unfortunate when one considers that the recent population trends of many farmland birds differ markedly between England and Scotland, with much more evidence of stable or increasing populations of some species in Scotland. In this context, a comparative

analysis might yield valuable insights into the relative effectiveness of AES management in Scottish and English agricultural landscapes, and the reasons for differences.

5.3.3 AES management for birds can deliver wider biodiversity benefits

Perhaps because of the availability of high quality population trend data from volunteer-based surveys such as the Common Birds Census (1962–2000) and its successor, the Breeding Bird Survey (since 1994), declines in farmland bird populations in the UK were detected early. Indeed, detailed research into the impacts of agricultural intensification on one species of economic interest—the grey partridge, see Chapter 13, this volume—began in the 1960s (Potts 1986). As a consequence, the evidence base to inform bird conservation measures in AES accrued early, and Grice et al. (2004) identified a need to test whether broader biodiversity conservation measures were being supported or hindered by the expansion of bird-focused conservation measures in AES. In response to this, RSPB and Natural England funded three parallel studies testing the species richness and numerical responses of a range of plant and invertebrate taxa to agri-environment management designed to benefit corncrakes, stone curlews,

Table 5.3 Summary of effects of agri-environment management targeted at single bird species on other taxa. Summarized from detailed results presented in MacDonald et al. 2012a,b and Wilkinson et al. 2012.

Target bird species	Management contrast	Other studied taxa benefiting from AES management in species richness, abundance, or activity-density	Other studied taxa reduced in species richness, abundance, or activity-density by AES management
Cirl bunting	Low-input spring barley (AES) versus conventional	Carabid beetles	None
Cirl bunting	No-input pasture (AES) versus conventional	Vascular plants; butterflies	None
Stone-curlew	Fallow plot in arable field (AES) versus conventional crop	Brown hare *Lepus europaeus*; vascular plants; butterflies; bumblebees	Predatory carabid beetles
Corncrake	Early and late margin cover (AES) versus conventional meadows	Butterflies; bumblebees; springtails; heteropteran bugs; homopteran bugs; spiders	Vascular plants
Corncrake	Delayed mowing (AES) versus conventionally cut meadows	Vascular plants; bumblebees; heteropteran bugs; homopteran bugs; beetles; hoverflies; caterpillars; sawfly larvae; Apocrita; mites; spiders	Springtails

and cirl buntings. These studies have all reported recently (MacDonald et al. 2012a, b; Wilkinson et al. 2012) and confirm that, across a wide range of measures designed to benefit three species of contrasting ecology, there is clear and general evidence of a wider beneficial impact (Table 5.3). Of course, conflicts and a need to prioritize AES objectives may occur in some specific circumstances. For example, implementation of wide grass margins to offer nesting and foraging resources for farmland birds may remove opportunities for rare arable flora that are better catered for using conservation headlands or fallow strips.

5.4 Future challenges for farmland bird conservation

5.4.1 Policy challenges

In 2001, in the wake of an epidemic of foot-and-mouth disease, the UK Government's *Policy Commission on the Future of Farming and Food* (2002) considered that the agricultural industry had become dysfunctional—unsustainable in every sense, detached from the rest of the economy, and also from the environment, which it continued to degrade. It laid much of the blame at the door of the CAP and its subsidies, and recommended that direct subsidy payments should be decoupled from production and ultimately phased out altogether, with subsidy support only paid under the principle that 'public money should be used to pay for public goods'. At that time, prospects for the continued expansion of agri-environment support for biodiversity

seemed good. However, between autumn 2006 and spring 2008, world cereal prices became much more volatile, after 25 years of relative stability (FAO 2012), with dramatic effects on policy thinking. In November 2008, the EU abolished compulsory set-aside to allow farmers to 'maximise their production potential' (European Commission 2008) and, by 2009, many influential commentators were identifying security of food supply as a global challenge of the twenty-first century as a result of the conjoint pressures on food production from human population and economic growth, globalization of food markets, and the impact of climate change (MacIntyre et al. 2009; Royal Society 2009); a *'perfect storm of global events'* as the then Chief Scientific Adviser to the UK Government put it (Beddington 2009).

Over the same time period, publication of the *Millennium Ecosystem Assessment* (2005) and subsequent initiatives, including *The Economics of Ecosystems and Biodiversity* (TEEB 2010) and the *UK National Ecosystem Assessment* (2011), had encouraged a more utilitarian perspective on biodiversity conservation amongst policy makers. This focused on the contribution of biodiversity to the range of ecosystem services that underpin sustainable agriculture, including pollination, biological pest control, soil health, nutrient cycling, and hydrological services (e.g. Power 2010; Norris 2012). There was recognition of the severe biodiversity losses already caused by agricultural intensification and hence that further increases in global agricultural production must represent a 'sustainable intensification' comprising increased yields, more efficient use of inputs, and without further adverse environmental impacts (e.g. Firbank 2009; Royal Society 2009; Foresight. Future of Food and Farming 2011). However, as many commentators have recognized, the integration of food production with sustainable delivery of public ecosystem services for which markets and valuation methods either do not exist or are in their infancy, is a huge challenge both scientifically and in policy terms (e.g. Robertson and Swinton 2005).

At landscape and national scales, there is some evidence of spatial congruence between biodiversity measures and the delivery of other ecosystem services priorities (e.g. Anderson et al. 2009). However, at the more intimate scale of AES management in intensive production landscapes in the UK, Bradbury et al. (2010) found only weak evidence, either that AES measures focused on bird conservation were likely to have benefits for other ecosystem services, or that management for those services is likely to be an efficient way to conserve farmland birds. They concluded that there is a clear risk that some management measures for birds may lose support if they are either neutral with respect to wider ecosystem service delivery or hinder it, especially because the cultural value of wild birds in their own right is poorly quantified. There is therefore a real risk that bird conservation on farmland becomes peripheral to a renewed preoccupation of agriculture and its scientific funding base with food production, and its reconciliation with mitigating and adapting to climate change and to the delivery of other, more readily measured and valued ecosystem services (Wilson et al. 2010). These pressures are being exacerbated at a time of European economic recession by strong pressure to reduce spending in an EU where the CAP budget still represents over 40% of total EU expenditure (European Commission 2011).

Ultimately, society will decide whether any further agricultural intensification is delivered without further environmental damage, whether public subsidy is used to deliver public goods and services, and whether a rich and thriving bird community remains a focus amongst environmental objectives. However, 50 years after Rachel Carson's *Silent Spring*, it is salutary to realize that the European Bird Census Council's 2012 update of its European farmland bird population indicator reveals that, in aggregate, farmland bird populations have lost 300 million individuals across Europe since 1980. To put this into perspective, it amounts to just over ten times the total British population of all 19 species included in the UK farmland bird indicator.

What might Rachel Carson make of these statistics? Given that bird populations are well established as robust indicators of wider biodiversity and ecosystem

health in agricultural systems (Gregory et al. 2005), it would seem important to continue to consider bird populations as a critical indicator of success or failure in achieving long-term sustainability of agricultural production (Wilson et al. 2010).

5.4.2 Scientific challenges

If we assume that bird conservation remains a core objective of AES and other forms of wildlife-friendly farming, then what questions do we need to tackle to ensure that conservation interventions are as effective as possible in halting and reversing losses of birds and other biodiversity?

One critical challenge is to close the gap between the scale at which the policy objectives underpinning AES are set (reversing national biodiversity losses) and the scale at which interventions are made and evaluations are undertaken (typically the field or farm scale). This point was made recently by Kleijn et al. (2011) and is a very important one. More studies are needed that build on examples, such as Perkins et al. (2011), which evaluate AES effectiveness at the farm scale and use the data to estimate the scale of AES delivery needed to deliver specified biodiversity targets nationally. To make such studies more robust, research should seek to understand the impacts of conservation initiatives on the demography and population dynamics of target species, as well as simple numerical responses.

A key element of scaling-up studies of AES effectiveness will be to understand the impact of landscape context on the effectiveness of AES delivered at the farm scale. There is growing evidence from studies of a variety of wildlife-friendly farming interventions, including organic farming (Gabriel et al. 2010; Rundlof et al. 2010; Smith et al. 2010; Chapter 6, this volume) and AES (Dallimer et al. 2010; Batary et al. 2011; Concepcion et al. 2012; Chapter 8, this volume), both that biodiversity benefits at the farm scale can depend on landscape context, and that interventions may have important benefits in landscapes beyond the farms on which they are implemented. Studies focused on individual species or assemblages of birds of conservation concern could be very influential in helping us to understand whether, for example, modifications to policy instruments to encourage farmers to work cooperatively to deliver additional landscape-scale benefits might be cost-effective. Such studies are especially important at a time of pressure on agri-environment budgets, which can result in too great a focus amongst policy makers on ensuring that schemes are kept 'simple' (e.g. Scottish Government 2010). As Armsworth et al. (2012) found in developing field-parameterized ecological–economic models of UK livestock farms, the administrative complexity that may accompany AES that offer evidence-based management options and spatial discrimination in payment rates and management options, is a price that must be paid if biodiversity benefits are to be maintained.

A further reason to take a landscape-scale perspective when considering the effectiveness of AES is to consider their potential role in contributing to what Lawton et al. (2010) described as the need for a 'step-change in nature conservation'—delivering a resilient and coherent ecological network of sites for nature conservation. Lawton et al.'s review is focused on England, but its conclusions resonate wherever landscapes are dominated by agricultural production. They note that sites designated, protected, and managed for nature conservation are often smaller and less well managed than they should be, and are increasingly isolated and degraded by the increasingly intensive management of the surrounding agricultural matrix. They argue that an ecological network fit to contribute to global targets to halt biodiversity loss will entail protecting more sites, increasing their size, improving their management, but also increasing their connectivity, and buffering them by improving the management and ecological diversity of the surrounding countryside. Management options offered by AES could make an important contribution to these last two objectives. For example, reviews by Vickery et al. (2004) and Donald and Evans (2006) of England's Environmental Stewardship Scheme identified a range of management measures, and the resources they provide, that could contribute to use of intensive agricultural landscapes by a wider variety of species. The effectiveness of such measures in increasing the resilience and coherence of ecological networks, as envisaged by Lawton et al. (2010), is likely to be all the greater if AES implementation advances from a farm-by-farm approach to more integrated delivery across catchments and landscapes (Wilson et al. 2009).

Benton et al. (2003) argued that loss of ecological heterogeneity at multiple spatial and temporal scales is a universal consequence of agricultural intensification, and that policy frameworks and management interventions should aim to re-create heterogeneity as a key principle of biodiversity restoration in agricultural systems. Since then, evidence has accumulated in support of this idea, both for bird and wider biodiversity conservation on farmland (e.g. Wilson et al. 2005; Schaub et al. 2010; Batary et al. 2011; Verhulst et al. 2011). Today, the argument is often expressed as whether

land 'sparing' or 'sharing' is the best approach to delivering nature conservation or other ecosystem services on land managed primarily for food production (Balmford et al. 2005). Sharing is usually taken to mean adopting less intensive farming systems to aid nature conservation or other ecosystem service objectives, but producing food less efficiently per unit area as a consequence, and therefore potentially converting more land to agricultural use. AES, low-intensity, high nature value farming systems, and organic farming could all be seen as examples from this end of the spectrum (e.g. Hodgson et al. 2010; Gabriel et al. 2013). Sparing involves concentrating food production into certain areas of land while making other areas available for delivery of other services. It is typically promoted as the optimum solution for nature conservation in contexts where habitats that might become converted to agriculture are of very high conservation value, such as tropical forests (Phalan et al. 2011). In essence, there is a continuum, with decisions being made about the spatial grain at which land is devoted to production or other purposes; an AES approach might be seen as land sharing when viewed from a regional or national perspective, but land sparing when viewed from the field or farm perspective. The optimum outcome can be determined if one knows the relative productivity of land for a given ecosystem service—whether devoted to that service or shared with others—and can value the range of different ecosystem services of interest (Firbank et al. 2011). A very neat empirical example of this is provided by Hodgson et al. (2010), who estimated in UK agricultural systems that if the productivity of organically farmed land fell below 87% of that of conventionally managed land, then the additional area of semi-natural habitat lost to farming to maintain production would outweigh the biodiversity benefits (measured as butterfly densities) on organic farmland. However, if the available non-farmed habitats for butterflies were simply uncultivated field margins, rather than patches of semi-natural habitat (implying a more intensive production landscape), then organic farming produced greater net benefits for butterflies even when yields were reduced to 35% of those on conventional farmland. In the context of this growing interest in the multiple services desired from agro-ecosystems, an essential research aim for farmland bird conservation is therefore to understand to what extent AES management for birds is providing other benefits and, equally, to what extent management designed to deliver other ecosystem services such as water quality, flood protection, carbon sequestration and storage, pollination, and soil health, on agricultural land can offer opportunities for bird conservation. For example, hydrological management ranging from management of in-field water tables to landscape-scale wetland creation to assist flood management and reduce diffuse pollution, may generate important opportunities to provide nesting and foraging opportunities for birds on farmland (Bradbury and Kirby 2006; Rhymer et al. 2010). However, reciprocal benefits cannot simply be assumed (Macfadyen et al. 2012), and more quantitative studies assessing the synergies and trade-offs, for birds, other biodiversity, and other ecosystem services, of agri-environment interventions, are needed. Nonetheless, perhaps Bateman et al. (2013) offer cause for optimism. This study combined spatially explicit ecosystem service modelling with valuation methods to predict that, while production-focused decision-making reduces overall ecosystem service values in agricultural landscapes, large net increases can be achieved if all ecosystem services are considered together in targeted planning, and that such an approach also predicts biodiversity conservation benefits.

5.5 Conclusions

Where management options are: (i) based on a sound understanding of causes of species decline; (ii) well-targeted to the right landscapes, farms, fields, and parts of fields; (iii) backed by good advisory support to farmers and evaluation studies to allow adaptive improvement of management over time; and (iv) delivered to a high proportion of the population of the target species, then AES can be spectacularly successful in reversing declines of farmland birds. The main limitation of these schemes on UK farmland is simply that implementation of the relevant management options is not at a sufficient scale or density to reverse decline of any but the rarest and most range-restricted species. More studies are needed which use evaluations of bird responses to AES implementation to quantify the levels of provision needed to reverse national population declines. More studies are also needed of the extent to which the benefits of AES management are landscape-context dependent, and whether policy measures to encourage farmers to work co-operatively across catchments and landscapes would increase the effectiveness of AES. A landscape-scale perspective is also needed to improve our understanding of the extent to which AES measures for bird conservation can contribute to building resilient and coherent ecological networks for nature conservation more generally. There is encouraging evidence that AES measures designed for bird conservation have many wider biodiversity benefits,

but the extent to which bird conservation measures in AES are compatible and efficient means that delivering other services from agro-ecosystems remains uncertain and more detailed quantitative studies are needed. Hydrological management, from the field scale to landscape-scale wetland creation, may provide particular opportunities for co-benefits. With farmland bird populations at their lowest ever recorded levels both in England and across Europe, Rachel Carson's call has not yet been heeded, and a post-industrial agriculture which recognizes the wider public benefits of agricultural management has not yet been achieved. Increased resources for AES within CAP budgets, or equivalent interventions, are essential, yet successive reforms of the CAP, including the latest in June 2013, have failed to achieve this. Nonetheless, the research efforts described above remain essential to ensure that whatever AES measures for bird conservation are resourced go on to deliver to their full potential as a key element of management of agricultural systems for public benefit.

A thing is right when it tends to preserve the integrity, stability and beauty of the biotic community. It is wrong when it tends otherwise – Aldo Leopold.

Acknowledgements

We thank Ruth Feber and David Macdonald for the invitation to contribute this chapter and for their support as editors, Cecilia Grundy for assistance with extracting and formatting references, and Allan Perkins for comments on drafts.

References

Anderson, B.J., Armsworth, P.R., Eigenbrod, F., Thomas, C.D., Gillings, S., Heinemeyer, H., Roy, D.B., and Gaston, K.J. (2009). Spatial covariance between biodiversity and other ecosystem service priorities. *Journal of Applied Ecology*, **46**, 888–896.

Armsworth, P.R., Acs, S., Dallimer, M., Gaston, K.J. Hanley, N., and Wilson, P. (2012). The cost of policy simplification in conservation incentive programs. *Ecology Letters*, **15**, 406–414.

Ausden, M. and Hirons, G.J.M. (2002). Grassland nature reserves for breeding wading birds in England and the implications for the ESA agri-environment scheme. *Biological Conservation*, **106** 279–291.

Baker, D.J., Freeman, S.N., Grice, P.V., and Siriwardena, G.M. (2012). Landscape-scale responses of birds to agri-environment management: a test of the English Environmental Stewardship Scheme. *Journal of Applied Ecology*, **49**, 871–882.

Balmford, A., Green, R.E., and Scharlemann, J.P.W. (2005). Sparing land for nature: exploring the potential impact of changes in agricultural yield on the area needed for crop production. *Global Change Biology*, **11**, 1594–1605.

Barnett, P.R., Whittingham, M.J., Bradbury, R.B., and Wilson, J.D. (2004). Use of unimproved and improved lowland grassland by wintering birds in the UK. *Agriculture, Ecosystems and Environment*, **102**, 49–60.

Batary, P., Baldi, A., Kleijn, D., and Tscharntke, T. (2011). Landscape-moderated biodiversity effects of agri-environmental management: a meta-analysis. *Proceedings of the Royal Society B-Biological Sciences*, **278**, 1894–1902.

Bateman, I.J., Harwood, A.R., Mace, G.M., et al. (2013). Bringing Ecosystem Services into Economic Decision-Making: Land Use in the United Kingdom. *Science*, **341**, 45–50.

Beddington, J. (2009). Food, energy, water and the climate: a perfect storm of global events? <http://www.bis.gov.uk/>, accessed 5 August 2012.

Benton, T.G., Vickery, J.A., and Wilson, J.D. (2003). Farmland biodiversity: is habitat heterogeneity the key? *Trends in Ecology and Evolution*, **18**, 182–188.

Bradbury, R.B., Hill, R.A., Mason, D.C., et al. (2005). Modelling relationships between birds and vegetation structure using airborne LiDAR data: a review with case studies from agricultural and woodland environments. *Ibis*, **147**, 442–452.

Bradbury, R.B. and Kirby, W.B. (2006). Farmland birds and resource protection in the UK: cross-cutting solutions for multi-functional farming? *Biological Conservation*, **129**, 530–542.

Bradbury, R.B., Kyrkos, A., Morris, A.J., Clark, S.C., Perkins, A.J., and Wilson, J.D. (2000). Habitat associations and breeding success of yellowhammers on lowland farmland. *Journal of Applied Ecology*, **37**, 789–805.

Bradbury, R.B., Stoate, C., and Tallowin, J.R.B. (2010). Lowland farmland bird conservation in the context of wider ecosystem service delivery. *Journal of Applied Ecology*, **47**, 986–993.

Bretagnolle, V., Villers, A., Denonfoux, L., Cornulier, T., Inchausti, P., and Badenhausser, I. (2011). Rapid recovery of a depleted population of Little Bustards *Tetrax tetrax* following provision of alfalfa through an agri-environment scheme. *Ibis*, **153**, 4–13.

Bright, J.A., Morris, A.J., Field, R.H., Cooke, A.I., Grice, P.V., Walker, L.K., Fern, J., and Peach, W.J. (2015). Higher-tier agri-environment scheme enhances breeding densities of some priority birds in England. *Agriculture, Ecosystems and Environment*, **203**, 69–79.

Buckwell, A. and Armstrong-Brown, S. (2004). Changes in farming and future prospects—technology and policy. *Ibis*, **146**, 14–21.

Butler, S.J., Bradbury, R.B., and Whittingham, M.J. (2005). Stubble height manipulation causes differential use of stubble fields by farmland birds. *Journal of Applied Ecology*, **42**, 469–476.

Carson, R. (1962). *Silent Spring*. Houghton Mifflin, Boston.

Chamberlain, D.E., Fuller, R.J., Bunce, R.G.H., Duckworth, J.C. and Shrubb, M. (2000). Changes in the abundance of farmland birds in relation to the timing of agricultural intensification in England and Wales. *Journal of Applied Ecology*, **37**, 771–788.

Concepcion, E.D., Diaz, M., Kleijn, D., et al. (2012). Interactive effects of landscape context constrain the effectiveness of local agri-environment management. *Journal of Applied Ecology*, **49**, 695–705.

Dallimer, M., Marini, L., Skinner, A.M.J., Hanley, N., Armsworth P.R., and Gaston, K.J. (2010). Agricultural land-use in the surrounding landscape affects moorland bird diversity. *Agriculture Ecosystems and Environment*, **139**, 578–583.

Davey, C.M., Vickery, J.A., Boatman, N.D. Chamberlain, D.E., Parry, H.R., and Siriwardena, G.M. (2010). Assessing the impact of Entry Level Stewardship on lowland farmland birds in England. *Ibis*, **152**, 459–474.

Donald, P.F. and Evans, A.D. (2006). Habitat connectivity and matrix restoration: the wider implications of agri-environment schemes. *Journal of Applied Ecology*, **43**, 209–218.

European Bird Census Council (2012). <http://www.ebcc.info>.

European Commission (2008). *Agriculture and Rural development. 'Health Check' of the Common Agricultural Policy*. <http://ec.europa.eu/>, accessed August 2012.

European Commission (2011). *CAP Post-2013: key graphs and figures*. <http://ec.europa.eu>, accessed August 2012.

Evans, A.D. and Green, R.E. (2007). An example of a two-tiered agri-environment scheme designed to deliver effectively the ecological requirements of both localised and widespread bird species in England. *Journal of Ornithology*, **148**, S279–S286.

Ewald, J.A., Aebischer, N.J., Richardson, S.M., Grice, P.V., and Cooke, A.I. (2010). The effect of agri-environment schemes on grey partridges at the farm level in England. *Agriculture Ecosystems and Environment*, **138**, 55–63.

FAO (2012) Food and Agriculture Organization of the United Nations. <http://www.fao.org/>.

Fewster, R.M., Buckland, S.T., Siriwardena, G.M., Baillie, S.R., and Wilson, J.D. (2000). Analysis of population trends for farmland birds using generalised additive models. *Ecology*, **81**, 1970–1984.

Firbank, L.G. (2009). It's not enough to develop agriculture that minimizes environmental impact. *International Journal of Agricultural Sustainability*, **7**, 1–2.

Firbank, L., Bradbury, R., McCracken, D., and Stoate, C. (2011). Enclosed Farmland. In *The UK National Ecosystem Assessment Technical Report*, pp. 197–240. UK National Ecosystem Assessment, UNEP-WCMC, Cambridge.

Foresight. The Future of Food and Farming (2011). *The Future of Food and Farming: Challenges and choices for global sustainability*. The Government Office for Science, London.

Fuentes-Montemayor, E., Goulson, D., and Park, K.J. (2011). Pipistrelle bats and their prey do not benefit from four widely applied agri-environment scheme prescriptions. *Biological Conservation*, **144**, 2233–2246.

Fuller, R.M. (1987). The changing extent and conservation interest of lowland grasslands in England and Wales: a review of grassland surveys 1930–1984. *Biological Conservation*, **40**, 281–300.

Gabriel, D., Sait, S.M., Hodgson, J.A., Schmutz, U., Kunin, W.E., and Benton, T.G. (2010). Scale matters: the impact of organic farming on biodiversity at different spatial scales. *Ecology Letters*, **13**, 858–869.

Gabriel, D., Sait, S.M., Kunin, W.E., and Benton, T.G. (2013). Food production vs. biodiversity: comparing organic and conventional agriculture. *Journal of Applied Ecology*, **50**, 355–364.

Gregory, R.D., van Strien, A., Vorisek, P., Meyling, A.W.G., Noble, D.G., Foppen, R.P.B., and Gibbons, D.W. (2005). Developing indicators for European birds. *Philosophical Transactions of the Royal Society B-Biological Sciences*, **360**, 269–288.

Grice, P., Evans, A., Osmond, J., and Brand-Hardy, R. (2004). Science into policy: the role of research in the development of a recovery plan for farmland birds in England. *Ibis*, **146**, 239–249.

Hodgson, J.A., Kunin, W.E., Thomas, C.D., Benton, T.G., and Gabriel, D. (2010). Comparing organic farming and land sparing: optimising (i) yield and (ii) butterfly populations at a landscape scale. *Ecology Letters*, **13**, 1358–1367.

Hole, D.G., Perkins, A.J., Wilson, J.D., Alexander, I.H., Grice, P.V., and Evans, A.D. (2005). Does organic farming benefit biodiversity? *Biological Conservation*, **122**, 113–130.

Hole, D.G., Whittingham, M.J., Bradbury, R.B., Anderson, G.Q.A., Lee, P.L.M., Wilson, J.D., and Krebs, J.R. (2002). Widespread local house sparrow extinctions. *Nature*, **418**, 931–932.

Kleijn, D., Rundlof, M., Scheper, J., Smith, H.G., and Tscharntke, T. (2011). Does conservation on farmland contribute to halting the biodiversity decline? *Trends in Ecology and Evolution*, **26**, 474–481.

Kleijn, D. and Sutherland, W.J. (2003). How effective are European agri-environment schemes in conserving and promoting biodiversity? *Journal of Applied Ecology*, **40**, 947–969.

Krebs, J.R., Wilson, J.D., Bradbury, R.B., and Siriwardena, G.M. (1999). The second silent spring? *Nature*, **400**, 611–612.

Lawton, J.H., Brotherton, P.N.M., Brown, V.K., et al. (2010). *Making Space for Nature: a review of England's wildlife sites and ecological network*. Report to Defra.

MacDonald, M.A., Cobbold, G., Mathews, F. et al. (2012a). Effects of agri-environment management for cirl buntings on other biodiversity. *Biodiversity and Conservation*, **21**, 1477–1492.

MacDonald, M.A., Maniakowski, M., Cobbold, G., Grice, P.V., and Anderson, G.Q.A. (2012b). Effects of agri-environment management for stone curlews on other biodiversity. *Biological Conservation*, **148**, 134–145.

Macdonald, D.W. and Johnson, P.J. (1995). The relationship between bird distribution and the botanical and structural characteristics of hedges. *Journal of Applied Ecology*, **32**, 492–505.

Macfadyen, S., Cunningham, S.A., Costamanga, A.C., and Schellhorn, N.A. (2012). Managing ecosystem services and biodiversity conservation in agricultural landscapes: are the solutions the same? *Journal of Applied Ecology*, **49**, 690–694.

MacIntyre, B.D., Herren, H.R., Wakhungu, J., and Watson, R.T., eds. (2009). *Agriculture at a Crossroads: International Assessment of Agricultural Science and Technology for Development Global Report*. Island Press, Washington, D.C.

Marren, P. (2002). *Nature Conservation*. Second ed. Harper Collins, London.

Millennium Ecosystem Assessment (2005). *Ecosystems and Human Well-being: Biodiversity Synthesis*. World Resources Institute, Washington, D.C.

Moorcroft, D., Whittingham, M.J., Bradbury, R.B. and Wilson, J.D. (2002). The selection of stubble fields by wintering granivorous birds reflects vegetation cover and food abundance. *Journal of Applied Ecology*, **39**, 535–547.

Moorcroft, D., Wilson, J.D., and Bradbury, R.B. (2006). The diet of nestling linnets *Carduelis cannabina* on lowland farmland before and after agricultural intensification. *Bird Study*, **53**, 153–162.

Morris, A.J., Bradbury, R.B., and Wilson, J.D. (2002). Determinants of patch selection by yellowhammers *Emberiza citrinella* foraging in cereal crops. *Aspects of Applied Biology*, **67**, 43–50.

Morris, A.J., Whittingham, M.J., Bradbury, R.B., Wilson, J.D., Kyrkos, A., Buckingham, D.L., and Evans, A.D. (2001). Foraging habitat selection by Yellowhammers (*Emberiza citrinella*) in agriculturally contrasting regions in southern England. *Biological Conservation*, **98**, 197–210.

Newton, I. (1986). *The Sparrowhawk*. Poyser, Calton, UK.

Newton, I. (2004). The recent declines of farmland bird populations in Britain: an appraisal of causal factors and conservation actions. *Ibis*, **146**, 579–600.

Norris, K. (2012). Biodiversity in the context of ecosystem services: the applied need for systems approaches. *Philosophical Transactions of the Royal Society B-Biological Sciences*, **367**, 191–199.

O'Brien, M. and Wilson, J.D. (2011). Population changes of breeding waders on farmland in relation to agri-environment management. *Bird Study*, **58**, 399–408.

Perkins, A.J., Maggs, H.E., Watson, A., and Wilson, J.D. (2011). Adaptive management and targeting of agri-environment schemes does benefit biodiversity: a case study of the corn bunting *Emberiza calandra*. *Journal of Applied Ecology*, **48**, 514–522.

Perkins, A.J., Whittingham, M.J., Bradbury, R.B., Wilson, J.D., Morris, A.J., and Barnett, P.R. (2000). Habitat characteristics affecting use of lowland agricultural grassland by birds in winter. *Biological Conservation*, **95**, 279–294.

Phalan, B., Onial, M., Blamford, A., and Green, R.E. (2011). Reconciling food production and biodiversity conservation: land sparing and sharing compared. *Science*, **333**, 1289–1291.

Policy Commission on the Future of Farming and Food (2002). *Farming and Food: A Sustainable Future*. Cabinet Office, London.

Potts, G.R. (1986). *The Partridge: Pesticides, Predation and Conservation*. Collins, London.

Power, A.G. (2010). Ecosystem services and agriculture: tradeoffs and synergies. *Philosophical Transactions of the Royal Society B-Biological Sciences*, **365**, 2959–2971.

Proffitt, F.M., Newton, I., Wilson, J.D., and Siriwardena, G.M. (2004). Bullfinch *Pyrrhula pyrrhula pileata* breeding ecology in lowland farmland and woodland: comparisons across time and habitat. *Ibis*, **146** (Suppl. 2), 78–86.

Pywell, R.F., Heard, M.S., Bradbury, R.B., Hinsley, S., Nowakowski, M., Walker, K.J., and Bullock, J.M. (2012). Wildlife-friendly farming benefits rare birds, bees and plants. *Biology Letters*, **8**, 772–775.

Rackham, O. (1986). *The History of the Countryside*. Dent & Sons, London.

Rhymer, C.M., Robinson, R.A., Smart, J., and Whittingham, M.J. (2010). Can ecosystem services be integrated with conservation? A case study of breeding waders on grassland. *Ibis*, **152**, 698–712.

Risely, K., Massimino, D., Jonston, A., Newton, S.E., Eaton, M.E., Musgrove, A.J., Noble, D.G., Procter, D., and Baillie, S.R. (2012). *The Breeding Bird Survey 2011*. BTO Reseach Report 624. British Trust for Ornithology, Thetford, UK.

Robertson, G.P. and Swinton, S.M. (2005). Reconciling agricultural productivity and environmental integrity: a grand challenge for agriculture. *Frontiers in Ecology and the Environment*, **3**, 38–46.

Robinson, R.A. and Sutherland, W.J. (2002). Post-war changes in arable farming and biodiversity in Great Britain. *Journal of Applied Ecology*, **39**, 157–176.

Royal Society (2009). *Reaping the Benefits: Science and Sustainable Intensification of Global Agriculture*. The Royal Society, London.

Rundlof, M, Edlund, M., and Smith, H.G. (2010). Organic farming at local and landscape scales benefits plant diversity. *Ecography*, **33**, 514–522.

Schaub, M., Martinez, N., Tagmann-Ioset, A., et al. (2010). Patches of bare ground as a staple commodity for declining ground-foraging insectivorous farmland birds. *PLOS ONE*, **5**, e13115.

Scottish Government (2010). *Mid-term Evaluation of Scotland Rural Development Programme*. Scottish Government, Edinburgh.

Siriwardena, G.M., Baillie, S.R., Buckland, S.T., Fewster, R.M., Marchant, J.H., and Wilson, J.D. (1998). Trends in the abundance of farmland birds: a quantitative comparison of smoothed Common Birds census indices. *Journal of Applied Ecology*, **35**, 24–43.

Siriwardena, G.M., Baillie, S.R., Crick, H.Q.P., Wilson, J.D., and Gates, S. (2000). The demography of lowland farmland birds. In N.J. Aebsicher, A.D. Evans, P.V. Grice, and J.A. Vickery, eds., *Ecology and Conservation of Lowland Farmland Birds*, pp. 117–133. British Ornithologists' Union, Tring, UK.

Smith, H.G., Danhardt, J., Lindstrom, A., and Rundlof, M. (2010). Consequences of organic farming and landscape heterogeneity for species richness and abundance of farmland birds. *Oecologia*, **162**, 1071–1079.

Sotherton, N.W. (1998). Land use changes and the decline of farmland wildlife: An appraisal of the set-aside approach. *Biological Conservation*, **83**, 259–268.

TEEB (2010). *The Economics of Ecosystems and Biodiversity: Mainstreaming the Economics of Nature: A synthesis of the approach, conclusions and recommendations of TEEB*. United Nations Environment Programme.

Tucker, G.M. (1992). Effects of agricultural practices on field use by invertebrate-feeding birds in winter. *Journal of Applied Ecology*, **29**, 779–790.

UK National Ecosystem Assessment (2011). *The UK National Ecosystem Assessment: Synthesis of Key Findings*. UNEP-WCMC, Cambridge.

Verhulst, J., Kleijn, D., and Berendse, F. (2007). Direct and indirect effects of the most widely implemented Dutch agri-environment schemes on breeding waders. *Journal of Applied Ecology*, **44**, 70–80.

Verhulst, J., Kleijn, D., Loonen, W., Berendse F., and Smit, C. (2011). Seasonal distribution of meadow birds in relation to in-field heterogeneity and management. *Agriculture, Ecosystems and Environment*, **142**, 161–166.

Vickery, J.A., Bradbury, R.B., Henderson, I.G., Eaton, M.A., and Grice, P.V. (2004). The role of agri-environment schemes and farm management practices in reversing the decline of farmland birds in England. *Biological Conservation*, **119**, 19–39.

Whittingham, M.J. (2007). Will agri-environment schemes deliver substantial biodiversity gain, and if not why not? *Journal of Applied Ecology*, **44**, 1–5.

Whittingham, M.J., Bradbury, R.B., Wilson, J.D., Morris, A.J., Perkins, A.J., and Siriwardena, G.M. (2001). Chaffinch *Fringilla coelebs* foraging patterns, nestling survival and territory density on lowland farmland. *Bird Study*, **48**, 257–270.

Whittingham, M.J., Devereux, C.L., Evans, A.D., and Bradbury, R.B. (2006). Altering perceived predation risk and food availability: management prescriptions to benefit farmland birds on stubble fields. *Journal of Applied Ecology*, **43**, 640–650.

Whittingham, M.J., Krebs, J.R., Swetnam, R.D., Vickery, J.A., Wilson, J.D., and Freckleton, R.P. (2007). Should conservation strategies consider spatial generality? Farmland birds show regional not national patterns of habitat association. *Ecology Letters*, **10**, 25–35.

Whittingham, M.J., Wilson, J.D., and Donald, P.F. (2003). Do habitat association models have any generality? Predicting skylark *Alauda arvensis* abundance in different regions of southern England. *Ecography*, **26**, 521–531.

Wilkinson, N.I., Wilson, J.D., and Anderson, G.Q.A. (2012). Agri-environment management for corncrake *Crex crex* delivers higher species richness and abundance across other taxonomic groups. *Agriculture Ecosystems* and *Environment*, **155**, 27–34.

Wilson, J.D., Evans, J., Browne, S.J., and King, J.R. (1997). Territory distribution and breeding success of skylarks *Alauda arvensis* on organic and intensive farmland in southern England. *Journal of Applied Ecology*, **34**, 1462–1478.

Wilson, J.D., Evans, A.D., and Grice, P.V. (2009). *Bird Conservation and Agriculture*. Cambridge University Press, Cambridge, UK.

Wilson, J.D., Evans, A.D., and Grice, P.V. (2010). Bird conservation and agriculture: a pivotal moment? *Ibis*, **152**, 176–179.

Wilson, J.D., Taylor, R., and Muirhead, L.B. (1996). Field use by farmland birds in winter: an analysis of field type preferences using resampling methods. *Bird Study*, **43**, 320–332.

Wilson, A., Vickery, J., and Pendlebury, C. (2007). Agri-environment schemes as a tool for reversing declining populations of grassland waders: Mixed benefits from Environmentally Sensitive Areas in England. *Biological Conservation*, **136**, 128–135.

Wilson, J.D., Whittingham, M.J., and Bradbury, R.B. (2005). Managing crop structure: a general approach to reversing the impact of agricultural intensification on birds? *Ibis*, **147**, 153–163.

CHAPTER 6

Does organic farming affect biodiversity?

Ruth E. Feber, Paul J. Johnson, Dan E. Chamberlain, Leslie G. Firbank,
Robert J. Fuller, Barbara Hart, Will Manley, Fiona Mathews,
Lisa R. Norton, Martin Townsend, and David W. Macdonald

Let us give Nature a chance; she knows her business better than we do.

Michel de Montaigne

6.1 Introduction

The second half of the twentieth century has seen enormous losses of farmland biodiversity in the UK. The consensus is that agricultural intensification is largely responsible (Chapter 1, this volume). For example, European countries with higher wheat yields tend to have more threatened or recently extinct arable weed species (Storkey et al. 2012). Similarly, plant species richness in six European countries has been shown to be negatively related to nitrogen input (Kleijn et al. 2009). For birds, population declines and range contractions are greatest in European countries which have the highest levels of production, with cereal yields indicating a suite of crop and livestock husbandry changes (Donald et al. 2001, 2006). These trends appear set to continue, given ever greater pressures on the land, exerted by the expansion of human populations and the need for food security, compounded by issues such as globalization of markets, biofuels, and climate change (Tilman et al. 2001). Non-cropped habitats on farmland, such as hedgerows, ponds, and patches of woodland, have also fared badly, with periods when they were aggressively removed to increase the cultivated area (Chapter 1, this volume). For example, in 1946, there were around 800 000 km of hedgerow in Britain, of which 1300–5500 km were destroyed annually between 1946 and 1970 (O'Connor et al. 1986). Since 1930, 7% of the UK's ancient woodland has been lost completely, while 38% has been replaced with plantations of non-native species (Spencer and Kirby 1992).

Intensification brings increases in the use of pesticides and fertilizers, changes in cropping patterns (such as a shift from spring sown to winter sown crops, and a reduction in the use of traditional crop rotations), land drainage, pasture improvement, and reduction or loss of quality of non-cropped habitats such as hedgerows (Fuller 1987; Vickery et al. 2001). While there is broad agreement that agricultural intensification has largely been responsible for the widespread declines of much farmland wildlife (the evidence for farmland birds is particularly strong: Chamberlain et al. 2000; Donald et al. 2001; Wilson et al. 2009), disentangling the underlying causes is challenging. Some studies have attempted this: for example, Geiger et al. (2010) measured 13 components of intensification and found that the use of insecticides and fungicides has had the most consistent negative effects on biodiversity. Chamberlain et al. (2000) speculated that there may have been a threshold for critical amounts of food and habitat, and that this threshold could explain the lag they observed between changes in land use and the effect on bird populations.

Agricultural production and biodiversity conservation place differing demands on the land, and have provoked much debate about farming 'systems'. The term 'system' has come to represent different philosophies about the long-term sustainability of food production and biodiversity conservation. The issue is not straightforwardly one of direct conflict between biodiversity and crop production—it is more subtle than

Wildlife Conservation on Farmland. Managing for Nature on Lowland Farms. Edited by David W. Macdonald and Ruth E. Feber.
© Oxford University Press 2015. Published 2015 by Oxford University Press.

that—but if an increasing percentage of primary production is channelled into crop production, then, all else being equal, biodiversity will suffer. Differences between 'land sharing' and 'land sparing' approaches have dominated much of the debate. At one end of the spectrum, land sparing advocates intensive management to maximize production in some areas while sparing others for the conservation of biodiversity. At the other end, land sharing implies the use of less intensive farming practice throughout much of the landscape, its philosophy being to encourage biodiversity as an integral part of, economically viable, farming (Green et al. 2005). Broadly speaking, agri-environment schemes (which offer financial incentives to farmers in the UK and Europe for restoring or managing habitat features for wildlife on farmland: Chapter 1, this volume) fall within this end of the spectrum, although there is much debate concerning their effectiveness e.g. Kleijn et al. (2006). Recently, the concept of sustainable intensification has gained momentum. Defined as a form of production wherein 'yields are increased without adverse environmental impact and without the cultivation of more land', it denotes an aspiration of what needs to be achieved (Garnet and Godfray 2012), and could be considered a useful framework for deciding which combinations of approaches might work best.

Organic farming is one form of low intensity farming which has attracted considerable attention. It relies on management using crop rotations (usually incorporating fertility-building grass leys), green manure, compost, and biological pest control. There is a greater emphasis on spring sowing of cereals (with associated overwinter stubbles). Fertilizers and pesticides are used, but they are strictly limited in type and amount compared with non-organic farming. Non-organic farming mainly uses manufactured fertilizers. These are not permitted on organic farms and animal manures are therefore important, hence many organic enterprises are mixed farms, having both arable and livestock. Certifying bodies hold lists of other products which may be applied, and prior permission may be necessary before some of these can be used. Organic farming is an internationally recognized system, with national standards for what can be labelled as 'organic' produce, based on the standards set by the International Federation of Organic Agriculture Movements (IFOAM), an umbrella organization established in 1972. In the UK, the Advisory Committee on Organic Standards (ACOS) sets the standards that organic farmers have to adhere to in order for organic status to be conferred. The certification is important—the increased cost of organic production has been largely offset for the farmer by the premium price that consumers have hitherto been willing to pay. The organic industry emphasizes that the system is 'holistic'—its standards taking animal welfare and the environment, as well as farm management *per se*, into account.

Organic farming and the 'extensification' that is part of its philosophy often comes at a cost to production. For example, Gabriel et al. (2010) found organic crop yields to be around 55% of those of non-organic farms in two areas of England. Yield differences between organic and non-organic farms depend on site characteristics (Seufert et al. 2012), but the generally lower yields from organic farming have led to doubts that it can make a meaningful contribution to resolving the wildlife/production conflict over large areas (Gabriel et al. 2010). How to balance areas of wildlife-friendly land versus productive land under different crop yield scenarios has been explored by Hodgson et al. (2010), who predicted that farming non-organically, while sparing land as nature reserves, would be better for butterflies than farming organically over the same area, as soon as the organic yield per hectare fell below approximately 90% of the non-organic yield.

The scale at which trade-offs should be measured is not straightforward, though. For example, what if increased local biodiversity is at the cost of displacing the impacts on biodiversity possibly to somewhere else where production is increased, with even more severe consequences? Furthermore, the landscape context is important. Some studies have demonstrated that the benefits to biodiversity from less intensive (usually organic) farms are greater in simple, less heterogeneous landscapes (e.g. Roschewitz et al. 2005; Rundlof and Smith 2006), while other work points to the benefits of concentrating support for lower intensity farms in areas that already retain higher levels of biodiversity (Kleijn et al. 2009); more analyses are needed across broad spatial scales and over the long term (Balmford et al. 2012).

An obvious approach for tackling questions about the effects of intensification is to make comparisons between farming systems which differ in their intensity, while controlling for landscape context. Organic farms therefore provide an opportunity for carrying out an observational experiment on farming intensification in the UK. If organic farms have fewer aspects typical of more intensive systems, is there any evidence that they benefit biodiversity? A further issue surrounding organic farming and biodiversity is whether forms of non-organic extensification can deliver any biodiversity benefits associated with organic management. There is a continuum of possibilities here that range from extensification of the entire farm area through, for example, low intensity grazing and various low input systems, to fine-scale land sparing. In the latter, intensive production over, say, 95% of the land area and a 5% sacrificial area where a farmer might create semi-natural habitat on less productive land would not necessarily support 'farmland' wildlife, but might be better for biodiversity than a land sharing approach, of which organic farming is arguably an example.

In the early 1990s, interest in the potential of organic farming for biodiversity conservation was growing. In 1994, WildCRU were invited to participate in a unique collaboration between the Research Councils. Biology, sociology, and economics were brought together to explore all aspects of organic farming systems, with the work based at HRH Prince Charles's organic farm in Gloucestershire (Cobb et al. 1999). WildCRU, together with colleagues from the Centre for Ecology and Hydrology (CEH), were responsible for quantifying biodiversity differences between organic and non-organic farms.

We decided to focus on two groups of invertebrates: butterflies and spiders. Butterflies have complex life cycles, comprising egg, larval, pupal and adult stages, and thus requiring larval foodplants (often species specific), a continuity of nectar sources for the adult stages during the spring and summer, and safe overwintering habitats. Their demands for specific microclimate, vegetation structure, and the co-occurrence of specific vegetation types suggest they can be useful indicators of some aspects of habitat quality on farmland (Erhardt and Thomas 1991). Butterflies respond rapidly to changes in plant communities (Feber et al. 1996) and, as nectar-feeders, are also likely to respond to the effects of farm management in similar ways to the wider group of wild pollinators, many of which are in severe decline (Potts et al. 2010). Butterflies are well-monitored in Britain and elsewhere, and could be considered as indicators of change in wider terrestrial invertebrates (Thomas 2005). The evidence for steep population declines in British butterflies is unequivocal: Fox et al. (2011) reported that 72% of butterfly species have decreased in abundance over 10 years and 54% have decreased in distribution at the UK level and, notably, the abundance of common butterflies had dropped by 24% over ten years. Loss of habitat due to urbanization, lack of woodland management, and intensification of farmland are all likely to have contributed to these declines. On farmland, key factors are likely to be loss of larval foodplants and adult nectar sources

through grassland improvement (Vickery et al. 2001), herbicide use and fertilizer application (Smart et al. 2000), and reduction in uncropped habitats (Feber and Smith 1995).

Our other chosen group, spiders, are beneficial predatory invertebrates and highly important biocontrol agents on farmland (Nyffeler and Sunderland 2003). Evidence from other studies suggested that they were likely to respond to different farmland features, such as architectural complexity of the vegetation (e.g. Baines et al. 1998; Chapter 3, this volume). Some aspects more common to organic management could plausibly increase the abundance of their prey—the use of manure for fertility, for example, and few or no inputs of pesticides (e.g. Haughton et al. 1999). Management of cropped, as well as uncropped, areas is likely to be important for spiders, many species of which can colonize crops rapidly and in great abundance when prey is plentiful. In arable areas, spiders will overwinter in uncropped habitats such as field margins and disperse from these areas into fields in the spring and summer (Lemke and Poehling 2002; Schmidt and Tscharntke 2005; Öberg and Ekbom 2006). Spiders are also affected by wider impacts such as landscape heterogeneity (Schmidt et al. 2005).

6.2 Impacts of organic farming on butterflies

To test whether butterflies differed in their abundance and species richness between the two farming systems, we studied 12 pairs of organic and non-organic farms across southern England. These farms were matched so as to eliminate geographic variation—the non-organic farms were neighbours of the organic farms and growing similar crops. Butterflies were monitored over three years on these farms (Feber et al. 2007). Recorders walked a measured fixed transect route which was divided into sections corresponding to crop and/or boundary type. Each transect included an area of organic and an area of non-organic farmland (transect lengths within each farm pair were broadly similar). For each section, all butterflies seen were recorded; those seen associated with the crop edge were recorded separately from those associated with the uncropped field boundary. Details of crop, boundary type, and vegetation were recorded for each section of the transect routes (see Feber et al. 2007 for details).

Over the three years, we recorded 28 species of butterfly (Britain's butterfly fauna comprises around 56 breeding species). Averaged over each season as a

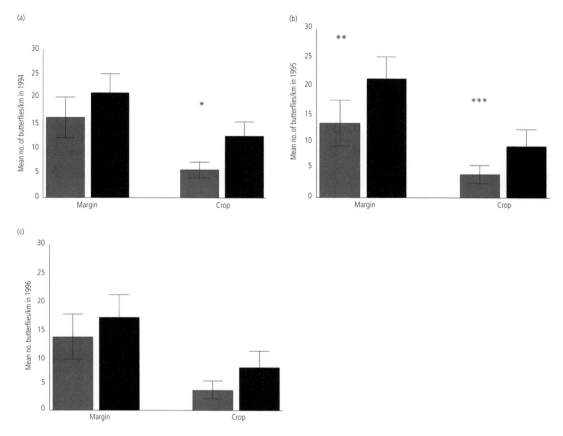

Figure 6.1 Effects of farming system (organic or non-organic) and habitat (margin or crop) on the mean abundance of butterflies per kilometre per transect in (a) 1994, (b) 1995, and (c) 1996. Solid bars: organic farmland, shaded bars: non-organic farmland. Significance of system effect indicated by asterisks. From Feber et al. (2007).

whole, we found that butterfly abundance was higher on organic compared to non-organic farms in all three years of the study (Fig. 6.1a,b,c, Table 6.1). Species richness followed a similar, but less striking, pattern, with more butterfly species on organic farms in each year, significantly so in 1994[1]. Most species had higher abundances of individuals recorded on organic than on non-organic farmland in most years, significantly so in at least one year for three of the less mobile species: large skipper *Ochlodes sylvanus*[2], common blue

Polyommatus icarus[3], and meadow brown *Maniola jurtina*[4]. The large and small white butterflies (often known as 'cabbage whites', considered to be pest species of brassica crops) are typically more mobile and organic farming did not increase their abundance (Feber et al. 2007).

Butterflies were consistently more abundant on un-cropped margins compared with the cropped area (Fig. 6.1), and there was also a tendency for the system effect to be more marked on the cropped area

[1] $t = 7.48$, d.f. $= 7$, $P < 0.001$: Feber et al. 2007.
[2] 1994: Mean abundance per km (S.E. in parentheses): 3.2 (0.98) on organic, 1.5 (0.66) on non-organic. $F_{(1,7)} = 6.17$, $P = 0.042$.

[3] 1994: Mean abundance per km (S.E. in parentheses): 7.4 (2.33) on organic, 2.5 (0.98) on non-organic, $F_{(1,7)} = 6.95$, $P = 0.034$.
[4] 1995: Mean abundance per km (S.E. in parentheses): 15.5 (1.54) on organic, 9.1 (1.07) on non-organic, $F_{(1,7)} = 21.51$, $P = 0.001$.

Table 6.1 ANOVA summary of the effects of farming system (organic or conventional) and habitat (margin or crop) on the mean abundance of butterflies (per kilometre, per transect) between 1994 and 1996. Analyses were performed on log-transformed data. Asterisks indicate level of significance of effect (exact P values also given). n = number of pairs of farms. From: Feber et al. (2007).

FACTOR	YEAR					
	1994 (n = 8)		1995 (n = 10)		1996 (n = 5)	
	$F_{(1,7)}$	P	$F_{(1,9)}$	P	$F_{(1,4)}$	P
System	9.22	* 0.020	31.46	*** 0.001	9.51	* 0.037
Habitat	13.00	** 0.009	22.83	*** 0.001	8.28	* 0.045
System x Habitat (interaction term)	5.16	ns 0.057	6.40	* 0.032	0.68	ns 0.450

compared with the uncropped margin[5]. The effect of system was therefore likely to be the result of differences in management of both uncropped and cropped habitats.

What might cause these differences in butterfly abundance and species richness? In intensively farmed landscapes, loss of plant diversity in hedge bottoms and grasslands has diminished the abundance and types of larval foodplants (many butterfly species, in their larval stage, are restricted to one or two species of plant) and reduced the availability of nectar sources for foraging adults (Feber and Smith 1995). Organic farms in our study had larger hedgerows and more perennial field-edge plant communities than non-organic farms, providing shelter and increased protection from pesticide application, as well as enhanced food resources; via these mechanisms they may have offered better habitat quality for farmland butterflies. Organic farms also differed from non-organic farms in terms of cropping regime. Because the use of artificial fertilizers is prohibited within organic systems, grass–clover leys usually form an integral part of the rotation to restore and maintain fertility. This increased proportion of grassland within organic farms is likely to have benefits for butterflies. Larvae of meadow brown *M. jurtina* and gatekeeper *Pyronia tithonus*, for example, are grass-feeders, and these species require undisturbed swards in which to overwinter in their larval stage. It seems likely that the greater temporal stability resulting from the maintenance of grass leys, which are in place for two or more years, the presence of more perennial field edge plant communities, and an increase in spatial heterogeneity at a landscape level sustained a greater diversity and abundance of butterflies.

[5] Indicated by a statistical interaction between the main effects of system (organic vs non-organic) and location (margin versus crop)—the effect was statistically significant in both 1994 and 1995.

6.3 Impacts of organic farming on spiders

On three of the farm pairs, we also tested whether effects of farming system could be detected on spider abundance and species richness, by sampling surface-active spiders from winter wheat fields at each of three of the sites, using pitfall trapping (Feber et al. 1998). Twelve pitfall traps were placed in a grid formation in each of the 18 fields under study, and were left out for a week in May and again in June. A number of vegetation measures were also recorded in each field. The methodology is described in full in Feber et al. (1998).

We identified 56 spider species from 8609 individuals in our pitfall trap samples, with most species belonging to the family Linyphiidae ('money' spiders, typical of agricultural land in the UK) (e.g. Fig. 6.2a). The Lycosidae (wolf spiders) were also well represented by *Pardosa* and *Trochosa* spp. (e.g. Fig. 6.2b).

Both the number of spiders captured and the species richness of spider samples were higher in organic than non-organic winter wheat fields (Fig. 6.3a,b). There was a pronounced difference between organic and non-organic fields in terms of understorey vegetation (both grasses and forbs), which was substantially more abundant on organic fields, and organic winter wheat was less dense than non-organic (fewer crop plants per square metre). Our most consistent finding was that there was an increased abundance and species richness of spiders in our samples, with increasing abundance of understorey vegetation within the crop, both overall and within each system, within each sampling session. For spiders, these system effects may have been mediated by the increased structural complexity within the crops on organic forms, increasing the opportunities for web-builders, and enhancing the availability of prey (Baines et al. 1998). The absence of agrochemicals and more complex crop rotations may also affect spider communities. Birkhofer et al. (2008)

Figure 6.2 Spiders commonly occurring on arable farmland in our study included (a) members of the Linyphiidae ('money spiders') such as *Bathyphantes gracilis* and (b) members of the Lycosidae ('wolf spiders') such as *Pardosa prativaga*. Photographs © Evan Jones.

found that organically managed grass–clover leys had higher spider activity density[6], and intensively farmed fields have been shown to have fewer spider species,

_____

[6] Activity density describes the number of individuals per unit captured by pitfall traps in a defined period (14 days for the study cited).

and lower activity of Lycosidae, than bio-dynamic fields (Gluck and Ingrish 1990).

6.4 Wider impacts of organic farming on biodiversity

A range of other studies have suggested that organic farming could benefit a wide range of taxa. There is evidence of positive effects of organic relative to non-organic for plants (e.g. Hyvonen et al. 2003; Gabriel et al 2006; Rundlof et al. 2010; Batáry et al. 2013), invertebrates (e.g. Holzschuh et al. 2008, Kragten et al. 2011), small mammals (e.g. Fischer et al. 2011), birds (e.g. Chamberlain et al. 1999, Kragten and de Snoo 2008, Smith et al. 2010), and bats (e.g. Wickramasinghe et al. 2003; Wickramasinghe et al. 2004). Meta-analysis of published studies using a range of methodologies and spatial scales (Bengtsson et al 2005) suggests that organic farming is associated with increased species richness and abundance of plants, predatory invertebrates, and birds and Hole et al.'s (2005) meta-analysis also provided clear evidence for a beneficial effect of organic farming on biodiversity. Tuck et al. (2014) found that, on average, organic farming increased species richness by about 30%. The response of animal communities to organic farming may partly depend on the trophic level of a taxonomic group (Birkhofer et al. 2014) and functional groups may be affected in different ways by organic farming (Batáry et al. 2012).

A shortcoming of many studies on the impacts of organic farming on biodiversity (reviewed by Hole et al., 2005), is that they have been limited in sample size or geographical scale. Furthermore, a common approach has been to pair out non-crop habitat differences between organic and non-organic farms and focus on system (within-field) effects alone, which prevents disentangling of the impacts of habitat features and field management on biodiversity. For example, the extent to which differences in biodiversity might be attributed to differences in the quality or quantity of uncropped habitat between organic and non-organic farms, or the extent to which field management might be the cause, remained poorly understood. This is important because, from a policy point of view, a key question remains—can non-organic farmers deploy elements of organic farms (such as amount or quality of uncropped habitats) which will result in biodiversity levels similar to those on organic farms? Or are there features intrinsic to the organic system (such as no artificial agrochemical input) that cannot easily be transferred to, or replicated within a non-organic system? As well as lack of pesticide and synthetic fertilizer use,

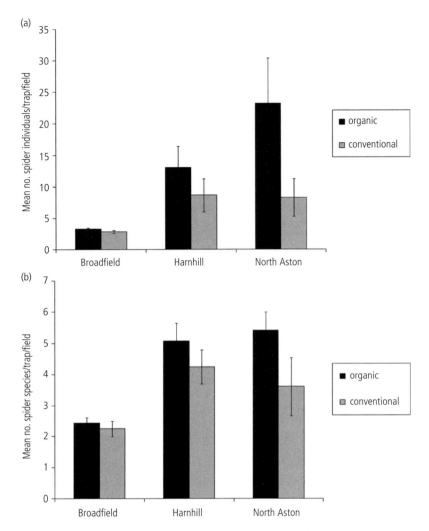

Figure 6.3 Mean number of (a) spider individuals and (b) spider species captured per trap per field on organic and non-organic (conventional) fields in June (see text for details). Re-drawn from Feber et al. (1998).

organic farms differ from non-organic farms in terms of a range of habitat variables and management practices (Norton et al. 2009), which vary in the extent to which they could be considered intrinsic to the system.

6.5 A large-scale multi-taxa study of organic farming

We used a multi-taxa, large-scale study of 89 pairs of farms (Fuller et al. 2005) to address these issues. Higher plants, spiders, ground (carabid) beetles, wintering birds, and bats were studied to represent a range

of trophic levels, niches, and ecological requirements. The project was of unprecedented scale for studies of this kind. Our project, a joint investigation with the British Trust for Ornithology (BTO), the Centre for Ecology and Hydrology (CEH), and the Royal Agricultural College (RAC), focused on the question: to what extent can biodiversity differences between organic and non-organic systems be related either to the amount and management of non-crop habitat, or to the differences in crop management?

The project had two primary aims. First, we aimed to assess the extent of differences in biodiversity between organic and non-organic farming systems over a wide geographical scale for a large sample of farms. Second, we asked if the extent to which observed biodiversity differences could be attributed to features intrinsic to organic systems, such as lack of artificial agrochemical use, or whether there were elements that could be used in non-organic systems to benefit biodiversity (non-cropped habitats for example). Figure 6.4 illustrates some hypothetical links between system (S), habitat (H), and biodiversity (B). Both the biodiversity and habitat axes are purely notional, and the habitat variables in our thought experiment vary independently of the biodiversity axis in question. In reality it is unlikely that any habitat can vary without there being

a causal link with some aspect of biodiversity. In the first three scenarios (a, b, and c), no system effect on biodiversity is expected, as there is no direct or indirect link between S and B. In scenario (d), we would observe a habitat effect, but system would not have any additional effect; in this case, H would appear to be a predictor of B, but only in a statistical model which does not take account of S. In (e), an effect of system is detected but habitat would not be a useful predictor in any model under this scenario, regardless of the inclusion of system. In (f), statistical models would detect an additive effect of system and habitat. Our paired design would detect the system effect in models not including habitat. But the system effect is not

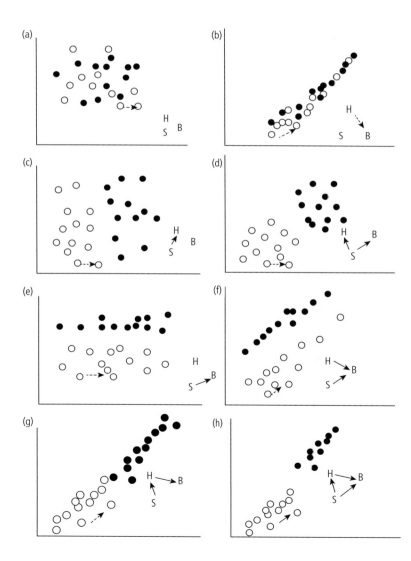

Figure 6.4 Schematic representation of a (not comprehensive) range of idealized scenarios ((a)–(h), see Section 6.5 for details) illustrating how different hypothesized causal links (solid arrows) between habitat (H), system (S) effects, and biodiversity (B) affect expected patterns. Arrows (inset) denote assumed causal links. Dotted arrow illustrates the likely effect of habitat enhancement on biodiversity for a non-organic farm under each scenario. Filled points represent organic farms, open points, non-organic farms.

explained by habitat—only in the final two scenarios is this wholly (g) or partially (h) true. Because the output of statistical models for scenario (g) could be close to those for (d), a full disentangling of the relationship between system, habitat, and biodiversity depends on understanding the *within-system* effects[7].

The possible links between system, hedgerow, and spiders provide a plausible example of how different statistical outputs are consistent with different patterns of cause and effect (bearing in mind the usual caveat for any associative model that correlation does not indicate causation). Consider the B axis as representing spider diversity while the H axis indicates a metric of increasing hedgerow complexity (it could be botanical or structural, or both). If there is a causal link between hedgerow and spiders, while system is irrelevant, we expect to observe scenario (b). In scenario (c), we visualize a situation where organic farmers have higher quality hedgerows but spider diversity does not respond to that quality. If organic farmers have higher quality hedgerows and, for reasons unconnected with hedgerow (lower pesticide impact, for example), also more diverse spiders, we expect to see pattern (d). A system effect on spiders independent of any hedgerow effect is described in (e). In (f) both system and hedgerow have an effect on spiders, in the absence of any link between system and hedgerow. In the final two scenarios, spider diversity is higher on organic farms *because* organic farms have hedgerows that favour spiders, the final scenario also having an additional direct system effect.

6.5.1 Site selection and sampling methods

For the purposes of this study, we focused on farms that had cereal fields, and we sampled both autumn-sown, referred to as 'winter' cereals (commonly grown on non-organic farms), and spring-sown, referred to as 'spring' cereals (more often grown on organic farms). Autumn sowing has greatly increased over the past 60 years while spring sowing has decreased, having a number of impacts on the farm environment, notably a significant reduction in overwinter stubbles. The organic farms in the study were: (1) at least 30 ha in area, (2) not highly fragmented holdings (i.e. where organic fields were interspersed with non-organic fields), (3) not predominantly agro-forestry or horticultural,

and (4) growing the 'right' crops in the 'right' years (spring cereal in 2000, and winter cereal in 2002 and 2003). Virtually all suitable organic farms in England growing relevant crop types (winter-sown wheat and spring cereals) at the time of the study were examined. The organic farms were paired with non-organic farms using a procedure that was purely geographical and not based on any attributes of either system. The selected study farm pairs were widely distributed, but there was a cluster to the east and south of Bristol (Fig. 6.5).

Plants and invertebrates were sampled at the 'field' scale; 89 pairs of cereal fields ('target fields') were sampled over three years. Both the spring cereal and the winter cereal fields were approximately equally divided between recently converted (<5 years) and old organic (>5 years). Plants were recorded in plots in the field boundary, the crop edge, and within the field. Spiders and carabid beetles were sampled using pit-fall trapping, before and after harvest. Eighteen traps per field sampled the crop and uncropped boundary habitats, with nine traps in each habitat. All spiders and carabid beetles captured were identified to species

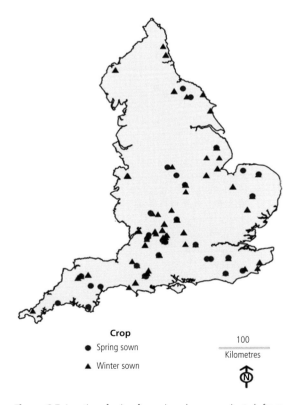

Crop
- Spring sown
▲ Winter sown

100
Kilometres

Figure 6.5 Location of pairs of organic and non-organic study farms.

---

[7] These scenarios are not, of course exhaustive; they do not, for example, allow for interaction between predictors. It is possible, for example, for the effect of habitat to be system specific.

level. Bats (Box 6.1) and birds were studied at the farm scale. Bird surveyors visited the target field monthly between October and February, recording birds seen during one walk around the perimeter of the target field, and one walk across it. Summer bat surveys were completed pre-harvest on 65 farm pairs between June and August in 2002 and 2003. Winter surveys of birds were carried out on 61 farm pairs on the target field and up to five adjacent fields. Birds were mapped on large-scale maps and individual records were subsequently allocated to habitat categories. The focus was on wintering, rather than breeding, birds because it was thought likely that in winter, flocking seed-eating birds in particular might be drawn into organic farms if these farms provided concentrations of seeds in the wider landscape.

6.5.2 How did organic farms differ from non-organic farms?

Our study quantified substantial differences between the two farming systems that might influence biodiversity (Fuller et al. 2005; Norton et al. 2009). Boundary density (km/ha) of all boundaries (including hedgerows) was higher on organic than non-organic farms (see Table 3 in Norton et al. 2009 for details). Hedgerows had fewer gaps and were larger on organic farms. They were also higher and wider than those surrounding non-organic fields. The number and diversity of trees and shrub species in the hedges were similar.

Organic farms had more grassland (as a proportion) compared to non-organic farms and organic target fields were smaller than their non-organic pairs (Norton et al. 2009). Organic and non-organic farms were similar sizes. They also had similar areas of woodland, permanent pasture, and numbers of ponds. There was no evidence that set-aside management differed (organic and non-organic set-aside were equally likely to be rotated or permanent).

Organic farmers tended to sow crops later than non-organic farmers and the crop rotations differed, with organic systems always including a grass ley as part of a cereal/vegetable rotation. Approximately a fifth of non-organic farms cropped continuously (set-aside excluded), but no organic farmers did this. Organic farms were more likely to include livestock (and a wider variety of types) and were more likely to use them on arable land, designated as such under the Arable Area Payments Scheme. Organic farmers cut their hedges less often and were more likely to use a traditional hedge management method (laying). More organic farms had agri-environment agreements (in addition

to the Organic Farming Scheme) than non-organic (Norton et al. 2009).

6.5.3 The effects of organic farming on diversity and abundance of plants, invertebrates, bats, and birds

Organic farming was mainly associated with positive effects on biodiversity, although there was substantial variation in the size of effects among taxonomic groups (Fuller et al. 2005). Species density and abundance were typically higher on organic farms, but patterns of diversity were less clear (Fuller et al. 2005). The direction of the effects was consistent, with all but one of the significant differences relating to higher diversity or higher abundance on organic farms compared to non-organic farms (Fig. 6.6). The largest and most consistent effects were for plants and the smallest for carabid (ground) beetles. Organic fields were estimated to hold 68–105% more plant species and 74–153% greater abundance of weeds (measured as cover) than non-organic fields, and cover of weeds was consistently higher at all distances into the crop (Fuller et al. 2005).

Chamberlain et al. (2010) looked in more detail at the bird effect illustrated in Figure 6.6. Of 16 species considered, none was more abundant on non-organic farms, and six out of 16 showed statistically significant effects. Variation in habitat abundance of the type detailed by Norton et al. (2009) was thought to be a plausible explanation for the effect, though there were no habitat effects for some species. Hedgerow density, the proportion of arable area at the farm scale (for stock

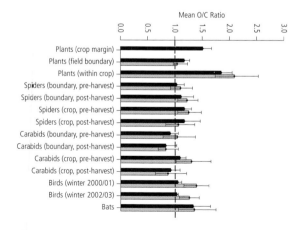

Figure 6.6 Effect of farming system on number of species (black bars) and abundance (grey bars) with confidence intervals. Dotted reference line at ratio = 1.0 indicates no system effect. No data for plant (crop margin) abundance.

Box 6.1 Bat activity on organic farms

Agricultural intensification has been linked with population declines of many bat species. One study of 24 farm pairs suggested bat diversity and abundance were lower on conventional than organic farms (Wickramasinghe et al. 2003), though most of these differences were for water-associated species. Whether any differences relate to farming system or to non-crop habitat remains unclear. We compared pre-harvest bat activity on 65 organic and non-organic farm pairs in the UK, and explored the effects of both farming system and habitat. Surveyors used methods comparable to those employed by the National Bat Monitoring Programme Field Survey (NBPFS). Activity of *Nyctalus leisleri*, *Nyctalus noctula*, and *Eptesicus serotinus* was identified using heterodyne bat detectors tuned to 25 kHz. Bat passes and feeding calls were counted for each 125 m transect section, at the end of which the detector was retuned to 50 kHz and numbers of *Pipistrellus* passes and feeding buzzes were counted for 1 min. Importantly, habitat was sampled in proportion to its availability in the landscape; this meant that any differences in abundance of features such as hedgerows was reflected in the sampling regime. The majority of bat activity was by *Pipistrellus* spp., and, for simplicity, we therefore consider all bat species together. The rate of encounters was converted to that expected over 3 km, for comparability with the monitoring methodology of the NBPFS.

As with our analyses of spiders and carabids, we compared alternative models and ranked them in terms of their ability to explain the observed data. First, we explored the links between bat abundance and the complexity of the landscape, measured at three different spatial scales, 1 km^2, 9 km^2, and 25 km^2, and asked whether the effect of farming system differed according to landscape type. The amount of land classified as 'open water' was also considered *a priori* as a likely predictor of bat activity. However, almost all values were well below 1% at all spatial scales and this variable may not reflect the availability of water features relevant for bats. Initial screening suggested that models which included this land class could not be distinguished from models excluding it.

As expected, there was strong evidence to link landscape complexity with bat activity: the model that included only farming system, pair ID, and year were supported much less than those that included a landscape variable. The model-averaged parameter estimate for the system variable across this suite of models was biased on the positive side of zero, but included zero, reflecting collinearity between system and landscape attributes and inflated standard errors of parameter estimates.

The organic farms tended to be found in landscapes with less arable cropping compared with non-organic farms (a local rather than large-scale effect given the paired design). The mean differences in arable crop abundances (organic minus non-organic) were −7.87% (CI—12.76 to −2.97) at the 1 km^2 scale, −3.54% (CI −6.68 to −0.41%) at the 9 km^2 scale, and −1.09% (CI −3.82 to 1.63) at the 25 km^2 scale. Although models including interactions between landscape complexity and farming system could not be ruled out as implausible, the effect sizes argue against there being any biologically important modification of landscape effects by farming system, or vice versa. The effects of landscape and system are illustrated in Box Fig. 6.1.1.

We went on to explore which aspects of farm structure affected bat activity, in addition to the landscape effect (or whether they were confounded with it, or with system). We also considered if any of the farming system difference not attributable to landscape could be explained by these variables. The potential predictors included were: number of ponds on the farm, a binary variable indicating whether stock was kept on the farm, the proportion of each transect where hedgerow was encountered, the proportion of each transect where water was encountered, the proportion of each transect where stock was encountered, the number of non-cropped habitats present on the farm, the proportion of woodland on the farm overall, and the proportion of pasture on the farm overall.

The predictors featuring in the most supported models were ponds, stock, and hedgerow. The presence of stock on the farm is clearly highest weighted by this metric (w (weight) = 0.53). Those for ponds and hedge were w = 0.26 and w = 0.09 respectively. Bat activity was higher on farms with stock, and the effect was more marked on organic farms (Box Fig. 6.1.2). The presence of stock was strongly associated with organic systems; while few organic farms surveyed contained no stock (9.2%) a substantial number of non-organic farms were stockless (43.1%). Hence models including the stock effect perform similarly to those with system alone.

The pond and hedge effects may also have been influential. As with livestock, the effect of both these features depended on the farming system. Bat activity increased with the number of ponds on organic farms, but not on non-organic farms, where the slope is negative (Box Fig. 6.1.3); for hedgerows the effect on organic farms was positive, whereas it was consistent with zero on non-organic farms (Box Fig. 6.1.4). Similar results were obtained regardless of whether an outlying farm was included in the

continued

Box 6.1 *Continued*

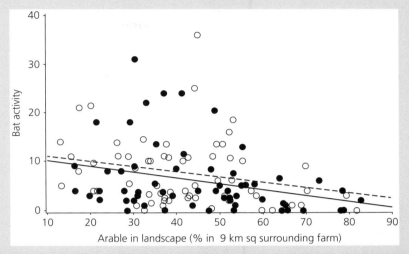

Box Figure 6.1.1 The effect of landscape structure and system on bat activity. Organic farms: open circles, non-organic: solid circles. Least squares regression lines fitted separately for each farm. Dotted line: organic, solid line: non-organic. Bat activity is number of passes expressed per 3 km.

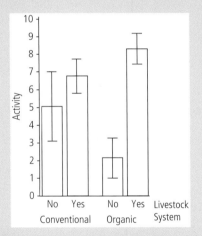

Box Figure 6.1.2 The association between the presence of stock on a farm and mean bat activity (+ /– SE).

analyses. System and landscape were each confounded with both habitats to some extent: organic farms tended to have more ponds than non-organic (mean = 2.70 SE = 0.43) compared with non-organic (mean = 2.50, SE = 0.41), and a higher density of hedgerows (see main text).

Overall, bat activity levels were low on all farms, and the differences in bat activity seen between non-organic and organic farms were modest. Our work, which sampled habitats in proportion to their availability, shows that much of the British landscape, and particularly that focused on cereal production, therefore offers rather unfavourable habitat for bats regardless of farming system. The mosaic of habitat surrounding the organic and non-organic farms differed, with the broader habitat around organic farms being more complex: some of the apparent benefits we attribute to organic farming could therefore be due to off-farm influences. On-farm features favourable to bats, particularly the presence of livestock, ponds, and hedgerow density were also, to some extent, confounded with farming system, as all were more common on organic than non-organic farms. Nevertheless, our data suggest that the 'quality' of these features in relation to bats was greater on organic than conventional systems. Organic farming therefore does not offer a simple panacea to the decline in bat populations, but the transfer of hedgerow and waterway management techniques to conventional farming, and an overall increase in the number of 'mixed' rather than purely arable enterprises would be likely to yield benefits.

continued

Box 6.1 *Continued*

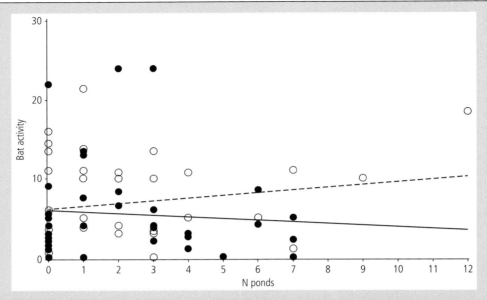

Box Figure 6.1.3 The association between the presence of ponds on a farm and mean bat activity. Organic farms: open circles, non-organic: solid circles. Dotted line: organic, solid line: non-organic.

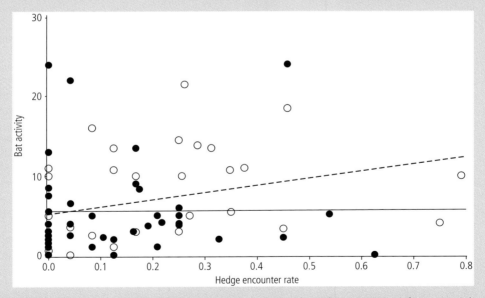

Box Figure 6.1.4 The association between transect hedgerow encounter rate and mean bat activity. Organic farms: open circles, non-organic: solid circles. Dotted line: organic, solid line: non-organic.

dove *Columba oenas* and jackdaw *Corvus monedula*), and the grass:arable ratio at the landscape scale (for woodpigeon *Columba palumbus* and jackdaw), were influential.

6.5.4 Local and landscape impacts on invertebrate groups

Here, we investigate in more detail the impacts of farming system and surrounding landscape on two of the taxonomic groups in this study: spiders and carabid (ground) beetles (Feber et al. in press). From results of other studies of the ecology of these two groups, one might predict complex and differing responses to organic farming. Hole et al. (2005) reviewed the evidence for an effect of organic management on biodiversity; they reported considerable evidence for a positive effect on both our target invertebrate groups. However, while the general pattern appeared to be one of higher abundance and diversity in organic fields, it was not universal across studies, and the caveats of small sample size and limited geographical scale remained.

Recent work has shown that landscape context influences the extent to which organic farms have impacts on biodiversity (Winqvist et al. 2012). For example, Schmidt et al. (2005) showed that the species richness of ground-dwelling spiders in crop fields was linked to large-scale landscape complexity irrespective of farming system; more spider species were recorded where the surrounding landscape had a higher proportionate area of non-cropped habitats. They attributed the effect to a higher availability of refuge and over-wintering habitats. The density of spiders responded instead to more local management practices including, in this case, organic farming.

Similar effects apply for carabid beetles. Local conditions such as vegetation and microclimate are important for carabids (Thiele 1977), but carabid assemblages are also strongly influenced by the quantity and arrangement of habitat elements at the scale of the landscape (Burel 1989; Millán de la Peña et al. 2003). Jonason et al. (2013) showed that, for carabid beetles, species richness (and weed seed predation by carabids) was influenced more by wider landscape context than local factors. Other taxonomic groups have shown similar effects: local plant species richness in arable fields is greatly influenced by processes operating at the landscape scale (Gabriel et al. 2005), and similar effects have been observed for arable weeds (Roschewitz et al. 2005). Butterfly abundance and species richness was significantly greater only on organic farms in homogeneous landscapes in Sweden

(Rundlof and Smith 2006; Rundlof et al. 2008), and Holzschuh et al. (2008) demonstrated that an increase in organic farming in the surrounding landscape increased bee species richness and bumblebee density at the local level.

Using data from the study of 89 pairs of organic and non-organic farms, we asked whether we could detect responses of spiders and carabid beetles to organic farming, and whether these differed between the two groups at local and landscape levels. We used data from the two crop types (spring and winter cereals) separately. Data from uncropped field margins and cropped field centres were analysed separately. Two responses were considered: abundance (the mean number of individuals in traps within the target field) and species richness of spiders and carabids in the trap samples. Spiders can be differentiated into two ecotypes based on their dispersal strategy. Hunting spiders, which include the wolf spiders (Lycosidae) (Fig. 6.2b), generally disperse by walking, and may be affected by local and landscape habitats differently compared with web-builders, which include the linyphiids (Fig. 6.2a). These spiders frequently disperse long distances by ballooning (Topping 1999; Schmidt and Tscharntke 2005). Some juvenile non web-builders also balloon, but the ecotypes are sufficiently distinct that we treated them separately.

For each response, we compared a series of candidate models using the model selection procedure of Burnham and Anderson (2002) and Anderson (2008). Relative model performance was assessed using the Akaike Information Criterion (AIC), adjusted for sample size (AIC_C). Competing models were ranked using Akaike weights. These are interpreted as the probability that the model in question is the 'best' model of the data set, among the series of models under consideration. The process accounts for model uncertainty, and the resulting estimates are 'unconditional'; they do not depend on any single model. Estimates of the relative importance of each predictor variable were provided by summing Akaike weights across all models in which that predictor variable occurred (Burnham and Anderson 2002). Our set of candidate models was based on competing and biologically plausible hypotheses. First we identified the best model from candidate models including system and two aspects of landscape structure. The percentage of arable land was used as a metric of landscape complexity, as for some previous studies. Models also included the percentage of woodland as a candidate predictor, given the likelihood that woodland will provide a diversity of refuge habitat for mobile species. We used the percentage of these land use categories at two different spatial scales (1 km² and 9 km²) around the focal farm.

Where a system effect was identified, either for responses where a system effect was previously reported (Fuller et al. 2005), or where a system effect was revealed here via its interaction with landscape (and therefore where a system effect may be confined to certain landscape types), we started with the model including system (and landscape where influential) and compared this model with models including habitat variables. For cropped area responses the habitat variable used was plant species diversity, based on previous work suggesting an influence (e.g. Schmidt et al. 2005), with plants providing both habitats and herbivorous prey species for predators. For field margin responses these were margin and hedge metrics. Interaction terms with system were included to allow for the possibility that the habitat effect differed with system.

We identified 131 species of spider from 29 377 individuals from the winter cereal sampling, and 7815 individuals of 77 species of spider from the spring-sown cereals. A total of 107 species of carabid were identified from 62 162 individuals from the winter-sown cereals, and 74 species from 19 313 individuals from the spring-sown cereals. More spiders and carabids were captured within the crop, compared with the uncropped field margin, particularly so for carabids. This was true for both spring and winter cereals and was consistent for both years where winter cereal was sampled. For example, in 2002, an average of 12.9 carabids (SE = 1.4) and 12.1 (SE = 1.10) spiders per trap were captured in the field margin (before harvest) compared with 34.0 (SE = 3.3) carabids and 14.8 (SE = 0.80) spiders in the cropped area. In 2003, the mean capture rates in the boundary were 20.5 (SE = 1.90) and 10.4 (SE = 0.92) respectively for carabids and spiders, compared with 35.3 (SE = 2.68) and 14.1 (SE = 1.22) in the cropped area. Trapping rates were between approximately 5–10 times higher before harvest compared with after harvest for both groups. The after-harvest trapping rates in the cropped area in 2002 were 8.54 (SE = 0.74) for carabids and 3.21 (SE = 0.41) for spiders, with rather similar rates observed in 2003.

6.5.4.1 Farming system effect varies with landscape type

We found that spiders in winter cereals were influenced by both farming system and landscape, but the patterns for hunting and web-building spiders were clearly different. For hunting spiders, farming system and landscape were both influential. There was very little evidence that the amount of woodland in the landscape was influential, but the amount of arable in the landscape clearly was (this predictor appeared in many of the best models), although its effect was not simple.

Spiders tended to be more abundant and species-rich on organic compared to non-organic farms; the overall farming system effect was particularly marked for hunting spiders in the cropped area before harvest (Fig. 6.7a), with our analysis suggesting a population average of 77% more individuals and 36% more hunting spider species on organic farms. In both crop and margin samples before harvest, the farming system effect depended on landscape type. The system difference—that is, that more hunting spiders were captured on organic compared to non-organic farms—was much more marked in non-arable landscapes (Fig. 6.7a).

This is clearly visible if we plot the response difference for each pair of farms against the extent of arable in the landscape. Almost all differences were positive for hunting spiders in the cropped area, indicating a positive organic effect, regardless of landscape (Fig. 6.7b). For hunting spiders in boundary samples (both abundance and species richness), the system effect was apparent only in complex landscapes (those comprised of less than approximately 40% arable; there was a clear system–landscape interaction); species richness followed a pattern similar to that of abundance, for hunting spiders (Fig. 6.7c,d).

After harvest, no effects of system or landscape were detected on hunting spiders in the cropped area. In the boundary, however, there were more hunting spiders on organic farms and, again, this effect was more marked in complex landscapes (Fig. 6.7e,f). In other words, there was an overall system effect (hunting spider abundance was higher on organic farms), while an interaction between system and landscape was again attributable to the system effect being more marked in complex landscapes.

In contrast to the hunting spiders, there was no evidence for any effect of farming system or landscape on web-building spiders in either before or after harvest samples (Table 6.2). The pattern in the cropped area before harvest (Fig. 6.8a,b) was therefore in marked contrast to that for the hunting spiders.

System effects on carabid beetles, where present, were not consistent in the two statistically significant findings (Fig. 6.9). First, there was a system effect on carabid abundance in two different directions. Before harvest, more carabids were captured in the cropped area on organic than on non-organic farms. Second, after harvest, fewer carabids were captured on organic

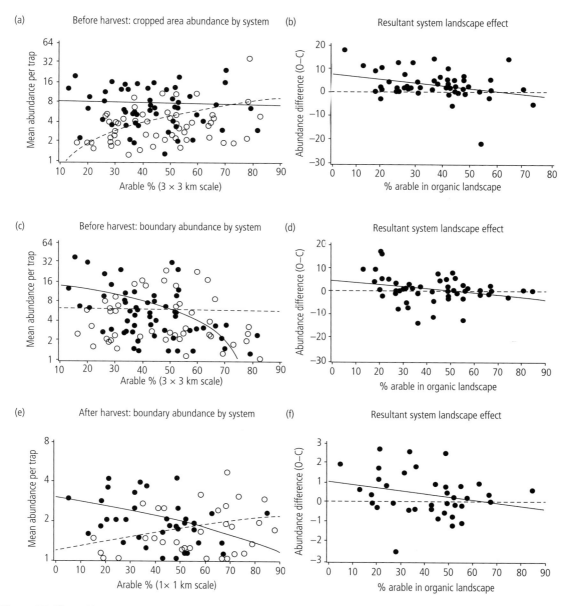

Figure 6.7 Effects of farming system and landscape on abundance of hunting spiders in winter cereals. Organic: solid lines, filled points; non-organic: dotted line, open points. Plots in left-hand panel (a,c,e): lines in plots are linear regressions plotted on log scale by system. Plots in right-hand panel (b,d,f): lines in plots are linear regression (solid line) and reference line (dotted at system difference = zero). Right-hand panel (b,d,f) extracts system effect from left-hand panel (a,c,e).

than non-organic boundaries, though the size of the effect was not large (Table 6.2). For carabids, there was no evidence for any landscape–system interactions, and the observed system effects were therefore not specific to any landscape type.

6.5.4.2 Does habitat explain the system effects?

The clearest system effect occurred for hunting spiders in the cropped area before harvest (Table 6.2). Can we explain this using the field-level habitat predictors? Models including field weed diversity were

Table 6.2 Model averaged parameter estimates for system and landscape effects in carabid beetle and spider models (winter cereal samples). The responses were square root transformed mean species counts per trap, and log mean numbers per trap. SR=Species richness. Numbers in bold are significant effects.

	Parameter estimate	SE	Effect size	LCI	UCI	Landscape effect	SE
Spiders before harvest							
Cropped area, Hunter abundance	0.578	0.093	**1.78**	**1.49**	**2.14**	0.220[§]	0.081
Cropped area, Hunter SR	0.31	0.05	**1.36**	**1.24**	**1.50**	0.067[§]	0.043
Boundary, Hunter abundance	0.085	0.114	1.09	0.87	1.36	−0.116[#]	0.108
Boundary, Hunter SR	0.039	0.066	1.04	0.91	1.18	−0.094[#]	0.057
Cropped area, Web-builder abundance	−0.066	0.096	0.94	0.78	1.13	none	
Cropped area, Web-builder SR	−0.043	0.056	0.96	0.86	1.07	none	
Boundary, Web-builder abundance	−0.023	0.088	0.98	0.82	1.16	none	
Boundary, Web-builder SR	−0.004	0.057	1.00	0.89	1.11	none	

[§] Main effects of arable land positive (more spiders and species in arable landscape), but with significant interaction indicating positive trend confined to conventional farms (Fig. 6.7a).
[#] Significant interaction, indicating negative trends in arable landscapes confined to organic farms (Fig. 6.7e).

	Parameter estimate	SE	Effect size	LCI	UCI	Landscape effect	SE
Carabids before harvest							
Cropped area, abundance	0.232	0.12	**1.26**	**1.00**	**1.60**	none	
Cropped area, species richness	0.045	0.059	1.05	0.93	1.17	−0.173[§]	**0.035**
Boundary, abundance	0.002	0.126	1.00	0.78	1.28	None	
Boundary, species richness	−0.012	0.007	0.99	0.97	1.00	−0.12[§]	0.004

[§] Fewer species in more arable landscapes.

	Parameter estimate	SE	Effect size	LCI	UCI	Landscape effect	SE
Spiders after harvest							
Cropped area, hunter abundance	0.118	0.143	1.13	0.85	1.49	none	
Cropped area, hunter SR	0.114	0.071	1.12	0.98	1.29	none	
Boundary, hunter abundance	0.165	0.08	**1.18**	**1.01**	**1.38**	0.113[§]	0.072
Boundary, hunter SR	0.156	0.066	**1.17**	**1.03**	**1.33**	−0.18[§]	0.075
Cropped area, web-builder abundance	−0.12	0.075	0.89	0.77	1.03	**−0.167**[#]	**0.069**
Cropped area, web-builder SR	−0.097	0.064	0.91	0.80	1.03	**−0.115**[#]	**0.047**
Boundary, web-builder abundance	0.008	0.08	1.01	0.86	1.18	None	
Boundary, web-builder SR	−0.095	0.053	0.91	0.82	1.01	**−0.092**[#]	**0.035**

[§] No overall effect of arable land extent, but significant interaction indicating a positive trend on conventional farms.
[#] Negative trends with increasing arable land in landscape.

	Parameter estimate	SE	Effect size	LCI	UCI	Landscape effect	SE
Carabids after harvest							
Cropped area, abundance	−0.103	0.145	0.90	0.68	1.20	None	
Cropped area, species richness	−0.066	0.072	0.94	0.81	1.08	none	
Boundary, abundance	−0.254	0.132	**0.78**	**0.60**	**1.00**	none	
Boundary, species richness	−0.177	0.082	**0.84**	**0.71**	**0.98**	none	

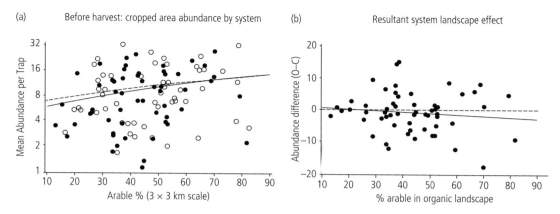

Figure 6.8 Effects of farming system and landscape on abundance of web-building spiders (winter cereal samples). Organic: solid lines, filled points; non-organic: dotted line, open points. Plot (a): line in plot is linear regression plotted on log scale by system. Plot (b): line in plot is linear regression (solid line) and reference line (dotted at system difference = zero). (b) extracts system effect from (a).

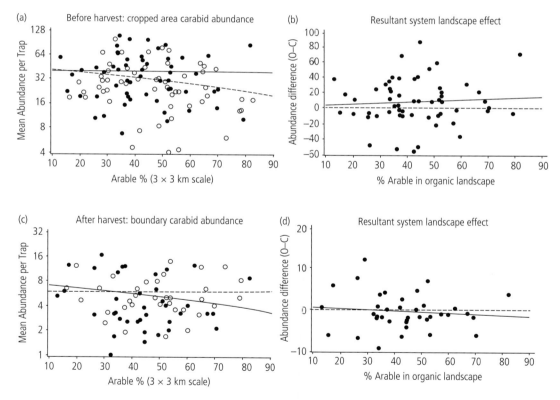

Figure 6.9 Effect of farming system and landscape on abundance of carabid beetles (winter cereal samples). Organic: solid lines, filled points; non-organic: dotted line, open points. Plots in left-hand panel (a,c): lines in plots are linear regressions plotted on log scale by system. Plots in right-hand panel (b,d): lines in plots are linear regression (solid line) and reference line (dotted at system difference = zero). Right-hand panel (b,d) extracts system effect from left-hand panel (a,c).

ranked higher than the 'baseline' system-arable model and the top models also included interaction between system and weed diversity. In other words, there was some evidence that weed diversity within the crop was influential. Separate modelling for organic and non-organic farms suggested field weed diversity was positively correlated with hunting spider abundance on non-organic farms, but not on organic (the parameter estimates were 0.10 SE = 0.049 and 0.02 (0.01) respectively). System and field weediness are highly correlated, i.e. organic fields are weedy, and the result is consistent with an upward trend where weed diversity is low (on non-organic farms), while no trend occurs over the range observed on organic farms (Fig. 6.10).

The weediness difference does not explain all of the system effect, as the overall system effect remains influential in models which include weed diversity. At the same time, the tendency for the within-pair spider abundance difference to be larger where the weed diversity difference is larger demonstrates that this variable contributes to the farming system difference, though this relationship is not strong (Fig. 6.11). A similar pattern was observed for hunting spider species richness in the cropped area (Fig. 6.12). There was no evidence that any non-cropped habitat—hedges, for example—had any effect.

There was little evidence from our results that differences in spider abundance and species richness between organic and non-organic farms were driven by differences in non-crop habitat. So the answer to one of our main questions—can the system difference be explained by differences in non-crop habitats?—is 'no'.

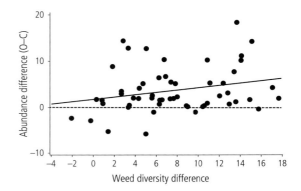

Figure 6.11 Farm pair difference in cropped area: hunting spider abundance and field weed diversity in winter cereal (r = 0.25, P = 0.06).

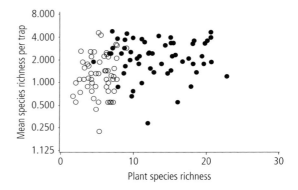

Figure 6.12 Hunting spider species richness and field weed diversity (winter cereal before harvest). Organic farms: solid circles, non-organic farms: open circles. Within system, there is no evidence for any association (r = 0.16, P = 0.22, r = 0.07, P = 0.61 for non-organic and organic respectively).

Both spiders and carabid beetles are surface-active predatory invertebrates which are abundant in arable fields, and play a potentially important role in crop pest control of arable crops (e.g. Lang et al. 1999). Their numbers and diversity are likely to indicate those of their prey, especially small invertebrates including mites, Collembola, and aphids. The observations reported here suggested that, for hunting spiders, the between-system difference may be related to weediness within the crop. This is likely to be due to the absence of herbicides in organic fields, resulting in larger weed populations, which provide structural complexity for the spiders.

We were unable to investigate the effects of agrochemical applications on spiders, as these were

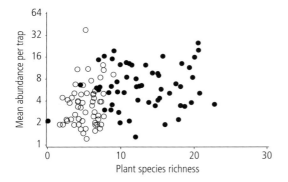

Figure 6.10 Hunting spider abundance and field weed diversity (winter wheat before harvest). Organic farms: solid circles, non-organic farms: open circles. Upward trend in non-organic farms, no trend among organic farms.

entirely confounded with management system. However, reduced abundance of predators with increased use of agrochemicals has been recorded; for example, Greig-Smith et al. (1991) found the densities of linyphiid spiders in areas receiving full pesticide inputs to be approximately 47% of those levels in reduced-input areas. Similar patterns were observed for staphylinid and coccinelid beetles (Vickerman 1992). Spiders and beetles on organic farms should not suffer from direct impacts of pesticide use, although they may be susceptible to non-organic farm practice occurring on neighbouring farms through spray drift or movement of individuals into pesticide treated areas.

Landscape context was influential in our analyses, with the impact of organic farming found to be greater in more complex landscapes. Schmidt et al. (2005) showed that high percentages of non-crop habitats in the landscape increased local species richness of spiders from 12 to 20 species, irrespective of local management and suggest that larger species pools are sustained in complex landscapes, where there is higher availability of refuge and overwintering habitats. Similarly Bergman et al.'s (2004) findings that butterfly assemblages are affected by the surrounding landscape at a large scale led to their conclusions that single-patch management might fail to maintain a diverse butterfly assemblage. Increasing some aspects of landscape complexity might be achieved by increasing the uptake of appropriate options within AES, which can have a positive impact on numbers of moths (Merckx et al. 2009) and numbers of bird species of conservation concern (Dallimer et al. 2010). It is clear that the interactions between invertebrate communities and local and landscape-scale factors are complex; our results for spiders, for example, contrast with those of Rundlof and Smith (2006), who showed that the beneficial effect of organic farming on butterflies was detectable only in intensively farmed homogeneous landscapes. Consideration of spatial scale (from local to landscape/region (Clough et al. 2005) and functional group (Clough et al. 2007)) are important for explaining patterns of invertebrate activity.

Dispersal ability of the species or group concerned is likely to be important. In our study, hunting spiders, which tend to have lower dispersal abilities, were affected more by organic farming and by surrounding landscape composition than web-builders, many species of which can disperse long distances. Dependence on overwintering sites and temporal and spatial closeness to prey populations may be key factors for the less dispersive species—increased landscape complexity may result in improved biocontrol by predator populations (Östman et al. 2001). Our biodiversity data support the conclusions reached by Gabriel et al. (2009), who developed models that show environmental factors associated with lower agricultural potential predispose conversion of farms to organic production, and that these factors naturally create regions with a high prevalence and intensity of organic farms. They argue that the most efficient conservation strategy for English farmland biodiversity would be to encourage the conversion of non-organic farms to organic production in regions where organic farming is already prevalent, and to maximize the intensity of production in areas where it is not. This raises the related question of whether the benefits of organic farming for biodiversity increase disproportionately as more of the landscape is managed organically.

Why did carabids in this study show a less consistent response to organic farming than spiders? At the farm scale, spatio-temporal dynamics of field and boundary overwintering carabid species show considerable variation both within fields and boundaries (Holland et al. 2005). Weedy areas are attractive to carabid beetles as sources of seed and invertebrate food, and conversion to organic farming may facilitate the movement of species more usually confined to field margins into cropped areas (Schröter and Irmler 2013); conversely, Holland et al. (2009) suggest that it may be possible for some types of vegetation cover to act as sink habitats for predatory invertebrates. Carabids may be particularly sensitive to elements of the landscape and less so to structural diversity at the field scale, using hedgerow networks and permanent elements of the landscape for dispersal and overwintering (Fournier and Loreau 2001). In our study, neighbouring farms were used and pairs within the same landscape were selected. However, the quality of the uncropped habitat did vary between farming systems, with larger and more sympathetically managed hedgerows occurring on organic farms. Pywell et al. (2005), in their study of beetles and spiders on arable land, concluded that hedgerows, rather than uncropped field margin habitats, provided the highest quality overwintering habitat for invertebrates, including staphylinid and carabid beetles, and spiders. Measures to conserve and enhance hedgerow habitats can be encouraged on both organic and non-organic farms, with important benefits for invertebrate and other farmland biodiversity. The long-term data set of Brooks et al. (2012) on carabid populations shows that, although carabids are in steep decline on their study area, the declines were less severe where hedgerows are managed for conservation. Lastly, our analysis using measures of species richness and abundance may have masked different impacts of agricultural management on the taxonomic and functional structure of carabid assemblages (Cole et al. 2012).

6.6 Conclusions and applied recommendations

The large sample size and wide range of data collected in our study provided a comprehensive assessment of differences in habitat and biodiversity between organic and non-organic arable farming systems throughout lowland England. We confirmed the findings of our earlier work, and that of others, that organic systems are associated with higher levels of biodiversity across a range of taxa (consistent with the review of Hole et al. (2005)). The most striking differences were for plants, where both species richness and cover of non-crop plants were consistently higher in organic fields (on average there were >80% more species within organic fields). We found that the difference in quality of non-crop habitats within farms, and the effects of surrounding landscape suggest that landscape attributes, non-cropped habitat, and crop management all affected biodiversity in ways that interacted and varied between taxa.

Habitats on organic and non-organic farms differed, and across a range of spatial scales. Organic farms had more grass and more non-crop plants within cereal crops, and more often had livestock. There was more hedgerow per unit area on organic farms, and organic hedges were cut less frequently, and were taller, wider, and had fewer gaps. While it might not be practical for non-organic farms to reduce field sizes (to those of organic farms) or to create more hedges, agri-environment scheme funding is available to support the improvement of hedge quality for both organic and non-organic farmers (Natural England 2009), with likely benefits for a range of taxa.

Management of non-crop habitat is not tied to any farming system. The tendency of organic farms to be more often mixed is more intimately bound up with the farming system. There are many ways in which the presence of stock can increase the potential habitats within a farm, for example through increasing the mix of arable and grassland, and therefore landscape heterogeneity, encouraging dense hedgerows as stock-proof barriers, and the input of dung, which encourages soil invertebrates and the bats and birds that feed upon them.

For spiders and carabids, the observed differences were largely confined to the cropped area (Table 6.2)

and were probably caused by factors associated with crop management, the direct effects of insecticide use being the most obvious candidate. The more diverse non-crop vegetation within the crop on organic farms, resulting from reduced inputs of herbicides and fertilizers, is also highly likely to be influential. There was some evidence for an effect of weed diversity on hunting spider abundance. The pattern for species richness fits that of scenario (d) in our Fig. 6.4, with no link between the habitat and biodiversity. The clearer effect for abundance, compared with species richness, points to insecticides as more likely to be responsible. The weed diversity abundance effect for non-organic farms, and not detectable in organic farms, is also consistent with a pesticide effect (as conventional pesticides are not used on organic farms). Better targeting of pesticide use within non-organic farms, and the possibility of field edge management with reduced pesticide inputs ('conservation headlands') could bring benefits; although little is known of how the scale over which such measures are put into place, and the length of time they are in place, affect the biodiversity benefits that accrue. We were unable to detect any effect of time since conversion on the system effect (Fuller et al. 2005), but scrutiny of community patterns may reveal more subtle side effects.

What are the main policy messages from our work? Our study showed that organic farming does have benefits for biodiversity across a range of taxa, although this study was not designed to consider relative agricultural productivity of the contrasting systems. Exactly which elements of organic farming can be transferred into non-organic systems for biodiversity enhancement remains uncertain. Some beneficial features of organic farms, in particular the quantity and quality of hedgerows, but also other features, such as increased plant diversity within crops (via, for example, conservation headlands), could be enhanced in all farming systems with appropriate support. Increasing the amount of grassland (and moves towards mixed farming systems) would be more difficult to achieve. We also found evidence of interactions among landscape and farming systems in their effects on species, highlighting the importance of developing strategies for managing farmland at the landscape scale for most effective conservation of biodiversity. The total area of organic farms relative to non-organic is small (currently c. 3.5% of English farmland is organic). If the benefit of organic farming is greater in some landscapes types, policy makers aiming to encourage conversion might need to consider regional targeting. At the moment, organic farms are concentrated in the more heterogeneous landscapes of south-western England where our results, for invertebrates at least, indicate the benefit of organic farming to biodiversity are likely to be greatest.

Acknowledgements

This work was funded by NERC and MAFF (now Defra). Martin Wolfe provided useful input at many stages of our work. We thank the farmers and land-owners who allowed us access to their farms and the Soil Association, especially Phil Stocker, and Organic Farmers and Growers for advice. We also thank Su Gough (project organization and design), Jim Dustow, Rick Goater, Dafydd Roberts, Anthony Taylor, James Bell, and Alison Haughton (data collection), Steve Gregory, Lawrence Bee, and Jonty Denton (invertebrate identification). Paul Johnson acknowledges the support of the Whitley Trust.

References

Anderson D.R. (2008). *Model Based Inference in the Life Sciences: a primer on evidence*. Springer, New York; London.

Baines, M., Hambler, C., Johnson, P.J., Macdonald, D.W., and Smith, H. (1998). The effects of arable field margin management on the abundance and species richness of Araneae (spiders). *Ecography*, **21**, 74–86.

Balmford, A., Green, R., and Phalan, B. (2012). What conservationists need to know about farming. *Proceedings of the Royal Society B*, **279**, 2714–2724.

Batáry, P., Sutcliffe, L., Dormann, C.F., and Tscharntke, T. (2013). Organic farming favours insect-pollinated over non-insect pollinated forbs in meadows and wheat fields. *PLOS ONE* 8(1): e54818. doi:10.1371/journal.pone.0054818.

Batáry, P., Holzschuh, A., Orci, K.M., Samu, F., and Tscharntke, T. (2012). Responses of plant, insect and spider biodiversity to local and landscape scale management intensity in cereal crops and grasslands. *Agriculture, Ecosystems and Environment*, **146**, 130–136.

Bengtsson, J., Ahnstrom J., and Weibull A.C. (2005). The effects of organic agriculture on biodiversity and abundance: a meta-analysis. *Journal of Applied Ecology*, **42**, 261–269.

Bergman, K.O., Askling, J., Ekberg, O., Ignell, H., Wahlman, H., and Milberg, P. (2004). Landscape effects on butterfly assemblages in an agricultural region. *Ecography*, **27**, 619–628.

Birkhofer, K., Fliessbach A., Wise D.H., and Scheu S. (2008). Generalist predators in organically and conventionally managed grass-clover fields: implications for conservation biological control. *Annals of Applied Biology*, **153**, 271–280.

Birkhofer, K., Ekroos, J., Corlett, B., and Smith, H.G. (2014). Winners and losers of organic cereal farming in animal communities across Central and Northern Europe. *Biological Conservation*, **175**, 25–33.

Brooks D.R., Bater, J.E., Clark, S.J., et al. (2012). Large carabid beetle declines in a United Kingdom monitoring network increases evidence for a widespread loss in insect biodiversity. *Journal of Applied Ecology*, **49**, 1009–1019.

Burel, F. (1989). Landscape structure effects on carabid beetles spatial patterns in Western France. *Landscape Ecology*, **2**, 215–226.

Burnham, K.P. and Anderson, **D.R.** (2002). *Model Selection and Multimodel Inference: a practical information-theoretic approach, 2nd ed*. Springer, New York, London.

Chamberlain, D.E., Fuller, R.J., Bunce, R.G.H., Duckworth, J.C., and Shrubb, M. (2000). Changes in the abundance of farmland birds in relation to the timing of agricultural intensification in England and Wales. *Journal of Applied Ecology*, **37**, 771–788.

Chamberlain, D.E., Joys, A., Johnson, P.J., Norton, L., Feber, R.E. and Fuller, R.J. (2010). Does organic farming benefit farmland birds in winter? *Biology Letters*, **6**, 82–84.

Chamberlain D.E., Wilson J.D., and Fuller R.J. (1999). A comparison of bird populations on organic and conventional farm systems in southern Britain. *Biological Conservation*, **88**, 307–320.

Clough, Y., Kruess, A., Kleijn, D., and Tscharntke, T. (2005). Spider diversity in cereal fields: comparing factors at local, landscape and regional scales. *Journal of Biogeography*, **32**, 2007–2014.

Clough, Y., Kruess, A., and Tscharntke, T. (2007). Organic versus non-organic arable farming systems: Functional grouping helps understand staphylinid response. *Agriculture, Ecosystems and Environment*, **118**, 285–290.

Cobb, R., Feber, R.E., Hopkins, A., et al. (1999). Integrating the environmental and economic consequences of converting to organic agriculture: evidence from a case study. *Land Use Policy*, **16**, 207–221.

Cole, L.J., Brocklehurst, S., Elston, D.A., and McCracken, D.I. (2012). Riparian field margins: can they enhance the functional structure of ground beetle (Coleoptera: Carabidae) assemblages in intensively managed grassland landscapes? *Journal of Applied Ecology*, **49**, 1384–1395.

Dallimer, M., Gaston, K.J., Skinner, A.M.J., Hanley, N., Acs, S., and Armsworth, P.R. (2010). Field-level bird abundances are enhanced by landscape-scale agri-environment scheme uptake. *Biology Letters*, **6**, 643–646.

Donald, P.F., Green, R.E., and Heath M.F. (2001). Agricultural intensification and the collapse of Europe's farmland bird populations. *Proceedings of the Royal Society B*, **268**, 25–29.

Donald, P.F., Sanderson, F.J., Burfield, I.J., and van Bommel, F.P.J. (2006). Further evidence of continent-wide impacts of agricultural intensification on European farmland birds, 1990–2000. *Agriculture, Ecosystems and Environment*, **116**, 189–196.

Erhardt, A. and Thomas, J.A. (1991). Lepidoptera as indicators of change in semi-natural grasslands of lowland and upland in Europe. In N.M. Collins, J. Thomas, eds., *The Conservation of Insects and their Habitats*, pp. 213–236. Academic Press, London.

Feber R.E., Bell, J., Johnson, P.J., Firbank, L.G., and Macdonald, D.W. (1998). The effects of organic farming on surface-active spider (Araneae) assemblages in wheat in southern England, UK. *Journal of Arachnology*, **26**, 190–202.

Feber, R.E., Johnson, P.J., Chamberlain, D.E. et al. (in press). *PLOS ONE*.

Feber, R.E., Johnson, P.J., Firbank, L.G., Hopkins, A., and Macdonald, D.W. (2007). A comparison of butterfly populations on organically and conventionally managed farmland. *Journal of Zoology*, **273**, 30–39.

Feber, R.E. and Smith, H. (1995). Butterfly conservation on arable farmland. In A.S. Pullen, ed. *Ecology and Conservation of Butterflies*. Chapman and Hall, London.

Feber, R.E., Smith, H., and Macdonald, D.W. (1996). The effects of management of uncropped edges of arable fields on butterfly abundance. *Journal of Applied Ecology*, **33**, 1191–1205.

Fischer, C., Thies, C., and Tscharntke, T. (2011). Small mammals in agricultural landscapes: opposing responses to farming practices and landscape complexity. *Biological Conservation*, **144**, 1130–1136.

Fournier, E. and Loreau, M. (2001). Respective roles of recent hedges and forest patch remnants in the maintenance of ground-beetle (Coleoptera: Carabidae) diversity in an agricultural landscape. *Landscape Ecology*, **16**, 17–32.

Fox, R., Brereton, T.M., Asher, J., et al. (2011). *The State of the UK's Butterflies 2011*. Wareham, Dorset, Butterfly Conservation; & Centre for Ecology & Hydrology, 16pp.

Fuller, R.M. (1987). The changing extent and conservation interest of lowland grasslands in England and Wales: a review of grassland surveys 1930–1984. *Biological Conservation*, **40**, 281–300.

Fuller, R.J., Norton, L.R., Feber, R.E., et al. (2005). Benefits of organic farming to biodiversity vary among taxa. *Biology Letters*, **1**, 431–434.

Gabriel, D., Carver, S.J., Durham, H., et al. (2009). The spatial aggregation of organic farming in England and its underlying environmental correlates. *Journal of Applied Ecology*, **46**, 323–333.

Gabriel, D., Roschewitz, I., Tscharntke, T., and Thies, C. (2006). Beta diversity at different spatial scales: plant communities in organic and conventional agriculture. *Ecological Applications*, **16**, 2011–2021.

Gabriel, D., Sait, S.M., Hodgson, J.A., Schmutz, U., Kunin, W.E., and Benton, T.G. (2010). Scale matters: the impact of organic farming on biodiversity at different spatial scales. *Ecology Letters*, **13**, 858–869.

Gabriel, D., Thies, C., and Tscharntke, T. (2005). Local diversity of arable weeds increases with landscape complexity. *Perspectives in Plant Ecology, Evolution and Systematics*, **7**, 85–93.

Garnett, T. and Godfray, C. (2012). *Sustainable intensification in agriculture. Navigating a course through competing food system priorities*. Food Climate Research Network and the Oxford Martin Programme on the Future of Food, University of Oxford, UK.

Geiger, F., Bengtsson, J., Berendse, F., et al. (2010). Persistent negative effects of pesticides on biodiversity and biological control potential on European farmland. *Basic and Applied Ecology*, **11**, 97–105.

Gluck, E. and Ingrish, S. (1990). The effect of bio-dynamic and conventional agriculture management on Erigoninae and Lycosidae spiders. *Journal of Applied Entomology*, **110**, 136–148.

Green, R.E., Cornell, S.J., Scharlemann, J.P.W., and Balmford, A. (2005). Farming and the fate of wild nature. *Science*, **307**, 550–555.

Greig-Smith, P.W., Frampton, G.K., and Hardy, A.R. (1991). *Pesticides, Cereal Farming and the Environment: The Boxworth Project*. HMSO, London.

Haughton, A.J., Bell, J.R., Boatman, N.D., and Wilcox, A. (1999). The effects of different rates of the herbicide glyphosate on spiders in arable field margins. *Journal of Arachnology*, **27**, 249–254.

Hodgson, J.A., Kunin, W.E., Thomas, C.D., Benton, T.G., and Gabriel, D. (2010). Comparing organic farming and land sparing: optimizing yield and butterfly populations at a landscape scale. *Ecology Letters*, **13**, 1358–1367.

Hole, D.G., Perkins, A.J., Wilson, J.D., Alexander, I.H., Grice, F., and Evans, A.D. (2005). Does organic farming benefit biodiversity? *Biological Conservation*, **122**, 113–130.

Holland, J.M., Birkett, T., and Southway, S. (2009). Contrasting the farm-scale spatio-temporal dynamics of boundary and field overwintering predatory beetles in arable crops. *BioControl*, **54**, 19–33.

Holland, J.M., Thomas, C.F.G, Birkett, T., Southway, S., and Oaten, H. (2005). Farm-scale spatio-temporal dynamics of predatory beetles in arable crops. *Journal of Applied Ecology*, **42**, 1140–1152.

Holzschuh, A., Steffan-Dewenter, I., and Tscharntke, T. (2008). Agricultural landscapes with organic crops support higher pollinator diversity. *Oikos*, **117**, 354–361.

Hyvonen, T., Ketoja, E., Salonen, J., Jalli, H., and Tiainen, J. (2003). Weed species diversity and community composition in organic and conventional cropping of spring cereals. *Agriculture, Ecosystems and Environment*, **97**, 131–149.

Jonason, D., Smith, H.G., Bengtsson, J., and Birkhofer, K. (2013). Landscape simplification promotes weed seed predation by carabid beetles (Coleoptera: Carabidae). *Landscape Ecology*, **28**, 487–494.

Kleijn D., Baquero R.A., Clough Y., et al. (2006). Mixed biodiversity benefits of agri-environment schemes in five European countries. *Ecology Letters*, **9**, 243–254.

Kleijn, D., Kohler, F., Báldi, A., et al. (2009). On the relationship between farmland biodiversity and land-use intensity in Europe. *Proceedings of the Royal Society B*, **276**, 903–909.

Kragten, S. and de Snoo, G.R. (2008). Field-breeding birds on organic and conventional arable farms in the Netherlands. *Agriculture, Ecosystems and Environment*, **126**, 270–274.

Kragten, S., Tamis, W.L.M., Gertenaar, E., Ramiro, S.M.M., Van der Poll, R.J., Wang, J., and De Snoo, G.R. (2011). Abundance of invertebrate prey for birds on organic and conventional arable farms in the Netherlands. *Bird Conservation International*, **21**, 1–11.

Lang, A., Filser, J., and Henschel, J.R. (1999). Predation by ground beetles and wolf spiders on herbivorous insects in a maize crop. *Agriculture, Ecosystems and Environment*, **72**, 189–199.

Lemke, A. and Poehling H.M. (2002). Sown weed strips in cereal fields: overwintering site and 'source' habitat for *Oedothorax apicatus* (Blackwall) and *Erigone atra* (Blackwall) (Araneae: Erigonidae). *Agriculture, Ecosystems and Environment*, **90**, 67–80.

Merckx, T., Feber, R.E., Riordan, P., et al. (2009). Optimizing the biodiversity gain from agri-environment schemes. *Agriculture, Ecosystems and Environment*, **130**, 177–182.

Millán de la Peña, N., Butt, A., Delettre, Y., Morant, P., and Burel, F. (2003). Landscape context and carabid beetle (Coleoptera: Carabidae) communities of hedgerows in western France. *Agriculture, Ecosystems and Environment*, **94**, 59–72.

Natural England (2009). *Agri-environment Schemes in England 2009: a review of results and effectiveness*. Natural England, Sheffield.

Norton, L., Johnson, P.J., Joys, A., et al. (2009). Consequences of organic and non-organic farming practices for field, farm and landscape complexity. *Agriculture, Ecosystems and Environment*, **129**, 221–227.

Nyffeler, M. and Sunderland, K.D. (2003). Composition, abundance and pest control potential of spider communities in agroecosystems: a comparison of European and US studies. *Agriculture Ecosystems and Environment*, **95**, 579–612.

Öberg, S. and Ekbom, B. (2006). Recolonisation and distribution of spiders and carabids in cereal fields after spring sowing. *Annals of Applied Biology*, **149**, 203–211.

O'Connor, R.J., Shrubb, M., and Watson, D. (1986). *Farming & Birds*. Cambridge University Press, Cambridge.

Östman, Ö., Ekbom, B., and Bengtsson, J. (2001). Farming practice and landscape heterogeneity influence biological control. *Basic and Applied Ecology*, **2**, 365–371.

Potts, S.G, Biesmeijer, J.C., Kremen, C., Neumann, P., Schweiger, O., and Kunin, W.E. (2010). Global pollinator declines: trends, impacts and drivers. *Trends in Ecology and Evolution*, **25**, 345–353.

Pywell, F.R., James, K.L., Herbert, I., et al. (2005). Determinants of overwintering habitat quality for beetles and spiders on arable farmland. *Biological Conservation*, **123**, 79–90.

Roschewitz, I., Gabriel, D., Tscharntke, T., and Thies, C. (2005). The effects of landscape complexity on arable weed species diversity in organic and non-organic farming. *Journal of Applied Ecology*, **42**, 873–882.

Rundlof, M., Bengtsson, J., and Smith H.G. (2008). Local and landscape effects of organic farming on butterfly species richness and abundance. *Journal of Applied Ecology*, **45**, 813–820.

Rundlof, M. and Smith, H.G. (2006). The effect of organic farming on butterfly diversity depends on landscape context. *Journal of Applied Ecology*, **43**, 1121–1127.

Rundlöf, M., Edlund, M., and Smith, H.G. (2010). Organic farming at local and landscape scales benefits plant diversity. *Ecography*, **33**, 514–522.

Schmidt, M.H., Roschewitz, I., Thies, C., and Tscharntke, T. (2005). Differential effects of landscape and management on diversity and density of ground-dwelling farmland spiders. *Journal of Applied Ecology*, **42**, 281–287.

Schmidt, M.H. and Tscharntke, T. (2005). Landscape context of sheetweb spider population dynamics in cereal fields. *Journal of Biogeography*, **32**, 467–473.

Schröter, L. and Irmler, U. (2013). Organic cultivation reduces barrier effect of arable fields on species diversity. *Agriculture, Ecosystems and Environment*, **164**, 176–180.

Seufert, V., Ramankutty, N., and Foley, J.A. (2012). Comparing the yields of organic and conventional agriculture. *Nature*, **485**, 229–232.

Smart, S.M., Firbank, L.G., Bunce, R.G.H., and Watkins, J.W. (2000). Quantifying changes in abundance of food plants for butterfly larvae and farmland birds. *Journal of Applied Ecology*, **37**, 398–414.

Smith, H.G., Dänhardt, J., Lindström, Å., and Rundlöf, M., (2010). Consequences of organic farming and landscape heterogeneity for species richness and abundance of farmland birds. *Oecologia*, **162**, 1071–1079.

Spencer, J.W. and Kirby, K.J. (1992). An Inventory of Ancient Woodland for England and Wales. *Biological Conservation*, **62**, 77–93.

Storkey, J., Meyer, S., Still, K.S., and Leuschner, C. (2012). The impact of agricultural intensification and land-use change on the European arable flora. *Proceedings of the Royal Society B*, **279**, 1421–1429.

Thiele, H. U. (1977). *Carabid Beetles in their Environments. A study on habitat selection by adaptation in physiology and behaviour. Volume 10 of Zoophysiology and Ecology*, Springer-Verlag, Springer Berlin Heidelberg.

Thomas, J.A. (2005). Monitoring change in the abundance and distribution of insects using butterflies and other indicator groups. *Philosophical Transactions of the Royal Society B*, **360**, 339–357.

Tilman, D., Fargione, J., Wolffe, B., et al. (2001). Forecasting agriculturally driven global environmental change. *Science*, **292**, 281–284.

Topping, C.J. (1999). An individual-based model for dispersive spiders in agroecosystems: simulations of the effects of landscape structure. *Journal of Arachnology*, **27**, 378–386.

Tuck, S.L., Winqvist, C., Mota, F. et al. (2014). Land-use intensity and the effects of organic farming on biodiversity: a hierarchical meta-analysis. *Journal of Applied Ecology*, **51**, 746–755.

Vickerman, G.P. (1992). The effects of different pesticide regimes on the invertebrate fauna of winter wheat. In: Greig-Smith, P.W., Frampton, G.K., Hardy, A.R., eds., *Pesticides, Cereal Farming and the Environment*. pp. 82–109. HMSO, London.

Vickery, J.A., Tallowin, J.T., Feber, R.E., Asteraki, E.A., Atkinson, P.W., Fuller, R.J., and Brown, V.K. (2001). Effects of grassland management on birds and their food resources, with special reference to recent changes in fertiliser, mowing and grazing practices on lowland neutral grasslands in Britain. *Journal of Applied Ecology*, **38**, 647–664.

Wickramasinghe, L.P., Harris, S., Jones, G., and Jennings, N.V. (2004). Abundance and species richness of nocturnal insects on organic and non-organic farms: Effects of agricultural intensification on bat foraging. *Conservation Biology*, **18**, 1283–1292.

Wickramasinghe, L.P., Harris, S., Jones, G., and Vaughan, N. (2003). Bat activity and species richness on organic and non-organic farms: impact of agricultural intensification. *Journal of Applied Ecology*, **40**, 984–993.

Wilson, J.D., Evans, A.D., and Grice, P.V. (2009). *Bird Conservation and Agriculture*. Cambridge University Press, Cambridge.

Winqvist, C., Ahnström, J., and Bengtsson, J. (2012). Effects of organic farming on biodiversity and ecosystem services taking landscape complexity into account. *Annals of the New York Academy of Sciences Issue*: *The Year in Ecology and Conservation Biology*, **1249**, 191–203.

CHAPTER 7

Farming for the future: optimizing farming systems for society and the environment

Hanna L. Tuomisto, Ian D. Hodge, Philip Riordan, and David W. Macdonald

Any sufficiently advanced technology is indistinguishable from Nature.
Karl Schroeder's revision of Clarke's Law.

7.1 Introduction

Wildlife conservationists cannot be other than pre-occupied with agriculture—it is one of the main drivers of global environmental change. Agricultural land, including arable land, grassland, and permanent pastures, occupies 38% (49 million km²) of the world's land area (FAOSTAT 2012), and it is estimated that a further 1.1 million km² of arable land will be needed in the developing countries by 2050 (Alexandratos and Bruinsma 2012).

Agriculture impacts wildlife both directly and indirectly. The direct impacts are more obvious, such as habitat loss and, sometimes, ensuing human–wildlife conflicts, but it also has immense indirect consequences for the wider environment through its contributions to climate change, nitrogen and phosphorus cycles (Box 7.1), global freshwater use, and chemical pollution.

Agriculture is facing the challenge of feeding the world's growing population while alleviating the environmental issues and facing binding resource constraints. The total environmental impact of agriculture is determined by three main factors: (i) the demand for agricultural products (which determines the quantity of agricultural production), (ii) the choice of farming practices that impacts on the intensity of agriculture, and (iii) the use of the 'rest of the land' that is not used for agriculture. Agricultural demand and agricultural intensity together determine the amount of land needed to meet agricultural demand, and in turn, the land area available for other uses. The use of the 'rest of the land' is an important factor to be included in any analysis

when environmental impacts of contrasting farming systems are compared, as that land area can potentially be used for providing environmental services such as habitats for wildlife or carbon sequestration.

To reduce total environmental impacts of agriculture, all of the three factors, demand, farming practices, and the use of the rest of the land, have to be addressed. In this chapter we concentrate on the farming practices by asking: what kind of farming systems can produce a given quantity of products with low environmental impacts when the alternative land use options of the rest of the land are taken into account? We start by comparing the environmental impacts of organic and conventional farming systems in Europe, and thereafter we explore the potential for integrated farming systems that combine the best practices from organic and conventional systems. In addition, we provide further information that can be used for designing policies that promote sustainable farming systems. We present the financial performance of the farming systems compared and we propose a novel method for weighting different environmental impact categories that allows a comparison of the overall environmental performance of the farming systems.

7.2 Comparing environmental impacts of organic and conventional farming

7.2.1 Introduction to organic farming

Organic farming aims at improving human and animal health, providing environmental benefits, and producing

Box 7.1 Plant nutrition: nitrogen and phosphorus cycles

Plants require 17 essential plant nutrients to complete a normal life cycle. Carbon (C) and oxygen (O) are absorbed from air, and hydrogen (H) from water. The remaining macro- and micro-nutrients are obtained from soil. Natural ecosystems recycle nutrients efficiently through food-web pathways that decompose biomass back to mineral nutrients. In agricultural systems, varying proportions of nutrients are applied as fertilizers, including six macronutrients: nitrogen (N), phosphorus (P), potassium (K), calcium (Ca), magnesium (Mg), and sulphur (S); and eight micronutrients: boron (B), chlorine (Cl), copper (Cu), iron (Fe), manganese (Mn), molybdenum (Mo), zinc (Zn), and nickel (Ni). Human interventions, especially, in nitrogen and phosphorus cycles are causing major environmental change (Rockström et al. 2009b).

Nitrogen cycle

Nitrogen is present in the environment in different forms, including nitrogen (N), ammonium (NH_4^+), nitrite (NO_2^-), nitrate (NO_3^-), nitrous oxide (N_2O), nitric oxide (NO), and nitrogen gas (N_2). The largest pool of nitrogen is in the Earth's atmosphere, 78% of which consists of nitrogen gas. In living organisms, nitrogen is needed to construct proteins. Plants can take up nitrogen only as nitrate or ammonium. Therefore, in natural ecosystems, nitrogen becomes available for plants through microbes that decompose organic matter or fix nitrogen from the atmosphere. When organic material is deposited in the soil, microbes convert organic nitrogen back into ammonium in a process called ammonification or mineralization. Atmospheric nitrogen can be fixed by freely living bacteria (e.g. *Azobacter*) or bacteria that live in legume root nodules (e.g. *Rhizobium*). Ammonia in soil is converted to nitrate in a process called nitrification. Bacteria (e.g. *Nitrosomonas* species) oxidize ammonium to nitrites, and other bacteria (e.g. *Nitrobacter* species) further oxidize nitrites into nitrates. The latter process is essential, as nitrites are toxic to plants. In a process called denitrification, soil bacteria (e.g. *Pseudomonas* and *Clostridium*) reduce nitrates back into nitrogen gas in anaerobic conditions. A powerful greenhouse gas, nitrous oxide, is produced during nitrification and denitrification processes.

Human intervention in the nitrogen cycle

Nitrogen is provided to agricultural soils through the use of nitrogen fixing legumes and application of organic or synthetic fertilizers. Synthetic nitrogen fertilizers are produced industrially by the Haber–Bosch process, in which ammonia is produced through a reaction of nitrogen and hydrogen gases. Methane from natural gas is used as a source of hydrogen and air as a source of nitrogen. The process is energy intensive as it requires high pressure and temperature.

Fertilizer application in agroecosystems has increased the nitrification and denitrification processes, and therefore, increased nitric oxide and nitrous oxide emissions to the atmosphere (Park et al. 2012). Furthermore, nitrification and livestock operations emit ammonia, which causes acidification. Increased nitrogen (wet and dry) depositions and leaching from agricultural fields impact on terrestrial and aquatic ecosystems, resulting in acidification of soils and waterways, changes in species composition, eutrophication of waterways, and contamination of groundwater.

Phosphorus cycle

Phosphorus is an essential plant nutrient as it is a component of adenosine triphosphate (ATP), which is needed for the conversion of light energy to chemical energy during photosynthesis in plant cells. Phosphorus most commonly exists in polyprotic phosphoric acid (H_3PO_4) in soil, but is more readily available for plants in dihydrogen phosphate (H_2PO_4). Phosphorus is the limiting element in most environments as it is released slowly from insoluble phosphates. Plants can increase phosphorus uptake by a mycorrhiza, a symbiotic association with soil fungi.

The phosphorus cycle is one of the slowest biochemical cycles: phosphorus cannot be found in gaseous form and phosphates only move quickly through plants and animals, weathering slowly from rocks and minerals. Small losses of phosphorus occur in terrestrial ecosystems through leaching and erosion. In soil, phosphorus is mostly in immobilized form and can be activated by plants and fungi.

Human impact on phosphorus cycle

Phosphorus fertilizers are made from phosphorus rock by dissolving it with nitric acid (HNO_3) to produce a mixture of phosphoric acid (H_3PO_4) and calcium nitrate ($Ca(NO_3)_2$), or by adding sulphuric acid (H_2SO_4) to process water-soluble phosphate (P_2O_5). Humans have intervened in natural phosphorus cycles by mining phosphorus rocks for use as fertilizer and transporting phosphorus around the globe in fertilizers, food, and feed. As a result, phosphorus concentration is high in some areas, which leads to phosphorus run-off to waterways from farms and from sewage effluents. This causes eutrophication and therefore reductions in fish and animal populations (Rockström et al. 2009b).

high quality food by relying on local resources, recycling, re-use, and efficient management of materials and energy, instead of using synthetic fertilizers, pesticides, and genetically modified crops. In organic farming, pest and weed control are based on preventative practices, such as carefully planned, versatile, crop rotations, and mechanical weeding. Instead of using synthetic nitrogen fertilizers, organic farmers include nitrogen fixing legumes in crop rotations. In the European Union (EU), organic farming is regulated according to the European Council Regulation No. 834/2007 (EC 2007), which sets the basis for national standards in the EU. All organic producers are inspected by organic inspection bodies, which may be private or government managed. Globally, organic farming occupies < 1% of the agricultural land (FAOSTAT 2012). The share of agricultural land under organic management in the EU-27 was 5.5% and in the UK 3.7% in 2011 (Eurostat 2013).

7.2.2 Meta-analysis comparing environmental impacts of organic and conventional farming

To provide comprehensive information about the wider environmental impacts of organic farming, we used a meta-analysis to evaluate the results of systematically selected peer-reviewed studies, comparing environmental impacts of organic and conventional farming in Europe (Tuomisto et al. 2012b).

Data for the meta-analysis were extracted from 71 studies that provided 257 quantitative measures of the environmental impacts of organic and conventional farming. We chose ten indicators to cover the major agricultural environmental impacts. In the studies, impacts were either reported per unit of field area, or per unit of product. We grouped the indicators as either Life Cycle Assessment (LCA) (ISO 14040 2006), or non-LCA, indicators. LCA indicators were those where all impacts occurring during the production chain from input production up to the farm gate were taken into account, whereas a non-LCA indicator only took into account the emissions occurring from the farming process. In our study, LCA indicators describe the magnitude of the final impact that may be caused by many pollutants, whereas non-LCA impacts are only emissions of particular pollutants. LCA is a powerful tool for comparing environmental impacts of different farming systems as it takes into account the whole production chain and considers many environmental impacts simultaneously.

Response ratios for each indicator—namely, how much better or worse was organic farming in comparison

to conventional farming—were calculated using the following formula:

$$[\text{Response ratio} = \text{impact of organic farming} / \text{impact of conventional farming} - 1].$$

Overall we found that, generally, organic farming has environmental benefits when measured per unit of field area, but when the impacts were measured per unit of product the benefits were reduced, or organic farming had higher negative impacts (Fig. 7.1). This was explained by higher land use requirement for producing a unit of product in organic systems.

Each environmental impact category included in the study, and our findings from the meta-analysis study for each impact, are described in the following sections.

7.2.2.1 Non-LCA indicators included in the study

Biodiversity. Since impacts of organic farming on biodiversity have been systematically reviewed before (Bengtsson et al. 2005; Hole et al. 2005), our meta-analysis only reviewed more recent literature (37 papers), published between 2003 and 2009, which compared biodiversity on organic and conventional farms. The recent studies supported the findings of earlier reviews. Most studies found that organic farming had positive impacts on species abundance and/or richness (see also Chapter 6, this volume), but 34% of the studies found negative impacts, mixed results, or no differences between the systems. In particular, species richness of non-crop plants, such as broadleaved weed species, and more rare and declining species, has been widely found to be greater in organic farms compared with conventional farms (Albrecht 2005; Roschewitz et al. 2005; Petersen et al. 2006; Gabriel et al. 2006; Romero et al. 2008). Some studies showed that the extent of impact of organic farming on biodiversity differed according to the landscape within which it was situated (Purtauf et al. 2005; Rundlöf and Smith 2006; Rundlöf et al. 2008; Kragten and de Snoo 2007; Piha et al. 2007). It was also found that organic farming alone, without additional practices (e.g. those included in agri-environmental schemes), was not adequate for conserving some bird species (Kragten and de Snoo 2007; Piha et al. 2007) or butterflies (Ekroos et al. 2008). The question as to whether conventional farming with specific targeted practices, such as creation and maintenance of non-crop habitats and field boundaries, can result in higher biodiversity than organic farming is yet to be fully answered.

Soil organic matter (SOM). SOM has a positive impact on many soil quality aspects, such as structure, erosion

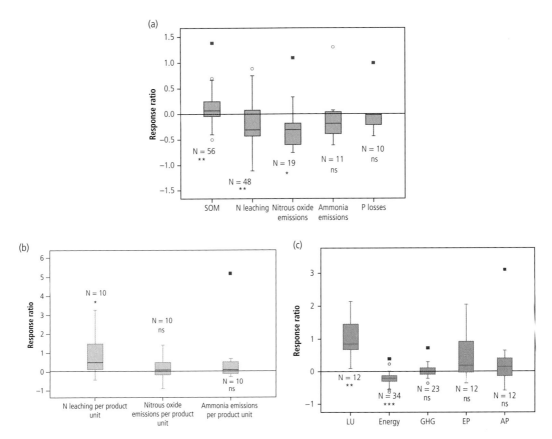

Figure 7.1 Response ratios for (a) non-LCA (life cycle assessment) impacts allocated per unit of land: soil organic matter (SOM), phosphorus (P) losses, nitrogen (N) leaching, nitrous oxide emissions, and ammonia emissions, and (b) non-LCA impacts allocated per unit of product: nitrogen (N) leaching, nitrous oxide emissions, and ammonia emissions, and (c) LCA impacts allocated per unit of product: energy use, greenhouse gas emissions (GHG), acidification potential (AP), eutrophication potential (EP), and land use (LU). Line through the box: median; upper and lower sides of the boxes: upper and lower quartiles; tiles: extreme values; o, ■: outliers 1.5–3 and over 3 box lengths from the upper or lower edge of the box, respectively; positive values: impacts from organic farming are higher; negative values: impacts from organic farming are lower; N = number of cases in the sample; ns = not significantly different from zero (Wilcoxon signed rank test P > 0.05); ***P < 0.001; **P < 0.01; *P < 0.05; Z-values: SOM = −2.457, N leaching = −2.643, Nitrous oxide emissions = −2.045, Ammonia emissions = −1.125, P losses = −0.674, LU = −2.521, Energy = −3.652, GHG = −0.880, EP = 0.000, and AP = −1.690. From Tuomisto et al. (2012c).

control, water retention, and long-term productivity; therefore, it is a sufficient indicator for describing the overall level of soil quality (Shepherd et al. 2002).

Our meta-analysis found 7% higher soil organic matter content in organic compared to conventional farms (Fig. 7.1a) The main explanations for higher SOM contents in organic systems were higher organic matter inputs, less intensive tillage, and inclusion of leys in the rotation (Canali et al. 2009; Quintern et al. 2006; Cardelli et al. 2004).

Nitrogen leaching. Nitrogen leaching causes contamination of ground water, eutrophication of waterways,

and also indirectly, nitrous oxide emissions (Box 7.1). Nitrogen leaching occurs when more nitrate is available in the soil than plants can use, at times of the year when water from rain, irrigation, or snowmelt moves through the soil into the groundwater. The level of nitrogen leaching is influenced by soil and weather conditions, and by management decisions such as choice of crop rotation, by type, timing, and amount of nitrogen fertilizer application, and by type and timing of cultivation (Shepherd et al. 2003).

Our meta-analysis found that nitrogen leaching was 31% lower from organic farming per unit of land area, but 49% higher when the results were compared per unit of product (Fig. 7.1a,b). The main explanation for lower nitrogen leaching levels from organic farming per unit of area was the lower levels of nitrogen inputs applied (Korsaeth 2008; Torstensson et al. 2006; Trydeman Knudsen et al. 2006; Hansen et al. 2000). Higher nitrogen leaching levels from organic systems per unit of land area were, in some cases, explained by poor synchrony between nutrient availability and crops' nutrient intake (Aronsson et al. 2007). In some cases, the use of overwinter cover crops in conventional systems was found to reduce nitrogen leaching, resulting in a leaching level that was lower than in organic farming (Torstensson et al. 2006).

Nitrous oxide emissions. Nitrous oxide is a greenhouse gas and, therefore, contributes to climate change. Nitrous oxide from agriculture originates mainly from application of nitrogen fertilizers, manure, and nitrogen fixing crops. Nitrous oxide is produced in soils aerobically during nitrification and anaerobically during denitrification.

Our meta-analysis found that nitrous oxide emissions were 31% lower from organic systems per unit of field area, but 8% higher when the impact was allocated per unit of product (Fig. 7.1a,b). The lower emissions from organic farming per unit of area were mainly due to lower overall nitrogen inputs in organic than in conventional systems.

Ammonia emissions. Agriculture, particularly livestock production, accounts for about 80% of ammonia emissions in Europe (EMEP 2008). Ammonia is produced when urea in urine and manure comes into contact with the enzyme urease, which is found commonly in bacteria and fungi inhabiting manure and soil. Thus animal housing, manure stores, and the spreading of manure to land, are the major sources of ammonia.

Our meta-analysis found 18% lower ammonia emissions from organic farming when measured per unit of field area and 11% higher emissions per unit of product (Fig. 7.1a,b). The lower emissions from organic farming per unit of area were mainly due to lower overall nitrogen inputs in organic than in conventional systems.

Phosphorus losses. Phosphorus losses contribute to eutrophication of waterways (Box 7.1). Many soils have large reserves of phosphorus, but often only 1% is available to crops (Shenoy and Kalagudi 2005). Phosphorus is added to soils as phosphate fertilizers, and recycled back into the soil from plant residues, agricultural wastes, and sewage sludge.

Phosphate reserves are a non-renewable resource and the accelerating use of this raw material will eventually lead to depletion. The existing stock is estimated to be able to sustain production for approximately 60 more years (Franz 2008). Therefore, enhanced recycling of phosphorus, minimizing phosphorus losses, and better utilization of soil phosphorus reserves are essential.

The median response ratio for phosphorus losses showed 1% lower emissions from organic systems (Fig. 7.1a); only one study found lower phosphorus losses from a conventional system (Aronsson et al. 2007). That result was due to incorporation of green manure, resulting in increased mineralization of crop residues in organic systems. The organic systems included in our meta-analysis had 55% lower total phosphorus inputs compared to conventional systems.

7.2.2.2 LCA indicators

Land use. The 'land use' impact category in this study relates to the area of land required for producing a unit of product output. Especially in organic systems, crop yield levels alone do not provide sufficient information about the total land requirements, since additional land is required for fertility-building crops.

Our meta-analysis showed that organic livestock production required 84% more land than conventional production (Fig. 7.1c). The greater requirement for land in organic systems was due to lower livestock and animal yields per unit of area. The average organic yields over all crops (including arable, ley, and horticultural crops) in the whole data were 75% (SD ±14%) of conventional yields. The main reason for lower organic crop yields identified in the studies was insufficient availability of nutrients (especially nitrogen), although some studies mentioned problems with weeds, diseases, or pests.

Energy use. Energy is used on farms directly in electricity and fuel oils, and indirectly in the manufacture and transport of fertilizers, pesticides, animal feeds, and in the manufacturing and maintenance of machinery. The production and distribution of mineral fertilizers account for 37% of the total energy input to agricultural production, and the production of pesticides accounts for approximately 5% (Deike et al. 2008).

Our meta-analysis showed that organic production required 21% less energy than conventional farming for producing the same amount of product (Fig. 7.1c). Higher energy inputs in conventional farming were mainly due to the high energy requirement for production and transport of inorganic fertilizers, especially synthetic nitrogen fertilizers. Only three studies out of 34 found higher energy use from organic systems, of which two cases were pork production (Basset-Mens and van der Werf 2005) and one potato production (Glendining et al. 2009). The higher energy use in those organic systems was explained by lower crop yields and lower productivity of the animals.

Greenhouse gas (GHG) emissions. The major GHG emissions from agriculture are carbon dioxide, methane, and nitrous oxide. GHG emissions are generally measured as carbon dioxide equivalents (CO_2-eq) over a 100-year time frame. The contribution of agriculture to the total global human-induced GHG emissions has been estimated to be *c.* 21% (11 054 MtCO$_2$-eq/ year) when both the direct and indirect emissions from agriculture are considered (Vermeulen et al. 2012). Vermeulen et al. (2012) estimated that direct emissions account for 55% of total agricultural emissions, including methane emissions from rice cultivation, enteric fermentation by ruminant animals, manure management, and burning of crop residues. Indirect emissions include production of farming inputs (5% of total agricultural emissions) and deforestation due to clearing forests for agricultural land (40% of total agricultural emissions). In the UK, direct emissions from agriculture account for around 9% of total UK GHG emissions when methane and nitrous oxide emissions from agriculture and carbon dioxide emissions from farming machinery are included (Defra 2011).

The median response ratio for GHG emissions in our meta-analysis was zero, which means that median emissions from organic and conventional farming were similar (Fig. 7.1c). There were clear differences in the median response ratios of GHG emissions between different product groups. Organic olive and beef had 6 and 10% lower GHG emissions respectively, whereas organic milk, cereals, and pork had 5%, 10%, and 40% higher GHG emissions respectively, when compared to conventional products. The emissions from olive production were closely related to the amount of fossil fuel used and that was found to be independent of the farming system (Kaltsas et al. 2007). Organic beef production was found to have lower GHG emissions compared to conventional, due to lower emissions from industrial inputs (Casey and Holden 2006).

Eutrophication potential. Eutrophication means enrichment of terrestrial and aquatic habitats with plant nutrients, which results in increased growth of plants and algae. The main agricultural sources are nitrate, phosphate, and ammonia. Eutrophication potential is quantified in terms of phosphate equivalents (Huijbregts and Seppälä 2001). Agriculture is the main contributor to eutrophication, accounting for 50–80% of the total aquatic nitrogen load and about 50% of the phosphorus load in Europe (EEA 2005). Nitrogen is more commonly the key limiting nutrient of marine waters, while phosphorus is the limiting factor in freshwaters (Smith et al. 1999).

Our meta-analysis found 20% higher eutrophication potential from organic farming per product unit (Fig. 7.1c). Eutrophication potential was generally lower in organic systems per unit area due to their lower nutrient inputs, but higher per product unit due to lower animal and crop yields, as compared to conventional systems. Some variation in the results was explained by differences in soil types between the organic and conventional systems compared.

Acidification potential. The major acidifying pollutants from agriculture are ammonia (NH_3) and sulphur dioxide (SO_2). Acidification potential is quantified in terms of SO_2 equivalents (SO_2-eq) (Seppälä et al. 2006). Acidifying pollutants impact on soil, ground and surface waters, biological organisms, and other materials, causing, for example, fish mortality, forest decline, and the erosion of building materials.

Our meta-analysis found 15% higher acidification potential from organic farming per product unit (Fig. 7.1c). Acidification potential was generally higher per unit of product from organic farming due to lower crop and animal yields. However, in some cases, organic farming had lower acidification potential per unit of product due to lower protein content in livestock feed (Cederberg and Mattsson 2000) or lower animal density in buildings, use of cover crops in winter, and use of solid manure (Basset-Mens and van der Werf 2005). However, conventional systems using targeted practices for reducing ammonia emissions, such as improvements in manure management, were found to have lower acidification potential than organic systems (Basset-Mens and van der Werf 2005).

7.2.3 Conclusions from the meta-analysis

Our results indicate that organic farming in Europe requires more land and often has higher environmental impacts per unit of product compared to conventional farming. As such, large-scale conversion to organic farming could provide environmental benefits only at the expense of reducing food production, if the difference in yields between organic and conventional farming remains. Given that the demand for food is increasing globally, there will be pressures to increase rather than reduce production per unit area (The Royal Society 2009). There is potential to increase yields in organic farming. Under carefully controlled management conditions, organic farming has the potential to achieve yields comparable with those in conventional farming (Jonsson 2004; Pimentel et al. 2005; Seufert et al. 2012). However, these controlled conditions may be impossible to achieve on commercial farms. Further research is needed to improve control strategies for weeds, pests, and diseases in organic systems, especially when reduced tillage is used. There is also a need for breeding both crops and animals that are best suited for organic farming, since many crop varieties and animal breeds currently used in organic farming have been developed for conventional farming systems (Wolfe et al. 2008).

The results of our meta-analysis also showed a wide variation between the impacts from different conventional farming systems. The major explanations for this variation between different studies were: (i) a wide range of different types of farming practices used, especially in conventional farming systems, (ii) differences in research methods (e.g. modelling study or experimental field investigation), and (iii) differences between products. Non-organic systems that used best practices for reducing environmental impacts, while producing high yields, tended to lead to the lowest environmental impacts (often even on the area basis) (Korsaeth 2008; Basset-Mens and van der Werf 2005; Torstensson et al. 2006). Thus, focusing on improving the environmental performance of conventional systems while integrating the elements achieving the best results might offer more scope for mitigating environmental impacts per unit of output than does organic farming.

7.3 Comparing environmental impacts and financial performance of organic, conventional, and integrated farming system models

7.3.1 Integrated Farm Management

Integrated Farm Management (IFM) is defined as 'a whole farm policy aiming to provide the basis for efficient and profitable production which is economically viable and environmentally responsible. It integrates beneficial natural processes into modern farming practice using advanced technology and aims to minimise the environmental risks while conserving, enhancing and recreating that which is of environmental importance' (Leake 2000). Crop rotations are designed to optimize pest, weed, and disease control. Biological and mechanical control methods are preferred over chemical ones, with pesticides only used as a last resort to avoid crop failures. Fertilizer inputs are planned based on crop requirements and soil analysis. Reduced and no-tillage are utilized for improving the soil structure where suitable. Organic farming relies on the natural processes while conventional farming relies more freely on modern technology: IFM aims to integrate both in order to achieve its targets. An organization called Linking Environment And Farming (LEAF) certifies integrated farming systems in the UK. LEAF was founded in 1991, and in April 2011 it had 2500 members in 47 countries (of which 1884 were in the UK) (LEAF 2012).

7.3.2 Farming system models

We used LCA for estimating and comparing energy use, GHG emissions, and biodiversity impacts of organic, conventional, and integrated crop farming system models (Tuomisto et al. 2012c, d). The integrated farming system models in the study were designed so as to combine the best practices for reducing environmental impact, including a versatile crop rotation, use of organic fertilizers, use of overwinter cover crops, use of pesticides only when needed, in order to avoid crop failures, integration of biogas production, and recycling of nutrients. The models did not represent average integrated farming systems, but rather were designed to enable comparison of the impacts of farming systems consisting of combinations of different farming practices. Therefore, this study does not provide information about the sustainability of the existing organic, conventional, and integrated farming systems, but aims to examine the potential impacts of different farming practices and systems in terms of energy use, GHG emissions, and biodiversity when alternative land use options are taken into account.

This exercise focused on arable-only farming systems, noting that the number of both organic and conventional farms without livestock is increasing in Europe (Stinner et al. 2008). Furthermore, due to the relatively high environmental impacts of livestock production as compared to crop production, the diet of the growing human population will increasingly have to be based on crops or alternative protein sources, such as cultured meat (Box 7.2) (Tuomisto and Teixeira de Mattos 2011), rather than livestock products (Kass et al. 2011). Therefore, farming systems, including organic farming, will possibly have to operate without livestock production in the future.

The functional unit (FU) towards which all of the inputs and impacts were proportioned was food crop output (in wet weights) of 460 t potatoes *Solanium*

Box 7.2 Cultured meat production

Livestock production contributes 18% to global greenhouse gas emissions, 33% to the global land use (FAO 2006), and 27% to the global water footprint (Mekonnen and Hoekstra 2011). Cultured meat (i.e. *in vitro* or lab-grown meat) is being developed in order to produce meat that is safer, healthier, and has fewer negative environmental impacts than conventionally produced meat. Cultured meat is produced by cultivating animal muscle cells *in vitro* without growing the whole animals. Currently, cultured meat production is in the research stage, but it has been estimated that commercial production could start within a decade.

Our study showed that the potential environmental impacts of cultured meat are substantially lower than those of conventionally produced meat (Tuomisto and Teixeira de Mattos 2011). We based our study on an assumption that cyanobacteria, produced in an open pond, were used as an energy and nutrient source for the cells that were grown in a bioreactor. Results showed that life-cycle-assessment-based GHG emissions, land use, and water use of cultured meat were approximately two orders of magnitude lower compared to those of conventionally produced European meat. Energy use for cultured meat production was approximately 40% higher than that of poultry, but lower than those of beef, sheep or pork. Cultured meat production could also have potential benefits for wildlife conservation, for two main reasons: it reduces pressure for converting natural habitats to agricultural land, and it provides an alternative way of producing meat from endangered and rare species that are currently over-hunted or over-fished for food. However, large-scale replacement of conventional meat production by cultured meat production may have some negative impacts on rural biodiversity due to the reduction in need for grasslands and pastures. In some hill areas, livestock has an important role in maintaining the open landscapes. Cultured meat could also reduce eutrophication impacts as its production has substantially lower nutrient losses to waterways compared to conventionally produced meat, since wastewaters from cyanobacteria production can be more efficiently controlled compared to run-offs from agricultural fields.

under a standard organic rotation in British lowland farming systems (Lampkin et al. 2008). Higher yielding systems required less land for producing the FU. The idea behind our simulations was that the land area not needed for production of food crops for the FU and green manure crops to sustain fertility was available for alternative uses. We considered three different alternative land uses for 'the rest of the land': cultivation of a biofuel crop *Miscanthus* energy grass, managed forest, and natural forest. In some of the systems, biogas was produced from green manure, cover crops, and straw. Biogas consists of mainly methane and carbon dioxide, produced by fermentation of organic material in anaerobic conditions. Biogas can be used in the same way as natural gas, for example, for generation of heat and electricity or for transportation fuel in gas powered vehicles. The energy produced from *Miscanthus*, wood, and biogas was assumed to replace the need for fossil fuels, and therefore, regarded as a negative energy input, reducing GHG emissions in the balance calculations.

The system boundaries took account of the production of farming inputs (e.g. fuels, fertilizers, and pesticides), machinery, buildings, the biogas production facility, field operations, and crop cooling and drying. We also considered soil nitrous oxide emissions that were calculated based on IPCC 2006 guidelines (IPCC 2006).

The model organic crop rotation was designed according to the recommendations for an arable organic farm that does not use external nitrogen inputs (Lampkin et al. 2008). The rotation was designed to be self-sufficient in nitrogen, consisting of: 1. grass-clover (GC); 2. potatoes; 3. winter wheat + undersown[1] over-winter legume cover crop (CC); 4. spring beans + CC; and 5. spring barley + undersown GC.

The model farming systems that we compared were:

1. *Organic farm without biogas production (O)*. The grass-clover, cover crops, and crop residues (CR) were incorporated into the soil. Ploughing was used.
2. *Organic farm with biogas production (OB)*. The grass-clover, cover crops, and crop residues (straw of wheat and bean crops) were harvested for biogas production. Ploughing was used.
3. *Conventional farm (C)*. Used mineral fertilizers and non-organic pesticides. No grass-clover, cover crops, or biogas production. Ploughing was used. Crop rotation consisted of potatoes, winter wheat, spring beans, and spring barley.

---

[1] Undersowing means that ley is established on the same field with the cereal crop after drilling the cereal seed. Once the cereal crop is harvested in autumn, the ley will be exposed.

tuberosum, 88 t winter wheat *Triticum aestivum*, 60 t field beans *Vicia faba*, and 66 t spring barley *Hordeum vulgare*, produced on a 100 ha farm. These crop outputs were determined by the yield output expected from 20 ha of organic field available for each crop

4. *Integrated farm (IF)*. The crop rotation and biogas production were similar to the OB system, but non-organic pesticides were used. Ploughing was used.
5. *Integrated farming special (IFS)*. As IF but instead of grass-clover ley, municipal biowaste was used as a fertilizer. Non-organic pesticides and no-tillage were used. Crop rotation consisted of potatoes, winter wheat + cover crop, spring beans + cover crop and spring barley.

The data used to populate the models of these alternative 'virtual realities' were retrieved from various publications, statistics, and data bases. The crop yields for each system are presented in Table 7.1, and the land use for food crops, green manure, and the rest of the land area that was used for *Miscanthus*, managed forest or natural forest, is presented in Table 7.2.

Table 7.1 Organic (O), conventional (C), and integrated farming special (IFS) crop yields (t wet weight ha⁻¹, variation in the brackets) and multiplication factors used for calculating organic biogas (OB) and integrated farming (IF) yields (factor multiplied by the organic yield).

	O[1]	OB[2]	IF[2]	C, IFS[1]
	t/ha	factor	factor	t/ha
Grass-clover	46 (44–48)	1.00	1.00	–
Potatoes	23 (14–29)	1.01	1.42	37 (34–40)
Winter wheat	4.4 (3.1–5)	1.09	1.39	7.9 (7.2–8.8)
Spring beans	3.0 (2–3.6)	1.00	1.15	4.0 (3.0–5.0)
Spring barley	3.3 (1.9–4.0)	1.10	1.25	6.0 (5.4–6.8)

[1] Moakes and Lampkin (2003–2009).
[2] Conversion factors based on data from (Cooper 2008; Deike et al. 2008; Delin et al. 2008).

Table 7.2 Land use for different purposes (ha). See Table 7.1 legend for land use categories.

	Food crops	Green manure	Rest of the land[1]
O	80.0	20.0	0.0
OB	76.3	20.0	3.7
C	49.6	0.0	50.4
IF	61.9	20.0	21.0
IFS	49.6	0.0	50.4

[1] *Miscanthus*, managed forest or natural forest.

7.3.3 Energy and GHG balances

Taking an overview of the performance of each simulated farming system, IFS had the lowest energy input and GHG emissions per unit of crop output (Fig. 7.2 a,b). When the opportunity costs of land use were taken into account, the system that emerged as most favourable depended on which of the various impact categories was given precedence. IFS system with *Miscanthus* production had the highest net energy production and GHG mitigation under the scenario where we assumed that the *Miscanthus* replaced oil in heating.

7.3.4 Biodiversity

The method for biodiversity assessment was adapted from De Schryver et al. (2010). Our approach was to assess the ecosystem damage of each system by comparing the fraction of species (specifically using vascular plants as an indicator) that was likely to disappear under each regime. The value of the indicator describes a change in vascular plant species richness within the occupied area, compared with the baseline. The baseline was taken as natural forest, because that is the land type that was assumed to have occurred naturally in the area without human intervention. The factors that characterized each different habitat type were determined by using data from the British Countryside Survey 2000 (Defra 2000).

In our study, IFS with natural forest had the lowest biodiversity loss index whereas conventional systems with *Miscanthus* and managed forest had the highest biodiversity loss indices (Fig. 7.2 c). Only integrated systems with natural forest had a lower biodiversity loss index than that of the organic systems.

7.3.5 Weighting factors

In order to compare the overall environmental performance of the different systems, the environmental impacts were weighted by using factors that took account of the relative importance of different impact categories (Tuomisto et al. 2012a). We developed novel weighting factors by extrapolating from a recent study that estimated the safe planetary boundaries for human life (Rockström et al. 2009a). In their study, Rockström et al. (2009a) argue that anthropogenic environmental change may push the global system disastrously beyond the boundaries of a stable environmental state. Maintenance of that stable state is essential for humanity and, to achieve this, Rockström et al. proposed a framework based on nine planetary boundaries that define the safe operating space for humanity. Those nine boundaries are: climate change, rate of biodiversity loss, interference with nitrogen and phosphorus cycles, stratospheric ozone depletion, ocean acidification, global fresh water use, change in land use, chemical pollution, and atmospheric aerosol loading. The weighting factors for each impact category in our study were generated by calculating the ratio between the current position and the estimated safe boundary based on the data from (Rockström et al. 2009b), which resulted in the following factors: climate change 1.31, biodiversity loss 10, nitrogen cycle 3.46, phosphorus cycle 0.82, ozone depletion 1.02, ocean acidification 1.05, global freshwater use 0.65, and land use 0.78.

When the total impact scores of the three farming systems were compared, the IFS system with natural forest had a markedly lower total impact than the others (Fig. 7.2 d). The reason for the higher impact scores of organic systems compared to some of the other systems was mainly the low availability of land for alternative land uses (e.g. for natural forest).

In all the scenarios, the highest impact was on the loss of biodiversity due to the high weighting factor for biodiversity.

7.3.6 Financial performance

We calculated the profitability of the three farming systems and simulated the financial feasibilities of some individual farming practices for reducing GHG emissions (Tuomisto et al., unpublished). The financial performance of the farming systems was calculated in terms of net margins: returns minus variable costs minus fixed costs. Variable costs included seeds, fertilizers, crop protection, and other crop costs. Fixed costs included paid labour, fuels, machinery repairs and depreciation, buildings depreciation, general farming costs, water, electricity, land expenses, insurance, rent, and interest payments. Returns included income from crops and the Single Payment Scheme. Data for the input prices and returns from crop production are presented in Table 7.3.

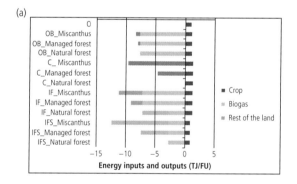

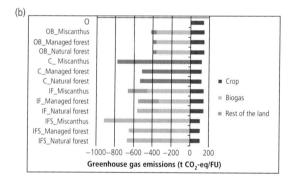

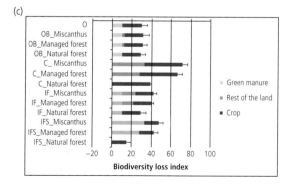

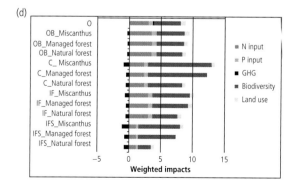

Figure 7.2 Results of the (a) energy inputs and outputs, (b) greenhouse gas emissions, (c) biodiversity loss score, and (d) weighted impacts of the farming systems studied. O = organic, OB = organic biogas, C = conventional, IF = integrated farming, IFS = integrated farming special. From Tuomisto et al. (2012b,d).

Table 7.3 Prices and returns from crop production. Source: (1) Moakes and Lampkin (2010) and (2) Nix (2009).

Source		
Conventional winter wheat	113 £/t	1
Conventional spring barley	110 £/t	1
Conventional potatoes	155 £/t	1
Conventional field beans	139 £/t	1
Organic winter wheat	214 £/t	1
Organic spring barley	184 £/t	1
Organic potatoes	289 £/t	1
Organic field beans	260 £/t	1
Single Payment Scheme	224 £/ha	2
Diesel	42 p/l	2
Labour	9.6 £/h	2
Nitrogen fertilizer	55 £/kg	2
P_2O_5 fertilizer	45 £/kg	2
K_2O fertilizer	60 £/kg	2

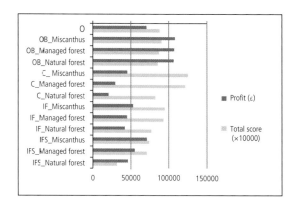

Figure 7.3 Comparison of the profitability and total impact score of the farming systems. O = organic, OB = organic biogas, C = conventional, IF = integrated farming, IFS = integrated farming special.

For natural forest, no costs or returns were assumed. For managed forest, it was assumed an average yearly harvest of 15 m³/ ha and an income of £13.57/ m³ for standing wood. Costs and returns for *Miscanthus* were based on Nix (2009). We assumed an average yield of 13.3 oven-dried tonnes (odt)/ha/year over 15-year crop longevity; a price of £60/odt, and an energy crop supplement of £28/ha. Variable costs amounted to £336/ha, including planting costs, fertilizers, sprays, and harvesting costs. The other fixed costs were included as for the arable crops described earlier. The yearly costs and returns of the whole biogas reactor were calculated based on data from Nix (2009) and Redman (2010).

When the profitability of the different systems was compared, the organic systems were the most profitable due to the organic price premium for the food crops (Fig. 7.3). Without the organic price premium, the organic systems would have been among the least profitable. Integrated systems were more profitable than conventional, due to lower input costs and inclusion of biogas production.

Comparing the environmental performance and profitability of the options for the uses for the 'rest of the land', the *Miscanthus* option came top in terms of profitability, GHG balances, and energy balances. Natural forest had the lowest profitability, but the highest biodiversity benefits.

Many of the farming practices that reduced GHG emissions also improved the profitability of the farms. Introducing biogas production into the organic system resulted in high GHG mitigation and a substantial increase in farm profit. Yield improvements also had high potential for GHG mitigation, with increased profit provided that the land area released from food crop production was used for *Miscanthus* production. Reduced tillage and no-tillage practices required less fuel, and therefore reduced both costs and GHG emissions. Replacing synthetic fertilizers with food waste digestate also provided a win–win situation, as GHG emissions from the manufacture of nitrogen fertilizers and the costs of buying them were avoided. The reduction of N_2O emissions by using nitrification inhibitors or slow release nitrogen fertilizers was the only option that reduced both GHG emissions and profitability, when it was assumed that the nitrification inhibitors reduced N_2O emissions from soils by approximately 38% and slow release fertilizers by 35% (Akiyama et al. 2010), and that the cost of fertilizers was increased by 50% (MacLeod et al. 2010) while the crop yields remained unchanged.

7.4 Conclusion

We are mindful that models are just that—models. Nevertheless, these desktop explorations of virtual reality can reveal helpful, and sometimes unexpected, insights. In this case, if nothing more, they provide a powerful didactic illustration of the joined-upness of everything to everything else in the context of environmental policy, farming systems, and biodiversity conservation. Insofar as our models reflect reality, they suggest that the highest environmental benefits might be achieved by developing integrated farming systems that use the best technologies for producing high yields with low negative environmental impacts, and simultaneously sparing land from agriculture for biodiversity conservation. Since land spared for biodiversity would have obvious susceptibility to being squeezed and eroded in the face of short-term, if short-sighted, financial gain, then policy instruments designed to foster this system of land use should have very sharp regulatory teeth for the guarding of different types of protected areas. Our findings also illustrate vividly the importance of taking into account the opportunity costs of land use when environmental impacts of different farming systems are compared. An obvious conclusion would be that land use should be planned so that the most productive land is used for agriculture, the areas most suitable for (and rich in) biodiversity should prioritize nature and in some cases be nature reserves, and carbon-rich soils should be left undisturbed. Macdonald et al. (2006) described the journey of a generation of conservationists from pioneering backwoodsmen to the besuited denizens of the corridors of political power. That journey could scarcely be more vividly illustrated than by those entering farmland conservation seeking a wistful pleasure in Constable's Haywain and fluttering fritillaries, who now find themselves musing on a plan for land-use policy that offers as enormous a political challenge as humanity has faced.

Acknowledgements

We are grateful to Holly Hill Charitable Trust for funding the project. We thank Kairsty Topp for helpful comments.

References

Akiyama, H., Yan, X.Y., and Yagi, K. (2010). Evaluation of effectiveness of enhanced-efficiency fertilizers as mitigation options for N_2O and NO emissions from agricultural soils: meta-analysis. *Global Change Biology*, **16**, 1837–1846.

Albrecht, H. (2005). Development of arable weed seedbanks during the 6 years after the change from conventional to organic farming. *Weed Research*, **45**, 339–350.

Alexandratos, N. and Bruinsma, J. (2012). *World Agriculture Towards 2030/2050—The 2012 Revision*. Agricultural Development Economics (ESA), The Food and Agriculture Organization of the United Nations (FAO), Rome, Italy.

Aronsson, H., Torstensson, G., and Bergstrom, L. (2007). Leaching and crop uptake of N, P and K from organic and conventional cropping systems on a clay soil. *Soil Use and Management*, **23**, 71–81.

Basset-Mens, C. and van der Werf, H.M.G. (2005). Scenario-based environmental assessment of farming systems: the case of pig production in France. *Agriculture, Ecosystems & Environment*, **105**, 127–144.

Bengtsson, J., Ahnstrom, J., and Weibull, A.C. (2005). The effects of organic agriculture on biodiversity and abundance: a meta-analysis. *Journal of Applied Ecology*, **42**, 261–269.

Canali, S., Di Bartolomeo, E., Trinchera, A., et al. (2009). Effect of different management strategies on soil quality of citrus orchards in Southern Italy. *Soil Use and Management*, **25**, 34–42.

Cardelli, R., Levi-Minzi, R., Saviozzi, A., and Riffaldi, R. (2004). Organically and conventionally managed soils: Biochemical characteristics. *Journal of Sustainable Agriculture*, **25**, 63–74.

Casey, J.W. and Holden, N.M. (2006). Greenhouse gas emissions from conventional, agri-environmental scheme, and organic Irish suckler-beef units. *Journal of Environmental Quality*, **35**, 231–239.

Cederberg, C. and Mattsson, B. (2000). Life cycle assessment of milk production—a comparison of conventional and organic farming. *Journal of Cleaner Production*, **8**, 49–60.

Cooper, J. (2008). *Yield differences between organic and conventional agriculture: causes and solutions*. The University of New Castle. Presentation at The 16th IFOAM Organic World Congress: Cultivate the Future 16.6.–20.6.2008.

De Schryver, A., Goedkoop, M., Leuven, R., and Huijbregts, M. (2010). Uncertainties in the application of the species area relationship for characterisation factors of land occupation in life cycle assessment. *The International Journal of Life Cycle Assessment*, **15**, 682–691.

Defra (2000). *Countryside Survey 2000*.

Defra (2011). *Greenhouse Gas Emission Projections for UK Agriculture to 2030* Economics Group, Defra.

Deike, S., Pallutt, B., and Christen, O. (2008). Investigations on the energy efficiency of organic and integrated farming with specific emphasis on pesticide use intensity. *European Journal of Agronomy*, **28**, 461–470.

Delin, S., Nyberg, A., Lindén, B., et al. (2008). Impact of crop protection on nitrogen utilisation and losses in winter wheat production. *European Journal of Agronomy*, **28**, 361–370.

EC (2007). *Council Regulation (EC) No 834/2007 of 28 June 2007 on organic production and labelling of organic products and repealing Regulation (EEC) No 2092/91*. **L 189**, Official Journal of the European Communities, pp. 1–23.

EEA (2005). *Source apportionment of nitrogen and phosphorus inputs into the aquatic environment*. European Environment Agency, Copenhagen.

Ekroos, J., Piha, M., and Tiainen, J. (2008). Role of organic and conventional field boundaries on boreal bumblebees and butterflies. *Agriculture, Ecosystems & Environment*, **124**, 155–159.

EMEP (2008). *Co-operative Programme for Monitoring and Evaluation of the Long-range Transmissions of Air Pollutants in Europe. Database of the national submissions to the UNECE*

LRTAP Convention maintained at EMEP. Available from: <http://www.emep-emissions.at/ceip/>.

Eurostat (2013). Certified organic crop area by crops products. E. European Commission.

FAO (2006). Livestock's long shadow –environmental issues and options. Food and Agricultural Organization of the United Nations, Rome.

FAOSTAT (2012). FAOSTAT Agricultural Data.

Franz, M. (2008). Phosphate fertilizer from sewage sludge ash (SSA). Waste Management, 28, 1809–1818.

Gabriel, D., Roschewitz, I., Tscharntke, T., and Thies, C. (2006). Beta diversity at different spatial scales: Plant communities in organic and conventional agriculture. Ecological Applications, 16, 2011–2021.

Glendining, M.J., Dailey, A.G., Williams, A.G., van Evert, F.K., Goulding, K.W.T., and Whitmore, A.P. (2009). Is it possible to increase the sustainability of arable and ruminant agriculture by reducing inputs? Agricultural Systems, 99, 117–125.

Hansen, B., Kristensen, E.S., Grant, R., Hogh-Jensen, H., Simmelsgaard, S.E., and Olesen, J.E. (2000). Nitrogen leaching from conventional versus organic farming systems—a systems modelling approach. European Journal of Agronomy, 13, 65–82.

Hole, D.G., Perkins, A.J., Wilson, J.D., Alexander, I.H., Grice, F., and Evans, A.D. (2005). Does organic farming benefit biodiversity? Biological Conservation, 122, 113–130.

Huijbregts, M. and Seppälä, J. (2001). Life cycle impact assessment of pollutants causing aquatic eutrophication. The International Journal of Life Cycle Assessment, 6, 339–343.

IPCC (2006). Guidelines for National Greenhouse Gas Inventories. Volume 4 Agriculture, Forestry and Other Land Use.

ISO 14040 (2006). Environmental management—Life cycle assessment -Principles and framework. International Organisation for Standardisation (ISO)

Jonsson, S. (2004). Öjebynprojektet –ekologisk produktion av livsmedel (in Swedish). SLU Institutionen for norrländsk jordbruksvetenskap, Öjebyn.

Kaltsas, A.M., Mamolos, A.P., Tsatsarelis, C.A., Nanos, G.D., and Kalburtji, K.L. (2007). Energy budget in organic and conventional olive groves. Agriculture, Ecosystems & Environment, 122, 243–251.

Kass, G.S., Shaw, R.F., Tew, T., and Macdonald, D.W. (2011). Securing the future of the natural environment: using scenarios to anticipate challenges to biodiversity, landscapes and public engagement with nature. Journal of Applied Ecology, 48, 1518–1526.

Korsaeth, A. (2008). Relations between nitrogen leaching and food productivity in organic and conventional cropping systems in a long-term field study. Agriculture, Ecosystems & Environment, 127, 177–188.

Kragten, S. and de Snoo, G.R. (2007). Nest success of Lapwings Vanellus vanellus on organic and conventional arable farms in the Netherlands. Ibis, 149, 742–749.

Lampkin, N., Measures, M., and Padel, S. (2008). 2009 Organic Farm Management Handbook, pp. 238. University of Wales, Aberystwyth.

LEAF (2012). LEAF Marque Facts. Available from: <http://www.leafuk.org/ >. [10/12/2012].

Leake, A. (2000). The development of integrated crop management in agricultural crops: comparisons with conventional methods. Pest Management Science, 56, 950–953.

Macdonald, D.W., Collins, N.M., and Wrangham, R. (2006). Principles, practice and priorities: the quest for 'alignment'. In D. Macdonald and K. Service, eds., Key Topics in Conservation Biology, pp. 271–290. Blackwell Publishing, Oxford.

MacLeod, M., Moran, D., Eory, V., et al. (2010). Developing greenhouse gas marginal abatement cost curves for agricultural emissions from crops and soils in the UK. Agricultural Systems, 103, 198–209.

Mekonnen, M.M. and Hoekstra, A.Y. (2011). The green, blue and grey water footprint of crops and derived crop products. Hydrology and Earth System Sciences, 15, 1577–1600.

Moakes, S. and Lampkin, N. (2003–2009). Organic Farm Incomes in England and Wales 2001/02–2007/08. Aberystwyth University, Aberystwyth.

Moakes, S. and Lampkin, N. (2010). Organic Farm Incomes in England and Wales 2008/09. Aberystwyth University, Aberystwyth.

Nix, J. (2009). Farm Management Pocketbook 2010, pp. 256. The Andersons Centre, Melton Mowbray.

Park, S., Croteau, P., Boering, K.A., et al. (2012). Trends and seasonal cycles in the isotopic composition of nitrous oxide since 1940. Nature Geosci, 5, 261–265.

Petersen, S., Axelsen, J.A., Tybirk, K., Aude, E., and Vestergaard, P. (2006). Effects of organic farming on field boundary vegetation in Denmark. Agriculture, Ecosystems & Environment, 113, 302–306.

Piha, M., Tiainen, J., Holopainen, J., and Vepsalainen, V. (2007). Effects of land-use and landscape characteristics on avian diversity and abundance in a boreal agricultural landscape with organic and conventional farms. Biological Conservation, 140, 50–61.

Pimentel, D., Hepperly, P., Hanson, J., Douds, D., and Seidel, R. (2005). Environmental, Energetic, and Economic Comparisons of Organic and Conventional Farming Systems. Bioscience, 55, 573–582.

Purtauf, T., Roschewitz, I., Dauber, J., Thies, C., Tscharntke, T., and Wolters, V. (2005). Landscape context of organic and conventional farms: Influences on carabid beetle diversity. Agriculture, Ecosystems & Environment, 108, 165–174.

Quintern, M., Joergensen, R.G., and Wildhagen, H. (2006). Permanent-soil monitoring sites for documentation of soil-fertility development after changing from conventional to organic farming. Journal of Plant Nutrition and Soil Science-Zeitschrift Fur Pflanzenernahrung Und Bodenkunde, 169, 564–572.

Redman, G. (2010). A Detailed Economic Assessment of Anaerobic Digestion Technology and its Suitability to UK Farming and Waste Systems. 2nd edition. The Andersons Centre, Melton Mowbray.

Rockström, J., Steffen, W., Noone, K., et al. (2009a). A safe operating space for humanity. Nature, 461, 472–475.

Rockström, J., Steffen, W., Noone, K., et al. (2009b). Planetary Boundaries: Exploring the Safe Operating Space for Humanity. *Ecology and Society*, **14**.

Romero, A., Chamorro, L., and Sans, F.X. (2008). Weed diversity in crop edges and inner fields of organic and conventional dryland winter cereal crops in NE Spain. *Agriculture, Ecosystems & Environment*, **124**, 97–104.

Roschewitz, I., Gabriel, D., Tscharntke, T., and Thies, C. (2005). The effects of landscape complexity on arable weed species diversity in organic and conventional farming. *Journal of Applied Ecology*, **42**, 873–882.

Rundlöf, M., Nilsson, H., and Smith, H.G. (2008). Interacting effects of farming practice and landscape context on bumble bees. *Biological Conservation*, **141**, 417–426.

Rundlöf, M. and Smith, H.G. (2006). The effect of organic farming on butterfly diversity depends on landscape context. *Journal of Applied Ecology*, **43**, 1121–1127.

Seppälä, J., Posch, M., Johansson, M., and Hettelingh, J.-P. (2006). Country-dependent characterisation factors for acidification and terrestrial eutrophication based on accumulated exceedance as an impact category indicator (14 pp). *The International Journal of Life Cycle Assessment*, **11**, 403–416.

Seufert, V., Ramankutty, N., and Foley, J.A. (2012). Comparing the yields of organic and conventional agriculture. *Nature*, **485**, 229–232.

Shenoy, V.V. and Kalagudi, G.M. (2005). Enhancing plant phosphorus use efficiency for sustainable cropping. *Biotechnology Advances*, **23**, 501–513.

Shepherd, M.A., Harrison, R., and Webb, J. (2002). Managing soil organic matter—implications for soil structure on organic farms. *Soil Use and Management*, **18**, 284–292.

Shepherd, M., Pearce, B., Cormack, B., et al. (2003). *An assessment of the environmental impacts of organic farming*. A review for Defra-funded project OF0405. <http://orgprints.org/6784/2/OF0405_909_TRP.pdf>.

Smith, V.H., Tilman, G.D., and Nekola, J.C. (1999). Eutrophication: impacts of excess nutrient inputs on freshwater, marine, and terrestrial ecosystems. *Environmental Pollution*, **100**, 179–196.

Stinner, W., Moller, K., and Leithold, G. (2008). Effects of biogas digestion of clover/grass-leys, cover crops and crop residues on nitrogen cycle and crop yield in organic stockless farming systems. *European Journal of Agronomy*, **29**, 125–134.

The Royal Society (2009). *Reaping the benefits: Science and the Sustainable Intensification of Global Agriculture*. The Royal Society, London, p. 72.

Torstensson, G., Aronsson, H., and Bergstrom, L. (2006). Nutrient Use Efficiencies and Leaching of Organic and Conventional Cropping Systems in Sweden. *Agron J*, **98**, 603–615.

Trydeman Knudsen, M., Sillebak Kristensen, I.B., Berntsen, J., et al. (2006). Estimated N leaching losses for organic and conventional farming in Denmark. *The Journal of Agricultural Science*, **144**, 135–149.

Tuomisto, H.L., Hodge, I.D., Riordan, P., and Macdonald, D.W. (2012a). Does organic farming reduce environmental impacts?—A meta-analysis of European research. *Journal of Environmental Management*, **112**, 309–320.

Tuomisto, H.L., Hodge, I.D., Riordan, P., and Macdonald, D.W. (2012b). Exploring a safe operating approach to weighting in life cycle impact assessment—a case study of organic, conventional and integrated farming systems. *Journal of Cleaner Production*, **37**, 147–153.

Tuomisto, H.L., Hodge, I.D., Riordan, P., and Macdonald, D.W. (2012c). Comparing global warming potential, energy use and land use of organic, conventional and integrated winter wheat production. *Annals of Applied Biology*, **161**, 116–126.

Tuomisto, H.L., Hodge, I.D., Riordan, P., and Macdonald, D.W. (2012d). Comparing energy balances, greenhouse gas balances and biodiversity impacts of contrasting farming systems with alternative land uses. *Agricultural Systems*, **108**, 42–49.

Tuomisto, H.L., Hodge, I.D., Riordan, P., and Macdonald, D.W. (Unpublished). Comparing profitability and environmental impacts of contrasting arable farming systems.

Tuomisto, H.L. and Teixeira de Mattos, M.J. (2011). Environmental impacts of cultured meat production. *Environmental Science & Technology*, **45**, 6117–6123.

Vermeulen, S.J., Campbell, B.M., and Ingram, J.S.I. (2012). Climate Change and Food Systems. *Annual Review of Environment and Resources*, **37**, 195–222.

Wolfe, M.S., Baresel, J.P., Desclaux, D., et al. (2008). Developments in breeding cereals for organic agriculture. *Euphytica*, **163**, 323–346.

CHAPTER 8

Landscape-scale conservation of farmland moths

Thomas Merckx and David W. Macdonald

When through the old oak forest I am gone,
Let me not wander in a barren dream
**John Keats, *On Sitting Down to
Read King Lear Once Again*.**

8.1 Scope of agri-environment schemes

Biodiversity has declined substantially throughout much of the European wider countryside. The most promising tools to reverse these declines are widely thought to be agri-environment schemes (AES) (Donald and Evans 2006). These governmental schemes provide financial rewards for 'environmentally friendly' methods of farmland management. However, AES do not always produce significant biodiversity benefits (Kleijn et al. 2006; Batáry et al. 2010). For example, in the UK, the broad and shallow 'Entry Level Stewardship' has often been unrewarding for wildlife (e.g. Davey et al. 2010, but see Baker et al. 2012), but, in many cases, the more targeted 'higher level' scheme has exceeded expectations (Jeremy Thomas, pers. comm.). Indeed, there is great scope for inventively designed AES to make a large impact on biodiversity conservation in regions where intensive agriculture has a dominant footprint; AES can be implemented over enormous areas of land and this matters because intensive agriculture is one of the main drivers of biodiversity declines worldwide (Donald et al. 2001; Benton et al. 2002; Green et al. 2005).

Globally, farmland covers about half of the potentially useable land (Tilman et al. 2001) with farmed crops feeding, dressing and, increasingly, fuelling the growing human population. However, land conversion to farming has brought destruction, degradation, and fragmentation of habitats, landscape homogenization, and pollution. It has not only destroyed the ecosystems converted to farmland, but often also reduced the ecosystem services (such as crop pollination, pest control, water retention, and soil protection) provided by the adjoining non-farmed land. Nevertheless, some biodiversity of the original ecosystems may be retained within farmland ecosystems, its amount heavily dependent on the spatial extent and degree of farmland intensification. Indeed, although species typically 'prefer' one ecosystem, they often occur in, and use resources from, neighbouring ecosystems (Pereira and Daily 2006; Dennis 2010). As such, many species may manage to persist within farmland systems, with at least some of them, such as the speckled wood *Pararge aegeria*, originally a woodland butterfly, adapting to these 'novel' ecosystems (Merckx et al. 2003). As a result, extensively farmed systems can often be characterized by flourishing biodiversity (e.g. chalk grasslands, the Iberian dehesa/montado); hence farmland, in general, has the potential to support biodiversity (Chapter 7, this volume), and all the more so when fostered by effective AES (Whittingham 2011).

Launched during the late 1980s, AES were conceived to reverse the severe declines in farmland biodiversity that were wrought by the techno-boom of agricultural intensification. They reflected a societal desire to restore biodiversity to farmland, and also, increasingly, recognition of the economic value of the ecosystem services they provide (Macdonald and Smith 1991). However, given that they are financed through tax-payers' money, it is essential to ensure AES are effective in delivering their goals (Kleijn et al. 2006; Macdonald et al. 2000, 2007). This could potentially be improved in a number of ways: for example, in regions

Wildlife Conservation on Farmland. Managing for Nature on Lowland Farms. Edited by David W. Macdonald and Ruth E. Feber.
© Oxford University Press 2015. Published 2015 by Oxford University Press.

characterized by marginal land or farmland abandonment, ecological restoration may be achieved more cost-effectively by including rewilding approaches, whether or not as part of effective and targeted AES supporting valuable, low-intensity farming practices (Warren and Bourn 2011; Monbiot 2013; Merckx and Pereira 2015). On intensive farmland, cost-effectiveness of biodiversity delivery through AES may be achieved by taking two vital steps: first, by identifying those elements of farmland that benefit biodiversity and can be integrated within intensive farming systems and, second, by managing them appropriately (Merckx et al. 2009a; Fuentes-Montemayor et al. 2011).

8.2 Why moths?

8.2.1 Indicators of farmland quality

With a particular focus on 'larger moths' (i.e. macro-moths), we set out to explore two ways in which the benefits of AES might be optimized. We chose macro-moths principally because of their ability to 'tell' us something about the state of the ecosystems of which they are part. Because of their fast generation turnover (i.e. one to several generations per year), their abundance as a group, their ecological diversity, their functional roles within ecosystems, and their species richness, macro-moths are considered a sensitive indicator group for biodiversity in terrestrial ecosystems (New 2004; Thomas 2005). In other words, they can be viewed as a relatively accessible 'miner's canary' for the health of other terrestrial insects on farmland. Although other groups may be good indicators too (e.g. Odonata diversity is indicative of both terrestrial health and the quality of aquatic habitat resources: Chapter 10, this volume), they are either far less abundant or species-rich, or more difficult to sample and identify (e.g. beetles, fungi). The attraction of moths to light and the fact that they are usually on the wing in high numbers mean that they can be sampled with relative ease using light traps. With around 2500 species in Britain, of which *c.* 900 are macro-moths, and with over 160 000 described and 500 000 estimated species worldwide, Lepidoptera (moths and butterflies) are a highly diverse group of insects that occupy a wide variety of habitats all over the world (Merckx et al. 2013). Their great variety in size, colour, and wing patterns makes the large majority of them easily identifiable. Also, moths are simply beautiful, and they are intriguing leaves of a rich and venerable phylogenetic tree (Mutanen et al. 2010), which has resulted in many fascinating evolutionary, ecological, and life-history aspects, such as pheromone mate-attraction, intricate host plant interactions, varied anti-predator responses, complete metamorphosis (from egg, over several larval stages and pupation, to the adult imago), and complex movement ecology (Young 1997; Chapman et al. 2010).

8.2.2 Population declines

This wealth deserves our attention and timely protection. All the more so since 62 species of macro- and micro-moths have become extinct in Britain during the twentieth century. Moreover, many more species are nationally threatened, and rapid, significant declines in abundance and distribution have been recorded for common and widespread macro-moth species that inhabit farmland in Britain. Of this last group, two thirds (227 of 337 species analysed) show a decreasing population trend over 40 years (1968–2007), with 61 having declined by at least 75% (Fox et al. 2013). For example, the figure of eight *Diloba caeruleocephala* was once a common and well-distributed woodland, hedgerow, and garden moth, feeding as a larva on hawthorn *Crataegus* spp. and blackthorn *Prunus spinosa*, but has declined by 96% over this period. The garden tiger *Arctia caja* has declined by 92% over the same 40 years (Fox et al. 2013). This species is coloured spectacularly and famous for its 'woolly bear' caterpillars that feature prominently in many childhood memories. It seems likely that these trends are part of a widespread loss in insect biodiversity in temperate-zone industrialized regions (macro-moths: Groenendijk and Ellis 2010; butterflies: Van Dyck et al. 2009; carabid beetles: Brooks et al. 2012), not to mention the unrecorded declines and extinctions of many specialist moths all over the world (Merckx et al. 2013).

8.2.3 Ecosystem services

These trends are of concern, as herbivorous macro-moth larvae—because of the huge numbers involved—are significant primary consumers and nutrient recyclers. Anyone who has heard the constant noise of falling 'frass' from winter moth *Operophtera brumata* larvae consuming leaves in temperate oak woodland, knows what we are talking about. Macro-moths are also key prey items, in all life-stages, for a wide range of other taxa (e.g. birds, bats, shrews, parasitoids, spiders, beetles); for example, it is estimated that blue tit *Parus caeruleus* chicks consume at least 35 billion caterpillars in Britain each year (Fox et al. 2006). Another significant ecosystem service to which moths contribute is pollination, with moths dominating both temperate and tropical flower-visitor faunas after dark (Devoto et al. 2011). For example, hawk moths (Sphingidae) are crucial to the pollination of many moth-pollinated plants, whose flowers co-evolved with their pollinators to produce a strong, sweet scent at night, and have long, tubular corollas which only allow long proboscises to reach the nectar produced, to fuel the high metabolic rates needed to power hovering flight; in fact very similar to hummingbird feeding behaviour (Darwin 1862).

8.3 Optimizing agri-environment schemes: field margins and hedgerow trees

Against this background, our AES study had two parts. First, we aimed to elucidate the effects on macro-moths of two prominent farmland elements, wide grassy field margins and hedgerow trees. Both of these farmland features provide habitat for moths, and their restoration and management can easily be implemented as options within AES. Second, we investigated whether implementing AES over larger, landscape-scale areas, rather than applying them to small, field-scale areas, had different consequences for macro-moth populations, and offered more scope for farmland conservation. We tackled these questions by conducting experiments in which we captured 311 species of macro-moth, ranging from the 'primitive' swift moths (Hepialidae) to the fan-foots (Herminiinae), a group of slender noctuid moths. This substantial data base allowed us to document patterns in their ecology on farmland, at a range of spatial scales, which we describe in this chapter. We contrasted species groups categorized with respect to feeding guild, mobility, and conservation status to elucidate the mechanisms underlying these patterns.

8.3.1 Wide field margins

Moths will make use of many farmland habitats, such as woodland, hedgerows, and scrub. Field margins, defined here as the uncropped strips of land which lie between the boundary feature (such as a hedge) and the field itself (whether arable or grass), can potentially provide nectar sources, larval foodplants, roosting and pupation sites, and protection from farm operations such as pesticide spraying. A first question we posed was whether field margin width affected macro-moth abundance and diversity. To this effect, we contrasted 24 sites characterized by wide (6 m) margins (current AES option; Defra/NE 2012) with 24 sites with

standard (1 m) margins (Fig. 8.1). All margins were well-established, tussocky, sown perennial grass strips of variable age, located next to hedges and machine-cut once every two or three years; they were ungrazed and unfertilized, although fertilizer may have drifted into the margin unintentionally. Wide field margins are a popular and important conservation tool, and their management to deliver biodiversity is rewarded by AES payments in a number of EU countries (Chapters 2 and 3, this volume). By early 2011, 67% of the utilizable agricultural area in England was under AES (Defra 2011), and grass/buffer strips on arable land were among the most popular (>116 000 km in 2009) scheme options (NE 2009).

8.3.2 Hedgerow trees

Hedgerow trees are solitary trees emergent from hedgerows. They characterize many European agricultural landscapes and are valuable because of the many ecosystem services they provide, such as shade for livestock, aesthetic value, carbon sequestration, and soil protection. For wildlife, they are a source of fruit and seeds. They also provide shelter from wind, as well as nesting and roosting sites for birds and bats, song posts, hiding places, and both food and mate location sites for many insects. Hedgerow trees can also support diverse invertebrate, lichen, and fungal communities, and they act as stepping stones for mobile organisms to move through otherwise typically bleak agricultural landscapes (Slade et al. 2013). In England, pedunculate oak *Quercus robur*, ash *Fraxinus excelsior*, and formerly elm *Ulmus procera* too, are by far the most common species, but around 20 million elm trees were lost from the English landscape through the Dutch elm fungal disease in the late 1960s. The recent arrival of the ash die-back fungal disease is now threatening to be as damaging to the ash population. From their most abundant in the eighteenth century, when they served as a vital source of timber, hedgerow tree numbers have declined dramatically as a consequence of field enlargements and the mechanical trimming of hedgerows, which drastically affects recruitment. There are now only an estimated 1.6 million, with annual recruitment only half the level required to maintain the current population (Defra 2010). A question we posed was whether local macro-moth abundance and diversity were affected by the presence of hedgerow trees. To this effect, we contrasted 24 sites next to a single open-grown hedgerow tree (minimum height: 15 m, usually pedunculate oak *Quercus robur*) versus 24 sites without any nearby tree (Fig. 8.1).

8.3.3 Joined-up approach

The effects on macro-moth abundance and species diversity were explored with respect to these two key farmland elements—field margins and hedgerow trees. Furthermore, mindful of the generally damaging effects on biodiversity of habitat fragmentation, we investigated whether the impacts on macro-moths of field margins and hedgerow trees differed when the two farmland elements were part of a wider landscape, managed specifically with conservation goals in mind. There is no current policy to encourage neighbouring farmers to join AES and thus increase connections between habitats. We tried a joined-up approach in two experimental areas (hereafter 'targeted' areas of our Upper Thames Project; Macdonald and Feber 2015: Chapter 14). This approach turned out to be successful in terms of uptake. For example, after only two years, the experimentally targeted areas had more conservation management of habitats such as hedges (c. 219 km of hedgerows under enhanced management versus 83 km in control areas). We asked whether and to what degree this targeting approach made a difference to the moths we recorded on field margins and next to hedgerow trees, by comparing their numbers in targeted and non-targeted areas.

8.3.4 Moth sampling

During four field seasons (2006–2009) we sampled moths on 16 predominantly arable farms; all located within a 1200 km^2 area of the lowland agricultural landscape of Oxfordshire, UK. Each farm contributed three sites to the total of 48 fixed sampling sites (Fig. 8.1a). In general, these 16 farms had fields characterized by having both standard and wide margins. Within each farm, the number of hedgerow trees per field margin varied from zero to one or more hedgerow trees. Nevertheless, the precise locations of all three sampling sites at a given farm were chosen so that they belonged to only one of four experimental groups (four farms per group), which differed in their combinations of hedgerow tree presence and field margin width: (i) hedgerow tree + wide margin; (ii) hedgerow tree + standard margin; (iii) no hedgerow tree + wide margin; (iv) no hedgerow tree + standard margin (Fig. 8.1b).

We sampled each farm 40 times in discrete fortnightly periods from mid May to mid October, once in each fortnightly period, and in random order within the period. We usually sampled three farms (i.e. nine sites) on any one night. The total of 240 trap nights and 1920 trap events resulted in a sample of almost 72 000 individuals, from 311 macro-moth species. This large quantity of sampled and identified farmland moth individuals, and the space- and time-wise intensity of sampling, makes our study unique. While the nocturnal lifestyle of moths makes them challenging to study, their well-known attraction to light means it is relatively easy to sample them. We used battery-run, portable heath pattern actinic light traps (6 W). These were operated from dusk

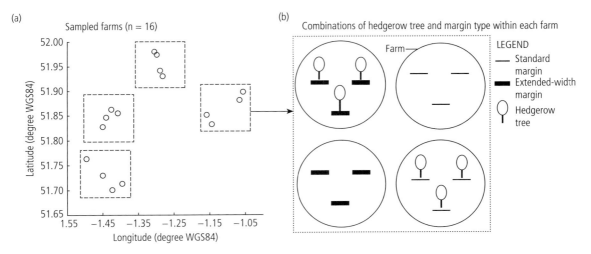

Figure 8.1 (a) Distribution of the 16 sampled farms (open circles) and (b) scheme of the sampling design within each farm (range of minimum, mean, and maximum distances between the three sites per farm: 60–470 m; 190–1290 m; 280–1930 m, respectively). From Merckx et al. (2012). Reproduced with permission from John Wiley & Sons.

until dawn, when the live sample in and on the trap was enumerated and identified to species level, except for five species-pair aggregates of essentially cryptic species. All sampling sites were positioned 1 m away from average-sized hedgerows (2–3 m high, 1.5–2.5 m wide), with both sides bordered by arable land. Sampling sites were at least 50 m away from hedgerow intersections and were at least 100 m apart, which prevented moth attraction radius interference (Merckx and Slade 2014). Traps were placed upon a standard-sized white sheet, which enhanced and equalized trap visibility and enabled us to include all individuals resting on the sheet. Sampling was conducted in similar, sufficiently favourable conditions to minimize bias due to differences in weather-related activity levels.

Apart from the main insights obtained for mostly common and widespread macro-moth species, our sustained trapping effort resulted in some rarer species too. We passed all of our records on to the National Moth Recording Scheme (<http://www.mothscount.org>), whose records help to provide a better picture of each species' national distribution. Perhaps our most fascinating find was the unexpected discovery, on arable farmland, of what is currently the largest known population of the pale shining brown *Polia bombycina*, a rare UK Biodiversity Action Plan (BAP) priority species (see Section 8.7).

8.3.5 Effects of hedgerow trees, wide field margins, and landscape on moths

Our findings highlighted the importance of hedgerow trees and wide field margins for moth conservation in the wider countryside, as both turned out to be beneficial for moth populations. The presence of hedgerow trees and wide margins each significantly increased macro-moth species numbers locally from *c.* 90 to *c.* 105 species on average, or by around 15% (Merckx et al. 2009a, 2012) (Fig. 8.2). Largest numbers of macro-moths were found at sites characterized by both wide margins and hedgerow trees, which was the result of additive rather than interactive effects. Such sites had, on average, 15% more individuals and ten more species than sites characterized by either one of these farmland elements (mean ± SE: abundance: 1286 ± 181 versus 1117 ± 90; species richness: 110 ± 6 versus 100 ± 4), and they had 33% more individuals and 26 more species than sites lacking both farmland elements (1286 ± 181 versus 965 ± 9; 110 ± 6 versus 84 ± 2.8, respectively).

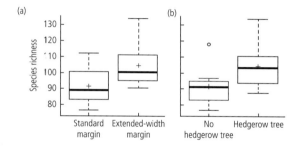

Figure 8.2 Main effects ($P < 0.05$) of the presence of (a) wide versus narrow margin, and (b) presence of a hedgerow tree at the local scale on macro-moth species richness. A solid line in the boxplots indicates the median, while a cross indicates the mean. From Merckx et al. (2012). Reproduced with permission from John Wiley & Sons.

The species most helped by wide field margins appeared to be the shoulder-striped wainscot *Mythimna comma*, a grass-feeding, common, widespread, but declining (72% national decline over 35 years) UK BAP species: 98% of all shoulder-striped wainscots were sampled in wide field margins. Other nationally declining macro-moth species particularly helped by the wide field margins' likely increase in foodplant quality and amount, and occurring in numbers four to eight times higher on wide versus standard margins, were (i) treble-bar *Aplocera plagiata*, feeding on St John's wort *Hypericum perforatum*, (ii) small phoenix *Ecliptopera silaceata*, a UK BAP species (77% decline over 35 years) feeding on willowherbs *Epilobium* spp., (iii) frosted orange *Gortyna flavago*, whose larvae feed and pupate internally in plant stems, (iv) small dotted buff *Photedes minima*, with tufted hair-grass *Deschampsia cespitosa* as foodplant, (v) feathered gothic *Tholera decimalis*, a grass-feeding UK BAP species (89% decline over 40 years), and (vi) grass rivulet *Perizoma albulata*, a UK BAP species (93% decline over 40 years) whose larvae feed on the hemiparasitic yellow rattle *Rhinanthus minor*. Both the barred rivulet *Perizoma bifaciata*, a local, though widely distributed, moth feeding on another hemiparasitic plant, red bartsia *Odontites vernus*, and the ghost moth *Hepialus humuli*, a species with a worrying national trend (62% decline over 40 years) feeding on the roots of grasses and herbs, occurred exclusively at wide field margins.

Nationally declining macro-moth species not feeding directly on herbaceous field margin species were also found in greater numbers on wide field margins. One possible explanation is that margins buffer hedgerows and trees from pesticide drift (Pywell et al.

2004). Examples of species that were between four and six times as abundant on wide compared to standard margins are: (i) buff arches *Habrosyne pyritoides*, feeding on bramble *Rubus* spp., (ii) beautiful hook-tip *Laspeyria flexula*, whose overwintering larvae feed on bark lichens, (iii) oak hook-tip *Watsonalla binaria*, a UK BAP (78% decline over 40 years) oak-feeding species, and (iv) pale eggar *Trichiura crataegi*, another UK BAP species (90% decline over 40 years), feeding on blackthorn and hawthorn. The declining, tree-feeding lunar-spotted pinion *Cosmia pyralina* occurred exclusively on wide margins.

The poplar grey *Acronicta megacephala*, fairly common and well distributed in England, is a powdery looking, greyish noctuid whose caterpillars feed on poplar *Populus* spp. leaves. Poplar grey appeared to be the species most helped by hedgerow trees—none of them poplars—, occurring almost twenty times more abundantly near hedgerow trees than at sites without hedgerow trees. Nationally (fairly) common and (fairly) well-distributed macro-moth species that were ten times more abundant at hedgerow trees, included: (i) white-spotted pug *Eupithecia tripunctaria*, a small geometrid with larvae on elder *Sambucus nigra* and wild angelica *Angelica sylvestris*, (ii) dingy shears *Parastichtis ypsillon*, whose nocturnal larvae feed on willow *Salix* spp., and (iii) clouded border *Lomaspilis marginata*, a nationally declining delicate geometrid with blackish and white markings, with sallow *Salix* spp. and poplars as foodplants.

Hedgerow trees seemed particularly important for some macro-moth species, 51 of which were only ever recorded at sites with hedgerow trees. These included: (i) pretty chalk carpet *Melanthia procellata*, an attractive but now nationally severely declining (88% decline over 40 years) geometrid, typical of hedgerows containing the climbing shrub traveller's joy *Clematis vitalba*, (ii) the nationally declining, campion *Silene* spp.-feeding, sandy carpet *Perizoma flavofasciata*, and (iii) a whole series of species more typical of woodland, such as satin beauty *Deileptenia ribeata*, large emerald *Geometra papilionaria*, pine hawk-moth *Hyloicus pinastri*, pale oak beauty *Hypomecis punctinalis*, olive *Ipimorpha subtusa*, and leopard moth *Zeuzera pyrina*.

We conclude that paying farmers to protect and establish more hedgerow trees could make a major contribution to halting the decline in moth diversity in 'farmscapes' typified by our study areas, as do existing AES payments for wide field margins. The likely main benefits of a higher density of hedgerow trees and wide field margins on farmland are (i) increased shelter for thermally constrained organisms in the otherwise barren and exposed environment (results from another experiment (see Section 8.4.1) appeared to confirm this), (ii) increased abundance and variety of foodplants, and (iii) increased provision of tree sap, nectar, pollen, and undisturbed roosting areas. Because moths are sensitive indicators for farmland biodiversity in general, and in particular for other terrestrial insects, we expect that these benefits would also benefit other insect groups, such as butterflies and bumblebees (Merckx et al. 2008; Goulson et al. 2011; Haaland et al. 2011).

In addition to the sheltering effect that hedgerow trees have on moths, they also provide larval feeding and female egg-laying resources, at least for some of the shrub and/or tree-feeding macro-moth species (hereafter called 'high-feeders'), although they do not for grass and/or herb-feeders ('low-feeders'). Also, it is likely that hedgerow trees exhibit a stronger attraction to high-feeders in terms of providing adequate roosting sites compared to low-feeders, as the latter may be more likely to roost in vegetation close to the ground (e.g. the highly abundant large yellow underwing *Noctua pronuba* and other common noctuid species, such as *Noctua*, *Agrotis*, *Xestia*, *Mythimna*, *Apamea*, and *Hoplodrina* spp.). In contrast to the stronger benefit of hedgerow tree presence for high-feeders, low-feeders did not benefit any more strongly than high-feeders from the presence of wide margins (Merckx et al. 2012), suggesting that both guilds benefited equally from the presence of wide margins. The positive effect of wide margins can be explained by the fact that they provide a relatively undisturbed breeding habitat and can act as buffer zones against the impact of agricultural chemicals on moth larvae and their host plants (Pywell et al. 2004). As such, this result indicates that wide margins may improve larval habitat quality, both for low-feeding larvae within the margins, and for high-feeding larvae in adjacent hedgerows and lower parts of hedgerow trees, by reducing exposure to pesticides and fertilizers. Another (complementary) explanation may be that floral resources are providing nectar to adults of both guilds. Increasing plant species richness of margins, and optimizing availability of resources through appropriate management, may have benefits for both larval and adult macro-moths, as is the case for other farmland Lepidoptera (Feber et al. 1996).

Although the observed local increase in macro-moth species richness is not, by itself, a measure of ecosystem functioning, evidence suggests that they are positively related (Hector and Bagchi 2007). Increased macro-moth richness is likely to provide a number of economically valuable ecosystem services. First, pollination success and pollination resilience are likely to be facilitated due to higher numbers of pollinating moth species (Devoto et al. 2011). Other pollinating taxa possibly also benefit from the presence of these farmland elements (Power and Stout 2011). Higher pollination success should not only benefit populations of wild plant species, but also increase fruit set and yields of insect-pollinated crops (Holzschuh et al. 2012). Second, moths are important prey for various taxa (see above), which may result in larger, and more stable, populations of a greater diversity of species at these higher trophic levels. In turn, this may provide better and cheaper crop pest control (Winqvist et al. 2011). The resulting increase in functional diversity of field margins and hedgerows will feed back into an improved ecosystem functioning of farmland as a whole via the affected ecosystem services (Cadotte 2011).

Our findings for macro-moths have important implications for AES policy and for delivering the best bangs-per-buck to the tax-payer who funds them. To the extent that we are correct in proposing macro-moths as model animals to inform biodiversity policy on farmland, the chief lesson of our findings is to advocate a policy shift from field- and farm-scale implementation of AES towards connective landscape-scale conservation. While hedgerow trees were found to benefit moths—c. 15% overall increase in species richness (Fig. 8.2) and c. 20% increase in overall abundance—the best results were in landscapes where we targeted farmers to join AES. In these areas, the abundance of moths was 60% greater at sites with a hedgerow tree than at sites without trees, and macro-moth species diversity was 38% greater at sites with a hedgerow tree (Fig. 8.3). We deduce that the mechanism underlying this striking result is that the higher proportion of land covered by AES in these areas resulted in the joining-up of habitat resources across the landscape. Thus, the *context* of the trees, and the *relationship* of protected areas to each other within a landscape are revealed to be important to delivering effective conservation. Our findings on macro-moths offer a foundation for policy thinking about wildlife on farmland such that, to preserve the biodiversity of agricultural landscapes, it is more effective to implement measures at spatial scales greater than those of individual fields; the most fruitful focus is on strengthening the diversity within landscapes and ecosystems as a whole. Although the financial and societal implications are far-reaching, the lesson of the macro-moths is that it could be beneficial for biodiversity on farmland to complement the entry-for-all approach to AES with a system that targets specific areas and/or landscapes in high nature conservation value farmland (Merckx et al. 2009a).

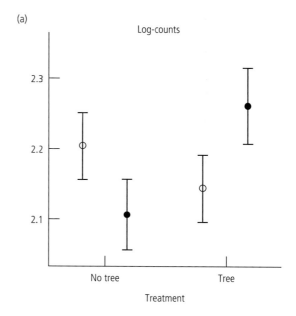

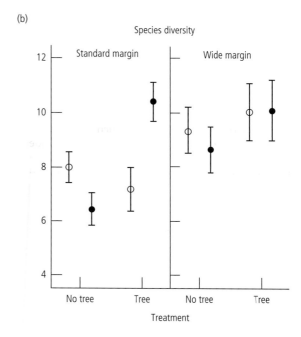

Figure 8.3 Fortnightly (a) individual moth counts (log $N + 1$) contrasting the effects of presence/absence of hedgerow trees on moth abundance, and (b) log-series α indexes of moth communities contrasting the effects of presence/absence of hedgerow trees and wide field margins on moth species diversity, in areas where farmers had (●) and had not (○) been targeted to apply for agri-environment schemes (error bars represent SE). From Merckx et al. (2009a).

Figure 8.4 (a) Male buff ermine *Spilosoma luteum*, captured with a Heath light trap (Actinic 6 W), which operates on the 'lobster-pot principle', whereby moths are drawn to an actinic tube secured vertically between baffles, fall unharmed down a funnel, and rest on the inside of the trap or on pieces of egg-tray provided. Traps were operated from dusk to dawn, when the live sample of selected species was marked (at first capture) by writing a unique number on the left forewing with a fine (0.4 mm), non-toxic, permanent waterproof marker and released *in situ* into nearby tall vegetation. (b) Same individual marked with a black number '6'. Photographs © Rita Gries.

Figure 8.5 Marked (red number '100') individual of drinker *Euthrix potatoria*, a common but nationally declining Eggar moth (Lasiocampidae). Both its common English name and scientific species name are derived from the larval liking for drinking drops of rain or dew on its foodplants, whereas the genus name refers to the hairy, thickset adults. The larvae too are densely covered in hairs, which gives them protection from being eaten by birds, except cuckoos. Photograph © Claire Mclaughlan.

8.4 Moth mobility: impacting local effects of farmland resources

8.4.1 Hedgerow trees

Motivated to better understand the mechanisms that underlay our finding that hedgerow trees were beneficial to the abundance and species diversity of macro-moths on farmland, we conducted a farm-scale mark-release-recapture (MRR) experiment (Fig. 8.4 and Fig. 8.5; Merckx et al. 2010a). We asked whether hedgerow trees increase moth numbers mainly because they provide shelter, which would be of strongest benefit to slender, sedentary macro-moths, or because they provide food resources for specific tree-feeding species only. If the latter hypothesis prevailed, hedgerow tree options within AES would favour significantly fewer species than if the former hypothesis prevailed, because hedgerow tree species would only benefit a specific suite of insect species adapted to use specific tree species as larval foodplants (or for adult sap-feeding). For example, hedgerow oak trees would only be beneficial to oak-feeders, such as the blotched emerald *Comibaena bajularia*, and not to other species, such as the small emerald *Hemistola chrysoprasaria*, a moth only feeding on traveller's joy. Furthermore, if the predominant importance of hedgerow trees to macro-moths was to provide larval foodplants, then the spatial abundance of not only hedgerow trees in general, but of specific tree species, would need to be taken into account to optimize the biodiversity gain from AES.

We opted for a MRR approach to find out how far individuals of different species travelled on farmland, as well as where the moths were, which gives a strong indication of species' habitat preferences. We were able to do this by individually and harmlessly marking moths with a pen, and by trying to recapture them by applying a continuous trapping effort during 33 nights. We marked and recaptured 23 pre-selected species of moth for which the larval foodplants are well known and which were assigned to one of two feeding guilds (i.e. 13 species of 'high-feeders' versus 10 species of 'low-feeders') using 20 fixed light-trap sampling points within five adjacent arable fields (Merckx et al. 2010a).

We found that hedgerow trees were significantly associated with increased adult moth numbers, even for eight shrub/tree-feeding species which did not feed on the tree species available in the hedgerows at our sampling points. We therefore deduced that the increased adult moth numbers associated with hedgerow trees were likely to have arisen, at least for these eight species, because of the shelter these trees provided in the typically exposed agricultural landscapes. We note that prominent trees do function as assembly points for adult mating in several insect species, providing a possible additional (or alternative) explanation for these higher moth numbers near hedgerow trees. Shelter is also the likely explanation for higher numbers of some 'low-feeders' near hedgerow trees, such as yellow shell *Camptogramma bilineata*, drinker *Euthrix potatoria*, treble lines *Charanyca trigrammica*, rustic shoulder-knot *Apamea sordens*, and shears *Hada plebeja*. This conclusion was corroborated by the observations that the hedgerow tree effect was (a) not significant for the two most mobile species of our set of high-feeders, i.e. scalloped oak *Crocallis elinguaria* and buff-tip *Phalera bucephala*, with mean covered distances above 550 m, and (b) strongest for less mobile species, such as many small geometrids (Merckx et al. 2010a). Sedentary species of macro-moth are believed to be more prone to convective cooling in typically exposed agricultural landscapes (Dover and Sparks 2000;

Pywell et al. 2004), so we predicted that they would benefit most from the additional shelter provided by trees (Dover and Sparks 2000). Macro-moths in this category included scorched carpet *Ligdia adustata*, a declining, small geometrid feeding on spindle, as well as the following geometrids feeding on traveller's joy: (i) small emerald, a UK BAP species (82% decline over 35 years), (ii) pretty chalk carpet, and (iii) small waved umber *Horisme vitalbata*. We found these species had covered relatively small mean distances (0–185 m), and were indeed associated, in large numbers (+ 73%), with hedgerow trees.

One plausible idea is that hedgerow trees may act as 'stepping stones' for some species, especially less mobile and woodland species, increasing the opportunity for moths, and other flying insects, to cross barren exposed farmland in search of resources, and facilitating dispersal between patches of semi-natural habitat. Tall hedgerows may serve a similar function; a study on Canadian farmland found 60% more macro-moth species, and in triple the abundance, near 20 m tall hedgerows compared to within fields, and this finding was more pronounced for two low-mobility (sub)families (i.e. Geometridae and Arctiinae) than for the generally more mobile noctuids (Boutin et al. 2011).

In the context of climate change, such stepping stones could facilitate the northward movements through farmland that may enable species to stay within their climatic envelope as the climate changes. This has been shown for butterflies (Hill et al. 2002; Menéndez et al. 2007) and would apply to the many moth species which have southern distributions but are struggling to expand their ranges northwards in response to climate change, due to low intrinsic dispersal capacity and/or the general hostility of the agricultural matrix. For example, although some macro-moths have shown substantial range shifts within Britain (up to 393 km between 1982 and 2009 for the red-necked footman *Atolmis rubricollis*; Fox et al. 2011), such species are very likely to be a small minority. Betzholtz et al. (2013) have recently analysed range margin shifts for all southerly distributed macro-moths and butterflies in Sweden (the analysis for macro-moths in Britain is currently in preparation; R. Fox pers. comm.). The Swedish study shows that 60% of the 282 analysed species had expanded their northern range margin between 1973 and 2010, yet it also shows a huge variation in expansion distance (min–max: 0–850 km; mean: 101 km). What is clear is that range shifts for the majority of thermally constrained insect species are probably happening already, and lagging behind to varying degrees depending on the species. Increasing the functional connectivity of 'farmscapes', helped by the establishment of more hedgerow trees providing more shelter, will help to mitigate these time lags. Although we believe that a higher density of hedgerow trees will benefit a majority of species by allowing them to move more easily through agricultural landscapes, we predict this will especially benefit woodland moth species, such as pale oak beauty, white-pinion spotted *Lomographa bimaculata*, slender brindle *Apamea scolopacina*, and oak hook-tip, to name just a few.

Given that most of Europe's intensive agricultural land was once dominated by forest, hedgerow trees are often the only remaining farmland element linking to this natural climax biotope. As such, it is not surprising that they make a large contribution to the ecological resilience of farmed landscapes. It is hence likely that hedgerow trees are keystone structures, with a disproportionate effect on ecosystem functioning given the small area occupied by any individual tree. Although proactive conservation management of hedgerow trees was not, until recently, rewarded financially in any EU country, it is now a recent addition to the set of general AES options within England (see also the new Scottish Rural Development Priorities), due in part to the results of our studies. These new AES options include the establishment of new hedgerow trees by tagging saplings, and the establishment of protective hedgerow tree buffer strips on both grassland and cultivated land (Defra/NE 2012). Expanding AES support to include these options should increase both the field- and landscape-scale supply of trees that not only provide shelter and stepping stones, but are also visually prominent enhancements of the landscape. Importantly, for the compatibility of food production and biodiversity conservation, the establishment of hedgerow trees, especially when using saplings already present within hedgerows, and the retention of hedgerow trees, have minimal costs to farmers, mainly related to the increased care that needs to be taken while trimming hedgerows, although mature hedgerow trees may very locally compete with crops for light and water.

As a result of the many ecosystem service benefits in return for these minimal costs, we believe that hedgerow trees are highly compatible with intensive farming systems. As such, hedgerow trees may be habitat resources that are spatially compatible within high-production areas, which are otherwise spatially separated from conservation areas in the 'land sparing' framework (Phalan et al. 2011). However, once a landowner opts to also provide hedgerow tree buffer strips, which are typically 6 m wide, we come conceptually closer to 'land sharing', where conservation and production are spatially integrated (Phalan et al. 2011). It is especially within a 'land sharing' context that foregone financial profit (cost of establishment and foregone crop of buffer strip) should be compensated.

Although it is impossible to give advice on a truly 'optimal' spacing of hedgerow trees, because such requirements differ from one organism to the other (Van Dyck 2012), we believe that AES advice on this (Defra/NE 2012)—two to three trees over 100 m of hedgerow—will be effective for increasing general farmland biodiversity and functional connectivity of agricultural landscapes. This advice is also in line with the suggestions for more, larger, better, and joined blocks of habitat given by the Lawton report (Lawton et al. 2010). Furthermore, as we have shown that the beneficial effect of hedgerow trees is mainly to be had from the shelter they provide, and only to a smaller extent from species-specific aspects, our only guideline is to use a variety of autochthonous tree species in line with what would be naturally on offer within the region.

8.4.2 Field margins

As with hedgerow trees, we also quarried into our findings regarding field margins, in order to try to uncover the mechanisms that caused their beneficial effects on the abundance and diversity of macro-moths. For example, these effects might stem from an enhanced habitat quality offered by the margins (e.g. a diverse mixture of native pollen/nectar-rich wildflowers; Haaland et al. 2011), or because the margins are configured as a ramifying *rete* which enhances connectivity at the landscape scale (e.g. Rundlöf et al. 2008). To explore these possibilities, we selected nine widespread moth species that, as larvae, mainly feed on various grasses and low-growing herbaceous plants. From the 1699 individuals that we marked, by far the three most common species, accounting for 73% of all observations, were the noctuid species heart and dart *Agrotis exclamationis* and large nutmeg *Apamea anceps*, and common swift *Hepialus lupulinus* (with root-feeding underground larvae like most 'swift' moths). We undertook a field-scale MRR light trapping study over 32 nights in four adjacent arable fields, which were bordered with hedgerows scattered with hedgerow trees throughout (Merckx et al. 2009b). Fields either had surrounding wide or standard-width margins and within each of the four fields, we sampled five sites: the field centre and one site at each margin; two of these margin sites were positioned near a hedgerow tree. Our goal was to discover whether the overall positive effect of grassy, wide field margins (reported previously) was dependent on species-specific mobility, as had transpired with the hedgerow tree effect. Although all nine species are common and widespread, two of them (white ermine *Spilosoma lubricipeda* (Fig. 8.6) and large nutmeg) are severely declining species, the latter having declined by 93% over 40 years (Fox et al. 2013).

Overall, light traps captured almost twice as many moths on field margins compared to field centres. For all but one species, field margins were more abundant in terms of moth individuals than field centres. For instance, all 56 trapped common footman *Eilema lurideola* were trapped in field margins. The only exception was the relatively mobile (see below) setaceous

Figure 8.6 White ermine *Spilosoma lubricipeda*—A common, widespread 'tiger' moth, which has nevertheless declined nationally by 70% over 40 years, and is hence listed as a UK Biodiversity Action Plan priority species (IUCN category: vulnerable). Its common English name refers to the black-spotted white fur, obtained from the winter skin of the stoat *Mustela erminea*, which is historically associated with royalty and high officials, whereas the scientific name refers to the bold spots on the abdomen of the adult and to the swift-footed, speedy gait of the larva, respectively. Photograph © Maarten Jacobs.

hebrew character *Xestia c-nigrum* (characterized by a dark and distinctive C-shaped mark in the centre of the forewings), which was actually 27% more abundant in field centres than in margins (average trap abundance: 6.75 versus 5.31 individuals, respectively). Moreover, wide field margins had significantly more moths overall (+ 40%) compared to standard-width field margins. For example, 78 brown-line bright-eye *Mythimna conigera* moths (relatively sedentary noctuids), were caught at wide field margins, compared to just seven on standard-width margins (four were caught in the field centres).

We also measured the abundance of nectar sources and found that flower heads were significantly more abundant (250–300%) on wide versus standard field margins, though not significantly so per unit area. Wide margins also offered a greater area and better quality of breeding habitat (relatively undisturbed larval habitat, increased larval food resources, foodplants, and larvae better buffered from agrochemicals: Pywell et al. 2004). Nectar sources were absent from cereal fields. Furthermore, there was no consistent difference in the nectar-producing characteristics of hedgerows that adjoined standard as opposed to wide margins, so this is unlikely to have confounded our results. Remarkably, the abundance of macro-moths at field centres was 60% higher for those fields bordered with wide margins than those bordered by standard margins. A plausible explanation is that the better habitat provided by wider margins increased the resource base and thus the abundance of moths that could spread from this source to the adjoining impoverished field centres. Some individuals captured at the field centres had indeed been caught and marked previously in a nearby field margin (Fig. 8.7), illustrating the tendency of these moths to make exploratory movements (Van Dyck and Baguette 2005).

In practical terms, our findings suggest that habitat quality of wide field margins could be improved by: (i) altering commercial seed mixes so that these contain more nectar-producing plants and more foodplants for Lepidoptera; and (ii) modifying their management from annual cutting (a common regime for grass and wildflower sown margins) to cutting once every two–three years (Kuussaari et al. 2007), which can be done by an annual rotation where half to a third of margins are cut every year. Summer cutting in particular should be avoided. Annual cutting, particularly in high summer, makes it difficult for moths, especially univoltine species, to complete a full life cycle, as is also the case for butterflies (Feber and Smith 1995).

Figure 8.7 Map showing the four study fields (grey); the two fields with wide field margins are outlined in bold. Sampling sites near a hedgerow tree are indicated with a square; sites lacking hedgerow trees are indicated with a circle. Observed individual movements (> 0 m) are contrasted between the species groups with opposite effects for the variable 'margin'. Individuals within the group of species where the statistical evidence for an effect of 'margin' was absent covered longer distances and were less frequently recaptured at the site of first capture than the group of species where the effect of 'margin' was stronger (bold dashed lines; large nutmeg *Apamea anceps*, setaceous hebrew character *Xestia c-nigrum*; N_{total} = 21; $N_{recaptured\ at\ site\ of\ first\ capture}$ = 4 versus slim dashed lines; treble lines *Charanyca trigrammica*, brown-line bright-eye *Mythimna conigera*, heart and dart *Agrotis exclamationis*, common footman *Eilema lurideola*, common swift *Hepialus lupulinus*; N_{total} = 43; $N_{recaptured\ at\ site\ of\ first\ capture}$ = 29, respectively). One of the bold dashed lines and one of the slim dashed lines cover the movements of two individuals each. From Merckx et al. (2009b).

Having discovered an interaction between a hedgerow tree effect and species-specific mobility, we found that differences in mobility among species had an impact on the effect of wide field margins too. The statistical evidence of our main effects was negatively correlated with the observed mobility of the nine species studied. For instance, species that were relatively mobile (seldom recaptured at site of capture; typical average distance covered between capture and first

recapture: 450–600 m), such as setaceous hebrew character and large nutmeg, were at least as abundant in field centres and standard field margins as they were in wide field margins. In contrast, relatively sedentary species (frequently recaptured at site of first capture; typical average distance: 50–300 m), such as treble lines and brown-line bright-eye, benefited from the presence of wide field margins (Fig. 8.7). These correlations between species mobility and the species-specific statistical strength of the effect of wide field margins, and hedgerow trees too, raise the possibility that the standard, field-scale uptake of AES may be effective only for less mobile species.

8.5 The need for landscape-scale implementation of habitat resources

We have seen that the variation in response of macromoths to the presence of hedgerow trees and wide field margins corresponded well to species mobility: the abundance of more sedentary moths increased near hedgerow trees and at wide margins, but not the abundance of more mobile moth species. We believe that these more mobile species, such as the severely declining large nutmeg, are not responding well, in terms of individual moth numbers, to these tree and margin resources because these extra resources were provided at a field or farm scale only. Mobile moths typically fly relatively large distances, up to several kilometres (Slade et al. 2013), and hence move around in search of resources on a landscape scale. For such mobile species, field-scale measures on an individual farm will consequently only make a relatively small, and often trivial, contribution to population levels (Fig. 8.8). As such, only sedentary species (i.e. species where the large majority of individuals move only a few hundred metres) are likely to benefit locally from a local, field-scale increase in habitat resources, such as the establishment of wide margins around an arable field. In order to benefit populations of relatively mobile species too, wide field margins and hedgerow trees will need to be established around a majority of fields in whole landscapes (see also Hambäck et al. 2007), rather than—as is currently still often the case—around single fields, diffusely scattered within inhospitable landscapes, composed mainly of fields with narrow margins and a low density of hedgerow trees (Fig. 8.8).

This principle most probably applies to other species groups and other semi-natural farmland elements, at least for structurally simple landscapes typified by low habitat heterogeneity (containing < 20% semi-natural habitat), such as intensive farmland. For instance, while

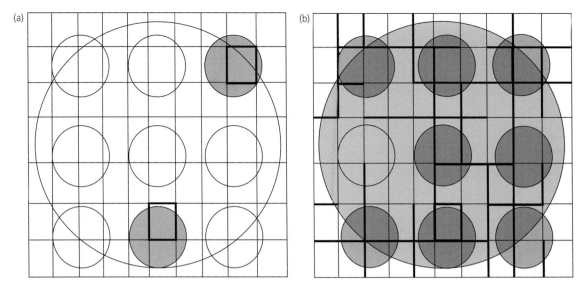

Figure 8.8 Schematic representation of two landscapes under intensive agricultural use, with fields represented as rectangles. Field margins covered by AES options (e.g. 6 m-wide strips) are shown in bold. Small circles represent populations of a sedentary species, whereas the large circle represents a population of a relatively mobile species. Filled circles represent populations benefiting from AES. We contrast (a) scattered, field-scale uptake with (b) landscape-scale uptake of AES. The first scenario does not meaningfully benefit the mobile species, and only benefits a couple of populations from the sedentary species, because whilst the two enhanced fields significantly improve overall habitat quality for two highly local populations of the sedentary species, they only represent a small proportion of the relatively large 'home range' of the mobile species. The second scenario does benefit the population of the mobile species, and also benefits all but one of the sedentary species' populations.

some sedentary woodland moth species, such as maiden's blush *Cyclophora punctaria*, black arches *Lymantria monacha*, and nut-tree tussock *Colocasia coryli*, do occur in small (a few hectares) farm woodlots, mobile woodland specialists, such as the lobster moth *Stauropus fagi*, appear to be restricted to larger (> 5 ha) woodland fragments only (Slade et al. 2013). Depending on their species-specific dispersal characteristics and corresponding differences in landscape-wide resource use, different species operate and experience the landscape at different spatial scales (Steffan-Dewenter et al. 2002; Van Dyck 2012) (Fig. 8.8). As a result, population densities of relatively mobile species will be affected by the surrounding landscape quality at large spatial scales only, whereas populations of low-mobility species will be affected at smaller scales (e.g. diameters of 6 km, 3 km, and 0.5 km for honeybees, bumblebees, and solitary wild bees, respectively; Steffan-Dewenter et al. 2002). This means that, while low-mobility species suffer more from locally adverse conditions than do mobile species, favourable local conditions within adverse landscape conditions will mostly benefit low-mobility species only (Tscharntke et al. 2005) (Fig. 8.8).

In line with these findings, Thomas (2000) and Rundlöf et al. (2008) concluded that relatively mobile butterfly and bumblebee species are more affected by habitat fragmentation than low mobility species (although species at the very high end of the mobility spectrum, like the small white *Pieris rapae* butterfly, are generally surviving well).

We have argued that macro-moths are revealing models for understanding patterns of animal communities on farmland and the ways in which AES may be tailored to deliver the best compromises for conserving biodiversity alongside food security. A key lesson from these model organisms is that AES should be devised to take account of the spatial scales at which populations of wider-countryside species use the agricultural matrix and the mosaic of semi-natural habitat within this matrix: although a field-scale uptake of AES options may bring significant benefits to low-mobility species, only a landscape-scale uptake is likely to benefit the whole set of wider-countryside species, inclusive of the high-mobility species (Fig. 8.8). This is important because, for agricultural landscapes characterized by intensive farming systems, mobile, generalist species are a key group in terms of ecosystem functioning for two reasons. First, plant-pollinator networks are highly dependent on the abundance of a core group of generalist species (Devoto et al. 2011) and, second, highly mobile, large-scale species influence food-web interactions more than small-scale species, as the latter are characterized by dispersal limitation (Tscharntke et al. 2005). In practical terms, we advocate that field margin options, and indeed other AES options, should be targeted and implemented at a landscape scale (rather than at the current, standard field or farm scale). By adopting our recommendations, AES options should not only benefit sedentary species, as they do now, but also more mobile species (Fig. 8.8).

In short, while small-scale AES may advantage less mobile species but not more mobile ones, landscape-scale AES will benefit both, and thus deliver more to the tax-payer and policy-maker. Delivering this outcome might involve, for example, encouraging contiguous farms to take up AES options in order to reduce habitat fragmentation and maximize habitat linkages, as in our experimental 'targeted' landscapes (Macdonald and Feber 2015: Chapter 14). This approach was also successfully implemented in the Chichester Plain, UK, where pro-active targeting of farms created a landscape-scale network of managed buffer strips along water courses, resulting in significant increases of the endangered water vole *Arvicola amphibius* (Macdonald et al. 2007; Dutton et al. 2008; Chapter 15, this volume). In England, 67% of the utilizable agricultural area was covered by AES in 2011. However, only 8.1% of the area was under the Higher Level Stewardship (HLS) scheme, which aims to deliver significant environmental benefits in priority areas (Defra 2011). Moreover, HLS remains discretionary, and may be less appealing to farmers because of the commitment to more complex environmental management.

In summary, we argue that a more effective repayment for investment by society in nature on farmland is likely to be delivered by the use of relatively simple existing and new AES prescriptions, but—importantly—implemented over large (landscape-scale) areas.

8.6 Landscape-scale impacts of agricultural intensification

Farms, while separate economic enterprises, are not ecological islands. To what extent does the fate of biodiversity on a farm, or the effectiveness of AES implemented on it, depend on the management of the neighbouring farms (Gabriel et al. 2010; Batáry et al.

2011)? Insofar as our findings about macro-moths pushed our perspective towards the scale of landscapes, and thus ecological patterns and dynamics that embrace several adjoining farms, we expected that AES options would have larger effects on biodiversity in settings typified by increasing intensification, simplification, and homogeneity of landscapes (Tscharntke et al. 2005). Moving on from the discovery that two elements of the 'farmscape' (i.e. wide field margins and hedgerow trees) were associated with increased abundance and diversity of macro-moths, we went on to test whether their effect was moderated by the amount of intensively managed agricultural fields in the surrounding landscape (Merckx et al. 2012). While we expected that a greater expanse of arable land would generally result in lower overall moth abundance and species richness, we predicted that the positive impact of wide field margins and hedgerow trees on both abundance and richness would be stronger in the context of a landscape more dominated by arable land.

Using a Geographic Information System (GIS, ArcMap 9.2), five circles (radii: 200 m, 400 m, 800 m, 1600 m, and 3200 m) were mapped around each of the three sampling sites of each of the 16 farms (Fig. 8.1). Using recent land-use data, we calculated the percentage of arable land within each of these circles. The five spatial scales were selected to cover roughly the extent of foraging movements for a gradient of low- to high-mobility species of macro-moths (Slade et al. 2013), whereas the variable 'percentage arable land' was chosen because it is considered to be a good indicator of the degree of agricultural intensification (Tscharntke et al. 2005).

Our original expectations were not fulfilled. The amount of arable land in the surrounding landscape did not affect overall species richness and abundance at any of the five spatial scales, and did not modify the effect of the local factors (i.e. there were no significant statistical interactions with hedgerow tree and wide margin). In practice, this means that wide field margins and hedgerow trees have an effect (increasing both overall abundance and species richness of macro-moths) no matter the degree of arable land cover on adjoining farms and irrespective of the spatial scale on which the landscape is viewed (at least for the range tested in our study—200–3200 m).

However, this unexpected generalization obscured different answers when our results were considered for species with different natural histories and, in particular, their conservation status. This became clear when we classified species into three classes based on

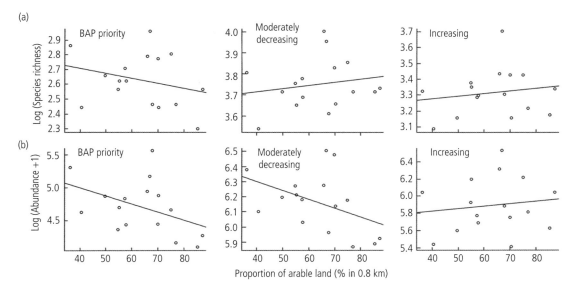

Figure 8.9 Effects of the cover of arable land in the landscape (% within 0.8 km radius) on (a) macro-moth species richness, and (b) macro-moth abundance, separately for severely declining species (i.e. UK BAP priority species), moderately declining species, and increasing species. From Merckx et al. (2012). Reproduced with permission from John Wiley & Sons.

national abundance trend data over 35 years (Conrad et al. 2006): (a) severe decline (70–99%; i.e. UK 'BAP' priority species, N = 44); (b) moderate decline (0–69%; i.e. 'declining' species, N = 106); and (c) increase (> 0%; i.e. 'increasing' species, N = 76). Typical examples for each of the three classes respectively are: (a) white ermine (Fig. 8.6); (b) drinker (Fig. 8.5); and (c) dingy footman *Eilema griseola*, which is nationally on the increase like most other moth species feeding on lichens/algae (Conrad et al. 2006). Species for which the national trend is unknown were not retained in the analysis. All three groups reacted positively in terms of abundance and species richness, whatever the degree of agricultural intensification, to wide margin and hedgerow tree presence. So, these farmland elements delivered a positive overall effect, and they did so at all spatial scales.

Our study showed that nationally declining species became less abundant with increasing levels of agricultural intensification in the surrounding landscape, especially so at the intermediate spatial scale of 0.8 km radius (Fig. 8.9b). Our study is the first to show that this may be the result of direct negative impacts of landscape intensification on these nocturnal insects. As such, this indicates that agricultural intensification may be the factor that explains most of the recent, extensive decline of many macro-moths, such as dusky thorn *Ennomos fuscantaria*, lackey *Malacosoma neustria*,

and garden tiger (98%, 93%, and 92% declines over 40 years in Britain, respectively) (Fox et al. 2013). These negative effects of agricultural intensification on the abundance of both nationally severely (UK BAP priority) and moderately declining species did translate into a negative effect with regard to species richness for the UK BAP priority species group only (Fig. 8.9a). The observation that agricultural intensification did not seem to affect the species richness levels for the nationally moderately declining species group may be due to considerable time lags between population declines and resulting local extinctions. We also show that the group of species that are nationally increasing is actually positively impacted with increasing levels of agricultural intensification (Fig. 8.9). This may possibly be a result of reduced competition with species that are declining or going extinct as a result of agricultural intensification.

8.7 Widespread versus localized species: the case of the pale shining brown

We have shown that general AES options are able to benefit widespread moth species, but there is little information on the extent to which rare, more localized, species may also benefit. During our main light-trapping experiment, we caught 88 individuals of the pale shining brown (Fig. 8.10) during the first year alone. This was an exciting, and all the more interesting, find as it meant the discovery of a population of a rare UK BAP priority species on farmland. The species was widely and well distributed in southern and southeast England until the mid 1970s but, since then, has undergone a massive decline. Since 2000, the vast majority of sightings have been from Salisbury Plain, an area known for its rich biodiversity, but the population is thought to be small. The Oxfordshire population we (re)discovered appears therefore to be the strongest currently known in Britain (Townsend and Merckx 2007).

The discovery provided a good opportunity to test whether AES options aimed at increasing general biodiversity would also benefit a highly endangered moth without species-specific tailoring. Since basic but much needed autecological (e.g. foodplant) information is lacking for this species, we tackled this by assessing the effects of wide field margins and hedgerow trees on the abundance of pale shining browns, both using light traps at the landscape scale and MRR

(also with light traps) at the farm scale (Merckx et al. 2010b). Based on the generalizations that we had previously demonstrated, we predicted that abundance of this rare species would be highest where field margins were wide and where hedgerows included emergent trees. We also expected that, if hedgerow trees conferred a positive effect, individual pale shining brown moths would be more likely to follow hedgerows than to cross exposed fields while on the move. These expectations were fulfilled: individuals were 8.5 times as abundant at sites with a hedgerow tree than at sites without a hedgerow tree (93 versus 11 individuals, respectively). Numbers were also higher at wide margins, but this was not statistically significant. No individuals were caught at field centres and, judging from the recorded movements, individuals may prefer to move within the sheltered space provided by hedgerows and hedgerow trees.

So, at least in the case of the rare pale shining brown, AES prescriptions designed to benefit the generality of farmland species, appeared to be beneficial to a rare and localized species too. This umbrella effect

Figure 8.10 Pale shining brown *Polia bombycina*—a rare and localized UK Biodiversity Action Plan (BAP) priority species. The population which is currently the largest in the UK was unexpectedly discovered—on farmland—during our light trap research programme. Photograph © Maarten Jacobs.

is encouraging, but obviously may not apply to many specialists because of their specialist life-history and habitat demands. Still, it would be helpful to explore which other rare UK BAP priority species fall inside the protective umbrella offered by specific AES options. For instance, the barberry carpet *Pareulype berberata* is limited to a few small sites in England. Barberry *Berberis vulgaris*, its foodplant, is a plant associated largely with hedgerows. Hence, well-designed hedgerow management options within AES, such as planting, protecting, and joining up stands of barberry, are likely to benefit remaining populations.

We are eager to promote conservation that is fit for purpose to protect priority endangered species, but the greater the extent to which several species can be protected by the same intervention, the more cost-effective it will be (a principle we have applied to widely different taxa; Macdonald et al. (2012)). There is scope for an integrative, biotope-focused approach going hand in hand with species-specific prescriptions where and when the precise ecological requirements of certain species of high conservation concern (and high charisma) are known.

8.8 Key recommendations for management and policy

In summary, what practical lessons for farmland conservation can be gleaned from our research? We propose the following:

1. Hedgerow trees and wide field margins are prominent features of farmland that can easily be integrated within effective AES. The presence and establishment of both hedgerow trees and wide field margins within intensive farmland offer substantial benefits to macro-moths overall, and are likely to benefit other flying insects too. This will positively impact on other trophic levels (e.g. birds/bats), on pollination levels, and other ecosystem services.
2. As the standard, field-scale uptake of AES options may be most effective for sedentary species, a move to a targeted, multi-farm, landscape-scale approach for AES implementation will additionally benefit more mobile species, which are a key group in terms of ecosystem functioning within agricultural landscapes.
3. A two-tier landscape restoration approach is needed, with increased investment in both (a) species-specific conservation measures for highly endangered specialist species with known resource requirements, and (b) biotope-specific conservation measures for mitigating declines in the large number of (once-) common and widespread species, as well as more localized species with poorly known ecological requirements.
4. We have shown that agricultural intensification results in negative effects on nationally declining and priority species, and that this is most pronounced at a spatially intermediate landscape scale (0.8 km radius). Our results hence suggest that the presence of wide field margins and hedgerow trees, promoted by AES targeting their implementation at this spatial scale, may help mitigate negative effects of agricultural intensification on macro-moths. A wide range of other taxa are dependent on macro-moths and may therefore benefit from these farmland features too. Nevertheless, taxa differ widely in their mobility. Consequently, measures mitigating biodiversity loss may need to be targeted at multiple spatial scales to maximize their effectiveness for multiple taxa.

Acknowledgements

This research was supported financially by the Esmée Fairbairn Foundation, within a project framework funded by Tubney Charitable Trust. We are grateful to Martin Townsend for assistance with fieldwork and moth identification, to Mark Parsons and Nigel Bourn (both at Butterfly Conservation) for general advice on moth conservation, to Ruth Feber, Mark Parsons, Eva Raebel, and Jeremy Thomas for commenting on and improving earlier drafts, and to all others who helped us in one way or another in carrying out the research.

References

Baker, D.J., Freeman, S.N., Grice, P.V., and Siriwardena, G.M. (2012). Landscape-scale responses of birds to agri-environment management: a test of the English Environmental Stewardship scheme. *Journal of Applied Ecology*, **49**, 871–882.

Batáry, P., Báldi, A., Kleijn, D., and Tscharntke, T. (2011). Landscape-moderated biodiversity effects of agri-environmental management: a meta-analysis. *Proceedings of the Royal Society of London B*, **278**, 1894–1902.

Batáry, P., Báldi, A., Sarospataki, M., Kohler, F., Verhulst, J., Knop, E., Herzog, F., and Kleijn, D. (2010). Effects of conservation management on bees and insect-pollinated grassland plant communities in three European countries. *Agriculture, Ecosystems & Environment*, **136**, 35–39.

Benton, T.G., Bryant, D.M., Cole, L., and Crick, H.Q.P. (2002). Linking agricultural practice to insect and bird populations: a historical study over three decades. *Journal of Applied Ecology*, **39**, 673–687.

Betzholtz, P.-E., Pettersson, L.B., Ryrholm, N., and Franzén, M. (2013). With that diet, you will go far: trait-based analysis reveals a link between rapid range expansion and a nitrogen-favoured diet. *Proceedings of the Royal Society of London B*, **280**, DOI: 10.1098/rspb.2012.2305.

Boutin, C., Baril, A., McCabe, S.K., Martin, P.A., and Guy, M. (2011). The value of woody hedgerows for moth diversity on organic and conventional farms. *Environmental Entomology*, **40**, 560–569.

Brooks, D.R., Bater, J.E., Clark, S.J., et al. (2012). Large carabid beetle declines in a United Kingdom monitoring network increases evidence for a widespread loss in insect biodiversity. *Journal of Applied Ecology*, **49**, 1009–1019.

Cadotte, M.W. (2011). The new diversity: management gains through insights into the functional diversity of communities. *Journal of Applied Ecology*, **48**, 1067–1069.

Chapman, J.W., Nesbit, R.L., Burgin, L.E., Reynolds, D.R., Smith, A.D., Middleton, D.R., and Hill, J.K. (2010). Flight orientation behaviors promote optimal migration trajectories in high-flying insects. *Science*, **327**, 682–685.

Conrad, K.F., Warren, M.S., Fox, R., Parsons, M.S., and Woiwod, I.P. (2006). Rapid declines of common, widespread British moths provide evidence of an insect biodiversity crisis. *Biological Conservation*, **132**, 279–291.

Darwin, C.R. (1862). *The Various Contrivances by which Orchids are Fertilised by Insects*. John Murray, London, UK.

Davey, C.M., Vickery, J.A., Boatman, N.D., Chamberlain, D.E., Parry, H.R., and Siriwardena, G.M. (2010). Assessing the impact of Entry Level Stewardship on lowland farmland birds in England. *Ibis*, **152**, 459–474.

Defra (2010). *Trends, long term survival and ecological values of hedgerow trees: development of population models to inform strategy*. Forest Research, Forestry Commission, report to the UK government Department for the Environment, Food and Rural Affairs, London.

Defra (2011). *Agri-environment performance report for Defra*. Data as at 31 March 2011: <http://www.naturalengland.org.uk/>.

Defra/NE (2012). *Entry Level Stewardship: Environmental Stewardship Handbook, fourth edition—January 2013*. UK government Department for the Environment, Food and Rural Affairs—Natural England.

Dennis, R.L.H. (2010). *A resource-based habitat view for conservation: butterflies in the British landscape*. Wiley-Blackwell, Oxford, UK.

Devoto, M., Bailey, S., and Memmott, J. (2011). The 'night shift': nocturnal pollen-transport networks in a boreal pine forest. *Ecological Entomology*, **36**, 25–35.

Donald, P.F. and Evans, A.D. (2006). Habitat connectivity and matrix restoration: the wider implications of agri-environment schemes. *Journal of Applied Ecology*, **43**, 209–218.

Donald, P.F., Green, R.E., and Heath, M.F. (2001). Agricultural intensification and the collapse of Europe's farmland bird populations. *Proceedings of the Royal Society of London B*, **268**, 25–29.

Dover, J.W. and Sparks, T. (2000). A review of the ecology of butterflies in British hedgerows. *Journal of Environmental Management*, **60**, 51–63.

Dutton, A., Edwards-Jones, G., Strachan, R., and Macdonald, D.W. (2008). Ecological and social challenges to biodiversity conservation on farmland: reconnecting habitats on a landscape scale. *Mammal Review*, **38**, 205–219.

Feber, R.E. and Smith, H. (1995). Butterfly conservation on arable farmland. In A.S. Pullin, ed. *Ecology and Conservation of Butterflies*. Chapman and Hall, London.

Feber, R.E., Smith, H., and Macdonald, D.W. (1996). The effects of management of uncropped edges of arable fields on butterfly abundance. *Journal of Applied Ecology*, **33**, 1191–1205.

Fox, R., Conrad, K.F., Parsons, M.S., Warren, M.S., and Woiwod, I.P. (2006). *The State of Britain's Larger Moths*. Butterfly Conservation and Rothamsted Research, Wareham, Dorset, UK.

Fox, R., Parsons, M.S., Chapman, J.W., Woiwod, I.P., Warren, M.S., and Brooks, D.R. (2013). *The State of Britain's Larger Moths 2013*. Butterfly Conservation and Rothamsted Research, Wareham, Dorset, UK.

Fox, R., Randle, Z., Hill, L., Anders, S., Wiffen, L., and Parsons, M.S. (2011). Moths count: recording moths for conservation in the UK. *Journal of Insect Conservation*, **15**, 55–68.

Fuentes-Montemayor, E., Goulson, D., and Park, K.J. (2011). The effectiveness of agri-environment schemes for the conservation of farmland moths: assessing the importance of a landscape-scale management approach. *Journal of Applied Ecology*, **48**, 532–542.

Gabriel, D., Sait, S.M., Hodgson, J.A., Schmutz, U., Kunin, W.E., and Benton, T.G. (2010). Scale matters: the impact of organic farming on biodiversity at different spatial scales. *Ecology Letters*, **13**, 858–869.

Goulson, D., Rayner, P., Dawson, B., and Darvill, B. (2011). Translating research into action; bumblebee conservation as a case study. *Journal of Applied Ecology*, **48**, 3–8.

Green, R.E., Cornell, S.J., Scharlemann, J.P.W., and Balmford, A. (2005). Farming and the fate of wild nature. *Science*, **307**, 550–555.

Groenendijk, D. and Ellis, W.N. (2010). The state of the Dutch larger moth fauna. *Journal of Insect Conservation*, **15**, 95–101.

Haaland, C., Naisbit, R.E., and Bersier, L.-F. (2011). Sown wildflower strips for insect conservation: a review. *Insect Conservation and Diversity*, **4**, 60–80.

Hambäck, P.A., Summerville, K.S., Steffan-Dewenter, I., Krauss, J., Englund, G., and Crist, T.O. (2007). Habitat specialization, body size, and family identity explain lepidopteran density-area relationships in a cross-continental comparison. *Proceedings of the National Academy of Sciences*, **104**, 8368–8373.

Hector, A. and Bagchi, R. (2007). Biodiversity and ecosystem multifunctionality. *Nature*, **448**, 188–191.

Hill, J.K., Thomas, C.D., Fox, R., Telfer, M.G., Willis, S.G., Asher, J., and Huntley, B. (2002). Responses of butterflies to 20th century climate warming: implications for future ranges. *Proceedings of the Royal Society of London B*, **269**, 2163–2171.

Holzschuh, A., Dudenhöffer, J.-H. and Tscharntke, T. (2012). Landscapes with wild bee habitats enhance pollination, fruit set and yield of sweet cherry. *Biological Conservation*, **153**, 101–107.

Kleijn, D., Baquero, R.A., Clough, Y., et al. (2006). Mixed biodiversity benefits of agri-environment schemes in five European countries. *Ecology Letters*, **9**, 243–254.

Kuussaari, M., Heliölä, J., Luoto, M., and Pöyry, J. (2007). Determinants of local species richness of diurnal Lepidoptera in boreal agricultural landscapes. *Agriculture, Ecosystems and Environment*, **122**, 366–376.

Lawton, J.H., Brotherton, P.N.M., Brown, V.K., et al. (2010). *Making space for nature: a review of England's wildlife sites and ecological network*. Report to Defra, UK.

Macdonald, D.W., Burnham, D., Hinks, A.E., and Wrangham, R. (2012). A problem shared is a problem reduced: seeking efficiency in the conservation of felids and primates. *Folia Primatologica*, **83**, 171–215.

Macdonald, D.W. and Feber, R.E., eds. (2015). *Wildlife Conservation on Farmland. Conflict in the Countryside*. Oxford University Press, Oxford.

Macdonald, D.W., Feber, R.E., Tattersall, F.H., and Johnson, P.J. (2000). Ecological experiments in farmland conservation. In M.J. Hutchings, E.A. John, and A.J.A. Stewart, eds., *The Ecological Consequences of Environmental Heterogeneity*, pp. 357–378. Blackwell Scientific Publications, Oxford.

Macdonald, D.W. and Smith, H.E. (1991). New perspectives on agro-ecology: between theory and practice in the agricultural ecosystem. In L.G. Firbank, N. Carter, J.F. Darbyshire, and G.R. Potts, eds., *The Ecology of Temperate Cereal Fields*, pp. 413–448. 32nd Symposium of the British Ecological Society. Blackwell Scientific Publications, Oxford.

Macdonald, D.W., Tattersall, F.H., Service, K.M., Firbank, L.G., and Feber, R.E. (2007). Mammals, agri-environment schemes and set-aside—what are the putative benefits? *Mammal Review*, **37**, 259–277.

Menéndez, R., González-Megías, A., Collingham, Y., Fox, R., Roy, D.B., Ohlemüller, R., and Thomas, C.D. (2007). Direct and indirect effects of climate and habitat factors on specialist and generalist butterfly diversity. *Ecology*, **88**, 605–611.

Merckx, T., Feber, R.E., Dulieu, R.L., et al. (2009b). Effect of field margins on moths depends on species mobility: field-based evidence for landscape-scale conservation. *Agriculture, Ecosystems and Environment*, **129**, 302–309.

Merckx, T., Feber, R.E., Mclaughlan, C., et al. (2010a). Shelter benefits less mobile moth species: the field-scale effect of hedgerow trees. *Agriculture, Ecosystems and Environment*, **138**, 147–151.

Merckx, T., Feber, R.E., Parsons, M.S., et al. (2010b). Habitat preference and mobility of *Polia bombycina*: are non-tailored agri-environment schemes any good for a rare and localised species? *Journal of Insect Conservation*, **14**, 499–510.

Merckx, T., Feber, R.E., Riordan, P., et al. (2009a). Optimizing the biodiversity gain from agri-environment schemes. *Agriculture, Ecosystems and Environment*, **130**, 177–182.

Merckx, T., Huertas, B., Basset, Y., and Thomas, J.A. (2013). A global perspective on conserving butterflies and moths and their habitats. In D.W. Macdonald and K.J. Willis, eds., *Key Topics in Conservation Biology 2*. John Wiley & Sons, Oxford, UK.

Merckx, T., Marini, L. Feber, R.E., and Macdonald, D.W. (2012). Hedgerow trees and extended-width field margins enhance macro-moth diversity: implications for management. *Journal of Applied Ecology*, **49**, 1396–1404.

Merckx, T. and Pereira, H.M. (2015). Reshaping agri-environmental subsidies: From marginal farming to large-scale rewilding. *Basic and Applied Ecology*, **16**: 95–103.

Merckx, T. and Slade, E.M. (2014). Macro-moth families differ in their attraction to light: implications for light-trap monitoring programmes. *Insect Conservation and Diversity*, 7, 453–461.

Merckx, T., Van Dongen, S., Matthysen, E., and Van Dyck, H. (2008). Thermal flight budget of a woodland butterfly in woodland versus agricultural landscapes: An experimental assessment. *Basic and Applied Ecology*, **9**, 433–442.

Merckx, T., Van Dyck, H., Karlsson, B., and Leimar, O. (2003). The evolution of movements and behaviour at boundaries in different landscapes: a common arena experiment with butterflies. *Proceedings of the Royal Society of London B*, **270**, 1815–1821.

Monbiot, G. (2013). *Feral: Searching for Enchantment on the Frontiers of Rewilding*. Penguin, UK.

Mutanen, M., Wahlberg, N., and Kaila, L. (2010). Comprehensive gene and taxon coverage elucidates radiation patterns in moths and butterflies. *Proceedings of the Royal Society of London B*, **277**, 2839–2848.

NE (2009). *Agri-environment Schemes in England 2009: a review of results and effectiveness*. Natural England, Sheffield, UK.

New, T.R. (2004). Moths (Insecta: Lepidoptera) and conservation: background and perspective. *Journal of Insect Conservation*, **8**, 79–94.

Pereira, H.M. and Daily, G.C. (2006). Modeling biodiversity dynamics in countryside landscapes. *Ecology*, **87**, 1877–1885.

Phalan, B., Onial, M., Balmford, A., and Green, R.E. (2011). Reconciling food production and biodiversity conservation: land sharing and land sparing compared. *Science*, **333**, 1289–1291.

Power, E.F. and Stout, J.C. (2011). Organic dairy farming: impacts on insect-flower interactions and pollination. *Journal of Applied Ecology*, **48**, 561–569.

Pywell, R.F., Warman, E.A., Sparks, T.H., et al. (2004). Assessing habitat quality for butterflies on intensively managed arable farmland. *Biological Conservation*, **118**, 313–325.

Rundlöf, M., Nilsson, H., and Smith, H.G. (2008). Interacting effects of farming practice and landscape context on bumblebees. *Biological Conservation*, **141**, 417–426.

Slade, E.M., Merckx, T., Riutta, T., Bebber, D.P., Redhead, D., Riordan, P., and Macdonald, D.W. (2013). Life history traits and landscape characteristics predict macro-moth responses to forest fragmentation at a landscape-scale. *Ecology*, **94**, 1519–1530.

Steffan-Dewenter, I., Münzenberg, U., Bürger, C., Thies, C., and Tscharntke, T. (2002). Scale-dependent effects of landscape structure on three pollinator guilds. *Ecology*, **83**, 1421–1432.

Thomas, C.D. (2000). Dispersal and extinction in fragmented landscapes. *Proceedings of the Royal Society of London B*, **267**, 139–145.

Thomas, J.A. (2005). Monitoring change in the abundance and distribution of insects using butterflies and other indicator groups. *Philosophical Transactions of the Royal Society of London B*, **360**, 339–357.

Tilman, D., Fargione, J., Wolff, B., et al. (2001). Forecasting agriculturally driven global environmental change. *Science*, **292**, 281–284.

Townsend, M.C. and Merckx, T. (2007). Pale shining brown *Polia bombycina* (Hufn.) (Lep.: Noctuidae) re-discovered in Oxfordshire in 2005 and 2006—a nationally significant population of a UK Biodiversity Action Plan priority species. *The Entomologist's Record and Journal of Variation*, **119**, 72–74.

Tscharntke, T., Klein, A.M., Kruess, A., Steffan-Dewenter, I., and Thies, C. (2005). Landscape perspectives on agricultural intensification and biodiversity—ecosystem service management. *Ecology Letters*, **8**, 857–874.

Van Dyck, H. (2012). Changing organisms in rapidly changing anthropogenic landscapes: the significance of the 'Umwelt'-concept and functional habitat for animal conservation. *Evolutionary Applications*, **5**, 144–153.

Van Dyck, H. and Baguette, M. (2005). Dispersal behaviour in fragmented landscapes: routine or special movements? *Basic and Applied Ecology*, **6**, 535–545.

Van Dyck, H., van Strien, A.J., Maes, D., and van Swaay, C.A.M. (2009). Declines in common, widespread butterflies in a landscape under intense human use. *Conservation Biology*, **23**, 957–965.

Warren, M.S. and Bourn, N.A.D. (2011). Ten challenges for 2010 and beyond to conserve Lepidoptera in Europe. *Journal of Insect Conservation*, **15**, 321–326.

Whittingham, M.J. (2011). The future of agri-environment schemes: biodiversity gains and ecosystem service delivery? *Journal of Applied Ecology*, **48**, 509–513.

Winqvist, C., Bengtsson, J., Aavik, T., et al. (2011). Mixed effects of organic farming and landscape complexity on farmland biodiversity and biological control potential across Europe. *Journal of Applied Ecology*, **48**, 570–579.

Young, M. (1997). *The Natural History of Moths*. Poyser, London, UK.

CHAPTER 9

Habitat use by vesper bats: disentangling local and landscape-scale effects within lowland farmland

Danielle Linton, Lauren A. Harrington, and David W. Macdonald

> A bat is beautifully soft and silky; I do not know any creature that is pleasanter to the touch or is more grateful of caressings, if offered in the right spirit.
>
> **Mark Twain**

9.1 Introduction

Bats belong to the order Chiroptera (literally 'hand-wing'), which contains over 1200 species within 19 families and 190 genera (Simmons 2005), making it the second largest mammalian order after Rodentia (Altringham 2011). Of the 20 or more British bat species, there are two horseshoe bats (the greater horseshoe bat *Rhinolophus ferrumequinum*, and the lesser horseshoe bat *R. hipposideros*) with all others belonging to the family Vespertilionidae (the vesper bats, or literally, 'evening' bats) (Dietz et al. 2009; Fig. 9.1).

Bats are considered to be indicators of habitat quality and environmental change (Jones et al. 2009) and, in 2008, were adopted as one of 24 UK Biodiversity Indicators (Defra 2013). But how sensitive are bats as indicators of agricultural intensification and what can be learnt from them about the effectiveness of agri-environment measures intended to ameliorate the negative impacts of intensive agriculture on bats and their prey?

Direct evidence of bat population declines linked to agricultural intensification is sparse, but there is evidence of agricultural change coinciding with range contractions in continental Europe (e.g. Daan 1980), and of declining numbers of bats at traditional roost sites in the UK (Stebbings 1995; Robinson and Sutherland 2002). Bat activity levels are comparatively low in arable-dominated areas (Walsh and Harris 1996a,b; Russ et al. 2003), and changes in agricultural trends in the UK, for example, the move from mixed farming towards specialization during the 1960s and 1970s (Robinson and Sutherland 2002), seems certain to have affected bat populations, particularly in eastern Britain where such changes were most widespread.

The term 'agricultural intensification' encompasses various processes likely to have had detrimental influences on bat populations: directly, through habitat loss and fragmentation (for example, hedgerow and tree removal, woodland clearance, drainage of wetland areas) or indirectly, through the negative impacts of increased pesticide use, loss of plant diversity, and altered crop rotations, on their invertebrate prey (Haysom 2003). In particular, the loss of unimproved pastures and associated declines in large invertebrates, such as dung beetles and cockchafers (important food for larger bat species), have no doubt restricted food supplies for foraging bats during critical periods. There are also concerns over the use of antihelminthic drugs (e.g. ivermectin) and their effect on invertebrates associated with the dung of treated livestock (Ransome 1996).

All bat species known to be resident within the UK exist within lowland agricultural landscapes (Wickramasinghe et al. 2003), and commute through, or forage in, farmland habitats to varying extents (Walsh and Harris 1996a, b). Bats can usually be detected in most farmland habitats, but the highest levels of activity tend to be associated with aquatic and uncropped habitats, such as streams and rivers, woodlands, and linear vegetation features (e.g. hedgerows) (Walsh

Wildlife Conservation on Farmland. Managing for Nature on Lowland Farms. Edited by David W. Macdonald and Ruth E. Feber.
© Oxford University Press 2015. Published 2015 by Oxford University Press.

Figure 9.1 A Natterer's bat *Myotis nattereri* in hand. This species belongs to the family Vespertilionidae, the vesper bats. Photograph © A.L. Harrington.

et al. 1995; Walsh and Harris 1996a, b; Russ and Montgomery 2002), although many bat species also utilize open habitats (e.g. Walsh and Harris 1996a; Vaughan et al. 1997a). Aquatic habitats, where emergent insects with aquatic larval stages are abundant, are important foraging areas for many bat species (Racey 1998; Lundy and Montgomery 2010) and woodland, in addition to providing foraging habitat for some species, provides roosting habitat for the majority of UK bat species (Mayle 1990). Hedgerows, treelines, and woodland belts provide habitat connectivity within highly fragmented agricultural landscapes, linking roosts and foraging areas (Limpens and Kapteyn 1991; Downs and Racey 2006).

Bats can be faithful to their foraging areas when a roost is lost (Brigham and Fenton 1986), and probably remain faithful to roosts when foraging areas are lost. All UK bats and their roosts are protected under a suite of national and European legislation[1]; but

foraging areas currently receive no statutory protection[2]. In south-west England, the 'Greater Horseshoe Bat Project' (Longley 2003) encouraged farmers and landowners around known roosts of the rare greater horseshoe bat to adopt a number of 'foraging habitat' management options through agri-environment scheme agreements, including arable reversion to grassland, managing grazing to encourage key prey invertebrates, and improving bat commuting routes through hedgerow restoration and tree planting (Longley 2003). The 'Landscapes for Lessers' project aimed to improve habitats for internationally important populations of the lesser horseshoe bat in Wales (Mapstone 2009). However, the design of scheme options specifically targeted at the conservation of vespertilionid bat species across the UK is currently less well developed, and

---

[1] <http://www.bats.org.uk/>; <http://www.naturalengland.org.uk/>, Accessed May 2013.

[2] With the exception of a small number of Special Areas of Conservation designated for the protection of bats listed in Annex II of the EC Habitats Directive, that encompass foraging areas as well as roost and hibernation sites; <http://jncc.defra.gov.uk/ProtectedSites/SACselection/SAC_species.asp>. Accessed May 2013.

existing options appear to be ill-suited to their habitat requirements (Fuentes-Montemayor et al. 2011).

Conservation measures for bats in the wider countryside are usually associated with hedgerow management options, especially planting of new hedges to reconnect existing linear features and increase habitat connectivity (Entwistle et al. 2001). Hedgerows are widely used by bats (for foraging and commuting, e.g. Boughey et al. 2011), but the influence of hedgerow density in the landscape is poorly understood (Verboom and Huitema 1997) and the effectiveness of increasing connectivity largely unknown. Quality of hedgerows and other linear vegetation features (e.g. hedgerow height, and the presence of mature standard trees) is also important (Pocock and Jennings 2008; Boughey et al. 2011). As foraging activity of bats can be

affected by weather conditions, such as ambient temperature, wind speed, and precipitation (Avery 1991; Park et al. 2000; Parsons et al. 2003; Ciechanowski et al. 2007), tall hedges and treelines may be even more important to bats during adverse weather (Verboom and Spoelstra 1999, Russ et al. 2003).

To improve the design of agri-environment schemes and habitat management plans for bat conservation, it is vital to know the habitat factors that influence bat activity patterns within agricultural landscapes. This is difficult because bats are highly mobile (most species typically travel several kilometres per night, e.g. Robinson and Stebbings 1997; Nicholls and Racey 2006a), therefore the scale at which habitat selection operates, and the extent to which habitat composition in the surrounding landscape influences local population

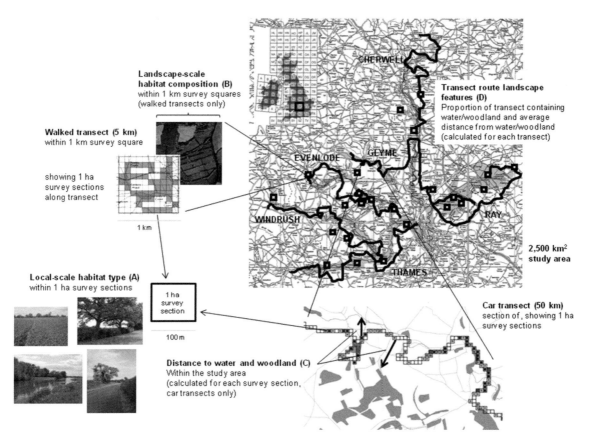

Figure 9.2 Study design and location of walked and car transects. Bat activity was assessed in relation to habitat type at the local scale (A) and habitat composition at the landscape scale (B); in an attempt to disentangle the relative importance of these different scales, we compared 1. Bat activity within a habitat type (A) among different habitat compositions (B), 2. Bat activity within a 1 ha survey section in response to distance from water or woodland in the wider landscape (C), and 3. Bat activity per car transect, as predicted according to local-scale habitat types available, among different transect route landscape features (D).

dynamics, are complex and not fully understood (Verboom and Huitema 1997; Lundy and Montgomery 2010; but see recent work by Bellamy et al. 2013).

In this chapter we describe studies carried out on lowland farmland of the Upper Thames Valley, UK, which aimed to provide information on distribution patterns and habitat use by vespertilionid (vesper) bats commuting and foraging within agricultural landscapes. The core of our work comprised over 180 bat detector surveys, for which we used two different methods: walked transects (each of 5 km), and car (driven) transects (each of 50 km). From these survey data we assessed the effects on bats of habitat type at the local (1 ha, i.e. 100 m x 100 m) scale, and habitat composition at the landscape (1 km²) scale (Fig. 9.2). In an attempt to disentangle the relative influence of these different scales, we then explored how the composition of the landscape influenced bat activity at the local scale by comparing: (1) bat activity within a habitat type (at the 1 ha level) among different landscape composition categories (at the 1 km² level), (2) bat activity (at the 1 ha level) in response to distance from the nearest water or woodland (within the 2500 km² study area), and (3) bat activity per 50 km car transect with predicted bat activity (based on local-scale habitat types available) among six different transect routes running through variable landscape features (Fig. 9.2). We addressed the issue of habitat connectivity, at both the local and landscape scales, by exploring the effect of the number of linear vegetation feature (hereafter LVF) connections (at the 1 ha level), and the effect of the density of LVFs (at the 1 km² level). We also assessed the influence of weather conditions on bat activity and behaviour.

To set the scene, we now briefly introduce the vesper bats found in UK lowland farmland.

9.2 Vesper bats on UK farmland

UK vesper bats include six genera: mouse-eared bats *Myotis*, house bats *Eptesicus*, noctule bats *Nyctalus*, pipistrelles *Pipistrellus*, barbastelles *Barbastella*, and long-eared bats *Plecotus*, each with one to eight (or more) species. Five species of *Myotis* bats are known to be present in the Upper Thames Valley: Daubenton's bat *Myotis daubentonii*, and Natterer's bat *M. nattereri* are widely distributed and abundant within the study area, whilst Bechstein's bat *M. bechsteinii*, whiskered bat *M. mystacinus*, and Brandt's bat *M. brandtii* appear (from records) to be scattered and infrequent. Three of the largest vesper bat species in the UK are known (from records of roost sites and injured or rescued bats) to be present within the study area: noctule *Nyctalus*

noctula is the most common species, while Leisler's bat *N. leisleri* and serotine *Eptesicus serotinus* are much scarcer. Two of the UK pipistrelles (common pipistrelle *Pipistrellus pipistrellus* and soprano pipistrelle *P. pygmaeus*) and the brown long-eared bat *Plecotus auritus* are common in the Upper Thames Valley; barbastelle *Barbastella barbastellus* is widespread and generally under-recorded (due to its low intensity calls and fast flight, Goerlitz et al. 2010). Nathusius' pipistrelle *P. nathusii* is rarely encountered, and the grey long-eared bat *P. austriacus* is not known to occur in the area. Two previously unrecorded bat species have been found in the UK in recent years: Alcathoe bat *M. alcathoe* (Jan et al. 2010) and Geoffroy's bat *M. emarginatus* (<http://www.naturalengland.org.uk>), and it is possible that other cryptic bat species remain undetected.

Considerable ecomorphological variation exists among these bats (Altringham 2003). Large bats with long narrow wings, such as the noctule bats (*c.* 40 g, with a 40 cm wingspan), are capable of fast efficient flight but have low manoeuvrability, and thus tend to forage by hawking flying insects in more open habitats, such as wetlands or grasslands (Walsh and Harris 1996a). At the other extreme, bats with relatively broad and short wings, such as the long-eared and *Myotis* bats, are capable of slow flight, even hovering, and are also highly manoeuvrable. These species can fly and forage within cluttered habitats, such as woodlands (Entwistle et al. 1996; Smith and Racey 2008), typically gleaning invertebrate prey from close to, or on the surface of, vegetation. Pipistrelles typically forage along the edge of vegetation features, and occupy a central position within this continuum (they are also the smallest, at *c.* 4–5 g, with a 20 cm wingspan). Many UK vesper bats forage near water, with aquatic prey appearing particularly important for Daubenton's bat (e.g. Dietz et al. 2006) and soprano pipistrelle (e.g. Sattler et al. 2007).

Vesper bats consume a vast number and a great variety of invertebrates: flying insects, such as cockchafers and dung beetles; small swarming insects, such as midges, mosquitos, craneflies, and small moths; aquatic invertebrates, such as caddisflies, stoneflies, and mayflies; as well as larger moths and spiders. Bechstein's bat and the brown long-eared bat feed predominantly on moths, whereas serotine bats feed mainly on beetles (especially ground, chafer, and dung beetles).

9.3 Research methods

9.3.1 'Counting' bats

To describe distribution patterns and habitat associations of any animal, it is necessary first to count them; bats, being aerial, fast-flying, and nocturnal (in the UK and mainland Europe), are particularly difficult to count. However, because bats are reliant on echolocation for spatial orientation, and location and capture of prey (e.g. Schnitzler et al. 2003), they can be detected via the ultrasonic calls that they emit[3].

The structure of echolocation calls varies substantially between bat species in accordance with their phylogeny (Jones and Teeling 2006) and size (Jones 1999), foraging strategy and dietary niche specialization (Siemers and Schnitzler 2004). Species can thus be identified by examination of recorded echolocation calls and the measurement of specific call parameters (see Russ 2012). However, within species (and between individuals), substantial variation, call plasticity, and modification of call structure also occur, depending on the surrounding environment and perceptual task. For example, the calls of noctules and serotines differ in structure when flying in the open but can be indistinguishable in cluttered environments. Foraging activity can be differentiated from other calls as 'feeding buzzes'—distinctive sequences of very quickly repeated echolocation pulses as the bat homes in on an insect (Kalko 1995, Kalko and Schnitzler 1998). Social behaviour, such as mother–infant, or male–female interactions, can also be inferred from the repertoire of non-navigational ultrasonic calls used for communication (Fenton 2003; Pfalzer and Kusch 2003), often audible to humans (unlike echolocation calls; Fenton 2003).

Technological advances have made continuous recordings across a range of frequencies (which enable simultaneous detection of several sympatric species), and subsequent analysis of multiple echolocation call sequences, feasible (Vaughan et al. 1997a). In our studies, we used continuous broadband, 'real-time' ultrasound recordings (using Ultra Sound Gate 116 bat detectors: Avisoft Bioacoustics, Berlin, Germany) to detect free-flying commuting and foraging bats. Individual, time-stamped echolocation calls within these recordings were geo-referenced with a spatial resolution of 5 m[2] and identified via sonogram analysis (BatSound version 3.31: Pettersson Electronik AB, Uppsala, Sweden).

Calls were classified to species level for common pipistrelle, soprano pipistrelle, brown long-eared bat,

and barbastelle, and to genus level for the *Myotis* bats. Bats from the genus *Nyctalus* (noctule and Leisler's bat), and *Eptesicus* (serotine) were grouped for analysis because, in many cases, it is not possible to distinguish between call sequences for these species. We refer to this latter group as the noctule-serotine group. Calls were identified as bat presence, feeding, or social behaviour, allowing us to describe not only general distribution patterns in relation to habitat, but also to identify those habitats used specifically for foraging, as well as habitats where social interactions occur.

One drawback of ultrasound detector-based survey data is the inability to distinguish between single passes from several individuals and several passes from a single individual, meaning that call counts do not relate directly to bat density or abundance (Hayes 2000). Another is that differences in call intensity and foraging strategy mean that not all bat calls are equally detectable (Obrist et al. 2004), rendering most interspecific comparisons invalid (Vaughan et al. 1997b). We, therefore, used 'bat activity' (quantified as the presence/absence of bat calls of each species/group within a survey unit, or total survey minutes within which a species/group was detected per transect) as a proxy for relative bat abundance, and refer specifically to differences within a species or group among habitat types.

9.3.2 Survey strategies

To describe bat activity across the farmscape, bats were recorded continuously along twenty-four 5 km walked transects (each within a 1 km² survey site) within the Upper Thames catchment, and six 50 km car transects along the River Thames and each of its major tributary rivers (the Cherwell, Evenlode, Glyme, Ray, and Windrush; see Fig. 9.2). Car transect survey methods were adapted from the BCT/MTUK 'Bats and Roadside Mammals Survey' protocol (Russ et al. 2006), and covered a variety of roadside habitats through predominately agricultural landscapes, as well as river corridors and woodlands. Car transects were driven at 24 km/hr and walked transects were covered at approximately 3.8 km/hr; car and walked transects together provided a balance between detail and coverage (Box 9.1).

To assess local (1 ha) scale habitat associations, both transect types were treated as a series of contiguous 1 ha survey sections (see Fig. 9.2), and bat calls were allocated to individual 1 ha survey sections (the survey unit) during analysis. To assess landscape (1 km²) scale habitat effects, bat calls were quantified per

---

[3] Most vespertilioniformes in the UK emit ultrasonic calls orally (exceptions are the two long-eared bats, belonging to the genus *Plecotus*—the 'whispering bats'—that emit calls nasally). See Fenton (1999).

walked transect. To account for variable weather conditions or seasonal effects, each walked transect was repeated five times, and each car transect repeated ten times, from late April to early October. In total, 120 walked transects and 60 car transects were completed over the summers of 2006 and 2007. The direction of travel along each transect was varied among repeats; all surveys commenced 30 minutes after sunset and continued for approximately two hours. For full details see Linton (2009).

Box 9.1 Recording bats at 15 mph—comparison of walked and car transects

Car transects are increasingly being used for monitoring bat activity, especially for establishing species inventories and providing information on spatial distribution in under-recorded localities. We compared data sets from car transects and walked transects and found that bats were recorded less frequently along car transects (22.2% survey events, n = 34160 total survey events) compared with walked transects (81.8% survey events, n = 5615, Box Fig. 9.1.1a). However, because of the speed with which each (1 ha) survey section was covered (c. 13 s for car transects, 101 s for walked transects), more bats were recorded on car transects (4204 presence/absence records generated over ten surveys of six car transects, compared with 2417 records from five surveys of 24 walked transects, Box Fig. 9.1.1b), over larger areas (3416 x 1 ha survey sections, versus 1123 x 1 ha survey sections), using less survey effort (60 survey nights versus 120 survey nights). However, we also found that *Myotis* bats were under-recorded along car transect routes (Box Fig. 9.1.1a,b), which may be related to bias in habitat types represented along road corridors, or may be due to behavioural responses to traffic noise and lighting (e.g. Berthinussen and Altringham 2012). We conclude that although car transects are undoubtedly an efficient and effective means of obtaining information on the presence of bats over a wide area, they should be used as only one component in a multi-method approach.

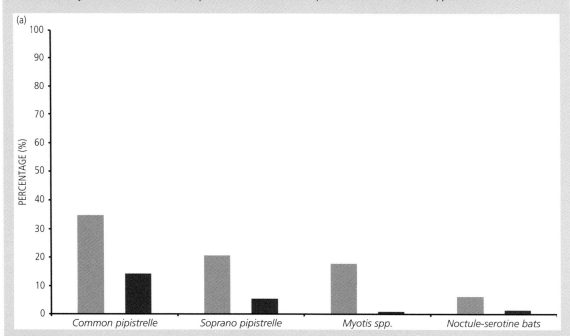

Box Figure 9.1.1 Bat detection in Upper Thames Valley farmland using walked transects (n = 24 x 5, grey bar in each pair) and car transects (n = 6 x 10, black bar in each pair) showing (a) the proportion of survey events during which each bat species/group was detected and (b) the total number of presence/absence records for each bat species/group (over all surveys, of all transects). A survey event is a single pass through a single 1 ha survey square.

continued

Box 9.1 *Continued*

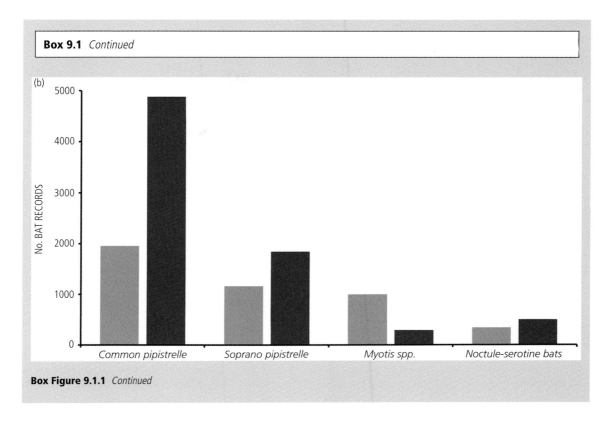

Box Figure 9.1.1 *Continued*

9.3.3 Habitat classification and environmental variables

Local-scale habitat type, field system, and connectivity, were described for each contiguous 1 ha survey section along transects, as the dominant habitat[4] within each hectare, the field type (classified as grassland, arable, or mixed farmland) within the survey section, and the number of linear vegetation feature (LVF) connections within each survey section containing LVFs.

Landscape (1 km²) scale habitat composition was described for each 1 km² survey site (for walked transects), and classified into one of eight habitat composition categories, according to the availability of surface water (> or < 1 km river), the amount of existing broadleaved woodland cover (> or < 25 ha), and connectivity (> or < 7 km LVFs) within each 1 km² survey site. Farm

[4] Habitat types included: broad-leaved woodland interior (WInt), woodland edge (WE), treeline (TREE), hedgerow (HEDGE), grassland field (GR), arable field (AR), rural (isolated buildings, industrial yards, and gardens), village (human settlements) and each of these habitat types adjacent to rivers (+R), or adjacent to streams (+S).

system within each 1 km² survey site was recorded as a broad descriptor of farmland landscape.

Within the wider (2500 km²) landscape, the location of all rivers or streams (water), and woodland blocks > 5 ha, were mapped, and the distance from each car transect survey section to the nearest water or woodland recorded (measured in ArcGIS: version 9.2, ESRI Inc., Redlands, CA). These spatial data were then used to describe each car transect, as a whole, in terms of the proportion of the transect that contained, and the average distance of the transect from, water or woodland.

To enable assessment of weather effects on bat activity, temperature, percentage cloud cover, and wind speed (ms⁻¹) were also recorded during each walked transect.

9.4 Survey results

Common pipistrelle, followed by soprano pipistrelle, were the most widespread and frequently detected bat species across all habitat types and for both walked transects (Table 9.1) and car transects. *Myotis* bats, and bats belonging to the noctule-serotine group, were also

Table 9.1 Bat activity detection rates for individual species/groups during 120 walked bat detector transects in the Upper Thames Valley, described as the number of 1 ha survey sections within which a species/group was recorded, or the number of survey minutes within which a species/group was recorded. A bat record is the presence of one species or group within a survey minute. Number of transects a species/group was detected on is a measure of distribution across the Upper Thames Valley, and number of surveys (one survey of one transect) during which a species/group was detected is a measure of consistency of activity through the summer survey season. n = 24 transects, each surveyed five times.

Species	No. 1 ha survey sections[1] (% total survey sections)	No. survey minutes (% total survey minutes)	% total bat records	No. transects	No. surveys
Common pipistrelle	867 (77.2)	3813 (24.6)	45.4	24	120
Soprano pipistrelle	590 (52.2)	2179 (14.0)	26.0	24	117
Myotis bats	567 (50.5)	1717 (11.1)	20.5	24	118
Noctule-serotine group	270 (24.0)	510 (3.3)	6.1	24	96
Brown long-eared bat	39 (3.5)	56 (0.4)	0.7	19	37
Barbastelle	79 (7.0)	109 (0.7)	1.3	17	47
Lesser horseshoe bat	5 (0.4)	7 (0.05)	0.08	1[2]	3
Nathusius' pipistrelle[3]	2 (0.2)	2 (0.01)	0.02	2	2

[1] Detection rates were similar for the six car transect surveys (each surveyed ten times): common pipistrelle were recorded in 2346 x 1 ha survey sections (68.7% total survey sections); soprano pipistrelle in 1082 (31.7%) survey sections; noctule-serotine group in 444 (13.0%) survey sections; and *Myotis* bats in 262 (7.7%) survey sections.

[2] This site was in close proximity to several known roosts.

[3] Probable recordings.

detected relatively frequently. Brown long-eared bats and barbastelle were detected infrequently but in all habitat types and in most 1 km² squares; lesser horseshoe bats and Nathusius' pipistrelle were detected rarely (in only two 1 km² survey sites and one 1 km² survey site respectively). The latter four were excluded from individual statistical analyses of habitat use due to low sample size, but data from these species were included in analyses of total bat activity.

9.4.1 Local-scale habitat effects

At the local scale, bat activity was significantly higher within terrestrial habitat types containing trees or woodland (woodland interior, woodland edge, and treelines) compared to habitats without trees (hedges, and grassland and arable fields), and higher in habitats adjacent to riparian corridors compared with those not adjacent to riparian corridors (and even more so for rivers than for streams, Fig. 9.3a). For all habitat types, bat activity was higher when bordered by grassland fields rather than arable fields (Fig. 9.3b), and for walked transects, which included grassland and arable 'open field' as a habitat type, bat activity was significantly lower in open arable fields than in any other terrestrial habitat type (Fig. 9.3a).

For car transects, bat activity was significantly higher where there were more LVF connections,

predominantly due to an increase in the number of pipistrelles (Linton 2009). Habitat distribution patterns overall were broadly similar for all bat species and groups, with the exception that soprano pipistrelle and the *Myotis* bats were more frequently recorded along rivers than along streams, and in woodland interior compared with either woodland edge or treelines; whereas common pipistrelle showed little difference between rivers and streams, and occurred more frequently in woodland edge and along treelines than in woodland interior. The noctule-serotine group also showed relatively little difference between rivers and streams.

When we compared habitat associations with habitat availability, we found that, while patterns of habitat preference were variable among species/groups and therefore difficult to generalize, all species/groups selected *against* hedgerows with no trees and open agricultural fields[5] (Fig. 9.4). Common pipistrelle

[5] Habitat selection per species/group was determined by calculating individual confidence intervals (CI), using χ^2 analysis and the z statistic, following Neu et al. (1974), around the observed proportions of bat activity per habitat type, and comparing with the proportion of habitat available within all 1 ha survey sections along all transects. Because of the number of simultaneous comparisons being made, a Bonferroni adjustment was made to each CI, with the overall confidence level set at 90% (Neu et al. 1974). For full details see Linton (2009).

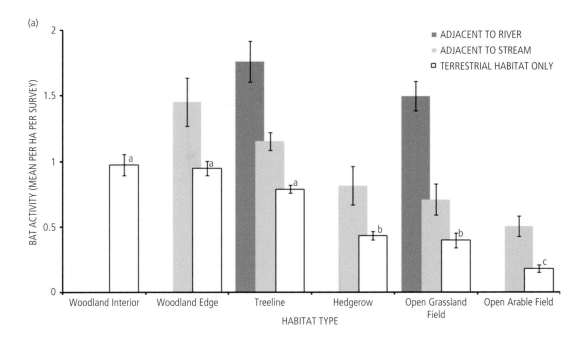

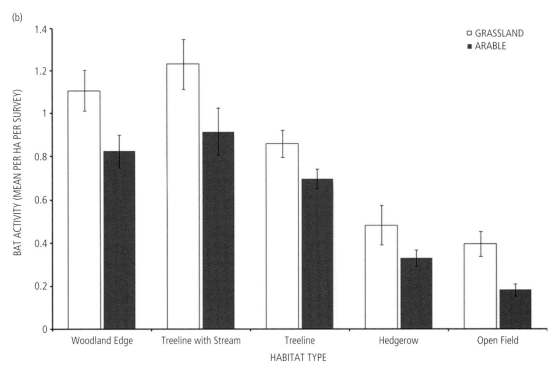

Figure 9.3 Local-scale bat activity across habitat types showing the influence of (a) trees and riparian corridors, and (b) adjacent field system. Bat activity differed significantly among terrestrial habitat types with different letters (a, b, and c), and among habitats adjacent to rivers, adjacent to streams, and equivalent terrestrial habitat types (for all habitat types > 20 x 1 ha survey sections). With the exception of treeline adjacent to stream, bat activity was significantly higher when bordered by grassland fields than arable fields (all = $p < 0.05$, open field = $p < 0.001$). Data are mean (± SE) bat activity within a 1 ha survey section per survey, shown for walked transects; a value of 1 indicates the presence of one bat species/group within a 1 ha survey section during each survey on average.

particularly selected for treelines, especially adjacent to streams (Fig. 9.4a) while soprano pipistrelle and *Myotis* bat activity were concentrated along river corridors and in woodland habitats (Fig. 9.4b,c). Considerable spatial overlap occurred among common and soprano pipistrelle, and the *Myotis* bats, particularly along river corridors (except along rivers through arable fields, Table 9.2).

Feeding buzzes and social calls of common pipistrelle, and feeding buzzes of soprano pipistrelle and *Myotis* bats, were widely distributed across habitat types, but particularly high proportions of these calls (relative to the number of calls of that species/group) were recorded at treelines adjacent to streams, or rivers, respectively (Fig. 9.4a,b,c). Along car transects, the distribution of common pipistrelle social calls was highly skewed, with 49.3% of 223 social calls occurring within human settlements. The proportion of feeding buzzes of all three species/groups, together with the proportion of soprano pipistrelle social calls, were low (below average/absent) in areas where only occasional or no trees were present in open agricultural fields and along streams in open fields, Fig. 9.4 a,b,c). For soprano pipistrelle, the proportion of feeding buzzes recorded was also low at woodland edge and treelines (Fig. 9.4b).

Noctule-serotine group bats demonstrated comparatively little habitat selection overall (Fig. 9.4d), but were frequently recorded during car transects in villages and industrial estates (with or without streams), particularly in areas with a high density of street lighting, where they foraged (Fig. 9.5).

9.4.2 Landscape-scale habitat effects

At a landscape (1 km^2) scale, common and soprano pipistrelle, and *Myotis* bat activity were higher where there was more water available in the surrounding area[6], and soprano pipistrelle and *Myotis* bat activity were also positively related to woodland cover[7]. Noctule-serotine group activity was highly variable within and among habitat composition categories, but was higher in grassland or in mixed farmland

landscapes than in arable landscapes[8]. In contrast to local-scale analyses, connectivity (LVF density per km^2) was not a significant factor influencing bat activity levels for any species/group.

The occurrence of common pipistrelle feeding buzzes and social calls was not related to any landscape-scale habitat variables, but soprano pipistrelle and *Myotis* (particularly Daubenton's) bat feeding buzzes were positively associated with the availability of water in the landscape[9], and the social calls of soprano pipistrelle were positively associated with the presence of more water and woodland[10].

9.4.3 Landscape-scale versus local-scale effects

Across local-level (1 ha) habitat types, soprano pipistrelle activity was higher in wetter landscapes (>1 km river within the 1 km^2 survey site) (Fig. 9.6). However, for all other bats, habitat composition of the landscape (including connectivity) had little or no effect on local-level activity. Further, although bat activity levels were significantly higher within survey sections containing water, and within sections containing woodland (i.e. within 100 m of water or woodland), relative to other sections not containing water/woodland, there was no effect of distance per se (from either water or woodland) beyond 100 m (Fig. 9.7).

Data from car transects revealed that, for most species or groups of bats, bat activity levels could be accurately predicted on the basis of local-level habitat types present along the transect (Fig. 9.8a). For soprano pipistrelles, however, activity levels were higher than predicted for the River Thames, and lower than predicted for the River Cherwell (Fig. 9.8b). These two rivers, and the respective landscapes surrounding them, differed quite substantially in terms of their

---

[6] Wetter (> 1 km river) versus drier (< 1 km river) sites, mixed model nested ANOVA, log-transformed data: common pipistrelle, $F_{1, 22} = 6.44$, $P = 0.019$; soprano pipistrelle, $F_{1, 22} = 17.60$, $P < 0.001$; *Myotis* bats, $F_{1, 22} = 5.10$, $P = 0.034$.

[7] Woodland effect, GLMM with repeated measures, log-transformed data: soprano pipistrelle, $F_{1, 16} = 5.14$, $P = 0.038$; *Myotis* bats, $F_{1, 16} = 9.32$, $P = 0.008$. The effect of water was also statistically significant for both pipistrelles and *Myotis* bats.

[8] Field system effect for noctule-serotine bats, GLMM with repeated measures, log-transformed data: $F_{4, 87} = 4.92$, $P = 0.001$. Field system did not have a statistically significant effect for any of the other bat species or groups.

[9] Water effect, GLMM with repeated measures, log-transformed data: soprano pipistrelle feeding buzzes, $F_{1, 16} = 11.16$, $P = 0.004$; *Myotis* bat feeding buzzes, $F_{1, 16} = 11.38$, $P = 0.004$; soprano pipistrelle social calls, $F_{1, 16} = 12.86$, $P = 0.003$. Woodland effect: soprano pipistrelle social calls, $F_{1, 16} = 10.37$, $P = 0.005$.

[10] The very few social calls recorded from *Myotis* bats during transects (n = 17) constituted too small a sample size to enable any further analysis. Social calls were, however, attributable to species; with recordings obtained from Daubenton's bat and Natterer's bat based on Pfalzer and Kusch (2003). Insufficient numbers of feeding buzzes (n = 16) and social calls (n = 11) were recorded from noctule-serotine bats to allow statistical analysis.

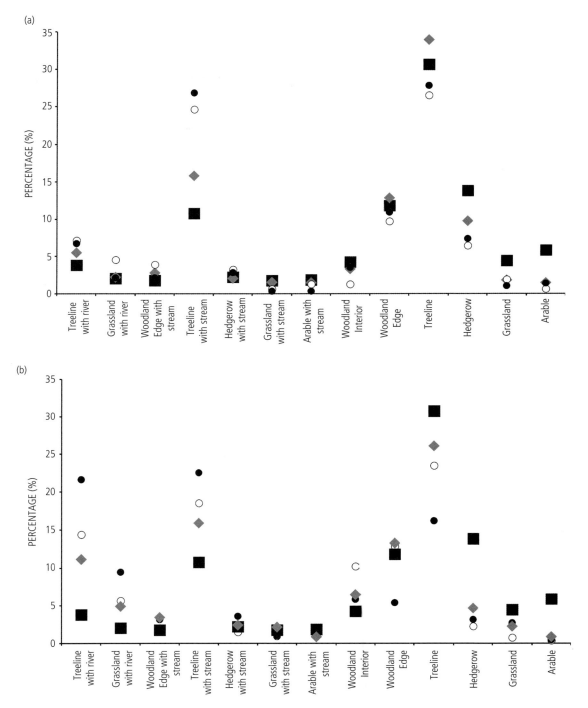

Figure 9.4 Habitat selection by (a) common pipistrelles, (b) soprano pipistrelles, (c) *Myotis* bats, and (d) noctule-serotine group, during walked transects. Proportion of bat activity within each habitat type (grey diamonds) relative to the availability of that habitat type (black squares); diamonds above the squares indicates selection for that habitat type, diamonds below the squares indicates selection against that habitat type. Proportions of feeding buzzes (closed circles) and social calls (open circles) are also shown. Habitat selection was similar for car transects.

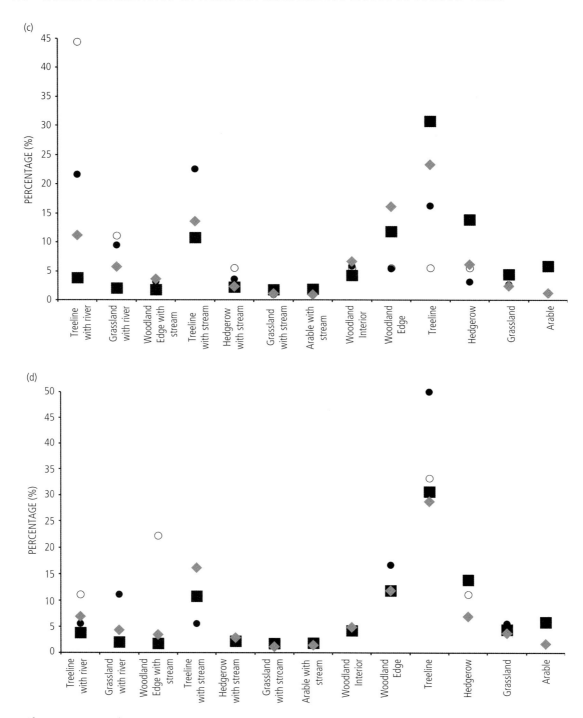

Figure 9.4 *Continued*

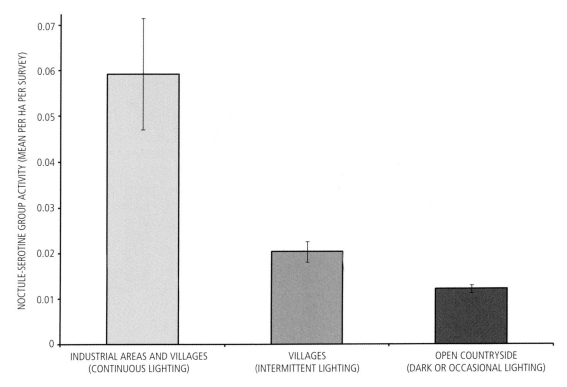

Figure 9.5 Noctule-serotine group bats were recorded significantly more frequently at industrial estates and well-lit sections of villages (n = 27), than in more poorly lit village areas (n = 530). They were recorded significantly more frequently at both these areas than in open countryside (where little or no lighting was present) (n = 1727) (Kruskal-Wallis test, H = 46.1, df = 2, P < 0.001). Data are mean (± SE) noctule-serotine group activity within a 1 ha survey section per survey, for ten repeat surveys (n = 6 transects), scored as 0 or 1 for presence/absence per ha per survey.

Table 9.2 Extent of spatial overlap among common pipistrelle, soprano pipistrelle, and the *Myotis* bats in farmland in the Upper Thames Valley, recorded along walked transects. Data are the number of 1 ha survey sections (and the proportion of total survey sections) within which all/both species or groups were recorded. Bats in the noctule-serotine group were not included due to their distinct foraging strategy (flying at height and covering large distances at speed) and the distance (> 50 m) over which their echolocation calls can be detected (compared with detection distances of 5–15 m for the other species/groups).

	All	common pipistrelle—soprano pipistrelle	common pipistrelle—*Myotis* bats	soprano pipistrelle—*Myotis* bats
All habitats	365 (32.5%)	532 (47.4%)	507 (45.1%)	384 (34.1%)
Terrestrial habitats[1]	216 (26.2%)	323 (39.2%)	326 (39.6%)	232 (28.2%)
Streams	84 (40.6%)	130 (62.8%)	108 (52.2%)	86 (41.5%)
River corridors[2]	60 (82.2%)	70 (95.9%)	61 (83.6%)	61 (83.6%)

[1] Including woodland.
[2] Excluding rivers in open arable fields.

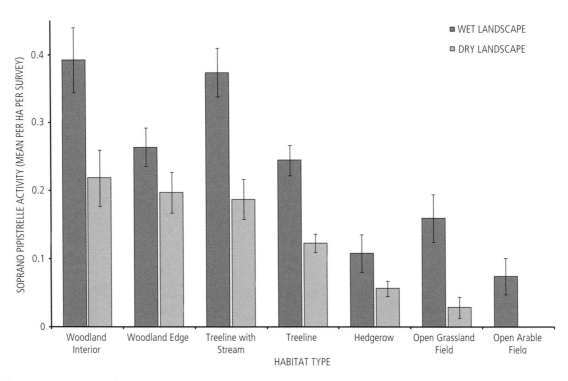

Figure 9.6 Soprano pipistrelle activity across local-level habitat types within wet (> 1 km river within 1 km²) and dry (< 1 km river) landscapes. Data are mean (± SE) bat activity within a 1 ha survey section per survey; Pair-wise comparisons were statistically significant in all cases (Linton 2009).

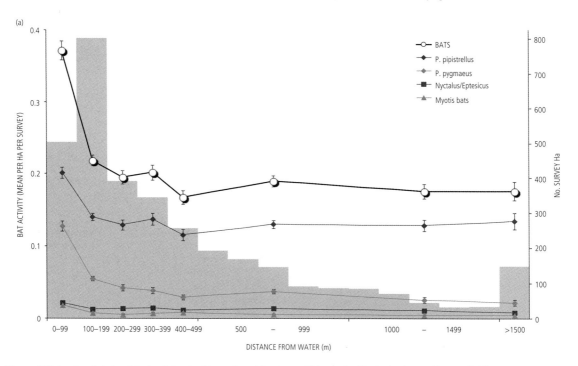

Figure 9.7 Local-scale bat activity in relation to distance from (a) water, and (b) woodland in the landscape. Distance 0–99 m represents survey squares that contain water or woodland (i.e. water or woodland occurs within 100 m, within the 1 ha survey section). Data are mean (± SE) bat activity within a 1 ha survey section per survey; a value of 1 indicates the presence of one bat species/group within a 1 ha survey section per survey on average. Sample size (number of survey squares) is shown by the grey bars and the scale is on the right-hand axis.

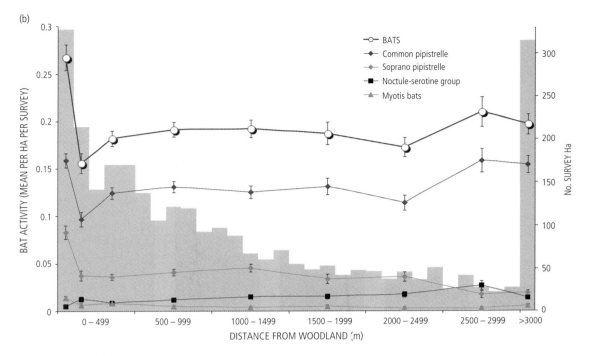

Figure 9.7 *Continued*

wetness and woodiness. The River Thames transect route was wet (35.5% one hectare survey sections contained water) and woody (6.4% one hectare survey sections contained woodland), and close to both water (mean distance = 332 m) and woodland (mean distance = 1.44 km). The River Cherwell route was relatively dry (9.1% one hectare survey sections contained water), had no woodland, and was further from water (mean distance = 507 m) and woodland (mean distance = 3.19 km) than the River Thames route.

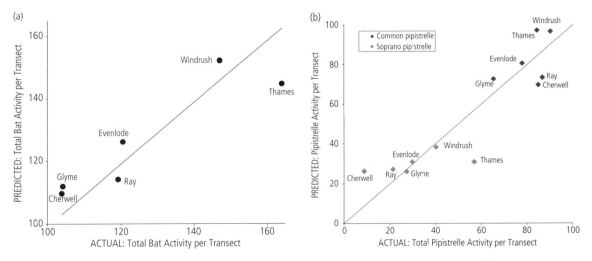

Figure 9.8 Actual versus predicted (a) total bat and (b) pipistrelle activity. Actual bat activity is mean total bat activity of ten surveys per car transect route. Regression line for all bats: $R^2 = 0.67$, df = 4, P = 0.047. Data analysed using cross-validation technique. Each transect route in turn was excluded from the sample data set. Expected levels of bat activity along the 'out-of-sample' test route were then predicted for each habitat type present along that route using 'mean actual bat activity per habitat type' data obtained from the five other 'in-sample' model routes. The overall predicted level of bat activity per transect route was then calculated as the sum of 'predicted bat activity per habitat type' for all habitat types present, based on the known local-level habitat composition of each transect route.

9.4.4 Environmental effects

Ambient night-time temperatures over the survey months ranged between 7 and 23 °C during the two years of the study and included periods of prolonged heat and drought, during July and September 2006, followed by extreme precipitation and extensive flooding across the majority of field sites during July 2007. Bat activity (for all species/groups except the *Myotis* bats) was generally higher when it was warmer[11], and (for *Myotis* bats and the noctule-serotine group) was reduced as wind increased[12]. Bats belonging to the noctule-serotine group appeared to be particularly sensitive to high winds and reduced their activity considerably at wind speeds over 1.5 m per second (Fig. 9.9). Feeding rates of common and soprano pipistrelles were higher at higher temperatures and (for common pipistrelles) with increased cloud cover, but (for soprano pipistrelles) was lower during windy weather[13].

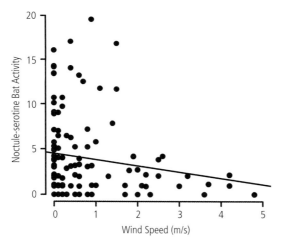

Figure 9.9 Noctule-serotine group activity in relation to wind speed, showing significant reduction in activity at wind speeds > 1.5 ms^{-1}. Activity level is the number of minutes during which noctule-serotine group activity was detected per survey. Wind speed was calculated as the mean of 10 measurements of maximum wind speed per survey. Activity at wind speeds < 1.5 ms^{-1} was significantly greater than activity levels > 1.5 ms^{-1} (one-way ANOVA, $F_{1, 118} = 6.19$, p = 0.014).

Feeding rates of *Myotis* (particularly Daubenton's) bats were apparently not affected by environmental conditions[14] but surveys were not conducted during periods of heavy rain.

9.5 Habitat associations, scale, and connectivity

Activity levels of vesper bats were significantly higher where there were trees and broad-leaved woodland, and also near water, at both local (1 ha) and (for the two pipistrelles, and the *Myotis* bats) landscape (1 km^2) scales. Vesper bats, although somewhat tolerant of variation in habitat quality, being rarely absent from even less favourable habitats (Walsh and Harris 1996a) and widely distributed across the Upper Thames Valley, are clearly sensitive to the presence of these non-agricultural habitats (woodland and water) within the farmscape. But at what scale are these habitat features important?

For most bat species/groups, activity was influenced more by local-scale habitat quality than by the composition of the surrounding landscape. For a particular local-level habitat type, there was no evidence of higher activity levels occurring when that habitat type occurred in wet or woody landscapes. Further, for local-level habitats that did not contain water or woodland, there was no benefit to being closer to water or woodland features in the landscape. Notably, the activity of most bat species/groups was as predicted based on local habitat availability despite each car transect route passing through quite different landscapes. In short, our results suggested that bats are able to utilize suitable local (1 ha) habitat patches regardless of the composition of the surrounding landscape matrix (see also Ober and Hayes 2008; Lundy and Montgomery 2010), but that poor or unsuitable habitat patches are used little (or not at all) even when surrounded by a high quality landscape (see also Bellamy et al. 2013).

The soprano pipistrelle appeared to be one exception to our general findings. Soprano pipistrelles are riparian and woodland specialists and, in contrast to the other more generalist bat species/groups, local-level activity of soprano pipistrelles in farmland appeared

---

[11] GLMM with repeated measures, log-transformed data: common pipistrelle, $F_{1, 87} = 5.94$, P = 0.017; soprano pipistrelle, $F_{1, 87} = 5.84$, P = 0.018; noctule-serotine group, $F_{1, 87} = 8.55$, P = 0.004.

[12] GLMM with repeated measures, log-transformed data: *Myotis* bats, $F_{1, 87} = 4.00$, P = 0.049; noctule-serotine group, $F_{1, 87} = 6.79$, P = 0.011.

[13] GLMM with repeated measures, log-transformed data: common pipistrelle feeding buzzes, $F_{1, 87} = 12.98$, P < 0.001; soprano pipistrelle feeding buzzes, $F_{1, 87} = 4.03$, P = 0.048. Cloud cover effect: common pipistrelle, $F_{2, 87} = 3.40$, P = 0.038. Wind effect: $F_{1, 87} = 7.75$, P = 0.007.

[14] Too few noctule-serotine group feeding buzzes (n = 16) were recorded for analysis.

to be influenced by the presence of water in the surrounding landscape (whereas common pipistrelles, and *Myotis* bats, despite similar local-level habitat requirements, are not; Lundy and Montgomery 2010).

If bats (other than soprano pipistrelles) are relatively tolerant of poor (dry and treeless) landscapes, the obvious question is: do suitable local habitat patches need to be connected? Our results suggest, somewhat unexpectedly, no. However, areas with very low connectivity did not exist in the Upper Thames landscape. Certainly, the dichotomy between wetter and drier sites, and between wooded and open sites, was far greater than the difference that existed between sites with variable densities of LVFs, and even the lowest densities of LVFs in our study area (*c.* 4.5 km/km^2) may have been above any (theoretical) critical connectivity threshold below which bat activity might be limited in lowland farmland. In other areas—such as East Anglia or Lincolnshire (arable-dominated landscapes where the density of LVFs is very low)—results might be different[15] and this warrants further study. Indeed, Frey-Ehrenbold et al. (2013) demonstrated such a threshold effect for *Myotis* bats (but not for pipistrelles or noctule-serotine group bats) in farmland in Switzerland, with very low levels of activity found in isolated habitats, but little difference in activity between 'poorly' connected and 'highly' connected habitats. That vesper bats responded positively to an increase in LVF connections at a local scale might be explained if connectivity were important for bats at the local scale, or if bats are influenced by some other attribute of LVF junctions (such as shelter), or if more LVFs simply provide more foraging habitat (and thus greater prey abundance). Considering the distances over which many bat species regularly forage, the latter two are most likely, and are non-mutually exclusive insofar as more complex habitat provides shelter for invertebrate prey, increased prey availability (e.g. Verboom and Spoelstra 1999), and shelter for foraging bats (and we will return to this point).

Habitat connectivity is an important factor in the biology of many species (Cushman et al. 2013), especially in highly fragmented agricultural landscapes (Donald and Evans 2006). However, for species such as bats that can fly between foraging patches, habitat *quality* may be more important than connectivity. Bat

activity was higher in landscapes with more woodland, and at a local scale the presence of trees appeared to be particularly important across a number of habitat types (to such an extent that bats actively selected against hedgerows with no trees). Indeed, bat activity along treeless LVFs (low hedges) in the Upper Thames Valley was equivalent to that recorded in open grassland fields (see Fig. 9.3). Apparent confusion in the literature regarding the importance of hedges to bats (see e.g. Russ and Montgomery 2002; Walsh and Harris 1996a, b) appears to be due to the distinction between treelines (which are selected by bats) and hedges (which may not be) (Boughey et al. 2011), and to precisely how 'hedge' is defined. A good quality hedge is one with a high density of mature emergent trees (Brandt et al. 2007; Boughey et al. 2011). Trees are beneficial to commuting and foraging bats as they provide shade at dusk and dawn, shelter during poor weather conditions, and feeding opportunities through the accumulation of high densities of aerial insects; and bats have been shown to prefer even isolated trees, relative to hedgerows without trees (Lumsden and Bennett 2005).

In contrast to the other vesper bats, the larger noctule-serotine group bats appeared not to be influenced by the amount of water or woodland within surveyed landscapes, but were influenced by farm type. These bats feed on larger invertebrates (mainly Coleoptera and Lepidoptera) found predominately in pastures (Vaughan 1997), and were recorded at significantly higher activity levels in pastoral landscapes and mixed landscapes compared to arable-dominated landscapes. At a local level, all other vesper bats selected against open agricultural fields (grassland or arable), and there was little or no evidence of feeding at streams running through them, whereas noctule-serotine group bats appeared to use open grassland fields in proportion to their availability, and demonstrated little evidence of local-scale habitat selection at all. However, because of the high and fast flight of noctule-serotine group bats (Jones 1995; Jensen and Miller 1999), and their high intensity calls which can be detected from distances of over 75 m (D. Linton, pers. obs.), it is difficult to associate these bats with specific fine-scale habitat types. They also forage around street lights (Rydell 1992), and were often recorded in well-lit villages and industrial estates along car transects, where they were apparently feeding on insects attracted to the lights. The only other bat to be frequently detected in villages was the common pipistrelle, which was probably due to the high availability of roost sites within buildings.

---

[15] This is a common problem in large-scale field studies—Fuentes-Montemayor et al. (2013). and Lentini et al. (2012) failed to detect an effect of water in the landscape on bat activity, which both authors attribute to a lack of variation among study sites.

9.6 Species-specific habitat use and niche partitioning

Although spatial distribution across different habitat types for all bat species and groups (with the exception of the noctule-serotine group) was broadly similar, species/group-specific differences were observed in the degree of preference exhibited for watercourses and woodland habitats, suggesting that some degree of niche differentiation exists among these sympatric species. Common pipistrelle was the most widespread and frequently encountered species across all habitat types, except along river corridors and within broad-leaved woodland interiors, where levels of soprano pipistrelle and *Myotis* bat activity were highest. It is possible that common pipistrelle might reduce activity adjacent to rivers or within woodland to avoid competitive interactions (Nicholls and Racey 2006a,b), but our study suggests that common pipistrelle are simply able to exploit a wider range of habitats (see also Davidson-Watts et al. 2006; Nicholls and Racey 2006b). Soprano pipistrelle and *Myotis* bats, in contrast, appear to be restricted to riparian and woodland habitat types (see also Boughey et al. 2011).

Differential habitat use is likely to be related to dietary preferences, as suggested by the higher proportions of feeding buzzes recorded along rivers (and in 'wet' landscapes) for soprano pipistrelle and *Myotis* bats, while the highest proportions of feeding buzzes for common pipistrelle were recorded along smaller streams (unaffected by landscape composition). Differences in roosting preferences may also contribute to observed differences in proportions of bat activity across habitat types, including the selection for woodland habitats by *Myotis* bats and soprano pipistelle (where activity levels were high but feeding rates were low), and to the selection for areas of human settlement by common pipistrelle.

Ecomorphological differences, foraging, and habitat use preferences, also lead to differences in activity patterns among the four main species groups (including the noctule-serotine group, Box 9.2). Whilst perhaps not driven by inter-specific competition, such temporal partitioning might facilitate co-existence, and thus the high levels of regional diversity (species richness) that occur (Aldridge and Rautenbach 1987; Altringham 2011; Neuweiler 2000; Findley 1993). Nevertheless, spatial and temporal overlap is high, certainly among bats detected in this study. Given the extent of overlap recorded, especially between the pipistrelles and the *Myotis* bats, as well as the

similarity in habitat selection between the soprano pipistrelle and the *Myotis* bats, further more detailed study of the interactions among these species would be insightful.

9.7 Environmental effects and implications for climate change

We found higher activity and feeding levels of all bats (with the exception of the *Myotis* bats) when it was warmer (and cloudy), but all species/groups were active even when summer temperatures fell to 7 °C, below which bat activity is usually severely reduced (Avery 1985). Bats appear to be highly tolerant of the temperature extremes that are likely to be experienced in the UK, and the influence of weather conditions on their foraging behaviour is most likely an indirect effect, reflecting the impact on their prey (Wickramasinghe et al. 2004). That feeding by the *Myotis* bats—predominantly Daubenton's bat—was not affected by environmental conditions, but that of pipistrelles was, is most likely explained by differences in their foraging ecology (Sherwin et al. 2013).

In the UK, windy weather seems to be more of a problem for bats than temperature. Noctule-serotine group bats, being adapted for fast flight in open environments (Altringham 2003), probably lack the manoeuvrability to fly well in high winds, and typically fly high in the sky (e.g. Jensen and Miller 1999) where they are most vulnerable to windy conditions. We found that these large bats were rarely active in high winds and the smaller pipistrelle and *Myotis* bats congregated along mature treelines or woodland edge habitats when it was windy—suggesting that the importance of LVF connections at the local scale may lie in their role as windbreaks for flying bats and insects in the, otherwise largely open, farmscape (see also Verboom and Spolestra 1999; Russ et al. 2003). We did not monitor bat activity in the rain due to equipment limitations, but rain can impact bats through noise (e.g. of raindrops on the water's surface), which can in turn interfere with echolocation (e.g. Rydell et al. 1999), and through the increased energetic costs of flying with wet fur (Voigt et al. 2011).

Bats are adaptable and opportunistic. The foraging we observed over flooded fields following the heavy rains in the UK in 2007 demonstrated that they are able to exploit temporary localized peaks in prey abundance. Nevertheless, their dependence on short-lived ectothermic prey (which is presumably highly sensitive to climate change, Wilson and Maclean 2011), means that bats are also likely to be vulnerable to

Box 9.2 Species differences in evening activity patterns

Combining all survey data and plotting detection rates (percentage of survey minutes containing echolocation calls) against time after sunset, revealed considerable differences among species/groups in evening activity patterns. Activity of common and soprano pipistrelles was comparatively stable across the late evening survey period, consistent with their intermediate emergence times and their tolerance to light (Altringham 2003). Noctule-serotine group activity was initially relatively high—these large, fast-flying, aerial hawking species typically emerge early (e.g. Catto et al. 1995; Shiel and Fairley 1999) to exploit abundant aerial insects early in the night—but declined steadily during the first few hours after dusk. *Myotis* bats exhibited late activity onset (consistent with their late emergence times, e.g. Swift 1997; Swift and Racey 1983), increasing sharply between 30–55 minutes after sunset when light levels were decreasing rapidly. No *Myotis* bats were encountered at light levels above 5.5 lux (Box Fig. 9.2.1). During the early evening, when it was

still relatively light, Daubenton's bats appeared to restrict their low flights over water to within the shadow of overhanging trees, but foraged along open stretches of river later when it was darker—their avoidance of high light levels most likely associated with predation risk (Jones and Rydell 1994; Rydell et al. 1996). Indeed, activity of *Myotis* bats was significantly lower during the full moon than during the new moon (on cloudless nights)—a phenomenon, termed 'lunar phobia', that is common among bats worldwide, and thought to be associated with avoidance of predators when risk is high (Saldaña-Vázquez and Munguía-Rosas 2013).

Our findings have important implications for the timing of bat surveys, to increase their efficiency and accuracy. We suggest that surveys should commence at or up to 30 minutes before sunset to ensure the peak in activity of noctule-serotine group bats is covered, and continue for at least 60 minutes after sunset, to avoid under-representation of *Myotis* bats (as recommended in Hundt 2012).

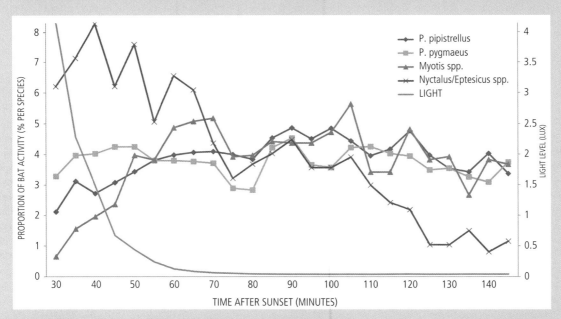

Box Figure 9.2.1 Activity of each bat species/group in relation to time after sunset, shown against evening light levels. All survey data combined (n = 180 surveys; mean 876 ± 45 SD minutes recorded per 5-minute period). Light levels measured at the start and finish of each 5-minute sampling period using a DT-131 (CEM) light meter.

climate change. Most research to date on the response of bats to environmental conditions (reviewed in Sherwin et al. 2013) has focused on hibernating bats (arousal frequency, emergence times, etc.), but there is comparatively little information on foraging bats, or of the implications of fluctuating food availability on long-term population dynamics. This issue warrants attention.

9.8 Farmland management for bats

Broadly speaking, all vesper bats favour, to some extent, water, trees and woodland, and grassland farmland, and this conclusion is supported by other studies in different areas (e.g. Bellamy et al. 2013). But what does this mean for farmland management? First, local-scale habitat creation, improvement, and protection are likely to be of greater benefit than creation of large-scale linkages (although we should be cautious about extrapolating beyond the 2500 km^2 study area, since any analysis involving scale is highly dependent on how 'landscapes' are defined, and the influence of landscape composition should be similarly assessed in other areas).

Second, habitats that are rare within the agricultural landscape—rivers, streams, treelines, and woodlands—should be created and managed to improve foraging opportunities for bats within agricultural ecosystems. Even artificial systems can be of value, such as sewage treatment works (Park and Christinacce 2006) and artificial water retention ponds (Stahlschmidt et al. 2012). It is worth noting, however, that most bat species/groups were at least occasionally recorded flying over open grassland and arable fields (≥ 50 m from any other habitat features), and were recorded feeding there, and thus that, within a given area, total bat activity over poor quality habitat can be high (albeit at low density), simply because there is a lot of it. For example, all 'tree-less' LVFs combined, along car transect routes, contained 9.4–22.2% of total bat activity (per species/group), whereas all LVFs with trees adjacent to rivers (a generally preferred but relatively rare habitat) contained only 2.4–7.5% of total bat activity.

Third, and probably most important, existing trees within the landscape should be better protected and new tree planting or allowance of tree growth into emergent standards along hedgerows ('hedgerow trees') to create treelines should be encouraged. Current emphasis (existing recommended practice) on reconnecting and planting low hedgerows without any emphasis on incorporating trees (Entwistle et al. 2001; Haysom 2003) is likely to be ineffective at benefiting foraging and commuting bats in all but the most isolated landscapes. A much greater difference was observed in bat activity levels between treelines relative to hedgerows without trees, compared to the differences found between grassland relative to arable fields. Therefore incentives for tree planting and for retention or encouragement of emergent standard trees along new or existing hedgerows should be of greater benefit to foraging and commuting bats than benefits derived from management of field margins for conservation, or even arable reversion to grassland (although the creation of pastures supporting high invertebrate abundance would benefit foraging bats belonging to the noctule-serotine group).

Fourth, bats need to be considered as a diverse group with highly species-specific preferences. Targeting of habitat restoration and enhancement works to benefit soprano pipistrelle, and the *Myotis* bats, for example, should be concentrated along riparian corridors (or around other waterbodies such as ponds and lakes). Woodland creation and improvement schemes should also be encouraged. Overall, as we explain for small mammals in Chapter 4, this volume, habitat heterogeneity is needed to foster the community of bats.

Finally, while short-term measures of bat activity are appropriate for large-scale, multi-species studies such as the one described here, long-term studies of population dynamics are sorely needed. Bats are long-lived and slow to reproduce and their populations would not be expected to reveal trends over the three-year duration of this study. For now, our results point to some simple management recommendations that could be implemented on farmland to benefit bats (and indeed other farmland biodiversity, e.g. Chapter 8, this volume), but only long-term commitment of the kind seen in our Wytham badger studies (Macdonald and Feber 2015: Chapter 4) will reveal the insights we need to predict (and act to prevent) changes in bat populations in response to climatic variation and broad-scale habitat alterations.

Acknowledgements

We thank our funders, Esmée Fairbairn Foundation and the People's Trust for Endangered Species, numerous field assistants for help with bat surveys, landowners for allowing access to their land, Oxfordshire and North Bucks Bat Groups for sharing their local bat knowledge, Jules Agate, John Russ, Chris Corben, and Stephen Ellwood for technical advice, and Philip

Riordan and the late Rob Strachan. John Altringham made helpful comments on an earlier version of this chapter. We are grateful to Ruth Feber for improving earlier drafts and Eva Raebel for editorial advice.

References

Aldridge, H.D.J.N. and Rautenbach, I.L. (1987). Morphology, echolocation and resource partitioning in insectivorous bats. *Journal of Animal Ecology*, **56**, 763–778.

Altringham, J.D. (2003). *British Bats*. New Naturalist Series 93, Harper Collins, London.

Altringham, J.D. (2011). *Bats: From Evolution to Conservation*. Oxford University Press, Oxford.

Avery, M.I. (1985). Winter activity in pipistrelle bats. *Journal of Animal Ecology*, **54**, 721–738.

Avery, M.I. (1991). Pipistrelle *Pipistrellus pipistrellus*. In: *Handbook of British Mammals* (eds. G. B. Corbett & S. Harris), pp 1248. Blackwell, Oxford.

Bellamy, C., Scott, C., and Altringham, J. (2013). Multiscale presence-only habitat suitability models: fine-resolution maps for eight bat species. *Journal of Applied Ecology*, 50, 892–901.

Berthinussen, A. and Altringham, J. (2012). The effect of a major road on bat activity and diversity. *Journal of Applied Ecology*, **49**, 82–89.

Boughey K, Lake I, Haysom K, and Dolman P. (2011). Improving the biodiversity benefits of hedgerows: How physical characteristics and the proximity of foraging habitat affect the use of linear features by bats. *Biological Conservation*, **144**, 1790–1798.

Brandt, G., Blows, L., Linton, D., Paling, N., and Prescott, C. (2007). Habitat associations of British bat species on lowland farmland within the Upper Thames catchment area. *Centre for Wildlife Assessment and Conservation e-journal*, **1**, 10–19.

Brigham, R.M. and Fenton, M.B. (1986). The influence of roost closure on the roosting and foraging behaviour of *Eptesicus fuscus* (Chiroptera: Vespertilionidae). *Canadian Journal of Zoology*, **64**, 1128–1133.

Catto, C.M.C., Racey, P.A., and Stephenson, P.J. (1995). Activity patterns of the serotine bat (*Eptesicus serotinus*) at a roost in southern England. *Journal of Zoology, London*, **235**, 635–644.

Ciechanowski, M., Zajac, T. Bitas, A., and Dunajski, R. (2007). Spatiotemporal variation in activity of bat species differing in hunting tactics: effects of weather, moonlight, food abundance, and structural clutter. *Canadian Journal of Zoology*, **85**, 1249–1263.

Cushman, S.A., McRae, B., Adriaensen, F., Beier, P., Shirley, M., and Wilson, K.A. (2013). Biological corridors and connectivity. In: *Key Topics in Conservation Biology* 2. Eds. D.W. Macdonald & K.J. Willis., pp. 384–404. Wiley-Blackwell, John Wiley & Sons Ltd., Oxford.

Daan, S. (1980). Long-term changes in bat populations in the Netherlands: a summary. *Lutra*, **22**, 95–118.

Davidson-Watts, I., Walls, S., and Jones, G. (2006). Differential habitat selection by *Pipistrellus pipistrellus* and *Pipistrellus pygmaeus* identifies distinct conservation needs for cryptic species of echolocating bats. *Biological Conservation*, **133**, 118–127.

Defra (2013). UK Biodiversity Indicators in Your Pocket 2013. Defra, London.

Dietz, M., Encarnação, J.A., and Kalko, E.K.V. (2006). Small scale distribution patterns of female and male Daubenton's bats (*Myotis daubentonii*). *Acta Chiropterologica*, **8**, 403–415.

Dietz, C., von Helversen, O., and Nill, D. (2009). *Bats of Britain, Europe and Northwest Africa*. A & C Black, London.

Donald, P.F. and Evans, A.D. (2006). Habitat connectivity and matrix restoration: the wider implications of agri-environment schemes. *Journal of Applied Ecology*, **43**, 209–218.

Downs, N.C. and Racey, P.A. (2006). The use by bats of habitat features in mixed farmland in Scotland. *Acta Chiropterologica*, **8**, 169–185.

Entwistle, A.C., Harris, S., Hutson, A.M., Racey, P.A., Walsh, A., Gibson, S.D., Hepburn, I., and Johnston, J. (2001). *Habitat Management for Bats*. Joint Nature Conservation Committee (JNCC), Peterborough. <http://jncc.defra.gov.uk/>page 2465 Accessed September 2013.

Entwistle, A.C., Racey, P.A., and Speakman, J.R. (1996). Habitat exploitation by a gleaning bat, *Plecotus auritus*. *Philosophical Transactions of the Royal Society of London B*, **351**, 921–931.

Fenton, M. B. (1999). Describing the echolocation calls and behaviour of bats. *Acta Chiropt*. **1**, 411–422.

Fenton, M.B. (2003). Eavesdropping on the echolocation and social calls of bats. *Mammal Review*, 33, 193–204.

Findley, J.S. (1993). *Bats: A Community Perspective*. Cambridge University Press. Cambridge.

Frey-Ehrenbold A., Bontadina F., Arlettaz R., and Obrist M.K. (2013). Landscape connectivity, habitat structure and activity of bat guilds in farmland-dominated matrices. *Journal of Applied Ecology*, **50**, 252–261.

Fuentes-Montemayor, E., Goulson, D., Cavin, L., Wallace, J.M. & Park, K. (2013). Fragmented woodlands in agricultural landscapes: the influence of woodland character and landscape context on bats and their insect prey. *Agriculture, ecosystems and environment*, **172**, 6–15.

Fuentes-Montemayor, E., Goulson, D., and Park, K. (2011). Pipistrelle bats and their prey do not benefit from four widely applied agri-environment management prescriptions. *Biological Conservation*, **144**, 2233–2246.

Goerlitz, H. R., ter Hofstede, H. M. Zeale, M. R. K. Jones, G., and Holderied, M. W. (2010). An aerial-hawking bat uses stealth echolocation to counter moth hearing. *Current Biology*, 20, 1588–1572.

Hayes, J.P. (2000). Assumptions and practical considerations in the design and interpretation of echolocation-monitoring studies. *Acta Chiropterologica*, **2**, 225–236.

Haysom, K. (2003). *Agricultural practice and bats: A review of current research literature and management recommendations*. Department for Environment, Food and Rural Affairs (Defra).

Hundt, L. (2012). *Bat Surveys: Good Practice Guidelines*, 2nd edition, Bat Conservation Trust, London.

Jan, C.M.I., Frith, K., Glover, A.M., Butlin, R.K., Scott, C.D., Greenaway, F., Ruedi, M., Franz, A.C., Dawson, D.A., and Altringham, J.D. (2010). *Myotis alcathoe* confirmed in the UK from mitochondrial and microsatellite DNA. *Acta Chiropterologica*, **12**, 471–483.

Jensen, M.E. and Miller, L.A. (1999). Echolocation signals of the bat *Eptesicus serotinus* recorded using a vertical microphone array: effect of flight altitude on searching signals. *Behavioral Ecology and Sociobiology*, **47**, 60–69.

Jones, G. (1995). Flight performance, echolocation and foraging behavior in noctule bats *Nyctalus noctula*. *Journal of Zoology, London*, **237**, 303–312.

Jones, G. (1999). Scaling of echolocation call parameters in bats. *The Journal of Experimental Biology*, **202**, 3359–3367.

Jones, G., Jacobs, D.S., Kunz, T.H., Willig, M.R., and Racey, P.A. (2009). Carpe noctem: the importance of bats as bioindicators. *Endangered Species Research*, **8**, 93–115.

Jones, G. and Rydell, J. (1994). Foraging strategy and predation risk as factors influencing emergence time in echolocating bats. *Philosophical Transactions of the Royal Society B.*, **346**, 445–455.

Jones, G. and Teeling, E.C. (2006). The evolution of echolocation in bats. *Trends in Ecology and Evolution*, **21**, 149–156.

Kalko, E.K.V. (1995). Insect pursuit, prey capture and echolocation in pipistrelle bats (Microchiroptera). *Animal Behaviour*, **50**, 861–880.

Kalko, E.K.V. & Schnitzler, H.-U. (1998). How Echolocating Bats Approach and Acquire Food. Pp. 197–204. In: T.H. Kunz & P.A. Racey (Eds.) *Bat Biology and Conservation*. Smithsonian Institution Press, Washington D.C.

Lentini, P.E., Gibbons, P., Fischer, J., Law, B., Hanspach, J., and Martin, T.G. (2012). Bats in a farming landscape benefit from linear remnants and unimproved pastures. *PLOS ONE*, **7**, e48201.

Limpens, H.J.G.A. and Kapteyn, K. (1991). Bats, their behaviour and linear landscape elements. *Myotis*, **29**, 39–48.

Linton, D.M. (2009). Bat Ecology and Conservation in Lowland Farmland. D.Phil Thesis, University of Oxford.

Longley, M. (2003). *Greater horseshoe bat Project 1998–2003*. English Nature Research Report, No. 532. English Nature, Peterborough. <http://publications.naturalengland.org.uk/publication/142004>. Accessed May 2013.

Lumsden, L.F. and Bennett, A.F. (2005). Scattered trees in rural landscapes: foraging habitat for insectivorous bats in southeastern Australia. *Biological Conservation*, **122**, 205–222.

Lundy, M. and Montgomery, I. (2010). Summer habitat associations of bats between riparian landscapes and within riparian areas. *European Journal of Wildlife Research*, **56**, 385–394.

Macdonald, D.W. and Feber, R.E., eds. (2015). *Wildlife Conservation on Farmland. Conflict in the Countryside*. Oxford University Press, Oxford.

Mapstone, L. (2009). *Landscapes for Lessers: Phase 2 Project Report*. Countryside Council for Wales (CCW) Science Report No. 896. Countryside Council for Wales, Bangor.

Mayle, B. A. (1990). A biological basis for bat conservation in British woodlands—a review. *Mammal Review*, **20**, 159–195.

Neu, C.W., Byers, C.R., and Peek, J.M. (1974). A technique for analysis of utilization-availability data. *Journal of Wildlife Management*, **38**, 541–545.

Neuweiler, G. (2000). *The Biology of Bats*. Oxford University Press, Oxford.

Nicholls, B. and Racey, P.A. (2006a). Contrasting home-range size and spatial partitioning in cryptic and sympatric pipistrelle bats. *Behavioural Ecology and Sociobiology*, **61**, 131–142.

Nicholls, B. and Racey, P.A. (2006b). Habitat selection as a mechanism of resource partitioning in two cryptic bat species *Pipistrellus pipistrellus* and *Pipistrellus pygmaeus*. *Ecography*, **29**, 697–708.

Ober, H.K. and Hayes, J.P. (2008). Influence of vegetation on bat use of riparian areas at multiple spatial scales. *Journal of Wildlife Management*, **72**, 396–404.

Obrist, M.K., Boesch, R., and Flückiger, P.F. (2004). Variability in echolocation call design of 26 Swiss bat species: consequences, limits and options for automated field identification with a synergetic pattern recognition approach. *Mammalia*, **68**, 307–322.

Park, K.J. and Cristinacce, A. (2006). Use of sewage treatment works as foraging sites by insectivorous bats. *Animal Conservation*, **9**, 259–268.

Park, K. J., Jones, G., and Ransome, R.D. (2000). Torpor, arousal and activity of hibernating greater horseshoe bats (*Rhinolophus ferrumequinum*). *Functional Ecology*, **14**, 580–588.

Parsons, K.N., Jones, G., and Greenaway, F. (2003). Swarming activity of temperate zone microchiropteran bats: effects of season, time of night and weather conditions. *Journal of Zoology, London*, **261**, 257–264.

Pfalzer, G. and Kusch, J. (2003). Structure and variability of bat social calls: implications for specificity and individual recognition. *Journal of Zoology, London*, **261**, 21–33.

Pocock, M.J.O and Jennings, N. (2008). Testing biotic indicator taxa: the sensitivity of insectivorous mammals and their prey to the intensification of lowland agriculture. *Journal of Applied Ecology*, **45**, 151–160.

Racey, P.A. (1998). The importance of the riparian environment as a habitat for British bats. In Dunstone, N., Gorman, M.L., eds., *Behaviour and Ecology of Riparian Mammals* (Symp. Zool. Soc. Lond. No. 71), pp.69–91. Cambridge University Press, Cambridge.

Ransome, R.D. (1996). *The management of feeding areas for greater horseshoe bats*. English Nature Research Report No. 174, English Nature, Peterborough.

Robinson, M.F. and Stebbings, R.E. (1997). Home range and habitat use by the serotine bat, *Eptesicus serotinus*, in England. *Journal of Zoology, London*, **243**, 117–136.

Robinson, R.A. and Sutherland, W.J. (2002). Post-war changes in arable farming and biodiversity in Great Britain. *Journal of Applied Ecology*, **39**, 157–176.

Russ, J. (2012). *British Bat Calls: A Guide to Species Identification*. Pelagic Publishing, Exeter.

Russ, J.M., Briffa, M., and Montgomery, W.I. (2003). Seasonal patterns in activity and habitat use by bats (*Pipistrellus* spp. and *Nyctalus leisleri*) in Northern Ireland, determined using a driven transect. *Journal of Zoology, London*, **259**, 289–299.

Russ, J., Catto, C., and Wembridge, D. (2006). *The Bats & Roadside Mammals Survey 2005: Final Report on First Year of Study submitted to The Bat Conservation Trust and the Mammals Trust UK*. The Bat Conservation Trust and the Mammals Trust UK, London.

Russ, J.M. and Mongomery, W.I. (2002). Habitat associations of bats in Northern Ireland: implications for conservation. *Biological Conservation*, **108**, 49–58.

Rydell, J. (1992). Exploitation of insects around street lamps by bats in Sweden. *Functional Ecology*, **6**, 744–750.

Rydell, J., Entwistle, A., and Racey, P.A. (1996). Timing of foraging flights of three species of bats in relation to insect activity and predation risk. *Oikos*, **76**, 243–252.

Rydell, J., Miller, L.A., and Jensen, M.E. (1999). Echolocation constraints of Daubenton's Bat foraging over water. *Functional Ecology*, **13**, 247–255.

Saldaña-Vázquez, R.A. and Munguía-Rosas, M.A. (2013). Lunar phobia in bats and its ecological correlates: A meta-analysis. *Mammalian Biology*, **78**, 216–219.

Sattler, T., Bontadina, F., Hirzel, A.H., and Arlettaz, R. (2007). Ecological niche modelling of two cryptic bat species calls for a reassessment of their conservation status. *Journal of Animal Ecology*, **44**, 1188–1199.

Schnitzler, H.U., Moss, C.F., and Denzinger A. (2003). From spatial orientation to food acquisition in echolocating bats. *Trends in Ecology and Evolution*, **18**, 386–394.

Sherwin, H.A., Montgomery, W.I., and Lundy, M.G. (2013). The impact and implications of climate change for bats. *Mammal Review*, **43**, 171–182.

Shiel, C.B. and Fairley, J.S. (1999). Evening emergence of two nursery colonies of Leisler's bat (*Nyctalus leisleri*) in Ireland. *Journal of Zoology, London*, **247**, 439–447.

Siemers, B.M. and Schnitzler, H.-U. (2004). Echolocation signals reflect niche differentiation in five sympatric congeneric bat species. *Nature*, **429**, 657–661.

Simmons, N.B. (2005). Order Chiroptera. In *Mammal Species of the World: A Taxonomic and Geographic Reference* (3rd edition), volume 1, D.E. Wilson and D.M. Reeder eds. pp. 312–529. Johns Hopkins University Press, Baltimore.

Smith, P.G. and Racey, P.A. (2008). Natterer's bats prefer foraging in broad-leaved woodlands and river corridors. *Journal of Zoology*, **275**, 314–322.

Stahlschmidt, P., Pätzold, A., Ressl, L., Schulz, R., and Brühl, C.A. (2012). Constructed wetlands support bats in agricultural landscapes. *Basic and Applied Ecology*, **13**, 196–203.

Stebbings, R.E. (1995). Why should bats be protected? A challenge for conservation. *Biological Journal of the Linnean Society*, 56 (Suppl. A), 103–118.

Swift, S.M. (1997). Roosting and foraging ehavior of Natterer's bats (*Myotis nattereri*) close to the northern border of their distribution. *Journal of Zoology, London*, **242**, 375–384.

Swift, S.M. and Racey, P.A. (1983). Resource partitioning in two species of vespertilionid bats (Chiroptera) occupying the same roost. *Journal of Zoology, London*, **200**, 249–259.

Vaughan, N. (1997). The diets of British bats (Chiroptera). *Mammal Review*, **27**, 77–94.

Vaughan, N., Jones, G., and Harris, S. (1997a). Habitat use by bats (Chiroptera) assessed by means of a broad-band acoustic method. *Journal of Applied Ecology*, **34**, 716–730.

Vaughan, N., Jones, G., and Harris, S. (1997b). Identification of British bat species by multivariate analysis of echolocation call parameters. *Bioacoustics*, **7**, 189–207.

Verboom, B. and Huitema, H. (1997). The importance of linear landscape elements for the pipistrelle *Pipistrellus pipistrellus* and the serotine bat *Eptesicus serotinus*. *Landscape Ecology*, **12**, 117–125.

Verboom, B. and Spoelstra, K. (1999). Effects of food abundance and wind on the use of tree lines by an insectivorous bat, *Pipistrellus pipistrellus*. *Canadian Journal of Zoology*, **77**, 139–140.

Voigt, C.C., Schneeberger, K., Voigt-Heucke, S.L., and Lewanzik D. (2011). Rain increases the energy cost of bat flight. *Biology Letters*, **7**, 793–795.

Walsh, A. and Harris, S. (1996a). Foraging habitat preferences of vespertilionid bats in Britain. *Journal of Applied Ecology*, **33**, 508–518.

Walsh, A. and Harris, S. (1996b). Factors determining the abundance of vespertilionid bats in Britain: geographical, land class and local habitat relationships. *Journal of Applied Ecology*, **33**, 519–529.

Walsh, A.L., Harris, S., and Hutson, A.M. (1995). Abundance and habitat selection of foraging vespertilionid bats in Britain: a landscape-scale approach. *Symposium of the Zoological Society of London*, **67**, 325–344.

Wickramasinghe, L.P., Harris, S., Jones, G., and Vaughan N. (2003). Bat activity and species richness on organic and conventional farms: impact of agricultural intensification. *Journal of Applied Ecology*, **40**, 984–993.

Wickramasinghe, L.P., Harris, S., Jones, G., and Vaughan Jennings, N. (2004). Abundance and species richness of nocturnal insects on organic and conventional farms: effects of agricultural intensification on bat foraging. *Conservation Biology*, **18**, 1283–1292.

Wilson, R.J. and Maclean, I.M.D. (2011). Recent evidence for the climate change threat to Lepidoptera and other insects. *Journal of Insect Conservation*, 15, 259–268.

CHAPTER 10

Local and landscape-scale management of Odonata

Eva M. Raebel, David J. Thompson, and David W. Macdonald

Thus the dragon-fly enters upon a more noble life than that it had hitherto led in the water, for in the latter it was obliged to live in misery, creeping or swimming slowly, but now it wings the air.

Swammerdam 1669

10.1 Introduction to Odonata

Odonata [Anisoptera (dragonflies) and Zygoptera (damselflies)], are among the most successful insects living on Earth: their ancestors (Protodonata) first appeared in the fossil record over 300 million years ago and, currently, around 6000 extant species of odonates have been described worldwide[1] (Corbet and Brooks 2008). Together with the Ephemeroptera (mayflies), odonates are considered to be among the oldest flying insects, and hence sustain a hemimetabolous metamorphosis (no pupal stage)[2] (Corbet and Brooks 2008).

10.1.1 Aquatic stage

Odonates have adapted to live in lentic (standing) and lotic (running) freshwater, and are present in rivers, streams, ditches, canals, lakes, ponds, bogs, and mires. Although well camouflaged within debris and mud, odonate larvae are preyed upon by frogs, fish, waterfowl, and other odonates, while, in turn, they are voracious predators of insect larvae, crustaceans, worms, snails, tadpoles, and even small fish (e.g. three-spined sticklebacks *Gasterosteus aculeatus*). As a consequence, they require a variety of niches within their aquatic habitat to provide food and shelter, and to facilitate co-existence of various species. Good water quality and the presence of aquatic plants are a must, but every species has different requirements; for example, some species are tolerant of nutrient-enriched water (e.g. broad-bodied chaser *Libellula depressa*), or drought (e.g. common darter *Sympetrum striolatum*), or are only associated with acid heathland (e.g. common hawker *Aeshna juncea*).

Eggs are oviposited in water (round-shaped exophytic eggs, e.g. darters) or in plant material/mud (elongated endophytic eggs; i.e. damselflies and hawkers), and hatch after 5–40 days in species with direct development, or after 80–230 days in species with delayed development (Corbet 1980). In the UK, most species lay eggs which hatch around 2–5 weeks after being laid and it is the larva which overwinters, but species with an egg diapause, such as emerald damselflies (Lestidae), some hawkers (Aeshnidae), and darters (Libellulidae), overwinter as eggs and hatch the following spring.

Time spent as larvae varies. Species can be bivoltine (two generations per year) in the tropics and warm temperate regions, while in temperate areas, species can be both univoltine (one generation per year), or partivoltine (one generation taking two (semivoltine) or more years) depending on latitude. In the UK, larval development typically lasts one or two years, but ranges from two to three months in emerald damselflies (Lestidae) to five years in the golden-ringed dragonfly *Cordulegaster boltonii*. Of course, the longer a species needs to spend underwater to develop, the more susceptible it will be to water quality and

---

[1] Belonging to 18 families of Zygoptera and 12 families of Anisoptera. Previously, a third suborder had been considered, the Anisozygoptera, from which only two species are extant. Current opinion (Corbet and Brooks 2008) is that species of this suborder are long extinct and the two remaining species are now included within the Anisoptera.

[2] As opposed to more recently evolved insects, such as Lepidoptera (butterflies and moths), which sustain a holometabolous, or complete, metamorphosis, including a pupa.

Wildlife Conservation on Farmland. Managing for Nature on Lowland Farms. Edited by David W. Macdonald and Ruth E. Feber.
© Oxford University Press 2015. Published 2015 by Oxford University Press.

Figure 10.1 Odonate larvae: (a) Emperor dragonfly *Anax imperator* larva. *A. imperator* is mostly found in large well-vegetated ponds, lakes, gravel-pits, and sometimes ditches and canals. This species is mainly semivoltine (one generation every two years) but can also be univoltine (one-generation per year), with both types of life cycle being able to occur simultaneously in the same pond. Photograph ©E. Raebel; (b) Azure damselfly *Coenagrion puella* larva. *C. puella* is a common species of small ponds and also found in streams; it can be univoltine (one generation per year) or semivoltine (one generation every two years) depending on environmental conditions, completing around 10–11 moults before emerging. It is generally found within aquatic vegetation. Photograph ©E. Raebel.

stochastic events (e.g. pollution incidents). Time-scale of development depends on food supply and on climate, taking longer in cooler areas than in warmer ones. However, some species such as the azure damselfly *Coenagrion puella* (Fig. 10.1b), can commonly have, in the UK, a simultaneously occurring univoltine and semivoltine population in the same pond.

Larvae undergo a series of moults (6 to 15 in the UK) before reaching the final larval stage. Once conditions of temperature and day length are favourable, final instar larvae (Fig. 10.1a,b) stop feeding and move to emergence areas (e.g. shallow areas of the pond with emergent vegetation to enable climbing, also tree trunks, or vegetation within a few metres of the bank), climbing out of the water and starting the process of metamorphosing into adults.

10.1.2 Terrestrial stage

After finding a suitable location, larvae will push the thorax, head, legs, and wings out of their larval skin (termed exuviae; Fig. 10.2) and, after a pause of about 30 minutes to harden their legs, they will withdraw the abdomen. In the UK, this process can take

from around one hour in damselflies to three hours in dragonflies, and during this time they are susceptible to predators, as well as to rain, wind, and temperature, that could hinder development. For example, Gribbin and Thompson (1990) reported 22% emergence mortality rates due to predation in the large-red damselfly *Pyrrhosoma nymphula*, and 6% due to weather-related factors. Mortality rates at this stage can be higher for populations that overwinter as final instar larvae (diapause), and have a synchronized emergence in spring ('spring species'; Corbet 1999; e.g. 20% of a population of 2714 azure damselflies emerged on a single day, Banks and Thompson 1985), compared to 'summer species' which show asynchronized emergence lasting throughout the summer. However, both emergence strategies have their own advantages: synchronized species find mates more easily as they have the whole population to choose from, while populations of asynchronized species may have fewer choices of mates but are less susceptible to climatic factors and local extinction.

Immature adults are referred to as tenerals in their first day after emergence, with their first flight—'maiden flight'—being long enough to find suitable

(a)

(b)

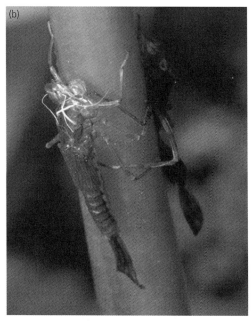

Figure 10.2 Odonate exuviae: (a) Four-spotted chaser *Libellula quadrimaculata* exuvia. After two years as larvae (present on ponds, pools, and ditches), final instar individuals emerge on marginal vegetation in late May and June. Photograph ©E. Raebel; (b) Azure damselfly *Coenagrion puella* exuvia. Emergence is synchronized (95% of the population at a pond emerges within three weeks; Brooks and Lewington 1997) between mid May and mid June. Photograph ©E. Raebel.

shelter away from water[3]. Tall grasses or trees in the terrestrial landscape near the emergence site are important for the maturing odonates, which need to harden their wings, strengthen their flight, and acquire full coloration, necessary to attract the opposite sex and breed. The immature period will last, in temperate species, from a few days to a few months, depending on climate and the duration of winter (Corbet 1999).

Once mature, odonates will return to water to breed. In Britain, the reproductive adult stage lasts for only a couple of weeks in small damselflies and about two months in large dragonflies (with most damselflies not living beyond a week and dragonflies not beyond 2–3 weeks[4]; British Dragonfly Society (BDS) 2013). It is in this aerial form that odonates will breed and disperse, and it is also as this colourful adult form that dragonflies and damselflies are most often encountered

(Fig. 10.3). Odonates feed and breed most actively from mid morning to mid afternoon, and will disappear to roosting sites in late evening or under rainy and windy conditions. At this stage, the management of the surrounding landscape will be important for their survival; for example, cutting or disturbance of roosting sites used by hundreds of individuals (e.g. in synchronized species) could be detrimental to a population. Landscape quality will also influence the rate of dispersal of individuals and their accessibility to feeding areas. Similarly, females spend most of their time away from water and only appear by waterbodies for mating[5].

10.1.3 Odonates as bio-indicators

This bipartite life cycle is the key element in using odonates as model organisms: the ecological processes affecting their life-history traits depend both on the quality of local waterbodies for growth, emergence,

[3] Some species will disperse at this stage, by flying vertically after emergence and letting themselves be transported by the wind.

[4] Mainly from accidents and predation, but also from starvation in bad weather, as they and their prey cannot fly (BDS 2013) without heat to thermoregulate and warm up their muscles.

[5] Refer to Corbet (1999) and Corbet and Brooks (2008) for a detailed description of the biology, life cycle, and general behaviour of odonates.

Figure 10.3 Odonate adults: (a) Coenagrionid azure damselfly *Coenagrion puella* (male). *C. puella* has limited dispersal ability (< 1.5 km), is non-territorial, and has a 'sit-and-wait' mating strategy where males sit by the breeding site and wait for females to arrive, adopting a scramble competition strategy. The flight season lasts until late August. Photograph ©X. Cervera; (b) Four-spotted chaser *Libellula quadrimaculata* (male). Males are highly aggressive and territorial at the breeding site. Males return to their preferred perch in between flights, chosen to have a good viewpoint of their territory from which to launch into rival males, prey, and females. At high densities, territories can break down and fights on the wing may be continuous. The flight season lasts until mid August. Photograph ©P. Watts; (c) Southern hawker *Aeshna cyanea* (male). *A. cyanea* is an aggressive territorial species, with males patrolling a territory on the wing while waiting for females to arrive at the breeding site. Males can time-share sites. *A. cyanea* is capable of hunting great distances away from water, especially along woodland rides. The flight season lasts from July to October. Photograph ©S. Cham.

and egg deposition for larvae and adults, and the quality and connectivity of the wider surrounding landscape for adult dispersal, feeding, and roosting. The renowned entomologist and worldwide odonate expert, Philip S. Corbet, emphasized that this dependence on both aquatic and terrestrial ecosystems, together with their position at the top of the food chain, renders odonates as a flagship group, whose patterns of abundance and diversity can reflect those of the wider community (Corbet 1999). Indeed, odonates have been found to be effective indicators of biotope quality (Clark and Samways 1996), have been identified as indicator-taxa of community richness of pond invertebrates (Briers and Biggs 2003) and lake vascular

---

Box 10.1   Odonate declines: global and UK threats

Odonates are declining worldwide due to the loss and de-terioration of their habitat. Clausnitzer et al. (2009) have assessed the status of odonates at a global scale, conclud-ing that one in ten species is threatened with extinction although, since 35% of species in their study were data deficient, the authors estimate that an overall 15% of all currently described odonate species should be considered as threatened. In the United States, almost two-thirds of known odonate species (441 species in 2005) are vulnerable (in-cluded in the country's Species of Greatest Conservation Need criterion; Bried and Mazzacano 2010), and a review of the status of European species of odonate revealed that, out of 137 native species, at least 15% of species are threatened and 24% of species have declining populations (Kalkman et al. 2010).

From the IUCN's Red List of Threatened Species of odon-ates worldwide[6] and the likely threats to those species, 77% (N = 1613) were related to habitat destruction and changes in management practices (residential and com-mercial development, agriculture and aquaculture, bio-logical resource use, and natural system modifications), while 12% of species are also threatened by water pollu-tion (Raebel 2010).

In the UK, out of a total of 46 British species recorded as long-term breeders, three are extinct, four are classified as 'Endangered', two as 'Vulnerable', and six as 'Near Threat-ened' (Daguet et al. 2008), with two species currently listed as Priority UK BAP species (Norfolk hawker *Aeshna isosceles* and southern damselfly *Coenagrion mercuriale*; JNCC 2013). Three species of Odonata became extinct in the 1950s: *Coe-nagrion scitulum* (due to catastrophic flooding), *Coenagrion*

armatum (habitat loss), and *Oxygastra curtisii* (due to a pol-lution incident) (BDS 2013).

British odonate declines can mainly be attributed to habitat loss and fragmentation, changes in farm management prac-tices (e.g. drainage, neglect, and infilling of ponds), increases in intensive agriculture and fishing with consequent pollution and eutrophication (e.g. increased use of agro-chemicals), and the spread of non-native aquatic species of other taxonomic groups (Daguet et al. 2008). Attempts to reverse this situation are hindered by the scarcity of financial incentives and policy frameworks to create, maintain, or improve waterbodies, and the lack of landscape-scale thinking about aquatic habitats (Declerck et al. 2006; Davies et al. 2009).

Climate change can also affect odonate species. Changes in phenology have already been described by Hassall et al. (2007), with British odonates advancing their flight period by a mean of 1.51 days/decade. The northwards shift at the range margin of all but three species of non-migratory Brit-ish odonates has also been attributed to a warmer climate (Hickling et al. 2005). This northwards shift also encourages species from continental Europe to arrive in the UK. The small red-eyed damselfly *Erythromma viridulum*, for example, arrived in the UK in 1999 and has now established itself, proceeding north-westerly by an average yearly expansion of 32 km per year (Watts et al. 2010). Indeed, the British Dragonfly Society mentions, in their list of UK species, three species of Zygoptera and ten species of Anisoptera as migrants or vagrants from Europe and the USA. These have been sighted in Britain but have not yet been recorded as breeding: for example, the win-ter damselfly *Sympecma fusca*. The consequences for native odonates of the arrival of new species are unclear.

---

plants (Sahlén and Ekestubbe 2001), and are represen-tative of aquatic carnivores (Thomas 2005). In addition, odonates are commonly used as ecological and evolu-tionary models (Córdoba-Aguilar 2008), and odonate distributions are a useful indicator of climate change responses by macroinvertebrates (Bush et al. 2013). However, it is this very same bipartite life cycle which makes them good bio-indicators that also increases their susceptibility to habitat change, as reflected in their declining worldwide status (Box 10.1). In this chapter, we describe our studies using odonates as a

model taxon to inform management recommendations for aquatic and terrestrial farmed landscapes.

10.2 The British agricultural landscape: odonates and ponds

There are currently 137 native species of odonate in Europe, from which 46 species have been recorded as breeding in Britain (20 damselflies and 26 dragon-flies; BDS 2013), together with at least 35 recorded migrants and vagrants[7], some of which have already attempted to breed and may eventually establish

---

[6] A total of 1994 species were classified in 2010 as: Extinct (2); Critically Endangered (56); Endangered (86); Vulnerable (119); Near Threatened (92); Data Deficient (607), and Least Concern (1032).

[7] Migrants: recorded regularly since 1970 and have at-tempted to establish breeding populations; vagrants: visiting species since 1998.

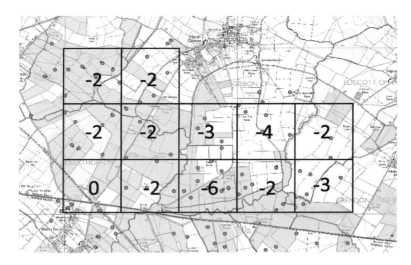

Figure 10.4 Pond loss in the River Ray catchment. Squares represent 1 km² and numbers within squares refer to the number of ponds that have been lost or gained by subtracting the number of ponds found in the field in 2006 from the number of ponds as given by the OS Map (Ordnance Survey 1998). Dots represent pond presence before loss.

breeding populations. Although British odonates have adapted to live in most freshwater habitats, ponds[8] are one of the most crucial; they are principal habitats of 35% of total British species, and a secondary habitat[9] of a further 38% (e.g. Table 10.1). Furthermore, ponds are distributed across the landscape and so can act as stepping-stones to maintain odonate populations and to encourage dispersal.

The global importance of integrated catchment conservation, encompassing not only rivers, but also lakes, channels, streams, ditches, and ponds, has already been highlighted (Pacini et al. 2013) but, in practice, government policies and conservation efforts have focused, until recently, primarily on running waters and large lakes. Monitoring of small waterbodies (although they comprise most of the water network), is patchy (Burns et al. 2013).

In Britain, ponds are the richest of the freshwater habitats, as measured by gamma diversity (total species diversity of macrophytes and macroinvertebrates; Williams et al. 2003). Ponds are able to sustain 10% more species and 50% more uncommon species than are other freshwater bodies: from 337 species of plants and invertebrates recorded in Williams et al.'s (2003) survey of UK freshwater habitats, ponds supported

71%, rivers 60%, streams 48%, and ditches 35%. Although British ponds were counted for the first time in the Countryside Survey of 1978, the quality of ponds was not monitored at a landscape scale (as opposed to surveying just one individual pond) until 1996, with the Lowland Pond Survey (Williams et al. 1998; Biggs et al. 2005). The most recently published pond survey, the Countryside Survey of 2007 (Carey et al. 2008), recorded a total of 482 000 ponds in Great Britain (excluding garden ponds). Although the number of ponds increased by 12.4% between 1998 and 2007 due to pond creation campaigns, there has been an estimated 50% pond loss during the last century (Carey et al. 2008), for which there are no records of their lost ecological quality and lost biodiversity (Williams et al. 2010). Our data have supported this estimate; a survey of a 6 x 4 km² area within the River Ray catchment for 126 ponds that were marked on an Ordnance Survey map revealed that 65% had been filled (Fig. 10.4). To address the loss of ponds in the countryside, the 'Million Ponds Project', led by the Freshwater Habitats Trust (formerly Pond Conservation) was launched in 2008. The project's target is to create one million high quality ponds by 2050, focusing on species of macroinvertebrates and plants that are priorities for national conservation (Freshwater Habitats Trust 2014a). One of the aims is to liaise with landowners implementing agri-environment schemes to help with pond creation options[10].

---

[8] For the purpose of our studies, ponds are defined as: 'water bodies between 1 m² and 2 ha in area which may be permanent or seasonal, including both man-made and natural water bodies' (Biggs et al. 2005).

[9] Secondary habitat refers to habitats used successfully for breeding after preference for other waterbodies (e.g. lake, canal, ditch, stream, river, bog, flush).

[10] Targeted schemes, such as the High Entry Level Schemes (2013) have a pond creation option.

In addition to pond loss, the quality of remaining lowland ponds deteriorated between 1996 and 2007, showing a 24% decrease in plant species richness, from an average of 10.2 to 8.3 emergent and submerged plant species per pond (Carey et al. 2008). In 2007, ponds were finally added as a Priority Habitat to the UK BAP, though only those that were classified as being of High Ecological Quality (UKBAP 2007). A pond qualifies as High Ecological Quality when it fulfils a set of criteria, as specified by the Freshwater Habitats Trust (2014b). For example, qualifying ponds could be in habitats of international importance, those supporting Red Data Book or UK BAP species, or ponds of high ecological quality with a PSYM[11] score $\geq 75\%$. Only 8% of British ponds have achieved Priority Habitat status because of their plant species (Carey et al. 2008), but it is estimated that 20% of the $c.$ 400 000 non-garden ponds in the UK meet some of the High Ecological Quality criteria (UKBAP 2008).

One of the criteria for a pond to qualify as a UK BAP Priority is that it should sustain 'exceptional assemblages of key biotic groups' (Freshwater Habitats Trust 2014b). These assemblages can comprise wetland plants (≥ 30 species) or macroinvertebrates (≥ 50 species), but ponds can also be of priority if they include species that meet the criteria required to be Sites of Special Scientific Interest (SSSI) (Freshwater Habitats Trust 2014b). Currently, SSSI guidelines are based on the presence of great-crested newts *Triturus cristatus,* or for ponds supporting an outstanding assemblage of odonates (JNCC 1998). In the Upper Thames region, and specifically in the area where our odonate study was conducted, ponds are currently considered to support an outstanding assemblage of odonates if they support 12 breeding species. Given the dynamic nature of current odonate distributions (i.e. climate change, north-western shift, new species), using set numbers of species to define outstanding assemblages is widely considered to be too static an approach in the face of changing reality. This prompted the British Dragonfly Society to recommend that designations be made if a rare species is present. The criteria for listing a pond as a priority under the UK BAP have not yet adopted this suggestion and, at the time of writing, the focus remains on demonstrating that a pond hosts an 'outstanding assemblage' of odonates.

[11] PSYM, Predictive SYstem for Multimetrics, is a method to assess the biological quality of still waters in England and Wales: plant species and invertebrate families are surveyed using a standard method and are combined to give an overall % value representing the waterbody's overall quality status (Biggs et al. 2000).

10.3 Reducing survey bias

Odonates on the wing are emblematic of joyous hot summer days, and Britain has a long history of recording odonates, with the British Dragonfly Society holding records from 1807 to the present day. Surveys are generally done by dedicated amateur odonatologists, members of the public, and environmental conservation agencies, and are used by national recording schemes and by national and local wildlife organizations to monitor reserve sites. They are also used by consultancy companies when assessing farmland biodiversity for agri-environment scheme (AES) application reports.

However, although these records are essential for assessing species distributions and changes in population trends, they are subject to coverage and recorder bias (Hassall and Thompson 2010). For example, records are commonly collected in areas that are easy to access (e.g. gardens, nature reserves); consequently, areas with restricted or limited access (e.g. in privately owned farmland or remote areas) are underrepresented. A further problem is that the presence of adults may not be indicative of successful breeding, since adults may be present, and even oviposit, at poor quality aquatic sites, where eggs may not develop and larvae may not reach the final instar (Corbet 1999). In spite of this, up to 96% of records held by the BDS are based on adults, as opposed to larvae (1.5%) or exuviae (3%) (N = 657 210), and continental European records are also based on presence of adults (Raebel et al. 2010).

Therefore, we set out to demonstrate the importance of exuvial surveys for recording odonate populations accurately, as they demonstrate unequivocal breeding success. While doing so, we aimed to assess which surveys were most appropriate for particular types of odonate species.

During three field seasons (2006–2008), we collected a total of 11 025 exuviae (e.g. Fig. 10.2a,b) of 17 species and undertook more than 348 surveys for adults (e.g. Fig. 10.3a,b,c) of 16 species at 29 ponds in the River Ray catchment, part of the Upper Thames (Table 10.1). The River Ray catchment (Buckinghamshire and Oxfordshire, UK) is a catchment of alkaline waters, comprising $c.$ 283 km^2 of lowland agricultural land, with clay as the main top substrate, and is located in a Nitrate Vulnerable Zone.

We found that adults were present at, and even oviposited in, all ponds, regardless of their quality (Raebel et al. 2010), and were found even at ponds that yielded no exuviae (Fig. 10.5a,b; Group 1). For example, in 2007, adult blue-tailed damselflies *Ischnura elegans*

Table 10.1 List of British species of Odonata found in the adult and exuviae surveys. 'A' refers to species found as an adult, 'E' as exuviae, 'P' refers to species for which ponds are the principal habitat, and 'S' to species for which ponds represent a secondary habitat (habitat preferences from Smallshire and Swash 2004).

Family	Species		Survey	Pond Habitat
ZYGOPTERA				
Calopterygidae	*Calopteryx splendens*	banded demoiselle	A	-
Lestidae	*Lestes sponsa*	emerald damselfly	A E	P
Coenagrionidae	*Coenagrion puella*	azure damselfly	A E	P
Coenagrionidae	*Enallagma cyathigerum*	common blue damselfly	A E	S
Coenagrionidae	*Erythromma najas*	red-eyed damselfly	A E	S
Coenagrionidae	*Erythromma viridulum*	small red-eyed damselfly	A E	P
Coenagrionidae	*Ischnura elegans*	blue-tailed damselfly	A E	P
Coenagrionidae	*Pyrrhosoma nymphula*	large red damselfly	A E	P
ANISOPTERA				
Aeshnidae	*Aeshna juncea*	common hawker	A E	P
Aeshnidae	*Aeshna mixta*	migrant hawker	A E	P
Aeshnidae	*Aeshna cyanea*	southern hawker	A E	P
Aeshnidae	*Aeshna grandis*	brown hawker	A E	P
Aeshnidae	*Anax imperator*	emperor dragonfly	A E	P
Libellulidae	*Libellula quadrimaculata*	four-spotted chaser	A E	P
Libellulidae	*Libellula depressa*	broad-bodied chaser	A E	P
Libellulidae	*Orthetrum cancellatum*	black-tailed skimmer	A E	S
Libellulidae	*Sympetrum striolatum*	common darter	A E	P
Libellulidae	*Sympetrum sanguineum*	ruddy darter	A E	P

were present at 27 out of 29 ponds, but we found their exuviae at only eight ponds. Therefore, the number of viable ponds had been overestimated, with presence of adults not being indicative of the healthy state of a pond, or of local pond management success.

With the aim of improving surveys, we also suggest that biases may arise if all species are surveyed, irrespective of their behaviour, according to the same set of rules. Species of odonates can show specific mating behaviour that will influence the pattern of their presence, and therefore the opportunity to record them, at a pond. For example, males of 'flier' territorial species (e.g. Aeshnidae: hawker dragonflies; Fig. 10.3c) will patrol breeding sites in flight to deter other males while waiting for females (Corbet and Brooks 2008). Aeshnidae guard territories in a hierarchical manner: a dominant male monopolizes the patrolling of the territory, while satellite/visiting males await his departure in order to take over his position (Corbet 1999).

Small ponds will only be able to hold one 'alpha' male at a time, which would show aggressive behaviour towards intruders. Corbet (1999) showed that southern hawker *Aeshna cyanea* (Fig. 10.3c) could visit a pond for several days, in 1–8 slots of 40 minutes each per day, allowing time-sharing with other individuals. Our study showed not only that records of adults at ponds cannot be used to estimate numbers for species exhibiting this mating behaviour (unless sufficient time is spent on site and adults are marked to avoid replication), but also that results of these surveys are not reliable indicators of a waterbody's suitability for their development. The bias will over-record males, and will tend to record lower numbers of adults than the total population emerging from that pond (Fig. 10.5b, Groups 2 & 3). For example, our records for adult hawkers (Aeshnidae) tended to remain below ten individuals per site, even when records of exuviae were up to 95 individuals per pond (Fig. 10.5b; Group 3).

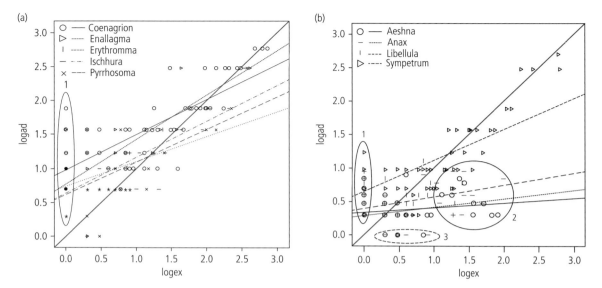

Figure 10.5 Adult abundance plotted against exuviae abundance at ponds for different genera of the odonate suborders: (a) Zygoptera: *Coenagrion*, *Enallagma*, *Erythromma*, and *Ischnura*; and (b) Anisoptera: *Aeshna*, *Anax*, *Libellula*, and *Sympetrum*. Abundances have been log transformed. 'Best fit' linear correlations are given for each genus, represented by different symbols. A reference line of a perfect linear correlation has been provided. Individual symbols can refer to more than one observation. Symbols grouped within a numbered ellipse are referred to within the text. From Raebel et al. (2010). With kind permission of Springer Science and Business Media.

In contrast, we found that abundance of adult and exuvial records coincided for species of 'perchers' (non-territorial Zygoptera (e.g. Fig. 10.3a) and territorial but percher *Libellula* (e.g. Fig. 10.3b) and *Sympetrum* spp.) at high population densities (Fig. 10.5a,b). In these species, males arrive at a pond and 'sit and wait' for females, adopting a strategy that leads to scramble competition (Corbet 1999). Under these circumstances, males spend most time making short flights between perches, which increase their conspicuousness to recorders.

Although extended exuvial surveys are costly and time consuming, we recommend that at least presence/absence of species should be ascertained through searches for exuviae when the aim is to deliver high quality biodiversity assessments (Raebel et al. 2010). Exhaustive exuvial surveys are essential when more detailed information (e.g. abundance) is needed, as is likely to be the case when dealing with protected species.

Monitoring using exuvial surveys can give an indication of the quality of both terrestrial and aquatic ecosystems at both small and larger spatial scales across farmland. In the remainder of this chapter, we show how we used our survey results (of all life-stages—larval, as represented by exuviae, and adults—,

and both sub-groups separately—damselflies and dragonflies—) to, first, determine high-quality habitats for odonates by identifying pond variables that benefited odonate species richness and abundance and, second, to investigate the relationships between odonate species richness at a pond and the land use of the landscape surrounding that pond at different spatial scales. We then assess the implications of our results for improving pond management, landscape management, and how these could relate to agri-environment scheme practice.

10.4 Farmland pond essentials for odonate abundance and richness: implications for agri-environment schemes

At the time of writing, in 2013, farmers receive a single-farm payment (Single Payment Scheme, SPS) for cross-compliance, which involves applying certain management practice standards in order to keep their land in good agricultural and environmental condition, regardless of production (NE 2013a; Chapter 1, this volume). Agri-environment schemes offer farmers the opportunity of further financial support for delivering additional environmental management on their land. Some 'broad and shallow' agri-environment schemes (those with basic management to improve the countryside as a whole e.g. Entry Level Scheme (ELS), introduced in England in 2005) include some options relevant to pond management: for example, the implementation of buffer strips and/or livestock fencing (NE 2013a,b) to mitigate pollution. 'Narrow and deep' schemes, which are local and targeted (e.g. Higher Level Scheme (HLS) introduced in England in 2005), support more complex management for priority habitats and species, and include options for maintenance and creation of ponds of 'high wildlife value' (NE 2013c). Our overall aim was to investigate what features of ponds determine high-quality habitats for odonates and to assess the effectiveness of general AES in providing those habitats, and how they might be improved.

We first needed to know how to optimize pond habitats for odonates, and tackled this by investigating which pond parameters encouraged species richness and abundance of odonates. There are readily available guidelines for the creation of good quality ponds (Pond Conservation 2008) and for ponds targeting odonates (BDS 2013). These have been derived from various studies that have found correlations between variables of lentic/lotic waterbodies and odonates. For example, odonate numbers have been shown to increase with greater levels of submerged and floating macrophytes (Painter 1998; Carchini et al. 2007), abundance of other macro-invertebrates (Foote and Rice Hornung 2005), and with certain types of channel substrate and wide berms (Rouquette and Thompson 2005). In contrast, odonate species and abundance can be lower with increased levels of turbidity (D'Amico et al. 2004), nutrient loads (Carchini et al. 2007), and cattle grazing (Foote and Rice Hornung 2005). Species-specific (either positive or negative) relationships have been found for shade, reed cover, bank height (Steytler and Samways 1995; Painter 1998; Hofmann and Mason 2005), presence of fish (Morin 1984), pond size, and pond age (Kadoya et al. 2004).

In our study, we used exuvial surveys of odonates (see Section 10.3) collected at a landscape scale (29 ponds) and used the Information Theoretic (IT) approach to investigate the effects of ten pond characteristics and their interactions (local scale: pond age; surface area variation; presence/absence of buffer, cattle, and fish; permanent or temporary pond; transparency; tree cover; vegetation type) on species richness and abundance of odonates, taking landscape (large scale: distance to nearest pond with exuviae, and

proof of breeding (exuviae) into account (Raebel et al. 2012a). The aim was to assess whether these variables were useful indicators of pond quality for odonates and whether they could become the basis of practical assessments in the field without the need for specialist equipment.

10.4.1 Pond vegetation and transparency: buffers and pond age

We found that ponds dominated by floating (e.g. broad-leaved pondweed *Potamogeton natans*) and submerged (e.g. water-milfoils: native *Myriophyllum* spp.) vegetation supported the highest abundance and species richness of exuviae (Table 10.2); two ponds with these features yielded 12 species[12] (e.g. Fig. 10.6a), and could therefore be classed as having outstanding assemblages. Vegetation provides refuge from predators, increases prey resources, can act as a cue for habitat selection (Thompson 1987; Foote and Rice Hornung 2005), and provides oviposition material for endophytic species (i.e. for damselflies and hawkers which lay eggs inside plant material).

Ponds with submerged and floating plants (see Table 10.2) were also the most transparent[13] (often with good water quality, although cloudy ponds can be of high conservation value for some macroinvertebrate species, or ponds can be transparent but polluted), with transparency increasing overall odonate abundance in our lowland farmland ponds (Fig. 10.7). For example, most turbid ponds (e.g. Fig. 10.6b) yielded an average of four individuals (N = 14), while most transparent ponds averaged 690 individuals (N = 9) (Fig. 10.7).

Invariably, well-vegetated, good quality ponds with higher species richness were surrounded by full buffer strips: these were unfertilized grass strips running the whole perimeter of a pond (as opposed to a partial buffer strip running around part of the perimeter, for example, to allow for cattle access) (Raebel et al. 2012a). We suggest that management that allows vegetation

---

[12] Azure damselfly *Coenagrion puella*, common blue damselfly *Enallagma cyathigerum*, large-red damselfly *Pyrrhosoma nymphula*, red-eyed damselfly *Erythromma najas*, small red-eyed damselfly *Erythromma viridulum*, blue-tailed damselfly *Ischnura elegans*, broad-bodied chaser *Libellula depressa*, four-spotted chaser *Libellula quadrimaculata*, migrant hawker *Aeshna mixta*, emperor dragonfly *Anax imperator*, common darter *Sympetrum striolatum*, and ruddy darter *Sympetrum sanguineum*.

[13] Water transparency; four classes: transparent > 0.5 m from bank; transparent on shallow bank only; low turbidity; high turbidity. See Raebel et al. (2012a) for details.

Table 10.2 Pond vegetation types and respective overall odonate abundance (N) and species richness (S). Vegetation types are based on the presence of major plant groups and their relative abundance: dominant (D: > 35%); abundant (A: 25–35%); frequent (F: 10–25%); occasional (O: <10%). From Raebel et al. (2012a). Reproduced with permission from John Wiley & Sons.

Type	Species composition and structure	N ± SE	S ± SE
Callitriche	*Callitriche* spp. (F); other spp. (O)	147.8 ± 82.1	5.0 ± 0.7
Emergent dicots	Emergent dicots: *Rorippa nasturtium-aquaticum, Apium nodiflorum, Veronica beccabunga, Mentha aquatica* (D); *Sparganium erectum* (A); *Alisma plantago-aquatica, Typha* spp. (O). (Ponds dried out at least once)	5.9 ± 3.5	0.7 ± 0.2
Floating	Floating spp. (*Potamogeton natans*) (D); emergent spp.: *Myosotis scorpioides, R. nasturtium-aquaticum, Typha* spp., *Callitriche* spp., *Nymphaea* spp. (A); *Lemna* spp. (F); Blanket weed, *Ceratophyllum demersum* (O)	278.2 ± 90.5	6.6 ± 0.6
Floating/ Submerged	Submerged spp. (e.g. *C. demersum, Elodea* spp.) (D); Blanket weed (A); floating vegetation: *P. natans* (D); *Typha, Callitriche* spp., *Nymphaea* spp., *M. scorpioides* (F); *Lemna* spp. (O)	466.0 ± 170.6	9.7 ± 0.6
Glyceria	*Glyceria fluitans, Typha* spp. (D); *S. erectum* (A); *Lemna* spp. (O)	54.8 ± 18.5	2.3 ± 1.5
Lemna	Any vegetation type with *Lemna* spp. (D)	39.1 ± 18.9	3.2 ± 0.7
Limited/ Absent	Floating and submerged spp. absent. Limited representation of species on marginal zone: *Callitriche* spp. (O)	0.0 ± 0.0	0.0 ± 0.0

Figure 10.6 Example of good quality versus poor quality pond in terms of odonate species richness. Pond (a) is an example of a good quality pond, yielding twelve species of odonate exuviae per season, an outstanding assemblage in the area. It is surrounded by lush vegetation, and most importantly, it supports floating (e.g. *Potamogeton natans*) and submerged (e.g. *Ceratophyllum demersum*) vegetation. Tree cover is confined to less than 10% of the pond, is partially buffered, but protected from cattle in its entirety. The pond contained an established population of Koi carp *Cyprinus carpio carpio*, but the presence of a large variety of mesohabitats, depths, and refuges created by the vegetation, did not affect odonates. Pond (b) is typical of a neglected pond, with no floating or submerged macrophytes, turbid water, and overgrown trees. No odonate exuviae were recorded. Photographs ©E. Raebel.

(b)

Figure 10.6 *Continued*

to grow around the pond perimeter will have benefits for the quality of the pond and its surroundings for odonates. Such grassy margins may also have positive impacts on other freshwater macroinvertebrates and terrestrial invertebrates, as discussed in Chapters 3 and 11, this volume.

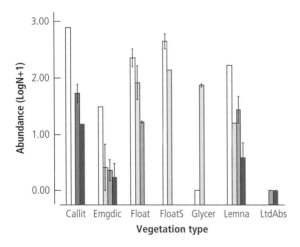

Figure 10.7 Effect of pond transparency and vegetation type on abundance of odonate exuviae (±SE) ($\log_{10}$). Bar colours refer to transparency levels, from white (highly transparent) to black (highly turbid). See Table 10.2 for description of vegetation types. From Raebel et al. (2012a). Reproduced with permission from John Wiley & Sons.

However, our findings also showed that 30% of ponds without exuviae were also fully buffered, suggesting that buffers are necessary but not always sufficient on their own to provide good quality pond habitats (e.g. when the pond does not have a clean water source, Feber and Macdonald 2013). For example, when ponds had limited/absent aquatic vegetation, they had no odonate species even when surrounded by a full buffer strip (full buffer: -0.75 ± 0.14, t-value$_{26}$ = -5.42, $P < 0.0001$) while ponds with floating vegetation always had high numbers of odonate species, even in the absence of buffer strips (absent buffer: 0.46 ± 0.07, t-value$_{18}$ = 6.66, $P < 0.0001$) (Raebel et al. 2012a). Current (2013) broad and shallow options for buffering in-field ponds demand 10 m wide buffers (e.g. ELS) but, although buffers were present at 66% of ponds under study, most were only established under cross-compliance, which required them to be a minimum of only 2 m wide (Raebel et al. 2012a). This is narrow in comparison to the recommendation of Davies et al. (2009), who conclude that pond buffers should be 25 m wide if they are to benefit freshwater biodiversity. We also suggest that the best results are likely to be achieved if buffers are positioned uphill of pond catchments to minimize contamination by runoff—this might be more cost-effective than deploying larger buffers or ones that are too small.

Pond age had an impact on odonate assemblages. Exuviae of all species combined were eight times more abundant at recently created ponds (< 3 years old) compared to well-established ones (Raebel et al. 2012a). Abundance of certain species (e.g. emperor dragonflies *Anax imperator*) can be greatest directly after colonization (in Brooks and Lewington 1997) due to the initial lack of competition, which increases with time. Although species richness was not higher at new ponds (non-pioneer species only settle at ponds at later seral stages), our data showed that exuviae of early colonizing species were already present in the second year after pond creation (e.g. univoltine species: blue-tailed damselfly *Ischnura elegans*, azure damselfly *Coenagrion puella* (Fig. 10.3a), common darter *Sympetrum striolatum*; and possibly univoltine in the study area: broad-bodied chaser *Libellula depressa*, emperor dragonfly *Anax imperator* (Fig. 10.1a)), with semivoltine species appearing in the third year (e.g. southern hawker *Aeshna cyanea* (Fig. 10.3c), red-eyed damselfly *Erythromma najas*, large-red damselfly *Pyrrhosoma nymphula*). This is encouraging news for habitat recreation and for the creation of new ponds[14] in the landscape, which can rapidly support odonates and be colonized by other organisms for which odonates are prey/predators, for example, amphibians. Indeed, amphibians were found in 11 ponds surveyed containing odonate exuviae (N = 18), but in only two ponds that did not yield odonates (N = 11).

10.4.2 Distance to the nearest viable pond

At a landscape scale, we found that distance to the nearest viable pond (a pond already supporting odonates, as distinct from distance to any pond, viable or not) was an important factor for maximizing odonate abundance and diversity. Indeed, species richness of exuviae fell by more than 40% when the nearest inhabited pond was more than 100 m away (Raebel et al. 2012a). Processes operating at broader scales, rather than local scales, in which the occupancy of a patch (pond) depends on the occupancy of a neighbouring patch, are essential to ensure dispersal and connectivity (e.g. Chapter 15, this volume). Our study suggested a distance of 100 m between ponds was required by odonates (but see Section 10.5.1 for differences in sub-orders) and this would most certainly suffice to sustain connectivity in populations of more mobile groups such as amphibians; for example, Sinsch et al. (2012) have estimated that a minimum distance of 600 m is required between breeding ponds to sustain populations of natterjack

toads *Epidalea calamita* (formerly *Bufo calamita*). Neighbouring ponds are also important for species such as great crested newts *Triturus cristatus*, whose metapopulations depend on movement in and between clusters of ponds throughout the year. Creation of nearby ponds at this distance would especially benefit aquatic plants and invertebrates that do not have an aerial stage, but that may depend on passive methods of dispersal (e.g. flooding, transportation by birds), and will ensure that 'pondscapes'[15] are maintained. For example, Niggebrugge et al. (2007) found that the occupancy of ditch habitat that was suitable for 20 British freshwater gastropods in marshes declined with an increase in the distance to the nearest occupied site; while Iversen et al. (2013) found that, for the European water beetle *Graphoderus bilineatus* (regionally extinct in the UK), the presence of local suitable habitat characteristics was not as important for its presence as the landscape connectivity between the available habitats.

Our findings suggest that AES might usefully offer farmers a premium for maintaining or creating more than one pond on the same farm, in order to increase connectivity, thereby helping to maintain biodiversity in pond networks. If AES payments were to be focused specifically on encouraging damselflies and dragonflies on farmland, then our results suggest that payments should reward a combination of pond options: creation of buffers that maximize coverage of the pond catchment (i.e. to minimize agro-chemical input), encouragement of complex vegetation structures with native submerged/floating vegetation, and maintenance of open pond surface areas for ovipositing, and creating and maintaining more than one pond within 100 m of each other (Raebel et al. 2012a).

10.5 Effects of multi-scale farmland management practices on odonate assemblages of ponds

There is mounting evidence that, to be effective, conservation needs to be planned at a landscape scale, and this has been argued for many taxa: birds (Wrbka et al. 2008); bumblebees (Rundlöf et al. 2008b); butterflies (Rundlöf et al. 2008a); mammals (Macdonald et al. 2007); moths (Merckx et al. 2009); grasshoppers (Marini et al. 2009); and freshwater macro-invertebrates and macrophytes (Davies et al. 2009) (and see Chapters 8, 11, and 14, this volume). Mindful of the dispersal

---

[14] Habitat recreation, such as creation of ponds, is usually included as optional capital works in targeted complex management schemes (e.g. HLS).

[15] Equivalent of landscape for lentic ecosystems (Baguette et al. 2013) where, in our study, the pondscape is a network of ponds connected by terrestrial corridors.

abilities of odonates and their utilization of the landscape surrounding waterbodies, we therefore related species richness of our adult and exuvial surveys to the quality of the wider landscape, by collecting information on water (number of ponds, percentage of running and still water), land-use (woodland, arable, grassland, urban), and AES (percentage land under different schemes, buffers)[16] over five different spatial scales[17] (see Raebel et al. 2012b). We measured directly the effects of these variables on both aquatic and terrestrial life stages of odonates.

10.5.1 Spatial scale: sub-order and life stage

We first explored the spatial scales at which species richness of odonates was affected by their evolutionary history—damselflies tend to fly short distances close to their natal pond, while dragonflies are known to disperse away from it by several kilometres (Banks and Thompson 1985; Conrad et al. 1999; Corbet 1999)—, and so we analysed our findings for the two sub-orders separately. Our results showed that adult and exuvial Anisoptera (dragonflies) were influenced by landscape at larger spatial scales (i.e. 1600 m) than were those of Zygoptera (damselflies: 100/400 m) (Raebel et al. 2012b). These distinctions reflect the relationship between dispersal behaviour and body size, such that larger species (Anisoptera) travel further than do smaller species (Zygoptera). This implies that current pond conservation strategies are most likely to benefit the less mobile damselflies, as they tend to focus on increasing pond quality and their immediate terrestrial surroundings (e.g. buffers) rather than including these conservation measures within landscape-scale approaches.

Second, we found that the presence of grassy buffer strips contributed to increased species richness of adult and exuvial odonates, but that Zygoptera benefited most from fully buffered ponds. Adult Zygoptera males need vegetation around a pond in which to wait for females, but Aeshnids (Anisoptera) wait for females on the wing, not needing surrounding vegetation on

which to perch. Furthermore, buffer strips bring better pond quality for larvae when coupled with an increase in aquatic vegetation and water quality.

In addition, although the number of ponds per unit area at different landscape scales did not affect the species richness of odonates (probably because a large number of ponds in our sample did not sustain odonate larvae), the amount of water[18] within 1600 m of a pond was strongly associated with increased species richness of exuviae, but only when ponds did not have a buffer strip surrounding them (Fig. 10.8). This suggested that ponds with buffers did not depend on water in the surrounding landscape to have high species richness of exuviae (Fig. 10.8 ellipse). However, 30% of ponds with buffer strips did not have any exuviae, suggesting that species richness in these ponds may depend on other factors, such as presence of floating/submerged vegetation within ponds (see Section 10.4.1; Raebel et al. 2012a).

The association of exuviae species richness and amount of nearby water was particularly pronounced for zygopterans ($F_{1,55} = 7.42$; $P = 0.009$). Most

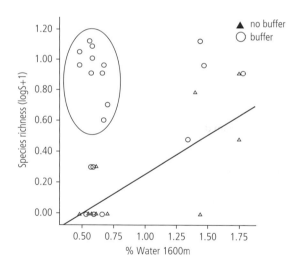

Figure 10.8 Log-transformed overall species richness of exuvial odonates in relation to percentage water at the 1600 m scale, for ponds with and without buffers. 'Buffer': ponds with a partial or full buffer. Line of 'best fit' has been given for ponds without a buffer. Symbols within ellipse are referred to within the text to highlight similarities. From Raebel et al. (2012b). Reproduced with permission from Elsevier.

---

[16] Landscape variables were obtained from the UK's Ordnance Survey (OS) digital database (Centre for Ecology and Hydrology), LandCover 2000, and Defra records (Defra, unpublished data).

[17] Landscape composition data were projected within a GIS environment and feature class data extracted at five circular spatial scales centred at each pond covering an area with a radius of 100 m, 200 m, 400 m, 800 m, and 1600 m. The maximum radius of 1600 m covered for the farthest estimated distance of 1.5 km travelled by common non-territorial damselflies (coenagrionids), as estimated by mark–recapture studies (Rouquette and Thompson 2007).

[18] Percentage of all still water (e.g. lakes, ponds) and running water (e.g. rivers, streams) within the chosen spatial-scale area, but excluding ditches and running water of < 1 m width.

zygopterans that we studied use other waterbodies (e.g. rivers, lakes, ditches) as secondary habitats and there is evidence that habitat connectivity is vital to facilitate gene flow and to maintain odonate populations (McCauley 2006). This reliance of damselflies (Zygoptera) on the amount of water in the landscape emphasizes the importance of connectivity between these waterbodies, especially of non-territorial species which are weaker fliers (e.g. some coenagrionids, such as azure damselfly; Fig. 10.3a). Damselfly dispersal largely depends on chance wandering rather than actively looking for new habitats, and they only tend to commute from roosting sites to ponds to breed, spending most of their time away from water (Conrad et al. 1999). We found that one of the land-use variables, the amount of woodland (especially a high percentage within 400 m), had a negative impact on adult coenagrionids (e.g. azure damselfly *Coenagrion puella*, blue-tailed damselfly *Ischnura elegans*, common blue damselfly *Enallagma cyathigerum*), which we interpret as arising because woodland makes the landscape less permeable to these species than does, for example, open pasture, potentially hindering movement, speed, and dispersal distance (Pither and Taylor 1998). However, for other highly mobile and larger species of damselfly, such as calopterygids, which are associated with running water, woodland within distances of < 500 m from the stream can increase movements away from the water (Jonsen and Taylor 2000), a similar scale to our study but with an opposite effect. This suggests that broad management should, of course, aim to benefit all species, but it should be tailored to accommodate the needs of mobile versus non-mobile species, as highlighted for moths in Chapter 8, this volume.

Finally, species richness varied between stages of the life cycle such that exuviae were affected by more landscape variables than were adults: explicable because larvae spend at least a year in the water and the quality of a waterbody will ultimately depend on its catchment, and this will allow for life cycle completion (presence of exuviae), while the effects of landscape management options were less important for adults because the habitat requirements for roosting and insect prey at this short stage (lasting only a few weeks) may be met more easily (Raebel et al. 2012b).

In summary, we recommend that effective conservation of odonates on farmland requires attention to habitats beyond waterbodies (especially for dragonflies, Anisoptera species). Management should integrate local and landscape-scale approaches, embracing all odonate species regardless of mobility levels and life-stages.

10.5.2 Broad and shallow versus targeted approaches to odonate conservation

Our purpose has been to reveal the science that will inform current and future management practices to contribute better to conserving biodiversity on farmland. We therefore used our results to challenge the efficacy of existing agri-environment scheme guidelines. The number of species of odonate exuviae (overall, and independently for Anisoptera and Zygoptera) decreased at ponds the greater the area of surrounding land that was entered into a broad and shallow scheme (i.e. ELS), and this decrease was most marked when land under this scheme occurred within 100 m of a pond (Fig. 10.9; Raebel et al. 2012b). In the UK, the broad and shallow scheme currently supports specific options intended to benefit pond life (e.g. buffering). However, the uptake of pond options remains low and other options—such as lowering fertilizer inputs and nurturing linear features—are perceived as easier and cheaper to implement and consequently are most popular, even when there are ponds on the farm (Dobbs and Pretty 2008; Hodge and Reader 2010). In our study area (N = 50 ELS agreements), uptake of options involving buffering in-field ponds in

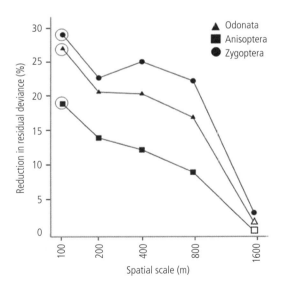

Figure 10.9 Scale-dependent effect of the uptake of Entry Level Scheme (ELS) on species richness of exuviae. Percentage drop in residual deviance of Poisson regressions between species richness and the proportion of ELS land at five spatial scales (significant: filled symbols; non-significant: empty symbols). Scales with largest drops in residual deviance are marked with circles. Modified from Raebel et al. (2012b). Reproduced with permission from Elsevier.

improved grassland and arable land was 2% and 4% respectively, while in contrast, there was a 46–52% uptake of options involving hedgerow management (both sides and one side of the hedge respectively), and 70% uptake of options related to low inputs of fertilizer and herbicides (Raebel et al. 2012b). There was no evidence to suggest that non-pond management options were of any benefit to ponds when undertaken around ponds for which pond management options had not been taken, so any benefits to odonates were probably due to other pond management practices, undertaken outside of agri-environment schemes. Indeed, many farmers undertake pond management work independently because of the personal value they place on them (Macdonald and Feber 2015: Chapter 14).

Table 10.3 Summary description and applications of recommendations for odonate conservation, with emphasis on farmland ponds. Responsibility for their application is suggested and the scale at which the implementation will yield benefits is noted: landscape (LAN) versus local (LOC) scales.

Recommendations	Description and applications	Responsibility/Scale	Scale
SURVEYING AND MONITORING			
1. Adult records to assess species distributions and changes in trends.	Broad general surveys undertaken by recording networks to monitor general patterns, to create atlases, and to assess general odonate diversity at specific sites. Indicative of landscape connectivity around a waterbody.	Recording networks, wildlife organizations, members of the public.	
2. Species-targeted surveys.	Adult, larvae, and/or exuvial surveys to be applied on a species-specific basis depending on territorial strategies and mobility of adults.	Senior recorders and experts within organizations. Volunteer recruitment to help wildlife organizations.	LOC
3. UKBAP, Red Data Book, and species of interest: exuvial surveys.	Targeting specific species of interest. Monitoring of sites of specific interest to assess which of the aquatic habitats allow for life cycle completion. Essential for single-species studies. Monitoring sustainability of emerging sites and focusing habitat management.	All organizations and studies focusing on specific and threatened species. Volunteer enrolment for surveys.	LOC
4. Rapid farmland pond surveying.	Farmers with ponds should assess their quality by rapid visual assessment (e.g. spring and summer: presence/absence of floating vegetation; transparency of the pond). This would be an indication of potential odonate sites and hence of good pond quality. Presence of amphibians, which can be at ponds of a completely different nature, should also be considered (e.g. with the habitat suitability index developed for the Great crested newt). This would ensure the effectiveness of taking up a buffer option.	This type of survey would be useful for advisors helping farmers apply for agri-environment schemes. AES should encourage uptake of pond buffer options when the application includes a potentially good quality pond.	LAN LOC
5. Monitoring of sites: farmland ponds.	Integration of 'evaluation programmes' within AES to find out which schemes/options are most effective (Kleijn and Sutherland 2003). Ponds of high quality to be monitored as part of AES.	AES policy and research.	
HABITAT CREATION AND ENHANCEMENT			
6. Pond creation	New ponds are colonized very rapidly. Creation of high quality ponds should be encouraged under targeted schemes and can be linked to Freshwater Habitats Trust's 'Million Ponds' project.	Pond creation options lie within targeted AES schemes. Appropriate advice and encouragement of uptake necessary.	LOC
7. Creation of new ponds near other viable ponds.	Location of ponds at a short distance from another viable pond will increase odonate species richness (< 100 m). The creation of new ponds with a source of clean water within the same farm should be encouraged and financially rewarded by AES.	Schemes to offer extra financial rewards when another viable pond is created/maintained within the same farm.	LOC LAN

continued

Table 10.3 *continued*

Recommendations	Description and applications	Responsibility/Scale	Scale
8. Encouragement of submerged and floating vegetation in ponds.	Encouragement of floating and submerged vegetation. Proven to be present on minimally impaired farmland ponds in this study and recommended by the British Dragonfly Society and the Freshwater Habitats Trust.	AES advice and encouragement. Ensure AES and local organizations use the correct advice by following certain guidelines. Policies in place to ban garden centres from selling invasive species, especially passing as native.	LOC
9. Encouragement of pond water quality on farms.	Good water quality, low levels of suspended sediment/algae and low pollution conditions. It encourages submerged and floating plants, which provide resources for odonates and their prey. It depends in part on the healthy state of the pond's catchment.	AES to continue and to encourage the uptake of options regarding low fertilizer inputs at a local and wetland scale.	LOC LAN
10. Targeting pond buffer locations.	Pond buffers located to strategically minimize run-off. Partial buffers and restricted cattle access is allowed. Target good quality ponds to increase effectiveness and avoid wasting financial resources.	AES policy. Encouragement of buffer uptake at broad and shallow schemes level.	LOC LAN
11. Water management: density of waterbodies.	Encouragement of lentic and lotic waterbodies within 1600 m of a pond increases odonate species richness. This is likely to benefit other taxa, depending on water features.	AES to encourage the uptake of water-related options.	LAN
12. Management at different spatial scales	Odonate subgroups, individuals at different life-stages, and species of different dispersal abilities will all be covered under this type of multi-scale management.	Environmental organizations, nature reserves, AES policy.	LOC LAN
13. Encouragement of targeted schemes uptake	Targeted schemes (e.g. HLS) provide the best benefits for odonates at all scales, and especially at large scales (1600 m). Uptake is currently low. Encouragement of uptake of targeted schemes by neighbouring farms could create medium-sized areas of high biodiversity quality, with odonate and pond hotspots.	AES policy.	LAN
14. Encouragement of pond-related options and options related to maintenance of surrounding fields.	Encouragement of options beneficial for ponds and hence odonates within broad-and-shallow schemes essential to connect areas of environmental high quality (within targeted schemes). In this manner, the farmed countryside can be a continuous landscape, with ponds located within the intervening matrix, and thus creating biodiversity benefits within and beyond farmland.	AES policy.	LAN

Nevertheless, the result that odonates appeared to do less well when in the vicinity of broad and general management conservation practices (i.e. ELS) merits exploration; besides pond neglect, one possible hypothesis is that harvesting dates dictated by these broad schemes are usually set with nesting birds in mind, resulting in the damage of odonate roosting sites[19]. This effect has been reported in other farmland organisms (e.g. Lepidoptera, Orthoptera; Marini et al. 2009). By leaving grass strips unmown, roosting sites could be better protected, and benefits

for other invertebrates are increased (Chapter 3, this volume).

More encouragingly, adult odonates were more abundant and diverse at ponds associated with 'narrow and deep' targeted schemes (i.e. HLS) which can deliver more specific conservation targets. In our study area, only 12% of agreements (of a total of 57) were targeted schemes, but three out of seven farmers joining those selected pond management options to maintain ponds of high quality value (Raebel et al. 2012b).

10.6 Recommendations

Our results have revealed the importance of local management in creating high quality ponds for breeding,

---

[19] Harvesting in June and July coincides with the peak season of adult coenagrionids and many other British species.

have shown what is important for creating them (vegetation, buffers, water quality), and have highlighted the importance of landscape-scale management and connectivity of habitats for odonate conservation. AES can foster connectivity if pond-related options are encouraged at a broad scale, within and among contiguous farms (e.g. to reduce distance between ponds). Maximum benefits would be accomplished by targeting functional clusters of species (e.g. by mobility) and focusing on restoration and quality management of specific areas (Sutherland 2002; Reid et al. 2007). This could be achieved by facilitating the uptake of targeted schemes, which have generally been considered by farmers as difficult to apply for (they are competitive and not open to all) and to implement with effective results.

Our results suggest that for odonates, as for moths (Chapter 8, this volume; Merckx et al. 2009) and water voles (Chapter 15, this volume), incentives should be available to encourage contiguous farms to link their activities under AES (see Macdonald and Feber 2015: Chapter 14). Specifically, we recommend that ponds are managed in an integrated way at the scale of the landscape by the 'broad and shallow' schemes, embracing species at all levels of mobility (and including other pond taxa, such as other macroinvertebrates and amphibians).

A summary of recommendations for odonate conservation are given in Table 10.3. This list combines our findings with previous knowledge of pond habitat creation for odonates (e.g. British Dragonfly Society), and provides guidance for future management of farmland ponds. The aim is to maintain or enhance odonate populations, as well as tentatively to offer advice on habitat creation. Recommendations can be used by farmland pond owners on an individual basis, but ultimately, need to be incorporated within the design of AES.

Acknowledgements

This research was financially supported by the Esmée Fairbairn Foundation. We express our gratitude to BBOWT and landowners who allowed us to carry out research on their land, and thank volunteers who offered their time to help with data collection, especially Robin Curtis and Charles Butt. We also thank the Freshwater Habitats Trust for helpful advice. Eva Raebel is grateful to Thomas Merckx for insightful discussions of results and to Paul Johnson for statistical advice.

References

Baguette, M., Blanchet, S., Legrand, D., Stevens, V.M., and Turlure, C. (2013). Individual dispersal, landscape connectivity and ecological networks. *Biological Reviews*, **88**, 310–326.

Banks, M.J. and Thompson, D.J. (1985). Emergence, longevity and breeding area fidelity in *Coenagrion puella* (L.) (Zygoptera: Coenagrionidae). *Odonatologica*, **14**, 279–286.

Biggs, J., Williams, P.J., Whitfield, M., Fox, G., and Nicolet, P. (2000). *Biological techniques of still water quality assessment. Phase 3 Methods development.* Environment Agency R&D Technical Report E110. Environment Agency. Bristol.

Biggs, J., Williams, P., Whitfield, M., Nicolet, P., and Weatherby, A. (2005). 15 years of pond assessment in Britain: results and lessons learned from the work of Pond Conservation. *Aquatic Conservation: Marine and Freshwater Ecosystems*, **15**, 693–714.

Bried, J.T. and Mazzacano, C.A. (2010). National review of state wildlife action plans for Odonata species of greatest conservation need. *Insect Conservation and Diversity*, **3**, 61–71.

Briers, R.A. and Biggs, J. (2003). Indicator taxa for the conservation of pond invertebrate diversity. *Aquatic Conservation: Marine and Freshwater Ecosystems*, **13**, 323–330.

British Dragonfly Society BDS (2013). *Dig a pond for dragonflies.* <http://www.british-dragonflies.org.uk/>

Brooks, S. and Lewington, R. (1997). *Field Guide to the Dragonflies and Damselflies of Great Britain and Ireland.* British Wildlife Publishing, Hampshire.

Burns, F., Eaton, M.A., Gregory, R.D., et al. (2013). *State of Nature report.* The State of Nature partnership.

Bush, A., Theischinger, G., Nipperess, D., Turak, E., and Hughes, L. (2013). Dragonflies: climate canaries for river management. *Diversity and Distributions*, **13**, 86–97.

Carchini, G., Della Bella, V., Solimini, A.G., and Bazzanti, M. (2007). Relationships between the presence of odonate species and environmental characteristics in lowland ponds of central Italy. *Annales de Limnologie-international Journal of Limnology*, **43**, 81–87.

Carey, P.D., Wallis, S., Chamberlain, P.M., et al. (2008). *Countryside Survey: UK Results from 2007.* NERC/Centre for Ecology and Hydrology, 105pp. (CEH Project Number: C03259).

Clark, T.E. and Samways, M.J. (1996). Dragonflies (Odonata) as indicators of biotope quality in the Kruger National Park, South Africa. *Journal of Applied Ecology*, **33**, 1001–1012.

Clausnitzer, V., Kalkman, V.J., Ram, M., et al. (2009). Odonata enter the biodiversity crisis debate: The first global assessment of an insect group. *Biological Conservation*, **142**, 1864–1869.

Conrad, K.F., Willson, K.H., Harvey, I.F., Thomas, C.J., and Sherratt, T.N. (1999). Dispersal characteristics of seven odonate species in an agricultural landscape. *Ecography*, **22**, 524–531.

Corbet, P.S. (1980). Biology of Odonata. *Annual Review of Entomology*, **25**, 189–217.

Corbet, P.S. (1999). *Dragonflies: Behaviour and Ecology of Odonata*. Harley Books, Colchester.

Corbet, P.S. and Brooks, S.J. (2008). *Dragonflies. The New Naturalist Library*. Harper Collins Publishers, London.

Córdoba-Aguilar, A., (ed.) (2008). *Dragonflies and damselflies. Model organisms for ecological and evolutionary research*. Oxford University Press, Oxford.

Daguet, C.A., French, G.C., and Taylor, P. (2008). The Odonata Red Data List for Great Britain. *Species Status*, **11**, 1–34. Joint Nature Conservation Committee, Peterborough.

D'Amico, F., Darblade, S., Avignon, S., Blanc-Manel, S., and Ormerod, S.J. (2004). Odonates as indicators of shallow lake restoration by liming: comparing adult and larval responses. *Restoration Ecology*, **12**, 439–446.

Davies, B., Biggs, J., Williams, P., and Thompson, S. (2009). Making agricultural landscapes more sustainable for freshwater biodiversity: a case study from southern England. *Aquatic Conservation: Marine and Freshwater Ecosystems*, **19**, 439–447.

Declerck, S., De Bie, T., Ercken, D., et al. (2006). Ecological characteristics of small farmland ponds: Associations with land use practices at multiple spatial scales. *Biological Conservation*, **131**, 523–532.

Dobbs, T.L. and Pretty, J. (2008). Case study of agri-environmental payments: The United Kingdom. *Ecological Economics*, **65**, 765–775.

Feber, R.E. and Macdonald, D.W. (2013). *Wildlife & Farming: Conservation on lowland farms*. Wildlife Conservation Research Unit, University of Oxford.

Foote, A.L. and Rice Hornung, C.L. (2005). Odonates as biological indicators of grazing effects on Canadian prairie wetlands. *Ecological Entomology*, **30**, 273–283.

Freshwater Habitats Trust (2014a). *Million Ponds Project. Pond Creation Toolkit*. <http://www.freshwaterhabitats.org.uk/>, accessed February 2014.

Freshwater Habitats Trust (2014b). *Priority Pond Criteria*. <http://www.freshwaterhabitats.org.uk/>, accessed January 2014.

Gribbin, S.D. and Thompson, D.J. (1990). A quantitative study of mortality at emergence in the damselfly *Pyrrhosoma nymphula* (Sulzer) (Zygoptera: Coenagrionidae). *Freshwater Biology*, **24**, 295–302.

Hassall, C. and Thompson, D.J. (2010). Accounting for recorder effort in the detection of range shifts from historical data. *Methods in Ecology & Evolution*, **1**, 343–350.

Hassall, C., Thompson, D.J., French, G.C., and Harvey, I.F. (2007) Historical changes in the phenology of British Odonata are related to climate. *Global Change Biology*, **13**, 933–941.

Hickling, R., Roy, D.B., Hill, J.K., and Thomas, C.D. (2005). A northwards shift of range margins in British Odonata. *Global Change Biology*, **11**, 502–506.

Hodge, I. and Reader, M. (2010). The introduction of Entry Level Stewardship in England: Extension or dilution in agri-environment policy? *Land Use Policy*, **27**, 270–282.

Hofmann, T.A. and Mason, C.F. (2005). Habitat characteristics and the distribution of Odonata in a lowland river catchment in eastern England. *Hydrobiologia*, **539**, 137–147.

Iversen, L. L., Rannap, R., Thomsen, P.F., Kielgast, J., and Sand-Jensen, K. (2013). How do low dispersal species establish large range sizes? The case of the water beetle *Graphoderus bilineatus Ecography*, **36**, 770–777.

JNCC (1998). *JNCC Guidelines for selection of Biological SSSIs—Chapter 19. Dragonflies*, pp. 284–287. Joint Nature Conservation Committee.

JNCC (2013). *UK BAP priority terrestrial invertebrate species: Insects. Joint Nature Conservation Committee.* <http://jncc.defra.gov.uk/>, accessed August 2013.

Jonsen, I.D. and Taylor, P.D. (2000). Fine-scale movement behaviors of calopterygid damselflies are influenced by landscape structure: an experimental manipulation. *Oikos*, **88**, 553–562.

Kadoya, T., Suda, S., and Washitani, I. (2004). Dragonfly species richness on man-made ponds: effects of pond size and pond age on newly established assemblages. *Ecological Research*, **19**, 461–467.

Kalkman, V.J., Boudot, J.P., Bernard, R., et al. (2010). *European Red List of Dragonflies*. Publications Office of the European Union, Luxembourg.

Kleijn, D. and Sutherland, W.J. (2003). How effective are European agri-environment schemes in conserving and promoting biodiversity? *Journal of Applied Ecology*, **40**, 947–969.

Macdonald, D.W. and Feber, R.E., eds. (2015). *Wildlife Conservation on Farmland. Conflict in the Countryside*. Oxford University Press, Oxford.

Macdonald, D.W., Tattersall, F.H., Service, K.M., Firbank, L.G., and Feber, R.E. (2007). Mammals, agri-environment schemes and set-aside—what are the putative benefits? *Mammal Review*, **37**, 259–277.

Marini, L., Fontana, P., Battisti, A., and Gaston, K.J. (2009). Agricultural management, vegetation traits and landscape drive orthopteran and butterfly diversity in a grassland-forest mosaic: a multi-scale approach. *Insect Conservation and Diversity*, **2**, 213–220.

McCauley, S.J. (2006). The effects of dispersal and recruitment limitation on community structure of odonates in artificial ponds. *Ecography*, **29**, 585–595.

Merckx, T., Feber, R.E., Dulieu, R.L., et al. (2009). Effect of field margins on moths depends on species mobility: Field-based evidence for landscape-scale conservation. *Agriculture, Ecosystems and Environment*, **129**, 302–309.

Morin, P.J. (1984). Odonate guild composition: Experiments with colonisation history and fish predation. *Ecology*, **65**, 1866–1873.

NE (2013a). *Entry Level Stewardship Handbook*, 4th ed. Natural England.

NE (2013b). *Organic Entry Level Stewardship Handbook*, 4th ed. Natural England.

NE (2013c). *Higher Level Stewardship Handbook*, 4th ed. Natural England.

Niggebrugge, K., Durance, I., Watson, A.M., Leuven, R.S.E.W., and Ormerod, S.J. (2007). Applying landscape ecology to conservation biology: Spatially explicit analysis reveals dispersal limits on threatened wetland gastropods. *Biological Conservation*, **139**, 286–296.

Ordnance Survey (1998). *Sheet 192, Buckingham & Milton Keynes*. Ed. 1:25 000. OS Explorer map. Ordnance Survey, Southampton.

Pacini, N., Harper, D., Henderson, P., and Le Quesne, T. (2013). Lost in muddy waters: freshwater biodiversity. In D.W. Macdonald and K. Willis, eds., *Key Topics in Conservation Biology 2*, pp. 184–203. John Wiley & Sons, Ltd., Oxford.

Painter, D. (1998). Effects of ditch management patterns on Odonata at Wicken Fen, Cambridgeshire, UK. *Biological Conservation*, **84**, 189–195.

Pither, J. and Taylor, P.D. (1998). An experimental assessment of landscape connectivity. *Oikos*, **83**, 166–174.

Pond Conservation (2008). *Ponds—A Priority Habitat*: best practice guidance for development control planning officers. Pond creation Toolkit Factsheet. <http://www.freshwaterhabitats.org.uk/>.

Raebel, E.M. (2010). *Ecology and conservation of dragonflies on farmland ponds*. PhD thesis, University of Liverpool.

Raebel, E.M., Merckx, T., Feber, R.E., Riordan, P., Macdonald, D.W., and Thompson, D.J. (2012a). Identifying high-quality pond habitats for Odonata in lowland England: implications for agri-environment schemes. *Insect Conservation and Diversity*, **5**, 422–432.

Raebel, E.M., Merckx, T., Feber, R.E., Riordan, P., Thompson, D.J., and Macdonald, D.W. (2012b). Multi-scale effects of farmland management on dragonfly and damselfly assemblages of farmland ponds. *Agriculture, Ecosystems and Environment*, **161**, 80–87.

Raebel, E.M., Merckx, T., Riordan, P., Macdonald, D.W., and Thompson, D.J. (2010). The dragonfly delusion: why it is essential to sample exuviae to avoid biased surveys. *Journal of Insect Conservation*, **14**, 523–533.

Reid, N.R., McDonald, R.A., and Montgomery, W.I. (2007). Mammals and agri-environment schemes: hare haven or pest paradise? *Journal of Applied Ecology*, **44**, 1200–1208.

Rouquette, J.R. and Thompson, D.J. (2005). Habitat associations of the endangered damselfly, *Coenagrion mercuriale*, in a water meadow ditch system in southern England. *Biological Conservation*, **123**, 225–235.

Rouquette, J.R. and Thompson, D.J. (2007). Patterns of movement and dispersal in an endangered damselfly and the consequences for its management. *Journal of Applied Ecology*, **44**, 692–701.

Rundlöf, M., Bengtsson, J., and Smith, H.G. (2008a). Local and landscape effects of organic farming on butterfly species richness and abundance. *Journal of Applied Ecology*, **45**, 813–820.

Rundlöf, M., Nilsson, H., and Smith, H.G. (2008b). Interacting effects of farming practice and landscape context on bumblebees, *Biological Conservation*, **141**, 417–426.

Sahlén, G. and Ekestubbe, K. (2001). Identification of dragonflies (Odonata) as indicators of general species richness in boreal forest lakes. *Biodiversity and Conservation*, **10**, 673–690.

Sinsch, U., Oromi, N., Miaud, C., Denton, J., and Sanuy, D. (2012). Connectivity of local amphibian populations: modelling the migratory capacity of radio-tracked natterjack toads. *Animal Conservation*, **15**, 388–396.

Smallshire, D. and Swash, A. (2004). *Britain's Dragonflies: A Field Guide to the Damselflies and Dragonflies of Britain and Ireland*. 1st ed. WildGuides Ltd, UK.

Steytler, N.S. and Samways, M.J. (1995). Biotope selection by adult male dragonflies (Odonata) at an artificial lake created for insect conservation in South Africa. *Biological Conservation*, **72**, 381–386.

Sutherland, W.J. (2002). Restoring a sustainable countryside. *Trends in Ecology and Evolution*, **17**, 148–150.

Thomas, J.A. (2005). Monitoring change in the abundance and distribution of insects using butterflies and other indicator groups. *Philosophical transactions of the Royal Society B*, **360**, 339–357.

Thompson, D.J. (1987). Regulation of damselfly populations: the effects of weed density on larval mortality due to predation. *Freshwater Biology*, **17**, 367–371.

UKBAP (2007). *Species and Habitat Review report 2007*, Annexes 4–6, pp.131. <http://www.ukbap.org.uk>.

UKBAP (2008). *Priority Habitat Descriptions*. <http://jncc.defra.gov.uk/>, accessed December 2013.

Watts, P.C., Keat, S., and Thompson, D.J. (2010). Patterns of spatial genetic structure and diversity at the onset of a rapid range expansion: colonisation of the UK by the small red-eyed damselfly *Erythromma viridulum*. *Biological Invasions*, **12**, 3887–3903.

Williams, P.J., Biggs, J., Barr, C.J., et al. (1998). *Lowland Pond Survey 1996*. Department of the Environment, Transport and the Regions.

Williams, P., Biggs, J., Crowe, A., Murphy, J., Nicolet, P., Weatherby, A., and Dunbar, M. (2010). *Countryside Survey: Ponds Report from 2007*. Technical Report No. 7/07. Pond Conservation and NERC/Centre for Ecology & Hydrology, 77pp. (CEH Project Number: C03259).

Williams, P., Whitfield, M., Biggs, J., et al. (2003). Comparative biodiversity of rivers, streams, ditches and ponds in an agricultural landscape in Southern England. *Biological Conservation*, **115**, 329–341.

Wrbka, T., Schindler, S., Pollheimer, M., Schmitzberger, I., and Peterseil, J. (2008). Impact of the Austrian Agri-Environmental Scheme on diversity of landscapes, plants and birds. *Community Ecology*, **9**, 217–227.

CHAPTER 11

Freshwaters and farming: impacts of land use and management on the biodiversity of rivers and ditches

Rosalind F. Shaw, Alison E. Poole, Ruth E. Feber, Eva M. Raebel, and David W. Macdonald

On either side the river lie
Long fields of barley and of rye,
That clothe the wold and meet the sky;

**Lord Alfred Tennyson, 'The Lady
of Shalott' (1832, revised 1842),
pt.1, l.1–5.**

11.1 The importance of freshwater habitats

Freshwater habitats such as rivers and lakes are vital resources for sustaining the human enterprise. As well as supplying water, they provide supporting, regulating, and cultural services (Millenium Ecosystem Assessment 2005; WWAP 2012). Pressures on freshwater systems are increasing as a result of increased human population growth and land use changes, such as urbanization and intensification of agriculture, and their degradation is at an all-time high (Darwall et al. 2008; Pacini et al. 2013). To address this, the EU's Water Framework Directive (WFD) aims to have all surface and groundwater bodies in Europe attaining 'good' ecological and chemical status by 2015 (European Union 2010).

Freshwaters provide habitat for a wide range of biodiversity. Plants found in freshwater range from the entirely aquatic but floating species (such as duckweeds *Lemna* spp.) and submerged species with roots and leaves underwater (such as curled pondweed *Potamogeton crispus*), to emergent plants (such as yellow flag *Iris pseudacorus*) which trap sediment between their stems and provide structural habitat for aquatic invertebrates. Many faunal groups are dependent on freshwater habitats. Some, such as aquatic worms and freshwater fish, inhabit water for their entire life.

Others, including several groups of aquatic invertebrates, such as mayflies (Ephemeroptera) (Fig. 11.1), stoneflies (Plecoptera), and dragonflies/damselflies (Odonata; Chapter 10, this volume) have life cycles that are completed partly outside the water. Mayflies, for example, start as underwater larvae known as nymphs and undergo several moults. Once adults, they fly for a very short period of time (between minutes and days) not feeding, only mating, before dying. This short adult life has lead to them sometimes being known as '24-hour flies'. Other groups of species use freshwater habitats for foraging and nesting; for example, birds will exploit the concentrated food source supplied by invertebrates that emerge from the water synchronously in large numbers, such as adult midges (Chironomids), while species such as lapwing *Vanellus vanellus* will find food in the soft, damp soil around watercourses (Bradbury and Kirby 2006). British bats regularly forage and sustain higher levels of activity along riparian corridors, with some species, such as Soprano pipistrelle and *Myotis* bats, being restricted to riparian (and woodland) habitats (Chapter 9, this volume). Water voles *Arvicola amphibius* burrow in riverbanks and use lush riparian vegetation for both food and shelter from predators (Chapter 14, this volume).

In spite of the importance of freshwater for biodiversity, the State of Nature report (State of Nature

Wildlife Conservation on Farmland. Managing for Nature on Lowland Farms. Edited by David W. Macdonald and Ruth E. Feber.
© Oxford University Press 2015. Published 2015 by Oxford University Press.

Figure 11.1 Mayflies, such as the green drake mayfly *Ephemera danica* pictured here, are indicators of good water quality. Photograph ©Steve Kett.

Partnership 2013) highlights that, in the UK, an alarming 57% of freshwater-related animals and plants (out of 1008 species assessed) have declined within the last 50 years. On a global level freshwater habitats are facing one of the greatest losses of any habitat (–2.4% per year, based on the decline of vertebrate populations: Balmford et al. 2002).

Key issues for biodiversity in freshwater bodies are the availability of habitat and the quality of the water. A major issue relating to land use is that of pollution. While geology, soil type, rainfall, and elevation all affect the basic chemical composition of water, the management of surrounding land is crucial in determining additional chemical loads and thus the water quality (Allan 2004; European Commission 2011). Discharge of pollutants from point sources such as sewage works and factories is strictly regulated by the EU (under the Urban Waste Directive and the Integrated Pollution Prevention and Control Directive), and is relatively easy to identify and control. However, diffuse agricultural pollution[1], particularly that which involves sediments, nutrients (such as nitrates and phosphates), and pesticides, has proved much more difficult to tackle. This is partly because pollutants generally enter the water system via soil, rather than directly, and also because

[1] Diffuse pollution is defined as pollution which comes from multiple, dispersed sources.

land draining into one water system may be managed by a large number of people, making it difficult to attribute pollution to any one of them (Scheierling 1995).

Agricultural pollutants affect biodiversity in a number of ways. For example, run-off from fields that have bare or poached (trampled) ground can increase river sediments, reducing light levels and primary productivity. By reducing macrophyte abundance and smothering gravel habitats, sediments can alter the macroinvertebrate community (Extence et al. 2011) and prevent spawning of fish such as salmon and trout (e.g. Greig et al. 2005). Organic waste, such as slurry, contains ammonia which can be toxic in the environment in this form, and is broken down into nitrite and nitrate by bacteria. Increases in nutrients such as nitrates and phosphates, commonly applied to crops to improve yield, lead to an initial increase in plant and animal biomass. If high levels of ammonia, nitrates, or phosphates enter the water system, this may lead to eutrophication: plants with floating leaves or mat-forming species increase, shading out the submerged plants below. As plants die they decompose, resulting in reduced oxygen levels as the bacteria breaking them down respire. Macrophytes are eventually lost and algae proliferate; oxygen levels become depleted and fish and invertebrates can no longer survive.

Diffuse pollution can have impacts on organisms throughout a watercourse. Diffuse pollutants can be difficult to measure; for example, pesticides or nutrients in run-off may only be measurable for a limited time, and only if heavy rainfall follows soon after application (Schäfer et al. 2007). However, an incidence of pesticide or chemical pollution that is not detected by water sampling, even a day after the event, can leave a footprint in the biotic communities. For these reasons, biotic communities—macroinvertebrates and macrophytes—are often used to monitor the long-term condition of freshwater systems, as they can reflect the pollution history of a site as well as being quick to respond to changes. For example, presence of the endangered but long-lived (up to 100 years) freshwater pearl mussel *Margaritifera margaritifera*, reflects past long-term good quality water. Loss of other species can indicate a pollution event—for example, exposure to the insecticide imidacloprid (a seed dressing with insecticidal properties, used commonly on a global scale) produces high mortality rates in the aquatic worm *Lumbriculus variegates* after ten days (Sardo and Soares 2010) and after just 12 hours in the caddis fly *Neureclipsis* sp. (Mohr et al. 2012). Similarly, freshwater shrimps *Gammarus pulex* can tolerate higher levels of ammonia than the large dark olive mayfly nymph *Baetis rhodani* (Williams et al. 1984). Such differences between species in their tolerance to pollutants (e.g. Williams et al. 1984) are important for monitoring purposes: they mean that indices can be developed to compare the biological communities of a site to what would be expected in an undegraded, high quality site of a similar type. In the UK, indices have been developed to do this for rivers (RIVPACS—River Invertebrate Prediction and Classification System; Wright et al. 2000) and still waters (PSYM—Predictive System for Multimetrics; Environment Agency and Pond Action 2002).

The extensive modification of freshwater habitats and the land surrounding them (in European countries and increasingly elsewhere; Pacini et al. 2013) has led to a critical need to understand the impacts of agricultural land use at different scales and how they can be mitigated. The UK has both regulatory and voluntary frameworks for mitigating the impacts of diffuse agricultural pollutants. For example, land managers farming within Nitrate Vulnerable Zones (areas which are vulnerable to water pollution from nitrates) are required to follow guidelines on nitrate application and storage of animal waste (Defra 2013). In priority catchments, where there is concern that agricultural pollutants are damaging important freshwater sites or reducing water quality, 'Catchment Sensitive Farming' (CSF) schemes have been created. These give free advice to farmers on how to reduce the agricultural pollutants entering watercourses (including nitrates but covering the entire range of pollutants) and competitive capital grants are available to specific holdings. Successful outcomes have been reported[2].

A key mechanism in the UK for encouraging environmentally sensitive agricultural land management is via agri-environment schemes (AES). A number of options within AES aim to reduce diffuse water pollution from agricultural activities. These include the creation of buffer strips next to watercourses, maintaining fencing next to watercourses (Natural England 2013a,b), management for nitrate run-off (e.g. Withers and Lord 2002), and support for ditch management, organic farming (Magbanua et al. 2010), and woodland conservation and restoration (Franken et al. 2007). UK AES are administered via payments[3] to land managers and

funded from tax revenue; it is, therefore, important that AES are optimized to deliver measurable ecological benefits. This cannot be determined without evaluating their effectiveness at improving the biological quality and conservation value of freshwater bodies.

In the remainder of this chapter we describe two large-scale studies which were designed to investigate how modifying land use at different scales can impact freshwater biodiversity. The purpose of the first study was to assess the ecological effectiveness of AES in a lowland river basin. We used aquatic macroinvertebrates as indicators of river health and conservation value, to assess whether, and at what scale, the implementation of AES options in the surrounding landscape improved river health. The second study focused on an important wetland feature on farms—farm ditches. Our aim was to explore how management of adjacent land, and management of the ditches themselves, affected their biodiversity.

11.2 Optimizing agri-environment schemes to improve conservation value of rivers

To conserve and improve freshwater biodiversity cost-effectively requires a full understanding of the relationship between catchment land use, improvement measures, and water quality, at multiple scales. Factors impacting at both catchment (e.g. soil type and origin) and local scales (e.g. habitat quality, flow stability, and riverbed sediment) are likely to be important (McRae et al. 2004; Vondracek et al. 2005; Wang et al. 2006). 'Distance from the river' has often been considered when looking at the influence of land use at the reach (length of river) scale, but studies at the catchment scale have tended not to focus on this aspect (e.g. Allan et al. 1997; Aspinall and Pearson 2000). However, if distance from the river is a factor in reach scale studies, why should it not also be important at the catchment scale? It is reasonable to assume that land closer to the river will have a greater effect on river health throughout the entire catchment than land further away. One could therefore hypothesize that in catchments, as for reaches, the effect of land use depends on its distance from the river.

The key objectives of our study were to determine which, and at what spatial scale, land use options could optimize the biological quality and conservation value of riverine macroinvertebrate communities within a catchment (Poole et al. 2013). We focused on UK AES options with potentially positive effects on the

---

[2] A CSF scheme targeted farmers within five catchments and successfully reduced the total annual pesticide load by 26% within that area (CSF Evidence Team 2011).

[3] For example, in 2013, farmers were being offered £36 per 100 m per year for managing a ditch of high environmental value in targeted schemes, £400 per year for 833 m of 12 m-wide buffer strip as part of a broad agreement, or £30 per 0.75 km of fencing as part of a broad agreement. For full details see Natural England (2013a) and Natural England (2013b).

Figure 11.2 Macroinvertebrate samples were taken from rivers in the Upper Thames study area. Photograph ©Rosie Salazar.

ecological quality of rivers: organic farming, deciduous woodland, and buffer strips designed to reduce run-off. Organic farming, for example, aims to produce food without synthetic fertilizers or pesticides, potentially reducing run-off of some pollutants, while streamside woodland has been shown to increase diversity of macro-invertebrates (Sweeney 1993) and reduce sedimentation. Grassy buffer strips can reduce the run-off of nitrogen and phosphorus from conventionally managed agricultural fields (Borin et al. 2004). Eighty per cent of the Upper Thames catchment was, at the time of our study, under AES; this high coverage with a variety of options allowed us to evaluate our results in terms of AES with and without river options (see Poole et al. 2013 for details). To assess river health, we took macroinvertebrate samples from rivers and streams throughout the catchment (Fig. 11.2) and analysed land-use cover within areas of various widths next to the river (10 m, 50 m, 100 m, 500 m, and 1000 m) and throughout the entire catchment[4] (see Poole et al. 2013).

Sixty sites throughout the study area were sampled a total of 159 times. We sampled macro-invertebrates during spring and autumn over three years, from 2009 to 2011. At each site, standard three-minute kick samples (where a surveyor stands in the river and kicks the sediment into a net held downstream) and a one-minute hand search (Murray-Bligh et al. 1997) were taken either by WildCRU staff, or by Environment Agency (EA) staff as part of their routine monitoring. Macroinvertebrates were identified to species level where possible and results used to calculate three indices of river health (Table 11.1).

[4] The land use cover in each area of different widths was calculated using ArcGIS (ESRI 2011) from Ordnance Survey data (woodlands), Natural England data (AES schemes and organic farming), and Environment Agency data (rivers), and the river catchment was calculated from a digital terrain model available from the Ordnance Survey (for further detail see Poole et al. 2013).

The first of these indices, developed in the 1970s, was the Average Score Per Taxon (ASPT; Table 11.1), a measure of the average pollution tolerance (and therefore pollution levels in the river) of species found in a sample (Hawkes 1998). For example, many species of mayflies (Ephemeroptera) are classed as the most highly sensitive taxa (scoring 10 where 10 is highly pollution sensitive and 1 is highly pollution tolerant). Highly pollution tolerant groups include the aquatic worms (Oligochaeta, score of 1) and non-biting midge larvae (Chironomidae, score of 2). The second and third indices we used, the Community Conservation Index (CCI) and the Proportion of Sediment-sensitive Invertebrates (PSI), have been developed more recently. CCI (Table 11.1) is a measure of how rare the invertebrates found in a sample are (Chadd and Extence 2004), and can be used to assess the influence of land use on aquatic macroinvertebrates of conservation concern. PSI (Table 11.1) is used to assess the amount of sediment in a river (Extence et al. 2011), a factor shown to be heavily influenced by farming (e.g. Allan et al. 1997). For example, the large dark olive mayfly *Baetis rhodani*, beloved of trout and trout fisherman alike, is highly sensitive to river sediments, as are Hydropsyche caddis flies, whose net-spinning behaviour and filter-feeding is affected by river sediments, which can clog their cases and increase mortality (Strand and Merritt 1997; Runde and Hallenthal 2000).

The number of macroinvertebrate taxa found at each site varied between 13 and 33 (mean 23 ± 5.0 SD). Overall, the average ASPT for invertebrates in the Upper Thames was 5.5 (very good quality), which is within the score range predicted for this region (4.8–6.5;

Hemsley-Flint 2000). However, although the mean ASPT for autumn and spring were similar (5.53 ± 0.08 SD and 5.60 ± 0.06 SD, respectively), the ASPT ranges of 3.69–6.57 in autumn and 4.48–6.42 in spring indicated that there were considerable differences between sites (from poor to very good quality; Table 11.1). There were also large differences in CCI between sites, which ranged from 2.8 (site of only common species or a community of low taxon richness) to 32.8 (supporting rarities or, in this case, a community of very high taxon richness), with sites scoring over 20 potentially being of national significance (Chadd and Extence 2004). The spring PSI in our study varied between 7.41 and 78.83, and in autumn between 4.17 and 72.36, ranging from heavily to slightly sedimented (Table 11.1; Poole et al. 2013).

Could land use explain these important differences? There was a positive relationship between the amount and location of woodland cover and our macroinvertebrate measures of river health. Pollution sensitivity in macroinvertebrates (as measured by ASPT) was most strongly correlated with the proportion of woodland within 500 m of the river (Fig. 11.3a), and the conservation value of the macroinvertebrates (as measured using CCI) was correlated with the proportion of woodland within both 100 m and 500 m (Fig. 11.3b). Neither was correlated with the proportion of woodland within smaller buffers. This meant, for example, that sites with more woodland in the surrounding catchment had higher proportions of macroinvertebrates that require non-polluted environments, such as pollution sensitive mayfly families (Ephemeroptera). This may be because an increase in woodland cover reduces the area of land which is a source of diffuse pollution and other stressors (farmland with/without AES river options, urban land; e.g. Davies et al. 2009) and, at shorter distances, not enough land is protected to produce this effect (Poole et al. 2013).

There was also a correlation between autumn PSI and the proportion of woodland within 100 m, suggesting that it is also important to retain woodland at a shorter distance from the river in order to limit sediment release of river banks in agricultural catchments, corroborating other studies (e.g. Larsen et al. 2009). In addition, the fact that the autumn, but not the spring, PSI sample was correlated with woodland, illustrates that sediment concentrations are seasonal and can be greater in the autumn (Owens and Collins 2006). Turbidity (suspended sediments), nitrate, and ammonia concentrations can also be higher in autumn than in summer and spring, probably as a consequence of the increased rainfall, and hence higher river flow, after the low flow period (Álvarez-Cabria et al. 2010). In addition, low vegetation cover in

Table 11.1 Macroinvertebrate indices and standard scores for individual sites.

Index	Scores
ASPT (Average Score Per Taxon): pollution quality	(> 5.4) very good quality; (4.81–5.4) good; (4.21–4.8) fair; (3.61–4.2) poor; (≤ 3.6) very poor
CCI (Community Conservation Index): conservation value of the macroinvertebrate community	(> 20) very high conservation value; (>15–20) high conservation value; (>10–15) fairly high; (>5–10) moderate; (0–5) low
PSI (Sediment-Sensitive Invertebrates): sediment in the river	(0–20) heavily sedimented; (21–40) sedimented; (41–60) moderately sedimented; (61–80) slightly sedimented; (81–100) naturally sedimented/unsedimented

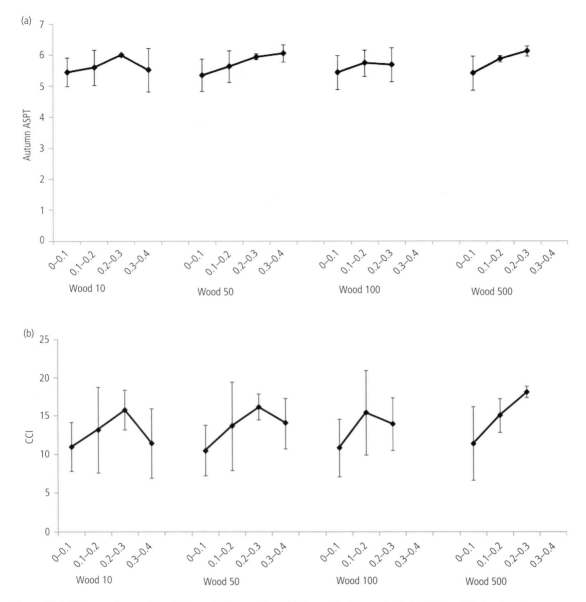

Figure 11.3 (a) Autumn Average Score Per Taxon (ASPT, n = 43) and (b) Community Conservation Index (CCI, n = 58) plotted against the proportion of woodland within areas of varying widths next to the river (in m) along the entire upstream length of the catchment (from Poole et al. 2013).

agricultural land in autumn may also result in increased sediment in watercourses (e.g. Steegen et al. 2000).

As well as the impact of woodland, we also found that, in spring, the PSI was positively correlated with the proportion of land within 500 m and 1000 m of the river in agri-environment scheme agreements which included river options (Fig. 11.4). This suggests that

increasing the implementation of AES 'river options' could help to reduce sediment run-off at the catchment level. Since the autumn PSI was correlated to the proportion of woodland within 100 m, our study demonstrates the importance of retaining woodland as well as AES river options to help reduce sedimentation levels throughout the year.

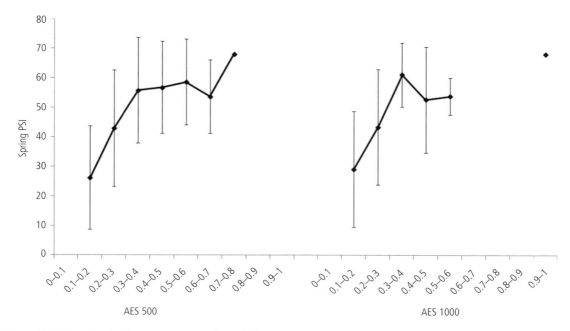

Figure 11.4 Proportion of Sediment-sensitive Invertebrates (PSI) plotted against the proportion of Agri-Environment Scheme (AES) river options within varying buffer widths (in m) along the entire upstream length of the catchment (n = 56). Reprinted from Poole *et al* 2013.

Numerous studies emphasize the importance of managing rivers at the catchment level (e.g. Harper et al. 1999; Larsen et al. 2009; Ormerod et al. 2010; Palmer et al. 2010b), but our findings were novel in demonstrating the importance of the distance from the river of habitats of conservation priority throughout the catchment. The goal of promoting macroinvertebrate conservation, as a proxy for protecting biodiversity and ecosystem services on lowland farms, was best served where there were high proportions of woodland within 100 m or 500 m of the river throughout the entire up-stream catchment. The results demonstrate the value of woodland beyond the immediate riparian corridor (considered to be much less than 100 m, e.g. Arnaiz et al. 2011) and supports calls for increasing, or at least maintaining, forest cover in the broader catchment (e.g. Harper et al. 1999). Current AES grants only allow for the creation of woodland up to 3 ha, in blocks no greater than 1 ha, and this option is only available under the narrow-and-targeted AES scheme (Higher Level Stewardship, HLS; Natural England 2013b), while larger grants can only be obtained from the English Woodland Grant Scheme (Forestry Commission 2013). AES river options within 100 m were associated with improved conservation of sediment-sensitive macroinvertebrates.

We therefore conclude that, if fostering biodiverse macroinvertebrate communities is the desired outcome, then efforts should be focused on preserving woodland within a 100–500 m buffer zone along the upstream length of the river, while retaining AES river options (Poole et al. 2013).

11.3 Land management and the biodiversity of ditches

Agricultural pollutants found in natural waterways, such as rivers and streams, are transported by water from the place they are applied, and one of the first places they are concentrated is often ditches. Ditches are man-made, linear, planform features (Williams et al. 2003), usually dug for drainage purposes or for storing water for irrigation. Ditches can range from permanently wet channels up to several metres deep and wide, or may be small channels of less than a metre, which are only wet during periods of high rainfall (JNCC 2005). The networks of ditches that collect and deliver water to larger watercourses have the potential to act as a biodiversity reservoir and to trap chemicals and sediments, thus functioning as a filter for water, improving the quality throughout the system (e.g. Kröger et al. 2009). Ditches are often actively managed by landowners, either by dredging (the removal of silt) or by cutting the vegetation, assumed to help maintain flow of water and drainage function of the ditch. Other management may include damming small sections to increase water retention (Aquilina et al. 2007). However, careful management is required to reduce negative impacts on both biotic communities and local hydrology—for example, high frequency of dredging may result in highly reduced biotic communities (Telfer 2000; Milsom et al. 2004) and damming may exacerbate flooding caused by heavy rainfall.

Much of what is known of life in British ditches comes from those with high conservation value in coastal grazing marshes (e.g. Drake et al. 2010) or those in intensively managed agricultural land that was once fenland (e.g. Foster et al. 1989; Painter 1999; Milsom et al. 2004 and references therein), both of which support rich remnants of marsh ditch flora and fauna communities. However, there are an estimated two kilometres of ditches in every kilometre square of intensively managed agricultural land in England (Brown et al. 2006). Of these, 43 430 km are managed under agri-environment scheme agreements (Natural England 2009), designed to encourage environmentally sensitive management of ditches by thinning of vegetation on a rotational basis (leaving 30–50% of vegetation undisturbed; Defra 2004), and careful and environmentally friendly dredging on short alternate sections of the bank, only when necessary and at certain times of the year (Natural England 2013a). Information on the aquatic communities of these wider countryside ditches, particularly seasonal ditches (those that regularly dry out, especially in summer), is rather more scarce (Biggs et al. 2007). We investigated how immediate land use and management of ditches affected both the water quality and the communities of plants and invertebrates found in them. Specifically we asked: does adjacent land use and ditch management affect the water quality of ditches? How does management of ditches and the surrounding landscape features affect the biotic communities of ditches?

11.3.1 Water quality of ditches in relation to land use and ditch management

To determine whether adjacent land use and ditch management affect the water quality of ditches, we sampled 175 ditches across 30 farms. We took water samples over two years, covering different seasons (late summer/autumn (August–October 2010); winter/

spring (February–April 2011); and early/mid-summer (May–July 2011; details in Shaw et. al in press). The amounts of key diffuse pollutants, particularly phosphates, ammonia, nitrates, and nitrites, together with pH, were recorded. We tested the relationship between water quality measures and the following factors: (1) the amount of adjacent arable land (recorded as a percentage of land on either side of the ditch under arable crop); (2) the presence of hedgerows or buffer strips (defined as non-cropped vegetation next to the ditch); and (3) management actions such as vegetation cutting and dredging.

The concentration of nitrate (NO_3^-) and phosphate differed depending on the time of year they were measured. We found that nitrate levels were highest in the winter/spring (February–April) sample[5]. Within Nitrate Vulnerable Zones, the recommendation is for organic manures (slurry/poultry or farmyard manures; sources of nitrogen) to be applied at the start of crop growth during the late winter/early spring period (February to April)[6] when nitrogen demand is at its greatest (Chambers et al. 2001; Defra 2010). This is meant to keep leaching to a minimum, because nitrogen is taken up by the crop rather than remaining in the soil. However, the risk of nitrates in run-off is closely linked with the rate at which nitrogen is released depending on the type of manure and, importantly, with rainfall events after application (Chambers et al. 2001).

Importantly, we found that the type of land use surrounding a ditch and the absence/presence of a buffer (such as that shown in Fig. 11.5) also had an effect on ditch water quality. This corroborates other studies showing that nitrate levels in leachate are higher in arable than in pasture land (e.g. ADAS 2007). Indeed, we not only found higher levels of nitrates in ditches that were surrounded by arable land, but we also detected a small increase of 0.01 mg/L NO_3^- with a 50% increase in the amount of arable land surrounding the ditch (i.e. an arable field on one side). We found, however, that nitrate levels in the water were significantly reduced as the width of vegetated buffer strip surrounding the ditch increased.

Overall, our results indicate first, that in arable land, vegetative buffers reduce some diffuse agricultural pollutants and, second, efforts to reduce nitrate

---

[5] Winter/spring 2011 mean of 16.8 mg/L NO_3^- ± 1.42 SE compared to 13.2 ± 1.70 SE and 9.4 ± 1.34 SE, summer/autumn 2010 and early/mid-summer 2011 respectively.

[6] Following the cross-compliance/AES closed period of September–end of January for the application of organic and manufactured nitrogen, depending on soil type.

Figure 11.5 Ditch on arable land with a grassy buffer strip on either side. Photograph © Rosalind Shaw.

concentrations in ditches should be targeted in winter and early spring (one of the key recommendations for protecting watercourses is that no manure is to be applied from October to the end of January, and green cover should be maintained on surrounding fields for as much of the year as possible; Defra 2009).

11.3.2 Biological communities of ditches

To determine which factors affected the species richness and community composition of biological communities in ditches, we tested the impacts of adjacent land use, water quality, and management as outlined above, and also ditch characteristics such as bank angle, ditch size, water depth (average from the three seasonal surveys), and the percentage of shade over the ditch from surrounding woody vegetation (for full list see Shaw et al. in press) on plant and invertebrate communities.

Plant communities of both the ditch bank and the ditch channel were surveyed using a 10 m vegetation quadrat placed midway between the end points of each ditch length (as per Palmer et al. 2010a, see Shaw et al. in press for more details) in summer

2010 and 2011. A total of 224 species of plant were recorded from ditch banks and 137 from the ditch channel surveys. Between one and fifteen plant taxa were recorded per ditch channel (mean 5.3 ± 3.10 SD). Fool's watercress *Apium nodiflorum* was the most frequently found wetland species, present in 22% of ditches. This species is typical of watercourses in nutrient-rich areas, in places so abundant that it sometimes requires control measures (Centre for Aquatic Plant Management 2005). On ditch banks, the most frequently found species were nettles *Urtica dioica* (present in 84% of ditches), brambles *Rubus fruticosa* (67% of ditches), and grasses, such as false oat-grass *Arrhenatherum elatius* (62% of ditches), and cocks foot *Dactylis glomerata* (54% of ditches).

In the spring of 2011, we sampled the macroinvertebrate communities of 49 ditches using a timed sweep netting approach (Palmer et al. 2010a). In total we recorded 190 taxa of aquatic invertebrates, between 9–35 taxa per ditch (mean 19.6 ± 5.76 SD), sustaining frequent groups such as chironomid larvae (found in 98% of ditches), oligochaete aquatic worms (82% of ditches), and the water beetle *Helophorus brevipalpis* (87% of ditches), but also near threatened species, such as the minute moss beetle *Hydraena palustris* (found in only two ditches), which is a specialist of temporary/semi-permanent, stagnant water with dense marginal vegetation, and occurs in isolated populations in Britain (Foster 2010).

Water depth and the level of shade over the ditch were the most important factors affecting channel plant and aquatic invertebrate communities (Shaw et al. in press). Water depth affected the plant communities of both ditch channels and ditch banks, with greater water depth leading to more species characteristic of open water, such as common duckweed *Lemna minor* in the channels, and wetland adapted species such as reed sweet-grass *Glyceria maxima* and reed canary-grass *Phalaris arundinacea* on the ditch banks. Aquatic invertebrate communities were most strongly affected by water depth, with species richness increasing in ditches with deeper water (Fig. 11.6). Ditches of an average water depth of less than 0.06 m had a mean taxonomic richness of 16 (± 2.01 SD), whereas in a water depth of between 0.2 and 0.57 m the mean taxonomic richness was 23 (± 6.5 SD). Families only found in deeper ditches included small square gill mayflies (Caenidae), burrowing mayflies (Ephemeridae), and humpless casemaker caddisflies (Brachycentridae).

For channel vegetation, species richness decreased with increasing shade (Fig. 11.7a) and the percentage cover of vascular plants also decreased with increasing shade (Shaw et al. in press). Increases in shade also led to a reduction in the taxonomic richness of aquatic invertebrates (Fig. 11.7b, Shaw et al. in press), potentially due to the lack of aquatic plants as substrate in highly shaded ditches. In other freshwater bodies, increased shade over ditches can be beneficial as it reduces fluctuations in water temperature; however, in narrow ditches, the shade provided by the banks themselves or riparian grasses may be enough to mediate temperature (Blann et al. 2002). Ditch bank vegetation was also affected by shade, with communities being dominated by shade tolerant species such as ivy *Hedera helix* and with a much higher percentage of bare ground and leaf litter—as shade increased by 10%, the amount of litter and bare ground increased by approximately 4.5%. Since increasing the vegetation cover of ditch banks has the potential to reduce the amount of pollutants and sediments entering the water (e.g. Rogers and Schumm 1991), controlling shade along ditches is important for both ditch biodiversity and water quality.

The invertebrate surveys also suggest that greater management of run-off is needed—ditches which had on average a higher, more alkaline pH (all ditches surveyed for macroinvertebrates had a pH of above 7)

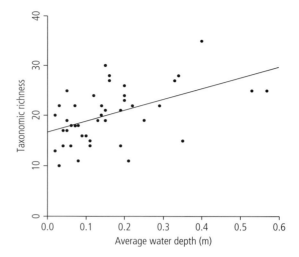

Figure 11.6 The impacts of average water depth of a ditch (measurements taken three times in summer 2010, winter/spring 2011, and spring/summer 2011) on the taxonomic richness of aquatic invertebrate species. From Shaw et al. (in press).

had a larger number of the families from the pollution intolerant Ephemeroptera, Plecoptera, and Trichoptera (EPT) groups. This potentially indicates that some ditches are fed more by ground water than run-off and are thus less polluted (ground water tends to have higher pH; Mason 1996).

11.3.3 Agri-environment schemes (AES) and ditches

We found little difference between ditches that were managed under an agri-environment scheme (AES) agreement and those that were not, in terms of the wildlife found within them (Shaw et al. in press). One contributory factor may be that AES options for ditches were generally designed for ditches of drained fenland, which are managed more frequently than those in the Upper Thames catchment—dredging cycles of 1–4 years and 4–10 years were reported for the Somerset Levels and Broads ESA respectively (McLaren et al. 2002). The Entry Level Stewardship (ELS) ditch management option requires that ditches are cleaned no more than once every five years, aimed at reducing the frequency of management, whereas ditches in our study were, on average, last dredged 15 years ago (although with a range between 1 and 55 years, Shaw et al. in press), indicating that in this area, high frequency of dredging is unlikely to be a problem.

(a)

(b)

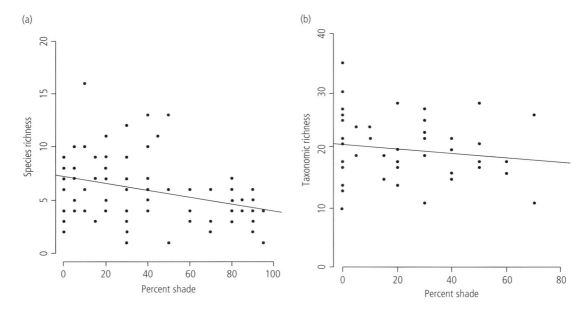

Figure 11.7 The impacts of the percentage of shade over the ditch channel on (a) the channel vegetation plant species richness and (b) the aquatic invertebrate community taxonomic richness. From Shaw et al. (in press).

Our study suggests that, in this lowland area, reducing the amount of shade improves the biodiversity of ditches. In narrow, often seasonal, ditches such as these, increased hedge trimming that reduces the amount of shade would be beneficial. Current best practice advice suggests that woody vegetation is good for stabilizing banks, reducing temperature fluctuations, preventing build up of vegetation, and providing leaf litter for aquatic invertebrates (Defra 2004). In the ditches we surveyed, given their narrow width, it is likely that low bank vegetation will provide sufficient shade for temperature regulation (Blann et al. 2002); lack of shade leading to wide temperature fluctuations is likely to be more of an issue in wide, deep ditches. Current AES prescriptions require infrequent trimming of hedges, which is appropriate for most hedges, but the management advice for a hedge is the same whether it occurs next to a ditch or not (Natural England 2013a). We suggest that AES options for combined hedge and ditch management could be modified to recommend more frequent trimming of hedges next to narrow ditches, if the aim is to preserve or enhance the aquatic biodiversity of the ditch.

We found that the species richness of ditches increased with water depth and suggest that management options could be considered that increase the depth and/or permanence of water in ditches. For

example, Avery (2012) discusses the use of barriers in ditches to retain water. However, the consequences of this practice for flooding in times of heavy rain need to be considered. Furthermore, some seasonal ditches may sustain uncommon temporary water invertebrate species that may not be present in any other (permanent) waterbody type (Williams et al. 2003). In line with broad recommendations for farmland management, habitat heterogeneity is the key—management that creates or maintains a mix of temporary and permanently wet ditches will provide habitat for the widest range of species.

11.4 Conclusions

Freshwater biodiversity, water quality, land use, and the ecosystem services delivered by freshwater systems are intricately linked. Freshwater biodiversity is not only worth conserving in its own right—despite its unglamorous name we would surely not want to lose the ditch dun mayfly *Habrophlebia fusca*—but it is also an indication of how well freshwater systems are able to meet human needs. The current policy framework surrounding water quality is directing attention to tackling the sources of pollutants rather than removing them downstream (Kay et al. 2009). We have identified two key scales at which land use and management can be manipulated to improve the ecological status of freshwater habitats in agricultural landscapes: the landscape scale, which considers the type of land use and its distance from the river, and the ditch scale, where appropriate management of this overlooked wetland habitat on farmland can deliver biodiversity gains. At both scales, woody vegetation was found to be an important influence on macroinvertebrate diversity, with greater woodland cover being correlated with increased biodiversity at large scales, but shading from hedgerows correlated with decreased biodiversity at the ditch scale. The mechanisms affecting macroinvertebrate biodiversity are likely to be different at these contrasting scales (Allan et al. 1997): high levels of shade caused by overgrown hedges at the ditch scale prevent the development of aquatic plant communities, thus reducing habitats for many groups of freshwater macroinvertebrates. At both scales, an increase in natural or semi-natural habitats such as woodland and vegetated buffer strips is likely to reduce the amount of agricultural pollutants, such as nitrates and phosphates, entering the water system by acting as a filter, and also by reducing agricultural activity immediately next to the watercourse. Manipulating management within the current framework of AES schemes has the potential to improve the quality of freshwater habitats within agricultural land.

The challenges of meeting our water needs in the future are only likely to increase. Our results suggest two areas where current management could be improved: (1) increasing the amount of woodland along river courses; and (2) improving management of ditch habitats by establishing grassy buffer strips alongside ditches, modifying hedge management practices, and creating a mosaic of ditches of various water depths. Targeting management to where it will have most impact can improve the effectiveness of mitigation measures, improving not only the biodiversity of freshwater habitats, but also ensuring the continued supply of clean, high quality water that we all require.

Acknowledgements

We would like to thank the farmers and land managers who gave permission for site access and provided information on ditch management (including the Earth Trust, Little Wittenham), and Joe Platt and Freya van Kesteren for water sample analysis. The ditch study was funded by Esmée Fairbairn Foundation. We thank the Rivers Trust and the Holly Hill Trust for sponsorship of the rivers project.

References

ADAS (2007). *Nitrates Consultation Supporting Paper D3: Diffuse nitrate pollution from agriculture—strategies for reducing nitrate leaching*. Defra, Norwich.

Allan, J.D. (2004). Landscapes and riverscapes: The influence of land use on stream ecosystems. *Annual Review of Ecology Evolution and Systematics*, **35**, 257–284.

Allan, J.D., Erickson, D.L., and Fay, J. (1997). The influence of catchment land use on stream integrity across multiple spatial scales. *Freshwater Biology*, **37**, 149–161.

Álvarez-Cabria, M., Barquín, J., and Juanes, J.A. (2010). Spatial and seasonal variability of macroinvertebrate metrics: Do macroinvertebrate communities track river health? *Ecological Indicators*, **10**, 370–379.

Aquilina R., Williams P., Nicolet, P., Stoate, C., and Bradbury, R. (2007). Effect of wetting-up ditches on emergent insect numbers. *Aspects of Applied Biology*, **81**, 261–262.

Arnaiz, O., Wilson, A., Watts, R., and Stevens, M. (2011). Influence of riparian condition on aquatic macroinvertebrate communities in an agricultural catchment in southeastern Australia. *Ecological Research*, **26**, 123–131.

Aspinall, R. and Pearson, D. (2000). Integrated geographical assessment of environmental condition in water catchments: Linking landscape ecology, environmental modelling and GIS. *Journal of Environmental Management*, **59**, 299–319.

Avery, L.M. (2012). *Rural Sustainable Drainage Systems (RSuDS)*. Environment Agency, Bristol.

Balmford, A., Bruner, A., Cooper, P., et al. (2002). Ecology—Economic reasons for conserving wild nature. *Science*, **297**, 950–953.

Biggs, J., Williams, P., Whitfield, M., et al. (2007). The freshwater biota of British agricultural landscapes and their sensitivity to pesticides. *Agriculture, Ecosystems & Environment*, **122**, 137–148.

Blann, K., Nerbonne, J.F., and Vondracek, B. (2002). Relationship of riparian buffer type to water temperature in the driftless area ecoregion of Minnesota. *North American Journal of Fisheries Management*, **22**, 441–451.

Borin, M., Bigon, E., Zanin, G., and Fava, L. (2004). Performance of a narrow buffer strip in abating agricultural pollutants in the shallow subsurface water flux. *Environmental Pollution*, **131**, 313–321.

Bradbury, R.B. and Kirby, W.B. (2006). Farmland birds and resource protection in the UK: Cross-cutting solutions for multi-functional farming? *Biological Conservation*, **129**, 530–542.

Brown, C.D., Turner, N., Hollis, J., et al. (2006). Morphological and physico-chemical properties of British aquatic habitats potentially exposed to pesticides. *Agriculture, Ecosystems & Environment*, **113**, 307–319.

Centre for Aquatic Plant Management (2005). *Information Sheet 28: Fools Watercress (Apium nodiflorum)*. Centre for Ecology and Hydrology, Wallingford.

Chadd, R. and Extence, C. (2004). The conservation of freshwater macroinvertebrate populations: a community-based classification scheme. *Aquatic Conservation-Marine and Freshwater Ecosystems*, **14**, 597–624.

Chambers, B., Nicholson, N., Smith, K., Pain, B., Cumby, T., and Scotford, I. (2001). *Making better use of livestock manures on arable land*. ADAS, Mansfield.

CSF Evidence Team (2011). *Catchment Sensitive Farming ECSFDI Phase 1 & 2 Full Evaluation Report*. Environment Agency, London.

Darwall, W., Smith, K., Allen, D., et al. (2008). Freshwater Biodiversity—a hidden resource under threat. In J.C. Vié, C. Hilton-Taylor, and S.N. Stuart, eds., *The 2008 Review of The IUCN Red List of Threatened Species*. IUCN, Gland, Switzerland.

Davies, B., Biggs, J., Williams, P., and Thompson, S. (2009). Making agricultural landscapes more sustainable for freshwater biodiversity: a case study from southern England. *Aquatic Conservation-Marine and Freshwater Ecosystems*, **19**, 439–447.

Defra (2004). *Farming and Watercourse Management Handbook*. The Stationary Office, Norwich.

Defra (2009). *Protecting our Water, Soil and Air: A Code of Good Agricultural Practice for farmers, growers and land managers*. The Stationary Office, Norwich.

Defra (2010). *Fertiliser Manual (RB209)*. The Stationary Office, Norwich.

Defra (2013). *Nitrate Vulnerable Zones*. <https://www.gov.uk/>, accessed June 2013.

Drake, C.M., Stewart, N.F., Palmer, M.A., and Kindemba, V.L. (2010). *The ecological status of ditch systems: an investigation into the current status of the aquatic invertebrate and plant communities of grazing marsh ditch systems in England and Wales. Technical Report*. Buglife—the Invertebrate Conservation Trust, Peterborough.

Environment Agency and Pond Action (2002). *A guide to monitoring the ecological quality of ponds and canals using PSYM*. Environment Agency, Peterborough.

ESRI (2011). *ArcGIS Desktop: Release 10*. Environmental Systems Research Institute, Redlands, CA.

European Commission (2011). On implementation of Council Directive 91/676/EEC concerning the protection of waters against pollution caused by nitrates from agricultural sources based on Member State reports for the period 2004–2007. <http://ec.europa.eu/>. Accessed online May 2013.

European Union (2010). *Introduction to the new EU Water Framework Directive*. <http://ec.europa.eu/>. Accessed online 27 July 2012.

Extence, C.A., Chadd, R.P., England, J., Dunbar, M.J., Wood, P.J., and Taylor, E.D. (2011). The assessment of fine sediment accumulation in rivers using macro-invertebrate community response. *River Research and Applications*, **29**, 17–55.

Forestry Commission (2013). *Woodland Creation Grant*. <http://www.forestry.gov.uk/>. Accessed online 12 May 2013.

Foster, G.N. (2010). A review of the scarce and threatened Coleoptera of Great Britain Part (3): Water beetles of Great Britain. *Species Status* 1. Joint Nature Conservation Committee, Peterborough.

Foster, G.N., Foster, A.P., Eyre, M.D., and Bilton, D.T. (1989). Classification of water beetle assemblages in arable fenland and ranking of sites in relation to conservation value. *Freshwater Biology*, **22**, 343–354.

Franken, R.J.M., Gardeniers, J.J.P., and Peeters, E. (2007). Secondary production of *Gammarus pulex* Linnaeus in small temperate streams that differ in riparian canopy cover. *Fundamental and Applied Limnology*, **168**, 211–219.

Greig, S.M., Sear, D.A., and Carling, P.A. (2005). The impact of fine sediment accumulation on the survival of incubating salmon progeny: Implications for sediment management. *Science of the Total Environment*, **344**, 241–258.

Harper, D.M., Ebrahimnezhad, M., Taylor, E., et al. (1999). A catchment-scale approach to the physical restoration of lowland UK rivers. *Aquatic Conservation-Marine and Freshwater Ecosystems*, **9**, 141–157.

Hawkes, H.A. (1998). Origin and development of the Biological Monitoring Working Party score system. *Water Research*, **32**, 964–968.

Hemsley-Flint, B. (2000). Classification of the biological quality of rivers in England and Wales. In J.F. Wright, D.W. Sutcliffe, and M.T. Furse, eds., *Assessing the Biological Quality of Fresh Waters*, pp. 55–70. Freshwater Biological Association, Ambleside.

JNCC (2005). *Common Standards Monitoring Guidance for Ditches*. JNCC, Peterborough.

Kay, P., Edwards, A.C., and Foulger, M. (2009). A review of the efficacy of contemporary agricultural stewardship measures for ameliorating water pollution problems of key concern to the UK water industry. *Agricultural Systems*, **99**, 67–75.

Kröger, R., Moore, M.T., Locke, M.A., et al. (2009). Evaluating the influence of wetland vegetation on chemical residence time in Mississippi Delta drainage ditches. *Agricultural Water Management*, **96**, 1175–1179.

Larsen, S., Vaughan, I.P., and Ormerod, S.J. (2009). Scale-dependent effects of fine sediments on temperate headwaters invertebrates. *Freshwater Biology*, **54**, 203–219.

Magbanua, F.S., Townsend, C.R., Blackwell, G.L., Phillips, N., and Matthaei, C.D. (2010). Responses of stream macroinvertebrates and ecosystem function to conventional, integrated and organic farming. *Journal of Applied Ecology*, **47**, 1014–1025.

Mason, C.F. (1996). *Biology of Freshwater Pollution*. Longman, Harlow.

McLaren, R., Riding, A., and Lyons-Visser, H. (2002). *The Effectiveness of Ditch Management for Wildlife in the Broads and Somerset Levels & Moors ESAs*. ADAS, Wolverhampton.

McRae, S.E., D. Allan, J., and Burch, J.B. (2004). Reach- and catchment-scale determinants of the distribution of freshwater mussels (Bivalvia: Unionidae) in south-eastern Michigan, USA. *Freshwater Biology*, **49**, 127–142.

Millennium Ecosystem Assessment (2005). *Ecosystems and human wellbeing: wetlands and water synthesis*. World Resources Insitute, Washington, DC.

Milsom, T.P., Sherwood, A.J., Rose, S.C., Town, S.J., and Runham, S.R. (2004). Dynamics and management of plant communities in ditches bordering arable fenland in eastern England. *Agriculture, Ecosystems & Environment*, **103**, 85–99.

Mohr, S., Berghahn, R., Schmiediche, R., et al. (2012). Macroinvertebrate community response to repeated short-term pulses of the insecticide imidacloprid. *Aquatic Toxicology*, 110–111, 25–36.

Murray-Bligh, J.A.D., Furse, M.T., Jones, F.H., Gunn, R.J.M., Dines, R.A., and Wright, J.F. (1997). *Procedure for Collecting and Analysing Macroinvertebrate Samples for RIVPACS*. The Institute of Freshwater Ecology and the Environment Agency.

Natural England (2009). *Agri-environment schemes in England 2009. A review of results and effectiveness*. Natural England, Peterborough.

Natural England (2013a). *Entry Level Stewardship: Environmental Stewardship Handbook*. Natural England, Peterborough.

Natural England (2013b). *Higher Level Stewardship Environmental Stewardship Handbook*. Natural England, Peterborough.

Ormerod, S.J., Dobson, M., Hildrew, A.G., and Townsend, C.R. (2010). Multiple stressors in freshwater ecosystems. *Freshwater Biology*, **55**, 1–4.

Owens, P.N. and Collins, A.J. ed. (2006). *Soil Erosion and Sediment Redistribution in River Catchments: Measurement, Modelling and Management*. CABI, Wallingford, UK.

Pacini, N., Harper, D., Henderson, P., and Le Quesne, T. (2013). Lost in muddy waters: freshwater biodiversity. In D.W. Macdonald and K.J. Willis, eds., *Key Topics in Conservation Biology 2*, pp. 184–203. Wiley-Blackwell, John Wiley & Sons Ltd., Oxford.

Painter, D. (1999). Macroinvertebrate distributions and the conservation value of aquatic Coleoptera, Mollusca and Odonata in the ditches of traditionally managed grazing fen at Wicken Fen, UK. *Journal of Applied Ecology*, **36**, 33–48.

Palmer, M., Drake, M., and Stewart, N. (2010a). *A manual for the survey and evaluation of the aquatic plant and invertebrate assemblages of ditches, Version 4*. Buglife, Peterborough.

Palmer, M.A., Menninger, H.L., and Bernhardt, E. (2010b). River restoration, habitat heterogeneity and biodiversity: a failure of theory or practice? *Freshwater Biology*, **55**, 205–222.

Poole, A.E., Bradley, D., Salazar, R., and Macdonald, D.W. (2013). Optimising Agri-Environment Schemes to improve river health and conservation value. *Agriculture Ecosystems & Environment*, **181**, 157–168.

Rogers, R.D. and Schumm, S.A. (1991). The effect of sparse vegetative cover on erosion and sediment yield. *Journal of Hydrology*, **123**, 19–24.

Runde, J.M. and Hallenthal, R.A. (2000). Effects of suspended particles on net-tending behaviors for Hydropsyche sparna (Trichoptera: Hydropsychidae) and related species. *Annals of the Entomological Society of America*, **9**, 678–683.

Sardo, A.M. and Soares, A.M.V.M. (2010). Assessment of the effects of the pesticide imidacloprid on the behaviour of the aquatic oligochaete *Lumbriculus variegatus*. *Archives of Environmental Contamination and Toxicology*, **58**, 648–656.

Schäfer, R.B., Caquet, T., Siimes, K., Mueller, R., Lagadic, L., and Liess, M. (2007). Effects of pesticides on community structure and ecosystem functions in agricultural streams of three biogeographical regions in Europe. *Science of the Total Environment*, **382**, 272–285.

Scheierling, S.M. (1995). *Overcoming Agricultural Pollution of Water: The Challenges of Integrating Agricultural and Environmental Policies in the European Union*. World Bank Techincal Paper No. 269, Washington DC.

Shaw, R.F., Johnson, P.J., Macdonald, D.W., and Feber, R.E. (in press). Enhancing the biodiversity of ditches in intensively managed UK farmland. *PLOS ONE*.

State of Nature Partnership (2013). The State of Nature in the UK and its Overseas Territories. RSPB, Sandy, Bedfordshire.

Steegen, A., Govers, G., Nachtergaele, J., Takken, I., Beuselinck, L., and Poesen, J. (2000). Sediment export by water from an agricultural catchment in the Loam Belt of central Belgium. *Geomorphology*, **33**, 25–36.

Strand, R.M. and Merritt, R.W. (1997). Effects of episodic sedimentation on the net-spinning caddisflies *Hydropsyche betteni* and *Ceratopsyche sparna* (Trichoptera: Hydrpsychida). *Environmental Pollution*, **98**, 129–134.

Sweeney, B.W. (1993). Effects of streamside vegetation on macroinvertebrate communities of White Clay Creek in eastern North America. *Proceedings of the Academy of Natural Sciences of Philadelphia*, **144**, 291–340.

Telfer, S. (2000). The influence of dredging on water vole populations in small ditches *Scottish Natural Heritage Commissioned Report F99LF13*.

Vondracek, B., Blann, K., Cox, C., et al. (2005). Land use, spatial scale, and stream systems: lessons from an agricultural region. *Environmental Management*, **36**, 775–791.

Wang, L., Seelbach, P.W., and Lyons, J. (2006). Effects of levels of human disturbance on the influence of catchment, riparian, and reach-scale factors on fish assemblages. In R.M. Hughes, L. Wang, and P.W. Seelbach, eds., *Landscape Influences on Stream Habitats and Biological Assemblages*, pp. 199–219. American Fisheries Society, Bethesda.

WWAP (World Water Assessment Programme) (2012). *The United Nations World Water Development Report 4: Managing Water under Uncertainty and Risk*. UNESCO, Paris.

Williams, K., Green, D., and Pascoe, D. (1984). Toxicity testing with freshwater macroinvertebrates: methods and application in environmental management. In D. Pascoe and R.W. Edwards, eds., *Freshwater Biological Monitoring*, pp. 81–91. Pergamon, Oxford.

Williams, P., Whitfield, M., Biggs, J., et al. (2003). Comparative biodiversity of rivers, streams, ditches and ponds in an agricultural landscape in Southern England. *Biological Conservation*, **115**, 329–341.

Withers, P.J.A. and Lord, E.I. (2002). Agricultural nutrient inputs to rivers and groundwaters in the UK: policy, environmental management and research needs. *Science of the Total Environment*, **282**, 9–24.

Wright, J.F., Sutcliffe, D.W., and Furse, M.T., eds. (2000). *Assessing the biological quality of freshwaters: RIVPACS and other techniques*. Freshwater Biological Association, Ambleside, UKAccess 2000.

CHAPTER 12

Local and landscape-scale impacts of woodland management on wildlife

Christina D. Buesching, Eleanor M. Slade, Thomas Merckx, and David W. Macdonald

The Wild Wood is pretty well populated by now; with all the usual lot, good, bad, and indifferent – I name no names. It takes all sorts to make a world.

The Wind in the Willows by Kenneth Graeme

12.1 Introduction

The typical 'farmscape' in Britain consists mainly of cropped fields and pastures, but also comprises a diverse range of other habitats, including semi-natural woodlands and hedgerows. Within many landscapes, these wooded elements are the only habitats reminiscent of the original forest ecosystem that developed after the last ice-age (Rackham 2003) and, as a result, many species still found in the wider British countryside depend on them for food (e.g. Merckx et al. 2012), shelter (e.g. Merckx et al. 2008), nesting opportunities (e.g. Macdonald et al. 2004), or dispersal (e.g. Slade et al. 2013). While single trees are sufficient sanctuaries or stepping stones for some species, others (including some invasive species, Box 12.1) need larger woodland patches or hedgerows and woodland corridors to facilitate their dispersal. Mitigation of forest fragmentation and appropriate woodland management are thus key topics in conservation-friendly farming (Buechner 1989; Buesching et al. 2008; Merckx et al. 2012; Slade et al. 2013), and in conservation work globally (Akcakaya et al. 2007; Cushman et al. 2013).

Depending on their mobility and dispersal patterns, the responses of different species and taxa to habitat management practices vary. Whereas some species are very sensitive to localized small-scale changes in their environment, others are affected predominantly by management at a landscape scale. Small mammals, for example, do not regularly traverse distances further than 100 m (Buechner 1989), and thus their local abundances are likely to respond sensitively to changing local habitat parameters. They are particularly sensitive to changes in cover from predators and in food availability due to habitat management and grazing pressure (Buesching et al. 2008; Buesching et al. 2011; Bush et al. 2012), making them a suitable model to study the impacts of small-scale habitat management. In Chapter 4, this volume, we have shown how, on farmland, habitat management at the local (i.e. field) scale affects wood mouse *Apodemus sylvaticus* behaviour and population densities. Similarly, the field-scale management of hedgerow trees and field margins has been shown to affect predominantly the abundance and species richness of less mobile moth species, whereas mobile moth species are likely to be more affected at the landscape scale (Chapter 8, this volume).

In this chapter, we will use these same taxonomic groups to look at the effects of different aspects of managing woodlands within the British 'farmscape'. First, we review how the impacts of management of individual woodlands affect small mammal distribution patterns and population densities. Then we investigate how the distribution, extent, and management of woodlands within the landscape influence the diversity, abundance, and movements of macro-moths.

Both taxa are of ecological importance in farmland ecosystems for three major reasons. Firstly, small mammals support a number of larger (and sometimes) endangered carnivores (e.g. stoats *Mustela erminea*, weasels *Mustela nivalis*, red foxes *Vulpes vulpes*, barn owls *Tyto alba*, tawny owls *Strix aluco* (Box 12.3), and buzzards *Buteo buteo* (see Macdonald and Feber 2015): Chapter 10; Harris et al. 2000), while moths

Wildlife Conservation on Farmland. Managing for Nature on Lowland Farms. Edited by David W. Macdonald and Ruth E. Feber.
© Oxford University Press 2015. Published 2015 by Oxford University Press.

are vital prey for bats (see Chapter 9, this volume), and birds (e.g. blue tits *Cyanistes caeruleus* eat a minimum of 35 billion caterpillars each year in Britain alone: Fox et al. 2006). Secondly, both taxa are good indicators for environmental health (small mammals: Harrington and Macdonald 2002; moths: Fox 2013), as well as for general habitat degradation and forest fragmentation (small mammals: Macdonald et al. 1998; moths: Summerville and Crist 2004). Thirdly, both taxa are implicated in woodland regeneration, as small mammals are important seed dispersers for many woodland tree species (e.g. Watts 1968), while some plants rely on moths for pollination (e.g. Devoto et al. 2011). In addition, they are well studied in Wytham (Buesching et al. 2010; Hambler et al. 2010).

Box 12.1 Aliens versus natives: can red and grey squirrels co-exist?

The North American grey squirrel *Sciurus carolinensis* was introduced into Britain in the late nineteenth century, and since its spread, the native red squirrel *Sciurus vulgaris* has experienced a dramatic loss of range. In general, when 'greys arrive, reds disappear'; according to the Forestry Commission (2013) there are only 140 000 red squirrels left in Britain, versus 2.5 million grey squirrels. Red squirrels are now restricted to a few island populations in the south of England, pockets in mid and north Wales, northern counties in England, and in Scotland, where an estimated 75% of all British red squirrels are found (Scottish Wildlife Trust 2013). Several factors have been involved in the ecological replacement of red squirrels by greys (Tomkins et al. 2003). The pattern of replacement has indicated that competition for habitats and food resources have played a major role (Usher et al. 1992). In addition, there is growing evidence that the squirrel-specific parapox virus has accelerated rates of competitive replacement in some areas (Rushton et al. 2006).

Interestingly, in the absence of the parapox virus, the native and alien species have been able to co-exist in some areas for prolonged periods. Since habitat composition may be the determinant of short- versus long-term co-existence, Jenny Bryce, leading a WildCRU team, aimed to ascertain similarities in patterns of woodland habitat use by both species. The work focused on the Craigvinean forest in Perthshire, Scotland, a commercial coniferous forest with riparian corridors of mixed broadleaves, where red and grey squirrels had co-existed for over 30 years at the time of our study in the late 1990s.

We radio-tracked 32 red and 34 grey squirrels and found a 77% interspecies dietary niche overlap based on their use of individual tree species, suggesting that the similarity of resources used was likely to preclude their co-existence (Bryce et al. 2002). However, due to the structure of woodlands in the landscape, the habitats within their home ranges overlapped by only 59%. Red squirrels selected home ranges within conifer areas (i.e. favouring Norway spruce *Picea abies* and Scots pine *Pinus sylvestris*), which were avoided by grey squirrels (population densities: 1.63 reds/ha versus 0.08 greys/ha), while grey squirrels used mixed conifer and broad-leaved habitats throughout the year, which red squirrels only used in autumn and winter (0.92 reds/ha and 0.88 greys/ha). Our results suggest that this partitioning of macrohabitats may have reduced interspecific competition between the species, allowing long-term (now 50+ years) co-existence in areas of extensive conifers with isolated patches of broad-leaved woodland. Hence, persistence in a wooded landscape is likely to depend on the extent and productivity of the patches of broadleaved woodland, with red squirrels being predicted to be out-competed at moderate grey squirrel densities (Wauters et al. 2002).

In addition, our studies revealed a near two-fold difference in body weight between the two species (red *c.* 300 g and grey *c.* 570 g; Bryce et al. 2001), and this difference in body mass can help to explain the greys' competitive advantage. By examining their relative daily energy expenditure (which increased with body mass), we estimated that one grey squirrel uses 1.65 times the energy of one red squirrel (Bryce et al. 2001). Social dominance in squirrels is driven by body size, hence we would expect grey squirrels to control access to, and consume relatively more of, the available food resources. According to some studies, these competitive advantages manifest through different rates of juvenile recruitment, rather than breeding rates or adult survival (Gurnell et al. 2004). Hence, conifer habitats with little or no broad-leaved woodland, which support relatively low densities of squirrels, may provide red squirrels with a refuge from competition with grey squirrels (Bryce et al. 2005). Interestingly, the mix of woodland types in the landscape is also thought to be important in determining disease transmission through habitat connectivity, and via squirrel population densities influencing encounter rates (White et al. 2014).

In conclusion, squirrels are relatively mobile species, but tend to stay close to least-cost corridors (Stevenson et al. 2013). Thus, it is habitat composition and connectivity at a landscape level, rather than overall forest extent, which are likely to be of crucial importance in this ongoing challenge from the introduced congener.

12.2 Local effects of deer browsing on small mammals

Wytham Woods are home to one of the longest-running small mammal studies in the world. Originated by H.N. Southern in 1943, and inspired by Charles Elton's (1924) on-site research into rodent ecology and population dynamics, small mammals have been live-trapped twice annually (at the end of May/beginning of June, and in December) at the same two sites, located in secondary deciduous woodland in Wytham's Great Wood, for over seventy years (Buesching et al. 2010). Capture-mark-recapture (CMR) techniques allow their numbers to be estimated, and reveal the effects of environmental factors on their population densities. Long-term monitoring of their abundance and distribution has thus contributed to a detailed understanding of their uses of, and preferences for, particular habitats. Such information, in turn, constitutes an important pre-requisite for the design of effective conservation and woodland management strategies.

Over the past 200 years, changes in climate, farming, forestry, and hunting practices, as well as the extinction of their natural predators (e.g. wolves and lynx: Sandom et al. 2013), have resulted in marked increases in deer numbers across the UK (see Box 12.2; Fig. 12.1), mainland Europe (Fuller and Gill 2001), and North America (McCabe and McCabe 1974). These have highlighted a suite of associated conservation issues ranging from intra-guild competition (Bartos et al. 2002), to parasite spread (Bindernagel and Anderson 1972), and human conflict (e.g. zoonotic diseases: Spielman 1994; agricultural damage: Putman and Moore 2002; Road Traffic Accidents: Putman 1997). The impacts of high densities of deer on woodland vegetation can be extreme (Morecroft et al. 2001; Savill et al. 2010). The shrub layer, especially bramble *Rubus fruticosa*, and many woodland forbs, are reduced by increased browsing pressure (Gill 2000), woodland plants are damaged by trampling (Gill and Beardall 2001), and forest regeneration is arrested (Kirby 2001). Thus, heavy deer grazing can alter the structure of the whole forest, with cascading effects on a wide variety of other species at all trophic levels (Fuller and Gill 2001). Invertebrate communities are affected by changes in forest vegetation; for example, the caterpillars of the peach blossom moth *Thyatira batis* feed almost exclusively on bramble in late summer (Pollard and Cooke 1994; Stewart 2001; Littlewood 2008; Merckx et al. 2012). The lack of cover from predators can lead to the elimination of ground-nesting birds, such as nightingales (Gosler 1990) and small mammal

populations may become less species diverse (Putman et al. 1989) and less abundant (Smit et al. 2001), resulting in fewer avian and mammalian predators at higher trophic levels (e.g. small mustelids: King 1985; tawny owls: Southern and Lowe 1982; Box 12.3).

Our long-term data set on small mammal populations in Wytham Woods, together with the evolving deer management situation described in Box 12.2, provided a unique opportunity to study not only the effects of high deer population densities on woodland rodents, but also the recovery of woodland ecosystems after the removal of known numbers of deer over a period of several years.

In woodlands throughout the UK and most of Europe, the two most common small mammal rodent species are the wood mouse *Apodemus sylvaticus* and the bank vole *Myodes glareolus* (formerly *Clethrionomys glareolus*). They occupy similar ecological niches throughout their range and constitute the prey base for many woodland predators (Flowerdew 1993). In both species, adults typically weigh between 15 and 25 g, and breed mainly between April and October, with

Figure 12.1 Roe deer *Capreolus capreolus*. ©A.L. Harrington.

Box 12.2 Deer management in Wytham Woods

Wytham Woods (total woodland area 425 ha: for a detailed description of the site see Savill et al. 2010; Buesching et al. 2010) are home to three species of deer: (1) the native roe deer *Capreolus capreolus* (Fig. 12.1), which, by the mid eighteenth century, had been extirpated from southern England by over-hunting, and only recolonized Oxfordshire in the 1970s (Ward 2005), supplemented by some imports from the continent (Macdonald and Burnham 2010); (2) the fallow deer *Cervus dama*, which was first introduced into royal hunting grounds by the Normans in the eleventh century and is now considered to be naturalized in England; and (3) the Asian muntjac *Muntiacus reevesi*, which was introduced from China and Taiwan to Woburn Abbey Park by the Victorians at the end of the nineteenth century, and was dispersed to other collections and menageries across southern England, from where animals escaped and spread widely, colonizing Wytham in the 1960s (Elton 1966).

Deer of all three species rely heavily on forest cover for resting sites but, if given the choice, roe and fallow deer graze preferentially in arable fields. The crop damage on farms adjacent to Wytham Woods forced Oxford University to erect a 2.5 m high deer fence along the 7 km perimeter of the Woods in 1987–1988, effectively fencing in the roe and fallow deer population, although the smaller muntjac could continue to migrate under, or even through, the fence. The absence of any natural predators, and only limited control of their numbers by culling, permitted steady increases in the population densities of deer in Wytham Woods throughout the 1990s until, in 1998, the combination of trampling, overgrazing and browsing, and shading by the dense canopy cover, had eliminated much of the ground cover vegetation (Morecroft et al. 2001), seriously affecting populations of birds (Gosler 1990) and small mammals (Flowerdew and Ellwood 2001), as well as the body condition of the deer themselves (Ellwood 2007). Thus, with the assistance of the British Deer Society, a rigorous culling programme was implemented in the winter of 1998/1999 (Perrins and Overall 2001), which is still in effect.

Empirical deer counts were first conducted in Wytham by Taylor and Morecroft (1997) using dung-count methods. The deer control strategy, however, facilitated the rare opportunity to calibrate different methods to estimate deer numbers by sampling before and after a known number of deer were removed from an effectively closed population (Ellwood 2007; Buesching et al. 2014). Between 1998 and 2003, WildCRU's Stephen Ellwood compared the statistical precision (reliability) and estimate accuracy (using known

population size determined by cohort analysis) of two dung-counting methods with observational ('distance') sampling, using a thermal imager at night, and concluded that distance sampling was more precise and accurate than dung counts for all three deer species (Ellwood 2007). However, as distance sampling requires expensive, specialized equipment not available to the average landowner, we also evaluated the accuracy of dropping counts. We found in a large-scale study over ten years that, with appropriate validation and calibration, faecal standing crop counts (Macdonald et al. 1998) can deliver accurate density estimates for larger deer species, such as roe and fallow deer, which can be calibrated against cull figures (Buesching et al. 2014). Faecal pellet group counts are, however, unsuitable for estimating densities of the small, pair-living, and territorial muntjacs, as they deposit their droppings in latrines rather than randomly throughout their range.

In Wytham, winter culling reduced the fallow population from an estimated 440 animals (103.5/100 ha) in 1998 to 35–40 individuals in 2003 (8.5/100 ha), and muntjac from *c.* 200 (47/100 ha) to 15 (3.5/100 ha). Interestingly, historically, the density of the roe deer population had been comparatively low in Wytham Woods, estimated at 2/100 ha, until the marked decrease in fallow, imposed by culling. This was followed by a sudden significant increase in roe numbers during the reproductive season of 2000 to almost double their previous densities (Savill et al. 2010). Thus, culling of roe commenced over the winter of 2000/2001.

As global warming is predicted to result in warmer and wetter conditions for most parts of the UK, we investigated the effects of population density and weather conditions on deer population dynamics using post-mortem examination of culled fallow and muntjac deer (Ellwood 2007). Our study showed that: (i) in both species, pregnancy rates were unaffected by population density, because most females became pregnant at their earliest opportunity; (ii) fallow fawns were smallest at high densities; and (iii) although fallow fawns and yearlings continued to grow, body condition of individuals of all ages deteriorated over the winter. This effect was exacerbated by high population density, but minimized in warm and wet winters. Muntjac, on the other hand, showed no weather dependency in their body condition (Buesching et al. 2010). Thus, as temperate woodlands are likely to support greater densities of deer in the future, due to predicted changes in weather patterns, present-day forest ecosystems could potentially be altered considerably (Newman and Macdonald 2013).

females establishing mutually exclusive breeding territories. However, their respective survival strategies differ significantly. Whereas the predominantly herbivorous and more sedentary bank vole relies on dense understorey for protection from predators (Fitzgibbon 1997), the wood mouse has well-developed olfactory, visual, and auditory senses, as well as considerable agility. These character traits allow flighty escape from predators (Hansson 1985), affording wood mice the freedom to forage for a more omnivorous diet, including seeds, berries, and up to 15% insects (Watts 1968). Nevertheless, while bank voles are active night and day under the protective shrub layer, wood mice are largely nocturnal, foraging under the cover of darkness (Flowerdew 1993).

Historically, the numbers of bank voles in Wytham Woods have always exceeded those of wood mice. However, while wood mouse numbers, albeit subject to inter-annual population cycles, stayed within the same density range over the past seven decades, bank voles showed a marked decline in their population density to approximately half their original numbers in the late 1980s, followed by a gradual increase from *c.* 2002 onwards (Flowerdew and Ellwood 2001; Buesching et al. 2010). While the decrease in bank vole numbers coincided with the substantial increase in deer numbers in the 1990s, after the erection of the deer fence in 1987 (see Box 12.2), their recovery appears to coincide with the regeneration of understorey and forest vegetation after the stringent deer control measures took effect (Buesching et al. 2010; 2014; Bush et al. 2012). Thus, small rodent numbers appear to be linked to the variation in deer-grazing pressure (Flowerdew and Ellwood 2001), and resultant changes in woodland vegetation (Morecroft et al. 2001; Buesching et al. 2011; Bush et al. 2012).

Empirical studies comparing small mammal communities in heavily deer-grazed versus deer-free areas within the same woodland are rare, as they are usually defeated by logistical challenges. In Wytham Woods, we were fortunate that the Environmental Change Network (ECN) established four deer exclosures of approximately 1 ha each in different parts of the Woods in 1998, specifically to study the effects of deer grazing on woodland vegetation (Morecroft et al. 2001). Each exclosure was delimited by a 2.5 m high deer fence, which excluded all deer whilst permitting free access to mice and voles. The absence of grazing pressure has led to a much denser understorey inside the exclosures than in the rest of the woodland, manifesting itself predominantly in increased bramble cover (Morecroft et al. 2001; Buesching et al. 2011).

Between June 2001 and September 2003, we compared the population densities of wood mice and bank voles inside the deer-free exclosures with those recorded in the open woodland subjected to deer grazing (Buesching et al. 2011). With the help of Earthwatch volunteers (see Macdonald and Feber 2015: Chapter 15) we carried out 16 live-trapping sessions, each lasting four days, during which we placed 50 Longworth traps (Fig. 12.2) inside one of the four deer exclosures and 50 traps in the surrounding woodland, rotating between the four different exclosures. Throughout the year, capture-mark-recapture estimates indicated that, on average, there were four times more bank voles than wood mice inside the exclosures, whereas the open woodland supported 1.6 times as many wood mice (Fig. 12.3a) as bank voles (Fig. 12.3b). This result illustrates neatly how strongly bank voles rely on thick vegetation, and how they benefit from reduced browsing pressure by deer. By contrast, wood mice use their acute senses and agility to detect, and evade, predators, allowing them to exploit also resources in the open woodland without jeopardizing their survival.

Figure 12.2 Longworth trap fixed in a tree. Photograph ©Karrie Langdon.

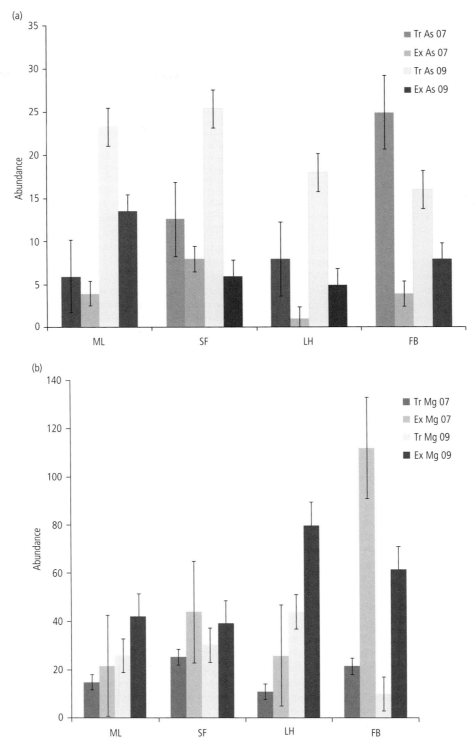

Figure 12.3 Comparison of population densities (± STD) of (a) wood mice (As) and (b) bank voles (Mg) in transects (Tr) and exclosures (Ex) in four different exclosures (Marley: ML; Swinford: SF; Lower Ash Hill: LH; Firebreak: FB) trapped in July (07) and September (09) (data from 2003). From Buesching et al. (2011). Reproduced with permission from Elsevier.

In contrast to previous studies (e.g. Yoccoz and Mesnager 1998), we found no difference in the sex ratio or age structure of either species between the open woodland and the exclosures, nor any differences in terms of body weight and/or reproductive status, which would equate to differences in reproductive fitness (Buesching et al. 2011). These results, combined with those of Hansson (1992), who found that in both species smaller females had greater reproductive fitness at high densities than did heavier individuals, provide further evidence that the commonly assumed positive correlation between body weight and reproductive fitness may not apply to cyclic microtine populations.

In the summer of 2010, we repeated this study (Bush et al. 2012). Although the densities of small mammals in different areas of Wytham Woods varied with the overall vegetation cover, our data confirmed that, after a decade of rigorous deer control, bank voles were spreading back into the open woodland, and had reached densities comparable to those inside the exclosures (i.e. between 10 and 30 animals/ha, depending on woodland area), with no significant differences recorded in population structure between grazed and ungrazed areas.

Both of these studies also provide some evidence for the previously suggested inter-specific competition, avoidance, and niche separation between wood mice and bank voles (Gipps 1985). While other studies propose that only bank voles will modify their behaviour to avoid competition with wood mice (Gipps 1985; Gurnell 1985), the results from our research in Wytham Woods suggested that wood mice might also change their behaviour. Wood mouse numbers have been reported to increase if deer are excluded (Smit et al. 2001), or if food resources are increased (Putman et al. 1989). However, under heavy grazing pressure in Wytham, wood mice occurred at approximately three times higher densities (3.37 ± 0.71 SE) in the open woodland than inside exclosures (Buesching et al. 2011).

Aside from predators, small mammal population densities are also heavily affected by food availability. The Habitat Saturation Hypothesis (Selander 1964) postulates that their populations are regulated by available resources in a density-dependent way (Plesner-Jensen 1996). Using behavioural observations and radio-tracking, our research at Wytham established that wood mice are not only capable of detecting and remembering rich food patches (Macdonald et al. 2006), but that the size of their territories correlates negatively with the richness of food resources.

We also found that wood mice and bank voles show a further degree of niche separation in their use of the three-dimensional forest structure (Buesching et al. 2008). In this study, half of the traps were placed on the ground and half were fixed in shrubs and trees between heights of 30 cm and 2.5 m (Fig. 12.2). The study confirmed that arboreality is common in small mammals: an average of one in five wood mice, and one in ten bank voles, were caught in trees. Although almost 90% of these arboreal animals were caught at heights of 50–100 cm above the ground, some reached heights of up to 2.2 m. While the majority (68%) of mice caught in trees were male, in bank voles, most arboreal individuals (72%) were female.

In both species, arboreality was correlated positively with population density. However, whilst bank voles were found to be arboreal only in July and August, wood mice were found above ground throughout the reproductive season, with males being caught in trees approximately twice as often as females. Dense intertwining vegetation, for instance provided by bramble, hawthorn *Crataegus monogyna*, hazel *Corylus avellana*, and elder *Sambucus nigra*, increased arboreality in both species, probably since these plants provide good branch networks for runways at the preferred vegetation height.

Although many hypotheses have been suggested to explain arboreality in small mammals (for review, see Buesching et al. 2008), we concluded that it can best be explained by a combination of factors. It is likely that only wood mice can afford to be truly arboreal, exploiting food resources (e.g. insects, berries) above ground, because only mice are sufficiently swift and agile to escape predators whilst running along thin branches. As the mating system of wood mice involves one male territory overlapping that of several females (female-defence polygyny: Tew and Macdonald 1994), it is likely that male territory holders use trees, not only during foraging trips, but also as direct routes and over-passes between different, and not necessarily contiguous, female territories. Bank voles, on the other hand, may be driven to arboreality only in times of high population densities and little food availability on the ground (i.e. in summer) to exploit (unripe) seeds and flowers above ground. Even then, they are unlikely to climb as high as wood mice due to their lack of agility and their reliance on dense vegetation for protection from predators (Buesching et al. 2008). Thus, reduction in the understorey layer, either through over-grazing by deer, or through well-established forest management practices, such as large-scale felling operations and brush cutting, has the potential to alter small mammal

---

Box 12.3 The effects of fluctuations in field vole populations on tawny owls *Strix aluco*

Fluctuations in prey populations affect abundance and population dynamics of their predators. For example, tawny owls *Strix aluco* predate small mammals (Southern and Lowe 1982) such as bank voles, wood mice, and field voles *Microtus agrestis* (see Chapter 4, this volume), as well as eating earthworms *Lumbricus terrestris*. Small mammal densities follow cyclical abundance patterns, and thus vary in their availability as prey for tawny owls on a seasonal as well as on an inter-annual basis, while earthworm distribution depends on soil micro-climate, and thus varies in availability as prey for tawny owls on a nightly basis (Macdonald 1976).

Studying tawny owls in Kielder Forest, Northumberland, in 1994 and 1995, a team led by WildCRU's Bridget Appleby investigated how changes in the abundance of field voles can affect the survival and parasite loads of tawny owls, and how parents can maximize their fitness by adjusting the sex ratio of their offspring in response to differences in resource availability (Appleby et al. 1997; 1999). During these two years, Appleby et al. determined the sex (through DNA fingerprinting of microsatellites in blood samples) of chicks from nests where all eggs hatched. Field vole abundance was estimated for each owl territory, based on the presence of fresh grass clippings in vole runs in March and June at 20 survey sites distributed throughout the study area. Although local field vole abundance fluctuates inter-annually following a three- to four-year cycle, these cycles can be asynchronous between neighbouring populations (Petty 1992), and indeed, during the research period at Kielder, field vole abundance increased in some parts of the forest, but simultaneously decreased in others. In 1994, vole abundance (and thus owl food) increased in all study territories, resulting in a low chick mortality between hatching and fledging of only 15% (N = 116). However, 1995 was a year of much lower vole abundance, and chick mortality increased to 31% (N = 70) (Appleby et al. 1997). Interestingly, female reproductive success (i.e. number of fledglings) was correlated with vole abundance in their respective territories only in 1995 (Appleby et al. 1997). The results therefore suggest that, in general, high abundance of prey (in 1994) allows even poorer individuals to breed successfully, while only high-quality individuals breed successfully when vole numbers are low (as in 1995).

As tawny owls are sexually dimorphic, with females weighing 20% more than males (Snow and Perrins 1998), rearing females implies a larger parental investment in acquiring food resources. Parents should thus produce more of the heavier sex when resources are most plentiful. Indeed, during both years, territories with higher abundances of field voles in June (when chicks fledge) supported a larger proportion of female chicks within a brood (Appleby et al. 1997). This appeared to be a flexible response to the prevailing supply of field voles in the territory in that year, rather than being a consequence of sex-biased mortality in eggs or nestlings. So, are owls able to predict future food supplies? Since tawny owls lay eggs in March, the results indicate that owls might be able to predict the future abundance of field voles, and respond by laying female-biased clutches, which are most likely to gain a long-term benefit when resources are good (Appleby et al. 1997). Possible explanations for this 'predictive ability' are the detection of pregnant and lactating vole females, the size of voles, which are larger on the increasing phase of multi-annual population cycles, and/or the detection of the voles' reproductive hormones (Appleby et al. 1997).

Furthermore, Appleby et al. (1999) also measured the prevalence of blood parasites in parent birds as well as chicks. They found that field vole availability had both a short-term effect on owls, in which birds suffering a decline in food abundance in their territories between the two study years showed an increase in parasite load, and a long-term effect, which was reflected in higher parasite burdens of adult owls that had experienced poorer food supply as chicks (Appleby et al. 1999). In conclusion, the results show that fluctuations in small mammal populations affect the quantity, quality, and sex of birds at higher trophic levels.

---

population structure, behaviour, and inter-specific interactions. As small mammals are important seed dispersers, forest regeneration may be impaired if rodent populations drop below a minimum level (Watts 1968), causing cascade effects throughout the ecosystem, and repressing population densities of many avian and mammalian predators (e.g. weasels, see Box 4.8 in Chapter 4, this volume; owls, Box 12.3).

12.3 Effects of woodland management on moths

English semi-natural, broad-leaved lowland woodlands, such as Wytham Woods, are often sheltered biotopes contained within highly exposed agricultural landscapes. Woodlands provide Lepidoptera (i.e. butterflies and moths) with habitat resources, including larval foodplants, nectar sources, and shelter (Usher and Keiller 1998; Summerville and Crist 2004; Dennis 2010). Nevertheless, many of these resources have been lost during the last half of the twentieth century, with the replacement of traditional coppicing through intensive forest management (Warren and Key 1991; Gorissen et al. 2004). This has led to losses of a significant proportion of butterflies from woodlands (van Swaay et al. 2006), including both woodland specialists and species of open woodland (Gorissen et al. 2004).

Our recent macro-moth conservation research has concentrated on farmscapes (see Chapter 8, this volume). However, populations of widespread macro-moth species are declining and the causes are likely to include factors other than management of farmland (Fox 2013). An estimated 60% of widespread moth species depend on semi-natural woodlands (Young 1997). Some species are exclusively restricted to dense woodland (e.g. the lobster moth *Stauropus fagi*, Fig. 12.6a, which received its English name from the remarkable crustacean-like appearance of the caterpillar), while others (e.g. rosy marbled *Elaphria venustula*) depend on open woodland complexes such as rides, heaths, glades, and scallops (Waring and Townsend 2009). However, these latter associations have rarely been quantified (see Merckx et al. 2012). The impacts of contrasting woodland management are only well understood for butterflies (Warren and Thomas 1992; Hodgson et al. 2009), with studies on moths being scarce and limited in terms of scale and re-sampling (Broome et al. 2011).

We conducted a large-scale study to explore whether woodland management, at the scale of the woodland patch, had the potential to halt and/or reverse macro-moth declines within woodlands for both rare and localized species of conservation concern and for widespread, yet declining species (Merckx et al. 2012). First, the study aimed to test how macro-moth abundance and species richness were affected by two common woodland conservation management practices, coppicing and ride management. Second, we assessed whether species groups of different conservation status reacted differently to these management practices, and if such variations depended on group/species ecology. Finally, we suggested clear recommendations for woodland conservation management that allowed for management practices that simultaneously catered for different ecological and conservation status groups of Lepidoptera.

The light-trap experiment was conducted in the Tytherley woodland landscape, a landscape of *c.* 17 000 ha, containing 98 woodland patches totalling 2500 ha of ancient semi-natural woodland, near Salisbury, UK. We compared the presence/absence, abundance, and species richness of macro-moths at 36 fixed trap sites located within the woodland. Trap sites were set within six experimental 'woodland management' treatments: hazel coppice (young: 1–2 yrs; medium: 3–6 yrs; old: 7–9 yrs), ride widths (wide: > 20 m; standard: < 10 m), and 'standard woodland' of non-coppiced, sheltered, high deciduous oak forest (at least 60 yrs old with some 300 yr-old individual trees). A total of 11 670 individuals from 265 macro-moth species were recorded, including 15 'Scarce/Red Data Book (RDB)' species (249 individuals), 38 'Common Severely Declining' species (891 individuals), and 90 'Common Declining' species (3564 individuals)[1].

Overall, numbers of both individuals and species of macro-moth were higher in the late-successional high deciduous forest biotope of 'standard woodland', which is characterized by higher levels of shelter, darkness, and humidity, compared to coppiced sites (Fig. 12.4). Shelter affects levels of convective cooling (Merckx et al. 2008), aiding the capacity of these endothermic insects to increase their temperature, and, as a consequence, their activity level, which may in part explain the higher overall abundance and species richness within sheltered woodland. Sheltered woodland habitats were especially important for scarce/RDB species (Fig. 12.5) such as the UK BAP species light crimson underwing *Catocala promissa*, and the nationally scarce great oak beauty *Hypomecis roboraria*, which was mirrored by the observation that 60% of species in this group (N = 15) were strict woodland species.

At the same time, we showed that coppicing and ride widening (> 20 m), which open up dense forest structures and which are practices often employed for butterfly conservation, are also indispensable woodland management tools to conserve other sets of macro-moth species. For instance, 18% (i.e. 49 species) of all our recorded species were fully restricted to sites with coppice/wide ride management.

The study also demonstrated that the mechanism behind the pattern of increased species richness at the woodland scale involved an increase in structural heterogeneity of plant species (e.g. in wide rides) and, hence, an increased micro-climatic and resource diversity for species with an affinity for more open biotopes. This increased habitat diversity especially benefited

---

[1] 'Common Severely Declining' and 'Common Declining' species show national declines > 69% and 0–69%, respectively, over a 35-year period (Conrad et al. 2006).

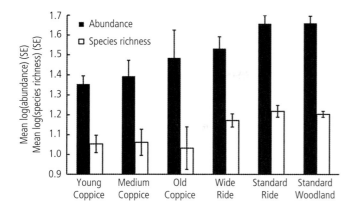

Figure 12.4 Overall macro-moth abundance (number of individuals) and species richness (number of species) ($\log_{10}$-transformed mean (SE)) for six experimental woodland management treatments. From Merckx et al. (2012). Reproduced with permission from Elsevier.

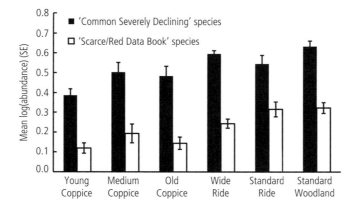

Figure 12.5 Abundance ($\log_{10}$-transformed mean (SE)) of two high profile conservation status species groups, for six experimental woodland management treatments. 'Scarce/RDB' species followed the overall trend with overall macro-moth abundance highest in standard rides and standard woodland and lowest in coppice and wide rides, while the 'Common Severely Declining' species, by contrast, were relatively abundant at wide woodland rides, while showing the smallest difference in abundance between the woodland management treatments. From Merckx et al. (2012). Reproduced with permission from Elsevier.

Common Severely Declining species, with ten (of 38) species of this group only found at open sites (e.g. figure of eight *Diloba caeruleocephala*). Wide woodland rides specifically benefited this group in terms of abundance (Fig. 12.5).

Based on these findings, we recommend a two-tier approach of zoning woodland conservation, which involves a combination of management practices that enlarge existing dense forest cores, while buffering those cores with lighter woodland zones (containing coppicing and wide woodland rides). This would allow for the differential value of successional stages for different ecological groups of Lepidoptera, and would

hence be beneficial for both woodland specialists and species of open habitats.

Importantly, this study also demonstrated that the size of the woodland patch surrounding coppiced plots was a key factor that positively affected abundance and species richness of nationally declining and severely declining species of moths in coppiced plots (Merckx et al. 2012). Thus, the woodland size/scale where the coppice management effort takes place matters to achieve maximum conservation outputs for moths. Our results suggested that larger woodlands offer more cost-effective opportunities to increase biodiversity through active coppice management.

12.4 Landscape-scale effects of woodland fragmentation on moths

Habitat connectivity, and in particular hedgerows and hedgerow trees, have been shown to be important for moth mobility at the field and farm scale (Merckx et al. 2010; Chapter 8, this volume). At the landscape scale, movement patterns of moth species associated with woodland are often restricted by forest fragmentation and the lack of functional corridors in agricultural landscapes. However, the impacts of fragmentation may also depend on the mobility of individual moth species (Merckx et al. 2010; Slade et al. 2013). We conducted the first investigation into the movement patterns of multiple species of moths on a landscape scale, by means of a large-scale capture-mark-recapture (CMR) study (Slade et al. 2013; see Macdonald and Feber 2015: Chapter 15). We examined how both life-history traits and landscape characteristics can be used to predict macro-moth responses to forest fragmentation, and the importance of woodland patch size, hedgerows, and single trees for movements of forest-associated moths within agricultural landscapes.

Enlisting the help of volunteers (Macdonald and Feber 2015: Chapter 15), we placed 44 six-watt actinic light traps across Wytham Woods, the surrounding forest fragments, and the landscape matrix in between, covering a 2 km radius from the centre of Wytham Woods. Captured moths from 89 species were individually marked on the hindwing with a unique identifying number using a permanent marker pen (Fig. 12.6a), and then released. This CMR experiment ran every night throughout most of June and July 2009. Nearly 15 000 moths were marked during the study, including several rare and UK BAP species, such as a healthy population of the nationally declining garden tiger *Arctia caja*, a species thought to be affected by climate change (Conrad et al. 2002).

Previous studies have shown that many moth species move at scales larger than the field scale (Merckx et al. 2009; 2010; Chapter 8, this volume), with individuals able to cover several kilometres a night. The longest movement distance we recorded was for a broad-bordered yellow underwing *Noctua fimbriata*, which was recaptured outside the study area, having moved 13.7 km in two months. However, the majority of species were predicted to move less than 500 m in a week, with the range increasing with wingspan. Wingspan, wing shape, adult feeding, and larval feeding guild all predicted macro-moth mobility. For moths with weak or strong forest affinity, those species with pointed wings moved further than those

with broad wings. Moreover, the most mobile species were those with pointed wings, rather than those with the largest wingspan (e.g. the weekly movement rate of green arches *Anaplectoides prasina*—48 mm wingspan, pointed wings—was 1207 m, whereas that of the garden tiger *Arctia caja*—70 mm wingspan, rounded wings—was 297 m). The three species with the largest predicted movement rate, i.e. lobster moth *Stauropus fagi* (Fig. 12.6a), green arches *Anaplectoides prasina* (Fig. 12.6b), and scarce silver-lines *Bena bicolorana* (Fig. 12.6c), were all forest species with pointed wings. Moths that fed as adults, and moths with shrub/tree-feeding larvae, had larger predicted movement rates than those that did not feed as adults or with grass/herb-feeding larvae. This pattern was similar for both forest specialists and ubiquitous species.

In terms of landscape structure, we found that both forest size and connectivity were important (Slade et al. 2013). Solitary trees and small fragments functioned as 'stepping stones', especially when their landscape connectivity was increased by being positioned within hedgerows or within a favourable matrix. Mobile forest specialists, such as the lobster moth, were most affected by forest fragmentation despite their high intrinsic dispersal capability. These species were confined mostly to the largest of the forest patches due to their strong affinity for the forest habitat, and thus were also heavily dependent on forest connectivity in order to cross the agricultural matrix. The majority of species with a strong forest affinity were first captured in the large fragments (> 5 ha) (94% of individuals and 89% of species), and were then only recaptured in the largest fragments, or rarely at hedgerow trees, and never in smaller fragments or at isolated trees (Fig. 12.7 a,c). Moreover, the results of this study suggest that forest fragments need to be larger than 5 ha, and need to have interior forest more than 100 m from the edge, in order to sustain populations of these forest specialists (Fig. 12.7).

The land use of the surrounding matrix was also found to have an effect on the number of both ubiquitous and forest specialist species using oak trees. Tall sward pastures were associated with higher moth numbers than short pastures and intensively managed arable crops. Reduced vegetation structure, fewer nectar sources, and reduced shelter, in grazed and arable fields, have been shown to be detrimental to many moth species (Littlewood 2008). Ubiquitous species, such as the nationally declining bright-line brown-eye *Lacanobia oleracea*, heart and dart *Agrotis exclamationis*, flame shoulder *Ochropleura plecta*, and flame *Axylia putris*, were relatively abundant at hedgerow oak trees in

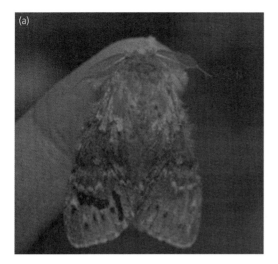

Figure 12.6 Mobile—note the pointed forewings—macro-moth species typical of forests: (a) lobster moth *Stauropus fagi* marked on the wing, the larvae feed on beech *Fagus*, oak *Quercus*, and several other trees. Photograph © E. Slade; (b) green arches *Anaplectoides prasina*, larvae feed on a number of plants, including bilberry *Vaccinium myrtillus*, honeysuckle *Lonicera*, and knotgrass *Polygonium*. Photograph © Maarten Jacobs; and (c) scarce silver-lines *Bena bicolorana* the larvae are typically found on oak. Photograph © Maarten Jacobs.

particular, suggesting that they are using hedgerows as corridors when crossing the agricultural landscape. In fact, hedgerow oaks approached small fragments in terms of abundance and species richness of species with weak to medium forest affinity (Fig. 12.7 a,c). Species with weak to medium forest affinity were also frequently captured at isolated oaks, and low recapture rates at the same tree suggest that they were being used as 'stepping stones', enabling movement across the landscape. Small forest fragments also seemed to be acting as 'stepping stones', and as a key habitat for many species with a medium forest affinity, such as blood-vein *Timandra comae*, lackey *Malacosoma neustria*, and pretty chalk carpet *Melanthia procellata*, all of which are severely declining in Britain (Conrad et al. 2006). Many farm woodland schemes conserve small

patches of forest (i.e. < 2 ha) within the agricultural landscape. Thus, while such schemes may not benefit true forest specialists, our results suggest that small forest fragments may provide key habitat resources for many other moth species.

Our study thus highlighted that both forest size and forest connectivity are important when considering how to conserve moth diversity in fragmented agricultural landscapes (Slade et al. 2013). Physical links in the landscape, such as hedgerows, may be important both for forest specialists and generalists, and such connectivity will become increasingly important in the light of climate change, as species try to move in order to stay within their climatic envelopes (Devictor et al. 2012). We suggest that increasing the landscape connectivity between patches of remaining forest should be a key

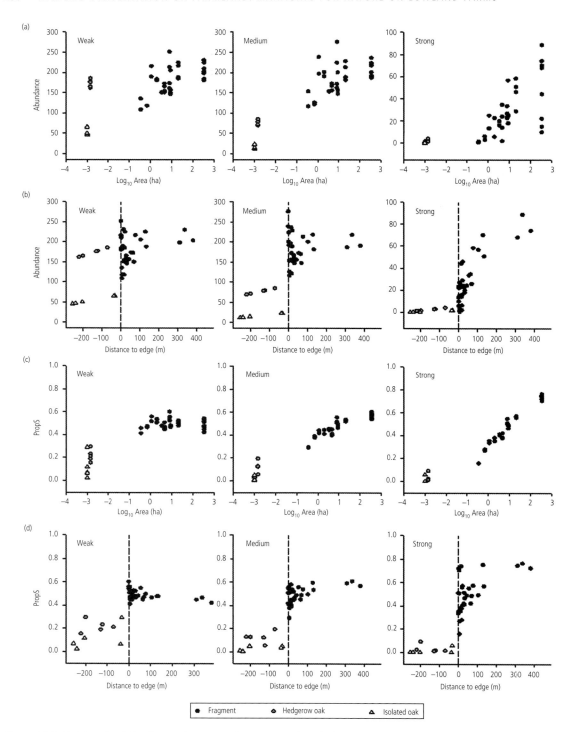

Figure 12.7 Model-predicted abundance and a species richness measure (PropS) of moths for three forest affinity classes (weak, medium, and strong), in relation to (a) and (c) forest area (hedgerow and isolated oaks were given a value of 0.0015 ha and 0.001 ha respectively) and (b) and (d) distance from forest edge. Note that in (a) and (b), the y-axis has a different range for the strong forest affinity class. In (b) and (d), positive values are the distance from within (i.e. trap is placed inside the forest) and negative values are the distance from outside (i.e. trap is outside the forest at solitary oak trees). Redrawn from Slade et al. (2013).

priority (e.g. for national agri-environment schemes, AES). The extent of forest cover in the UK, although very small, is now increasing (Mason 2007) and AES, combined with farm business models that include sustainable forest revenue streams, have the potential to be used to increase connectivity and create forest networks at the landscape level, leading to ecologically functional landscapes. However, in order to maintain the full complement of species within the landscape, and in particular forest specialists, it is necessary to maintain both connectivity and large patches of suitable forest habitat.

12.5 Forest management in the 'farmscape'

Our small mammal and moth research portfolios highlight some key conservation issues for the management of woodlands and the intervening matrix in the 'farmscape'. To maintain forest ecosystem functionality within smaller woodland patches, grazing pressure on forest regeneration needs to be controlled through rigorous deer population management, thus benefiting forest flora and ground cover. Encouraging a well-developed three-dimensional forest structure with sufficient bramble and hawthorn cover provides small mammals and ground-nesting birds with cover from avian and mammalian predators, while simultaneously providing food and nesting opportunities for a variety of species (Buesching et al. 2008).

Pure forest specialists can be accommodated through the implementation of a two-tier management approach for larger woodlands (Merckx et al. 2012), which serves to create core areas of dark, sheltered, and humid forest, while the buffering of these core areas with lighter zones of more open woodland helps to provide a variety of resources to those species which cannot satisfy all their requirements of food, shelter, and nesting opportunities in our increasingly intensified countryside. However, caution must be exercised in establishing too many rides through contiguous woodlands, as there is strong evidence that they may act as movement and dispersal barriers for small mammals (Buesching et al. unpublished data)

At the landscape scale, maintaining sufficient functional connectivity between woodland patches (e.g. through hedgerows, hedgerow trees, solitary field trees, small woodlots, and favourable ground cover; Slade et al. 2013) facilitates species dispersal—and thus re-population of those areas in the 'farmscape', which, due to land use and agricultural practices, have become depauperate.

Acknowledgements

We would like to thank Chris Newman and the many volunteers from the Earthwatch Institute and the HSBC Climate Partnership Programme, who helped with small mammal trapping, deer dropping surveys, and moth trapping. We thank Jenny Bryce, Bridget Appleby and Stephen Ellwood for their work described in this chapter. The landscape-scale MRR study of woodland moths was funded by the HSBC Climate Partnership Programme through the Earthwatch Institute. The study on woodland management for moths was funded by Defra (Project CR0470: Understanding the role of woodland management in the conservation of UK BAP moths). We thank Butterfly Conservation for valuable support, and M. Townsend and M. Botham for assistance with fieldwork and moth identification. Carolyn King and Andrew Gosler provided valuable comments on an earlier draft of this chapter. We thank Ruth Feber for help with the development of this chapter and Eva Raebel for editorial support.

References

Akcakaya, H.R., Mills, G., and Doncaster, C.P. (2007). The role of metapopulations in conservation. In D.W. Macdonald and K. Service, eds., *Key Topics in Conservation Biology*, 64–84. Blackwell Publishing, Oxford.

Appleby, B.M., Anwar, M.A., and Petty, S.J. (1999). Short-term and long-term effects of food supply on parasite burdens in tawny owls, *Strix aluco. Functional Ecology*, **13**, 315–321.

Appleby, B.M., Petty, S.J., Blakey, J.K., Rainey, P., and Macdonald, D.W. (1997). Does variation of sex ratio enhance reproductive success of offspring in tawny owls (*Strix aluco*)? *Proceedings of the Royal Society Biological Science*s, **264**, 1111–1116.

Bartos, L., Vankova, D., Miller, K.V., and Siler, J. (2002). Interspecific competition between white-tailed, fallow, red, and roe deer. *Journal of Wildlife Management*, **66**, 522–527.

Bindernagel, J.A. and Anderson, R.C. (1972). Distribution of the meningeal worm in white-tailed deer in Canada. *Journal of Wildlife Management*, **36**, 1349–1353.

Broome, A., Clarke, S., Peace, A., and Parsons, M. (2011). The effect of coppice management on moth assemblages in an English woodland. *Biodiversity and Conservation*, **20**, 729–749.

Bryce, J., Cartmel, S., and Quine, C.P. (2005). Habitat use by red and grey squirrels: results of two recent studies and implications for management. *Forestry Commission Information Note 76*. Forestry Commission, Edinburgh.

Bryce, J.M., Johnson, P.J., and Macdonald, D.W. (2002). Can niche use in red and grey squirrels offer clues for their apparent coexistence? *Journal of Applied Ecology*, **39**, 875–887.

Bryce, J.M, Speakman, J.R., Johnson, P.J., and Macdonald, D.W. (2001). Competition between Eurasian red and introduced Eastern grey squirrels: the energetic significance of body mass differences. *Proceedings of the Royal Society of London B*, **268**: 1–6.

Buechner, M. (1989). Are small-scale landscape features important factors for field studies of small mammal dispersal sinks? *Landscape Ecology*, **2**, 191–199.

Buesching, C.D., Clarke, J.R., Ellwood, S.A., King, C., Newman, C., and Macdonald, D.W. (2010).The Mammals of Wytham Woods. In: P.S. Savill, C.M. Perrins, K.J. Kirby, and N. Fisher eds., *Wytham Woods: Oxford's Ecological Laboratory*. Oxford University Press, Oxford.

Buesching, C.D., Newman, C., Jones, J.T., and Macdonald, D.W. (2011).Testing the effects of deer grazing on two woodland rodents, bankvoles and woodmice. *Basic and Applied Ecology*, **12**, 207–214.

Buesching, C.D., Newman, C., and Macdonald, D.W. (2014). How dear are deer volunteers: the efficiency of monitoring deer using teams of volunteers to conduct pellet group counts. *Oryx*, 1–9.

Buesching, C.D., Newman, C., Twell, R., and Macdonald, D.W. (2008). Reasons for arboreality in wood mice *Apodemus sylvaticus* and bank voles *Myodes glareolus Mammalian Biology*, **73**, 318–324.

Bush, E.R., Buesching, C.D., Slade, E.M., and Macdonald, D.W. (2012). Woodland recovery after suppression of deer: cascade effects for small mammals, wood mice (*Apodemus sylvaticus*) and bank voles (*Myodes glareolus*). *PLOS ONE*, **7**, e31404.

Conrad, K.F., Warren, M.S., Fox, R., Parsons, M.S., and Woiwod, I.P. (2006). Rapid declines of common, widespread British moths provide evidence of an insect biodiversity crisis. *Biological Conservation*, **132**, 279–291.

Conrad, K. F., Woiwod, I. P., and Perry, J. N. (2002). Longterm decline in abundance and distribution of the garden tiger moth (*Arctia caja*) in Great Britain. *Biological Conservation*, **106**, 329–337.

Cushman, S.A., Mcrae, B., Adriaensen, F., Beier, P., Shirley, M., and Zeller, K. (2013). Biological corridors and connectivity. In D.W. Macdonald and K. Willis, eds., *Key Topics 2*, pp. 384–404. John Wiley & Sons, Ltd., Oxford..

Dennis, R.L.H. (2010). *A Resource-based Habitat View for Conservation—Butterflies in the British Landscape*. Wiley-Blackwell, Oxford.

Devictor, V., van Swaay, C., Brereton, T., et al. (2012). Differences in the climatic debts of birds and butterflies at a continental scale. *Nature Climate Change*, **2**, 121–124.

Devoto, M., Bailey, S., and Memmott, J. (2011). The 'night shift': nocturnal pollen-transport networks in a boreal pine forest. *Ecological Entomology*, **36**, 25–35.

Ellwood, S.A. (2007). *Evaluating deer monitoring methods and the density dependence and independence of skeletal size and body condition of fallow and muntjac deer in a UK lowland wood*. D.Phil. thesis, Department of Zoology, University of Oxford.

Elton, C.S. (1924). Periodic fluctuations in the numbers of animals: their causes and effects. *British Journal of Experimental Biology*, **2**, 119–163.

Elton, C.S. (1966). *The Pattern of Animal Communities*. Methuen, London.

Fitzgibbon, S.D. (1997). Small mammals in farm woodlands: the effect of habitat, isolation and surrounding land use patterns. *Journal of Applied Ecology*, **34**, 530–539.

Flowerdew, J.R. (1993). *Mice and Voles*. Whittet Books, London.

Flowerdew, J.R. and Ellwood, S.A. (2001). Impacts of woodland deer on small mammal ecology. *Forestry*, **74**, 277–287.

Fox, R. (2013). The decline of moths in Great Britain: a review of possible causes. *Insect Conservation and Diversity*, **6**, 5–19.

Fox, R., Conrad, K.F., Parsons, M.S., Warren, M.S., and Woiwod, I.P. (2006). *The State of Britain's Larger Moths*. Butterfly Conservation and Rothamsted Research, Wareham, Dorset, UK.

Fuller, R.J. and Gill, R.M.A. (2001). Ecological impacts of increasing numbers of deer in British woodland. *Forestry*, **74**, 189–192.

Gill, R. (2000). *The Impact of Deer on Woodland Biodiversity*. Forestry Commission Information Note, HMSO, UK.

Gill, R.M.A. and Beardall, V. (2001). The impact of deer on woodlands: the effects of browsing and seed dispersal on vegetation structure and composition. *Forestry*, **74**, 209–218.

Gipps, J.H.W. (1985). The behaviour of bank voles. *Symposium of the Zoological Society London*, **55**, 61–87.

Gorissen, D., Merckx, T., Vercoutere, B., and Maes, D. (2004). Changed woodland use and butterflies. Why did butterflies disappear from woodlands in Flanders? *Journal for Landscape Ecology and Environmental Science in Flanders and the Netherlands*, **21**, 85–95.

Gosler, A.G. (1990). The Birds of Wytham—An Historical Survey. *Fritillary*, **1**, 29–74.

Gurnell, J. (1985). Woodland rodent communities. *Symposium of the Zoological Society London*, **55**, 377–411.

Gurnell, J., Wauters, L.A., Lurz, P.W.W., and Tosi, G. (2004). Alien species and interspecific competition: effects of introduced eastern grey squirrels on red squirrel population dynamics. *Journal of Animal Ecology*, **73**, 26–35.

Hambler, C, Wint, G.R.W., and Rogers, D.J. (2010). *Invertebrates*. In P S Savill, C M Perrins, K.J. Kirby, and N Fisher, eds., *Wytham Woods: Oxford's Ecological Laboratory*, pp. 109–144. Oxford University Press, Oxford.

Hansson, L. (1985). The food of bank voles, wood mice and yellow-necked mice. In: J. R. Flowerdew, J. Gurnell, and J.H.W. Gipps eds., *The Ecology of Woodland Rodents: Bank Voles and Wood Mice*, pp. 141–165. Oxford Scientific Publications, Oxford.

Hansson, L. (1992). Fitness and life-history correlates of weight variations in small mammals. *Oikos*, **64**, 479–484.

Harrington, L.A. and Macdonald, D.W. (2002). *A review of the effects of pesticides on wild terrestrial mammals in Britain*. WildCRU, Oxford.

Harris, S., McLaren, G., Morris, M., Morris, P., and Yalden, D. (2000). Abundance/mass relationships as a quantified basis for establishing mammal conservation priorities.

In A. Entwhistle and N. Dunstone., eds., *Priorities for the Conservation of Mammalian Diversity: Has the Panda Had its Day?* Cambridge University Press, Cambridge.

Hodgson, J.A., Moilanen, A., Bourn, N.A.D., Bulman, C.R., and Thomas, C.D. (2009). Managing successional species: modelling the dependence of heath fritillary populations on the spatial distribution of woodland management. *Biological Conservation*, **142**, 2743–2751.

King, C.M. (1985). Interactions between woodland rodents and their predators. *Symposium of the Zoological Society London*, **55**, 219–247.

Kirby, K. J. (2001). The impact of deer on the ground flora of British broadleaved woodland. *Forestry*, **74**, 219–229.

Littlewood, N.A. (2008). Grazing impacts on moth diversity and abundance on a Scottish upland estate. *Insect Conservation and Diversity*, **1**, 151–160.

Macdonald, D.W. (1976). Nocturnal observations of tawny owls *Strix aluco* preying upon earthworms. *Ibis*, **118**, 579–580.

Macdonald, D.W. and Burnham, D. (2010). *The State of Britain's mammals a focus on invasive species*. WildCRU for PTES, Oxford.

Macdonald, D.W. and Feber, R.E., eds. (2015). *Wildlife Conservation on Farmland. Conflict in the Countryside*. Oxford University Press, Oxford.

Macdonald, D.W., Mace, G., and Rushton, S. (1998). *Proposals for future monitoring of British mammals*. HMSO, London.

Macdonald, D. W., Newman, C., Dean, J., Buesching, C. D., and Johnson, P. J. (2004). The distribution of Eurasian badger, *Meles meles*, setts in a high-density area: field observations contradict the sett dispersion hypothesis. *Oikos*, **106**, 295–307.

Macdonald, D.W., Tew, T.E., Todd, I.A., Garner, J.P., and Johnson, P.J. (2006). Arable habitat use by wood mice (*Apodemus sylvaticus*). 3. A farm-scale experiment on the effects of crop rotation. *Journal of Zoology*, **250**, 313–320.

Mason, W. L. (2007). Changes in the management of British forests between 1945 and 2000 and possible future trends. *Ibis*, **149**, 41–52.

McCabe, R.E. and McCabe, T.R. (1974). Of slings and arrows: an historical introspection. In: L.K. Halls ed. *The White-tailed Deer, Ecology and Management*. Stackpole Books, Harrisburg.

Merckx, T., Feber, R.E., Dulieu, R.L., et al. (2009). Effect of field margins on moths depends on species mobility: field-based evidence for landscape-scale conservation. *Agriculture, Ecosystems & Environment*, **129**, 302–309.

Merckx, T., Feber, R.E., Hoare, D.J., et al. (2012). Conserving threatened Lepidoptera: Towards an effective woodland management policy in landscapes under intense human land-use. *Biological Conservation*, **149**, 32–39.

Merckx, T., Feber, R.E., McLaughlan, C., et al. (2010). Shelter benefits less mobile moth species: The field-scale effect of hedgerow trees. *Agriculture, Ecosystems & Environment*, **138**, 147–151.

Merckx, T., Van Dongen, S., Matthysen, E., and Van Dyck, H. (2008). Thermal flight budget of a woodland butterfly in woodland versus agricultural landscapes: an experimental assessment. *Basic & Applied Ecology*, **9**, 433–442.

Morecroft, M.D., Taylor, M.E., Ellwood, S.A., and Quinn, A.A. (2001). Impacts of deer herbivory on ground vegetation at Wytham Woods, central England. *Forestry*, **74**, 251–257.

Newman, C. and Macdonald, D.W. (2013). *The Implications of climate change for terrestrial UK Mammals*. Terrestrial biodiversity climate change impacts report card technical paper. Living with environmental change partnership. http://www.lwec.org.uk/.

Perrins, C. M. and Overall, R. (2001). Effect of increasing numbers of deer on bird populations in Wytham Woods, central England. *Forestry*, **74**, 299–309.

Petty, S. J. (1992). The ecology of the tawny owl Strix aluco in the spruce forests of Northumberland and Argyll. Unpublished Ph.D. thesis, Open University.

Plesner Jensen, S. (1996). Juvenile dispersal in relation to adult densities in wood mice *Apodemus sylvaticus*. *Acta Theriologica*, **41**, 177–186.

Pollard, E. and Cooke, A. S. (1994). Impact of muntjac deer *Muntiacusreevesi* on egg-laying sites of the white admiral butterfly *Ladoga camilla* in a Cambridgeshire wood. *Biological Conservation*, **70**, 189–191.

Putman, R.J. (1997). Deer and road traffic accidents: options for management. *Journal of Environmental Management*, **51**, 43–57.

Putman, R.J., Edwards, P.J., Mann, J.E.E., Howe, R.C., and Hill, S.D. (1989). Vegetational and faunal change in an area of heavily grazed woodland following relief of grazing. *Biological Conservation*, **47**, 13–32.

Putman, R.J. and Moore, N.P. (2002). Impact of deer in lowland Britain on agriculture, forestry and conservation habitats. *Mammal Review*, **4**, 141–164.

Rackham, O. (2003). *The Illustrated History of the Countryside*, new ed. Weidenfeld and Nicolson, London.

Rushton, S.P., Lurz, P.W.W., Gurnell, J., et al. (2006). Disease threats posed by alien species: the role of a poxvirus in the decline of the native red squirrel in Britain. *Epidemiology and Infection*, **134**, 521–533.

Sandom, C., Donlan, C.J., Swenning, J.-C., and Hanson, D. (2013). Rewilding. In: D.W. Macdonald and K. Willis, eds., *Key Topics in Conservation Biology 2*. Wiley-Blackwell, Oxford.

Savill, P.S., Perrins, C.M., Kirby, K.J., and Fisher, N. (2010). *Wytham Woods: Oxford's Ecological Laboratory*. Oxford University Press, Oxford.

Scottish Wildlife Trust (2013). <http://scottishwildlifetrust.org.uk/>.

Selander, R. K. (1964). *Speciation in wrens of the genus Campylorhynchus*. University of California Press.

Slade, E.M., Merckx, T., Riutta, T., Bebber, D.P., Redhead, D., Riordan, P., and Macdonald, D.W. (2013). Life-history traits and landscape characteristics predict macro-moth responses to forest fragmentation. *Ecology*, **94**, 1519–1530.

Smit, R., Bokdam, J., den Ouden, J., Olff, H., Schot-Opschoor, H., and Schrijvers, M. (2001). Effects of introduction and exclusion of large herbivores on small rodent communities. *Plant Ecology*, **155**, 119–127.

Snow, D.W. and Perrins, C. (1998). *The Birds of the Western Palearctic, Concise edition*. Oxford University Press, Oxford.

Southern, H.N. and Lowe, V.P.W. (1982). Predation by tawny owls (*Stix aluco*) on bank voles (*Clethrionomys glareolus*) and wood mice (*Apodemus sylvaticus*). *Journal of Zoology*, **198**, 83–102.

Spielman, A. (1994). The emergence of Lyme disease and human babesiosis in a changing environment. *Annals of the New York Academy of Sciences*, **740**, 146–156.

Stevenson, C.D., Ferryman, M., Nevin, O.T., Ramsey, A.D., Bailey, S., and Watts, K. (2013). Using GPS telemetry to validate least-cost modelling of grey squirrel (*Sciurus carolinensis*) movement within a fragmented landscape. *Ecology and Evolution*, **3**, 2350–2361.

Stewart, A.J.A. (2001). The impact of deer on lowland woodland invertebrates: a review of the evidence and priorities for future research. *Forestry*, **74**, 259–270.

Summerville, K.S. and Crist, T.O. (2004). Contrasting effects of habitat quantity and quality on moth communities in fragmented landscapes. *Ecography*, **27**, 3–12.

Taylor, G. and Morecroft, M.D. (1997). *Estimating summer 1997 deer populations in Wytham woods, via faecal accumulation method*. Unpublished internal report to Environmental Change Network, Wytham.

Tew, T.E. and Macdonald, D.W. (1994). Dynamics of space use and male vigour amongst wood mice, *Apodemus sylvaticus*, in the cereal ecosystem. *Behavioral Ecology and Sociobiology*, **34**, 337–345.

Tomkins, D.M., White, A.R., and Boots, M. (2003). Ecological replacement of native red squirrels by invasive greys driven by disease. *Ecology Letters*, **6**, 189–196.

Usher, M.B., Crawford, T.J., and Banwell, J.L. (1992). An American invasion of Great Britain: the case of the native and alien squirrel (*Sciurus*) species. *Conservation Biology*, **6**, 108–115.

Usher, M.B. and Keiller, S.W.J. (1998). The macrolepidoptera of farm woodlands: determinants of diversity and community structure. *Biodiversity and Conservation*, **7**, 725–748.

Van Swaay, C.A.M., Warren, M.S., and Lois, G. (2006). Biotope use and trends of European butterflies. *Journal of Insect Conservation*, **10**, 189–209.

Ward, A.I. (2005). Expanding ranges of wild and feral deer in Great Britain. *Mammal Review*, **35**, 165–173.

Waring, P. and Townsend, M. (2009). *Field Guide to the Moths of Great Britain and Ireland, Second ed*. British Wildlife Publishing, Oxford.

Warren, M.S. and Key, R.S. (1991). Woodlands: Past, present and potential for insects. In N.M. Collins and J.A. Thomas, eds., *The Conservation of Insects and their Habitats*. The Royal Entomological Society of London, Academic Press Ltd., London.

Warren, M.S. and Thomas, J.A. (1992). Butterfly responses to coppicing. In G.P. Buckley, ed. *The Ecological Effects of Coppice Management*, pp. 249–270. Chapman and Hall, London.

Watts, C.H.S. (1968). The foods eaten by wood mice (*Apodemus sylvaticus*) and bank voles (*Clethrionomys glareolus*) in Wytham Woods, Berkshire. *Journal of Animal Ecology*, **37**, 25–42.

Wauters, L.A., Gurnell, J., Martinoli, A., and Tosi. G. (2002). Interspecific competition between native Eurasian red squirrels and alien grey squirrels: does resource partitioning occur? *Behavioral Ecology and Sociobiology*, **52**, 332–334.

White, A., Bell, S. S., Lurz, P. W., & Boots, M. (2014). Conservation management within strongholds in the face of disease-mediated invasions: red and grey squirrels as a case study. *Journal of Applied Ecology*, **51**(6), 1631–1642.

Yoccoz, N.G. and Mesnager, S. (1998). Are alpine bank voles larger and more sexually dimorphic because adults survive better? *Oikos*, **82**, 85–98.

Young, M. (1997). *The Natural History of Moths*. Poyser Natural History, UK.

CHAPTER 13

Improving reintroduction success of the grey partridge using behavioural studies

Elina Rantanen, David W. Macdonald, Nick W. Sotherton, and Francis Buner

A grain in the balance will determine which individual shall live and which shall die - which variety or species shall increase in number, and which shall decrease, or finally become extinct.

Charles Darwin, *The Origin of Species*

13.1 Introduction

The transformation in status of the grey partridge *Perdix perdix*, from Britain's main shooting quarry to the UK Red List of Conservation Concern, exemplifies many of the impacts of farming change on wildlife in the twentieth century. From its evolution in the temperate grassland ecosystems of the steppes, the grey partridge expanded its range to inhabit open arable landscapes across much of the western Palearctic. Fossil evidence shows that the species has inhabited the UK since after the last ice age, with the lowland arable areas in the south, east, and midlands of Great Britain traditionally being the strongholds of grey partridge populations (Robinson 2009). This iconic farmland bird, immortalized in Christmas rhyme, was abundant on lowland farmland, with around two million birds shot annually between 1870 and 1930 (Tapper 1992). The Holkham Estate game books, for example, provide a glimpse into the species' past, recording that, in a single day in 1905, a bag of 1671 grey partridge were shot on the Warham beat in Norfolk[1]. Yet a hundred years on, the UK population size has dwindled to around 73 000 breeding pairs[2] (Baker et al. 2006; Baillie et al. 2013), with numbers of grey partridge having decreased by 88% between 1967 and 2006, leading to the species being added, in 2009, to the Red List of UK

birds[3]. In the meantime, the size of the national game bag has continued to rise, sustained by the release on to farmland, each year, of an estimated 35 million non-native common pheasants *Phasianus colchicus* and six million non-native red-legged partridges *Alectoris rufa* (GWCT 2012; Macdonald and Feber 2015: Chapter 9). Grey partridge are still shot, sometimes inadvertently, when they are mistaken for red-legged partridge, with potentially severe consequences for their populations (Aebischer and Ewald 2004).

A suite of factors relating to the post-war intensification of farmland are believed to have been largely responsible for the decline of the grey partridge (Baillie et al. 2013). Widespread hedgerow removal in the years after World War II, and leading up to the 1970s, reduced nesting cover for the partridge, making the birds more vulnerable to predation, leading to greater losses of nests, lowering hen survival rate during incubation, reducing winter survival, and hence decreasing population productivity (Potts 1986; Potts and Aebischer 1995; Tapper et al. 1996; Aebischer and Ewald 2004). Adult grey partridges are mainly granivorous, but chicks require protein-rich insect food (aphids, plant bugs, insect larvae, beetles) for successful development. Over the same time period, increased use of herbicides and insecticides had the effect of lowering chick survival rates, by reducing the availability of invertebrates in

---

[1] <http://www.holkham.co.uk/>.
[2] The grey partridge's decline has also been sharp in the rest of Europe (Baillie et al. 2013).

[3] Species with the highest conservation priority and needing urgent action (Eaton et al. 2009).

Wildlife Conservation on Farmland. Managing for Nature on Lowland Farms. Edited by David W. Macdonald and Ruth E. Feber.
© Oxford University Press 2015. Published 2015 by Oxford University Press.

cereal crops (Potts 1980, 1986; Rands 1985). Post-war Britain also saw a move towards reduced predator control and far fewer gamekeepers (Martin 2011).

In 1995, in recognition of the species' severe decline, the grey partridge was designated a priority under the UK Biodiversity Action Plan (BAP). In order to meet the BAP target, the UK Government, together with the Game & Wildlife Conservation Trust (GWCT), launched a recovery programme for the grey partridge. Under this programme, the GWCT monitors grey partridge numbers on UK farmland and provides land management advice for farmers to help to restore grey partridge populations (Aebischer and Ewald 2004). There have been encouraging examples of increases in pair densities (i.e. 66% increase on average over two years) on estates where the availability of hedges, grass margins, and conservation headlands (selectively sprayed edges of arable fields to encourage broadleaved plants and invertebrates), as nesting and brood-rearing habitats have been improved (Aebischer and Ewald 2004). Financial incentives for managing key habitats such as these for grey partridges and other wildlife, and other options such as planting game cover crops, have come from UK government agri-environment schemes (Vickery et al. 2004; Aebischer and Ewald 2010). Aebischer and Ewald (2010) demonstrated that grey partridge pair densities were over twice as high on study sites managed under agri-environment schemes compared to sites on conventional arable land.

As well as focusing on farm habitat improvements, one aim of the grey partridge recovery programme has been to investigate the possibility of increasing grey partridge populations locally through releases of birds bred in captivity (Buner et al. 2011). Releases of captive-bred birds with the aim of re-establishing a viable population, as opposed to releases solely for shooting, could help restore populations of native game birds in areas from which they have largely disappeared, provided that the key habitats for the species have also been restored and maintained (Buner and Aebischer 2008; Buner et al. 2011). Such game bird releases, carried out with the long-term goal of increasing the size of the local population for sustainable shooting, share similar objectives to those of wildlife reintroductions, that is, to re-establish or re-stock wild populations through releases of captive-bred animals (IUCN 1998; Chapter 14, this volume).

Traditional release methods of grey partridges, dating back to the nineteenth century, include releases of captive-bred pairs in spring, and family groups, called coveys, in the autumn, and even fostering captive-born

grey partridge young to wild parents that have lost their brood (Browne et al. 2009; Buner et al. 2011). In the UK and elsewhere, there have been numerous attempts to re-establish or re-stock wild grey partridge populations through releases of captive-bred birds (e.g. Rands and Hayward 1987; Brun and Aubineau 1989; Dowell 1990; Putaala et al. 2001; Parish and Sotherton 2007; Buner and Schaub 2008; Buner et al. 2011). However, the survival and breeding success of released captive-bred birds have been poor in comparison with wild grey partridges (e.g. Rands and Hayward 1987; Dowell 1990; Putaala et al. 2001; Buner et al. 2011; Rymešová et al. 2013). For instance, Buner et al. (2011) reported breeding success rates of 17.4–44.0% for released, captive-bred grey partridges compared to 49.6–65.6% for wild birds. Similarly, Rymešová et al. (2013) reported poor mating success of commercially reared individuals compared to wild partridges (both released) due to their early mortality: in their study, unpaired captive birds died within seven days after release, which was the time it took for wild males to form pairs. Indeed, high mortality and/or low breeding success are general problems also in wildlife reintroductions, where the success rates of re-establishing viable populations have also been low (Griffith et al. 1989; Wolf et al. 1996; Fischer and Lindenmayer 2000). Since there is a vast knowledge on the habitat requirements and management practices necessary to achieve and sustain viable habitat niches for the reintroduction of partridges (a primary condition for successful reintroductions), behavioural deficiencies resulting from captive breeding, and the lack of studies on this matter, are considered one of the main causes of reintroduction failures (Curio 1996; Snyder et al. 1996; Wallace 2000; Mathews et al. 2005; Macdonald 2009), which often occur during the first days and weeks of the establishment phase (Moehrenschlanger et al. 2013).

In the UK, farms have bred game bird strains in captivity for at least ten generations (Dowell 1990). The consequences of this for affecting anti-predator behaviour (Box 13.1) may be an important contributory factor leading to the low success rates of grey partridge releases and may explain why, more widely, there have been so many failures when reintroducing species that are naturally subjected to considerable predation in the wild (Snyder et al. 1996). In the UK, grey partridges are predated mainly by foxes *Vulpes vuipes* and raptors such as the sparrow hawk *Accipiter nisus* (Potts 1986; Tapper et al. 1996; Watson et al. 2007). The high breeding potential of the grey partridge helps compensate for predation; hens nest in their first summer (in May–June), their average clutch size of 15–16

Box 13.1 Behavioural differences between captive-bred and wild animals

Behavioural deficiencies in captive-bred animals originate from the fact that there are fundamental differences between the captive environment and the conditions in the wild (Moehrenschlager et al. 2013). For instance, in the wild, predation acts as one of the strongest agents of natural selection in prey species, whereas captive animals are generally bred in non-rich, predator-sheltered environments, which exclude selective forces of considerable evolutionary consequences (Caro 2005), and this may lead to relaxed selection in, for example, predator avoidance behaviour (Price 1999; McPhee 2003; Håkansson and Jensen 2005).

Anti-predator behaviour can be at least partly innate (Caro 2005) and thereby not immediately lost in captivity. However, the absence of predation in the captive environment may, over generations, increase variation in the performance of innate anti-predator behaviours in the captive population, which may decrease their general effectiveness in protecting the released population from predation (McPhee 2003). For example, just one generation of hatchery rearing wild Atlantic salmon *Salmo salar* is sufficient to select for maladaptive responses to predators under natural conditions (Jackson and Brown 2011). Anti-predator behaviour has direct consequences for survival and breeding success (Caro 2005), and any deterioration in this behaviour would therefore negatively affect the prospects of a successful reintroduction. Håkansson and Jensen (2005) determined that captive populations of the red junglefowl *Gallus gallus* obtained from different captive environments showed significant differences in anti-predatory responses, but that these differences diminished after just four generations exposed to the same simulated predator attacks (Håkansson and Jensen 2008).

Overall, a general problem in captive breeding for reintroductions is domestication, i.e. adaptations to the captive environment, developed both through the habituation of

the captive animal during its lifetime of development and experience, and through genetic changes in its progeny over generations (Price 1984). Domestication through selection may be difficult to prevent, since those individuals that are more tolerant of the captive conditions will have a reproductive advantage over less tolerant animals. For instance, stress is detrimental for breeding (Moburg 1991), and therefore animals that are less emotionally distressed would have a fitness advantage in captive conditions that often inflict psychological stress on the animals[4] (Moorhouse et al. 2007; Morgan and Tromborg 2007; Chapter 14, this volume). Traditionally, artificial rearing methods for game birds have been designed for intensive rearing of large numbers of birds (e.g. brooder houses with heat lamps for up to 100 birds; Game Advisory Booklet 1964), rather than for their natural behavioural development, and these same rearing practices are still widely used. Moreover, captive breeding usually involves regular contact with humans, and low fearfulness of humans would therefore be favoured by selection in the captive breeders (Price 1999). However, once released into the wild, low emotional reactivity becomes a disadvantage in the face of predators (Boissy 1995). For example, captive Persian fallow deer *Dama mesopotamica*, reared in predator-free facilities with reduced human presence and released into the wild, showed a higher flight and flush distance and use of cover over time under predatory attack than captive animals reared in facilities that were constantly visited by the public (Zidon et al. 2009). Furthermore, after 200 days, no animals from the population receiving high numbers of visitors when captive-bred had survived, compared with an 80% survival rate from the population with reduced human visits (Zidon et al. 2009), implying that levels of anthropogenic contact at the captive breeding stage can influence the success of reintroductions.

---

eggs is the largest among all birds, and they lay replacement clutches if the first one is predated (Potts 1986). Furthermore, wild grey partridges obviously seek to avoid predation. The hen nests in tall cover to avoid detection, and the parents use alarm calls to alert their chicks of predators (Potts 1986). In autumn,

the parents, their young, and additional adults form groups, or so-called coveys, of around 12 individuals, whose collective vigilance increases their predator detection, before eventual pairing in January–February for a new breeding season. Parental example is therefore important for fine-tuning innate anti-predator behaviour in grey partridges (Potts 1986; Beani and Dessi-Fulgheri 1998), and artificially reared birds are thus at a clear disadvantage.

Wild grey partridges also perform vigilance behaviour (adopting an upright position with the head stretched out: Beani and Dessi-Fulgheri 1998) for,

---

[4] Animal welfare is linked to re-introduction success: stress of a captive animal can arise from handling and transportation, but a captive animal that is not prepared for life in the wild is more likely to suffer post-release stress than its wild peers, and therefore more likely to die, as reviewed by Harrington et al. (2013).

on average, 43% of their time (Watson et al. 2007), and crouching behaviour to avoid detection (Dowell 1990). Although captive-bred grey partridges do still perform innate anti-predator behaviour towards a raptor model in behavioural tests, parent-reared birds crouched (Dowell 1990; Anttila et al. 1995) and froze (Beani and Dessi-Fulgheri 1998) more often and for a longer duration than artificially reared birds. In addition, wild grey partridges concentrate their feeding around dawn and dusk (Jenkins 1961), fly to their roosting site to avoid leaving a scent trail on the ground (Tillmann 2009), and tend to roost overnight in open fields away from field edges, probably reducing the risk from predators, such as foxes, that hunt along such linear features at that time (Dowell 1990; Harris and Woollard 1990). In contrast, released captive-bred grey partridges tend to walk to their roosting sites rather than flying, and remain mostly within 10 m of field edges at dusk/dawn/overnight, increasing their encounters with predators (Dowell 1990; Harris and Woollard 1990).

If grey partridge releases are to be a significant tool in grey partridge conservation, pre- and post-release behavioural studies would help to identify the nature and extent of such behavioural problems, leading to the development of better rearing techniques and release methods. In this chapter we describe our studies on reintroduced grey partridges, in which we have sought to identify reasons for previous failures in grey partridge releases and to address behavioural issues hindering success in reintroductions. Our ultimate aim was to provide new and practical information on the behaviour of captive-bred grey partridges that could improve rearing methods and post-release management for reintroductions of this and similar species (Rantanen et al. 2010a, b). In particular, we focused on the following behavioural research questions:

1. Could released grey partridges be using their habitats maladaptively, leading to an ecological trap (preference of low-quality habitats which would cause reduced survival)?
2. How much time do released grey partridges spend being vigilant for predators, and how does it compare with previous knowledge on wild grey partridge vigilance rates?
3. Do released grey partridges display maladaptive roosting behaviour, and how frequently?
4. What are the post-release survival rates of captive-bred grey partridges, and could they be predicted by testing their temperament?

13.2 Behavioural studies of grey partridge

13.2.1 Rearing and release methods

We established four study sites in Oxfordshire and east Gloucestershire, on arable farms growing winter and spring cereals, oilseed rape, and beans. All were participating in agri-environment schemes and provided the key habitats for grey partridges, i.e. nesting and brood-rearing habitats (hedges and field margins around each field), as well as overwintering cover (hedges, planted game covers) and predator control. Grey partridges were not hunted on these farms and, as opposed to intensive commercial shooting estates, there were no more than eight days of shooting a year on any of the sites, which limited the probability of disturbance or of accidental shooting of grey partridges mistaken for pheasants or red-legged partridges. We applied two traditional, commonly used release methods for grey partridges: pair releases in the spring (April), and covey releases in the autumn (October–November), as employed previously by the Game & Wildlife Conservation Trust (Buner et al. 2011). We did not attempt the fostering of captive-bred young to wild parents due to the lack of sufficient wild grey partridges on the study sites.

The birds were obtained in July as three-week-old chicks from a typical game farm where the grey partridges had been bred for at least seven generations and hand-reared in artificial conditions in brooder houses. For the spring releases, adult birds were kept over winter in large outdoor pens, as mixed-sex flocks of approximately 100 birds, where they were allowed to pair-up naturally. The autumn coveys each consisted of a brood of 9–16 young that had been fostered to an adult pair at the age of three weeks, after which the chicks and the foster parents had been housed in the same pen outdoors until the releases in mid October and early November (for more details on the rearing methods, see Rantanen et al. 2010a,b).

Before the releases, four to five randomly chosen juvenile birds in the autumn coveys and all the females of the spring pairs were radio-tagged using 10 g necklace radio-transmitters with mortality sensors (Fig. 13.1). The 10 g radio-tags accounted for 2.6% of the average weight of grey partridges (390 g; Robinson 2009). For further details on the tagging methods, see Buner et al. (2011). Parish and Sotherton (2007) had previously detected no difference in survival rates between radio-tagged and non-tagged released grey partridges, giving us grounds to expect that the survival of the radio-tagged birds in this study would represent that of all released birds.

Figure 13.1 Radio-tagged grey partridge *Perdix perdix* male. Photograph © F. Buner.

Between mid October and early November in 2006 and 2007, 19–20 autumn coveys were released each year on the four study sites, four or five coveys per site. Coveys were put into one of five separate release pens, each located next to tall vegetation (e.g. game cover or field margin with tall grass at least 50 cm in height; Fig. 13.2). Food, water, and shelter were provided in these pens, where each covey was kept for an acclimatization period of four days before release. We used this 'soft release' method with the coveys in autumn in order to facilitate settlement, as the dispersal distances of autumn coveys have been nearly three times greater than those of spring pairs (1.4 km versus 0.5 km on average) in previous releases (Buner et al. 2011). In April 2007, 70 pairs were released on two study sites (30 or 40 pairs on each site) using the 'hard release' method, i.e. each pair was released directly after transporting to the site, along field edges (approximately 100 metres apart and next to crops providing cover). For full details on the release methodology, see Buner and Aebischer (2008) and Buner et al. (2011).

13.2.2 Settlement and habitat use by released grey partridges

In order to identify possible ecological traps, we studied intensively the activity ranges, habitat use, and survival of released partridges, by radio-tracking the locations of the released pairs and coveys two to four times per day for the first two weeks after the releases, a critical period for their survival (Dowell 1990). The radio-tagged birds were located by 'homing-in'[5] (Kenward 2001).

---

[5] We followed the radio-signal to each tagged individual so as to yield the most accurate locations.

Figure 13.2 An autumn covey of grey partridges in a release pen, situated next to a hedge and an area of game cover (maize). Photograph © E. Rantanen.

For each recorded location, we categorized the habitat, e.g. crop (cereals, beans, or oilseed rape), grass field (pastures, silage grass, or set-aside), field margin, woodland (young or mature). In autumn, after harvest, some habitat categories were different: e.g. game cover (small areas of tall vegetation, usually strips of maize, planted in spring to provide cover for game birds after harvest). We then determined the availability of these habitats from maps (Fig. 13.3). Radio-tracking enabled us to determine activity ranges for each released pair and covey, which we could overlay on the habitat maps to plot the habitat compositions of these areas (Fig. 13.3). Habitat preferences by the released grey partridge pairs and coveys were then analysed using compositional analysis (see Aebischer et al. (1993) for details), a widely used method to test for whether individual animals use different habitats more or less in proportion to their availability, i.e. whether any habitat was actively chosen or avoided.

Our results showed that released grey partridges tended to settle onto the release sites (Fig. 13.3). In

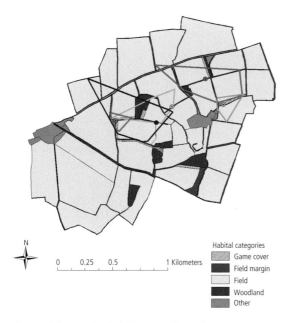

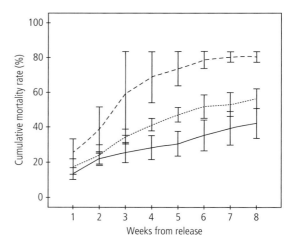

Figure 13.4 Graph depicting the cumulative mortality rates of released radio-tagged grey partridges during the first eight weeks following the releases in autumn 2006 (the middle dashed line), spring 2007 (the top dashed line), and autumn 2007 (the solid line). From Rantanen (2009).

Figure 13.3 Example of a habitat map of a study site in autumn. The circles represent the release points of a covey and the polygons represent these coveys' activity ranges over the first two weeks after release. From Rantanen (2009).

2007, the autumn coveys disbanded in January and this was followed by pairing, as is normal in wild grey partridges, while in autumn 2006, the coveys broke up and the birds started pairing exceptionally early, in October–November, for an unknown reason. The mortality rates were higher in spring (80% by eight weeks after release) than in autumn (40–50% by eight weeks after release; Fig. 13.4), suggesting that releasing coveys in autumn would be a better method for grey partridge reintroductions than spring pair releases. Released grey partridges were killed both by mammalian and avian predators, and higher proportions were predated by mammalian predators, both in spring (72%) and in autumn (67%).

We found that pairs of grey partridges released in spring used the crops and field margins most frequently (Fig. 13.5) and significantly more than expected from their availability (Rantanen et al. 2010a). Furthermore, their survival in spring was influenced positively by the use of crops, but negatively by the use of field margins, suggesting that field margins could serve as ecological traps for released grey partridges (Rantanen et al. 2010a). This means that released grey partridges were attracted to a habitat which probably hosted concentrated predator activity (Morris

and Gilroy 2008). In autumn, the birds released as coveys (family groups) in autumn used game covers more than expected from their availability, and habitat use did not affect their survival. Our findings flag the need in reintroduction projects to be alert to such ecological traps that could undermine conservation efforts (Moehrenschlager et al. 2013; Chapter 16, this volume).

13.2.3 Post-release vigilance behaviour

Released, captive-bred grey partridges may be less wary of predators than are wild-bred birds, thus being less vigilant and thereby more exposed to predation risk. To investigate this, we recorded the vigilance behaviour and time budgets in the released grey partridge coveys, both as group vigilance rates and as time budgets of individual behaviour, including vigilance and feeding.

The coveys were located by radio-tracking until they were in view, and the number of vigilant birds and the total number of birds in view were recorded once every five minutes during the observation session (see Rantanen et al. 2010b for details). We estimated visually the distance from the centre of the covey to the nearest habitat providing cover (vegetation taller than the height of the birds of c. 20–30 cm). Where possible, during the five minutes between each scan sampling,

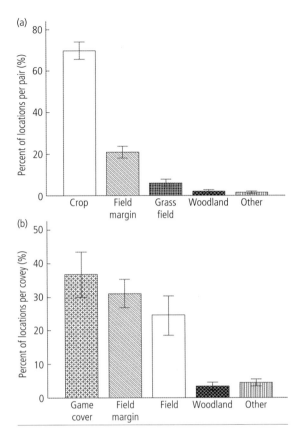

Figure 13.5 Graphs indicating the average frequencies with which released grey partridges used each habitat category in spring (a) and in autumn (b). From Rantanen et al. (2010a). Reproduced with permission from John Wiley & Sons.

average) or as individuals (2–7% of time spent vigilant on average; Rantanen et al. 2010b) compared with known vigilance behaviour of wild grey partridges (43% of time spent vigilant on average; Watson et al. 2007). Individuals spent 30–50% of their time feeding, and they tended to feed at this rate throughout the day (Fig. 13.6; Rantanen et al. 2010b) instead of having concentrated feeding bouts around dawn and dusk as do wild grey partridges (Potts 1986). The poor vigilance and the pattern of continuous feeding behaviour were likely to increase vulnerability to predators in the released grey partridges.

13.2.4 Post-release roosting behaviour

To investigate the extent of maladaptive roosting behaviour, we observed released grey partridge coveys from sunset until 30 minutes after, to establish whether the birds flew to their roosting site. The observer was then positioned so that the birds were either in view or, if in high vegetation, their possible flight to roost could be seen. We counted the number of calls the covey made per minute, and recorded whether they took flight or stayed on the ground. After dark, the roost locations chosen by the released coveys and pairs were determined by radio-tracking.

Our observations showed that released grey partridges were variable and inconsistent in performing the adaptive roosting behaviour typical of wild grey partridges. There were cases when the birds behaved maladaptively by walking instead of flying to their roost site (in 32.5% of cases), and by roosting close

the behaviour[6] of a randomly chosen individual bird was recorded on a dictaphone continuously for one minute. From these recordings, we calculated the average proportion of vigilant birds in the coveys and the average proportion of the total observation time the individual birds spent carrying out each behaviour, with particular focus on vigilance and feeding.

The released grey partridges were not sufficiently vigilant either as groups (3–5% of birds vigilant on

[6] The recorded behaviours were: vigilant (standing upright with neck out-stretched, scanning the surroundings); standing (sitting with the head raised above the level of the back); feeding (neck stretched towards the ground, pecking the ground or vegetation); preening (sitting with the head moving the beak through feathers); moving (walking or running); inactive.

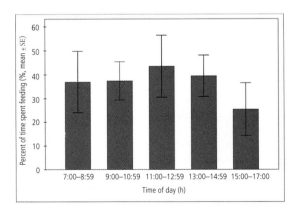

Figure 13.6 Graph showing the average percentages of time released grey partridges in the autumn coveys spent feeding at different times of day. From Rantanen et al. (2010b). Reproduced with permission from Elsevier.

to field boundaries (i.e. 26–32% of all roost locations recorded over different years were within a 20 m zone from field boundaries). However, we observed that the released birds did tend to roost on average 30 m further from field boundaries over time, suggesting that the surviving birds learned to roost further away from areas with higher predator activity. The variation and the inconsistency observed in adaptive roosting behaviour suggest that captive breeding had impaired these behaviours, and the observed maladaptive roosting behaviours could contribute to the high predation rates witnessed in these and previous releases.

13.2.5 Impact of temperament on survival

Conservation research and practical approaches to reintroductions are generally standardized to measure outcomes of success at the population level. However, Moehrenshlager et al. (2013) highlight that reintroduction success could be improved by understanding animal behaviour at the individual level, suggesting that high mortality during the establishment phase (first days/weeks) depends more on the disruption to the individual's environment than being due to large-scale ecological processes. We explored the link between temperament (i.e. consistent behavioural tendencies in individuals) and survival in reintroductions by investigating the post-release survival rates of captive-bred grey partridges, and whether they could be predicted by testing their temperament. Levels of exploration in a novel environment are considered to express an individual animal's fear of novelty (Boissy 1995), and

we investigated whether these levels could serve as an indicator for bold or fearful temperament.

Before the covey releases in autumn 2007, we recorded the exploratory behaviour of individual birds in a novel environment (Box 13.2), a new pen (Box Fig. 13.2.1). We also tested if, and how, exploratory behaviour changed in the presence of a predator model, in this case a stuffed fox, and if the exploratory behaviour was consistent across situations and over time. We then monitored the survival of these individuals after release using radio-telemetry, as the possible effect of temperament on survival may be linked to susceptibility to predation. Less fearful birds could be too incautious in exploring a new and potentially dangerous environment, whereas more fearful birds could be less willing to explore for food and therefore become weak and less able to escape from predators.

After each release, we used radio-tracking to monitor survival in the released birds. When the radio signal had changed to indicate that a tag had been stationary for 12 hours or more, we recovered the tag and recorded remains or signs found with it. Scattered plucked feathers were regarded as raptor kills, whereas wing or tail feathers neatly snapped from the base or tooth marks on the tag indicated a mammalian predator (Thirgood et al. 1998). A buried carcass was regarded as a fox kill (Macdonald 1976; Parish and Sotherton 2007). The estimated time between the death of the bird and recovering the tag was between 12 and 30 hours, leaving a possibility that the remains found were left by a scavenger rather than the initial predator. However, Parish and Sotherton (2007) reported

---

Box 13.2 Temperament tests

Tests of exploration behaviour in a novel environment, reflecting temperament, were conducted on individual grey partridges when the young were 10–15 weeks old. The test was performed on 106 individuals from 20 coveys, and each individual was marked with an aluminium leg ring bearing a unique number. The test pen (Box Fig. 13.2.1) represented a novel environment and was therefore of different size, shape, and colour from the rearing pens. The top of the pen was enclosed with a flexible nylon net (mesh size 2 cm) preventing injury from possible flights inside the pen. Two sides of the test pen were covered from outside with 1.5 metre-high wooden boards to hide the observers. A video camera

was set up at the other end of the test pen behind boards with a 10 x 20 cm hole for a view into the pen. An observer operated the video camera, staying silent and motionless behind the cover boards. Each individual was filmed in the test pen for five minutes, and afterwards released back into their home pen. We designed the test procedure to cause the birds as little stress as possible. This same test was repeated on 30 randomly chosen birds (three birds per ten coveys) 4–5 weeks later to test for consistency in the behaviour of individual birds.

We also tested, on some individuals, how exploratory behaviour would change in the presence of a predator model

continued

Box 13.2 *Continued*

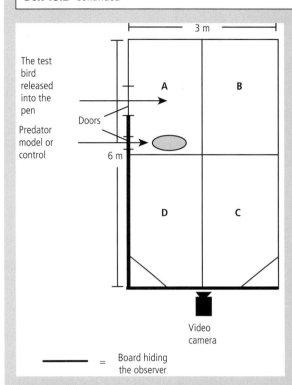

Box Figure 13.2.1 Diagram of the test pen representing a novel environment in the pre-release temperaments tests on the captive-bred grey partridges. From Rantanen (2009).

versus a control object. Following the first test as described above, one of two objects, either a predator model (stuffed fox) or control (clear 1.5 litre plastic bottle), was inserted into the pen through a covered hole on one side of the pen. The bird was then filmed for three minutes, after which the model or the control object was removed and the bird was allowed to settle for one minute. Then, the other of the two objects was inserted into the pen through the hole and the bird was filmed for another three minutes, after which the second object was removed. The order of the two objects was randomized. After the test, the birds were released back into their home pen.

The video recordings from the novel environment and the predator model tests were analysed from a TV screen by counting the number of steps taken by the tested animal as well as recording the number of different pen zones visited per 30-second period. These analyses gave the following measures: the average per-minute stepping rates and the average per-minute zone-visiting rates describing the level of exploration performed by the individual. Number of pecks as displacement behaviour and flights as escape behaviour performed by the individuals were also counted. For statistical analyses, the behavioural variables that correlated positively between the first tests and the repeat tests were combined by principal component analysis (PCA) into one principal component which was used as a measure of exploration, with high scores reflecting active, bold behaviour and low scores passive, shy behaviour.

that, on similar lowland farmland in the UK, scavengers generally did not discover test carcasses for five to eight days, so we consider that the remains of our tagged birds were unlikely to have been scavenged before we recovered them.

Our pre-release behavioural observations on individual grey partridges in the novel environment showed that the captive stock consisted of individuals with dispositions that ranged from fearful to bold, which correlated with their reactions to the predator model. This would imply that the individual's fearfulness is reflected in its behaviour in threatening situations. However, statistical tests showed that we could not predict post-release survival from the animal's behaviour in the pre-release tests: we did not find any significant difference other than a weak trend when

we compared the results of the temperament tests between the birds that died during the first 30 or 60 days after release and the birds that survived beyond this period (Figure 13.7a,b).

13.3 Implications for captive breeding and reintroductions

Our observations revealed maladaptive behaviour in released captive-bred grey partridges, which may contribute to the poor success experienced in grey partridge reintroductions. Released birds preferred habitat that may actually be bad for their survival. We also demonstrated how captive breeding in grey partridge may weaken innate anti-predator behaviours, such as vigilance and flying to roost. Since

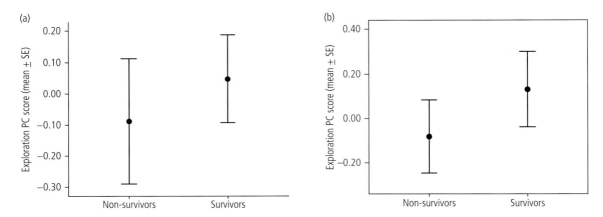

Figure 13.7 Comparison of average exploration scores from the novel environment test between the released grey partridges that were the non-survivors and survivors of the first 30 days (a) and the first 60 days (b) from release. From Rantanen (2009).

anti-predator behaviour such as vigilance is connected with emotional reactivity and fearfulness in animals (Boissy 1995), the low vigilance rates of released grey partridges support Price's (1999) general conclusion that domestication during captive breeding would produce stock with low emotional reactivity. In addition, the pre-release experiment indicated that some of the captive grey partridges were highly exploratory in a novel environment, suggesting low fearfulness in these individuals.

Selective breeding could make the captive population more fearful, by excluding from the rearing environment the factors that originally favoured the breeding of less emotionally reactive individuals, or at least minimizing their influence. This would generally mean tackling the sources of psychological stress in the captive conditions in order to cater for the more fearful individuals. For example, zoos have been using so-called environmental enrichment to stimulate and improve psychological well-being in captive animals (Shepherdson 1998). Such techniques could help to accommodate shy and fearful individuals better within the captive environment, for instance, by providing more shelter for them so that they feel less exposed, and thus possibly decreasing the domestication effect which favours low emotional reactivity.

Some of our results also support the notion that protection from predation in captive breeding leads to variation and inconsistency in anti-predator behaviour in the captive-bred animals (Anttila et al. 1995; McPhee 2003; Håkansson and Jensen 2005), because anti-predator strategies such as flying to roost at sunset and roosting in the open were observed variably

and inconsistently among the released grey partridges. The same individuals were observed flying to roost on one evening and walking on another, or roosting in the middle of a field on one night while roosting near the field boundary with cover on another. While the latter behaviours could be considered maladaptive in terms of predator avoidance, the same individuals appeared to be quite capable of behaving appropriately by flying to roost and roosting away from the field boundaries on other occasions. So when the appropriate innate behaviour has been preserved despite captive breeding, experience may reinforce it after release. Indeed, the roosting study showed that the tendency of the released birds to roost at increased distances from field boundaries over time was associated with a decline in their weekly mortality rates.

On the other hand, inexperience may also lead to maladaptive day-time use of habitats and to an ecological trap, as was demonstrated by the habitat preferences study (Rantanen et al. 2010a). Day- and night-time use of habitats by the released birds may be difficult to influence by training or by modifying the rearing environment, unless it is possible to use vast outdoor enclosures and perform simulated predator attacks in the 'risky' habitats. There are examples of anti-predator training by using wild-caught, experienced animals to demonstrate the proper anti-predator behaviour to captive-bred animals, which have resulted in higher post-release survival among the released, trained individuals (Shier and Owings 2007; Gaudioso et al. 2011). However, we cannot recommend catching wild grey partridges, and, where they are present, it is much better to foster captive-bred young to a wild pair (Buner and Aebischer 2008; Buner et al. 2011) and let captive-bred young learn the appropriate anti-predator behaviour from a wild bird that way.

In terms of grey partridge release methods, our results support the recommendations of Buner et al. (2011) of autumn releases of coveys being the preferred method over pair releases in spring, if fostering captive-bred young to wild parents is not possible because of low wild grey partridge numbers in the area, as was the case with our study. Buner et al. (2011) found that the fostering method resulted in highest post-release survival rates compared to releases of coveys in autumn and pairs in spring, and recommended that, once the number of free-living adult birds has been boosted by releases of autumn coveys, fostering should then be the favoured release method. Nevertheless, the quality of the autumn birds in the initial releases is still important, as it will determine their survival and ability to serve as appropriate parents to the fostered young.

13.4 Conclusions

Our findings prompt several practical suggestions for the future of grey partridge management.

For planning of releases and post-release management, the lower post-release mortality rates in autumn suggest that grey partridges should be released as coveys, rather than as pairs in spring when the mortality rates were unsustainable (Rantanen et al. 2010a). Indeed, wild grey partridges also have better survival during the covey period in the autumn than the nesting period in the spring, possibly due to the benefits of group vigilance (Rymensova et al. 2012). Inadvertently, we also demonstrated the importance of covey stability to these autumn releases of grey partridges. We observed two autumn releases; soon after the first, the coveys disintegrated and pairing started exceptionally early, whereas after the second, the coveys remained intact for much longer. It appeared that stable coveys were more vigilant, more likely to fly to roost, and suffered lower mortality, which suggests that rearing practices should promote strong social bonds in grey partridge coveys and avoid early pairing if possible. In addition, coveys should not be released too late in the autumn and hence too close to the period of pairing, so that the released birds can have the protection of their covey at least through the first few weeks after release when mortality rates are highest. Furthermore, the habitat preference for game covers by autumn coveys suggests that these habitats could induce settlement onto the release sites, where supplementary feeding and predator control could be concentrated to decrease overwinter mortality (Rantanen et al. 2010a).

We found that released grey partridges fed throughout the day. However, if permitted by animal welfare regulations, it may be beneficial to restrict the feeding times in grey partridge rearing to dawn and dusk (Rantanen et al. 2010b), thus mimicking the bimodal feeding routine of wild birds that has probably emerged as optimal from a trade-off between energy gain and predation risk (McNamara et al. 1994). Therefore, this protocol could teach the birds to be less occupied with feeding during the day, and thereby less vulnerable to day-time predators. Flying to roost could be promoted by selective breeding and/or releases of coveys consistently showing pre-roost 'restlessness' (calling and fluttering their wings) which could be more likely to fly to roost. Finally, the release of parent-reared coveys (young hatched and reared by their parents) as opposed to artificially reared birds (as described and used in the studies above) is likely to improve post-release survival as a result of more natural behaviour (Buner and Schaub 2008).

We hope that this chapter also demonstrates the fruitful contribution that behavioural research can make to improving the effectiveness of wildlife reintroductions. Every species, game or otherwise, has a set of crucial behavioural traits which may be impaired by captive breeding, and it raises welfare questions regarding the ethics of releasing with impaired survival abilities (Harrington et al. 2013). Caution should be applied in reintroductions to make sure that returning species have not lost their original performance (if compared to the original wild population) due to being in captivity (Moehrenschlager et al. 2013), such as with changes in morphological traits (e.g. increased reproduction in captive butterflies in detriment of flight capacity if compared to wild individuals; Lewis and Thomas 2001). Game birds bred for conservation should be seen and treated much less as domestic animals and much more as wild animals, because they need to be able to survive and flourish in the wild. Moreover, wildlife reintroductions and game bird releases for conservation have not only similar goals but also similar problems associated with the consequences of captive breeding, and we suggest that sharing knowledge and experiences between reintroductions and game science should benefit both interests.

Acknowledgements

We thank the land owners who allowed us to carry out the research on these study sites, and we give special thanks to all field assistants who helped with releases and collecting data. We are grateful to Phil Riordan, Dawn Burnham, John Quinn, and Sandra Baker for ideas and methodological guidance, and to Paul Johnson for statistical advice. This study was funded by the generous donations from the late Mr Fred Packard and the late Mr Hugh van Cutsem. We also received funding from various donors through the Game & Wildlife Conservation Trust. Ruth Feber and Eva Raebel gave helpful advice and input to the chapter.

References

Aebischer, N.J. and Ewald, J.A. (2004). Managing the UK Grey Partridge *Perdix perdix* recovery: population change, reproduction, habitat and shooting. *Ibis*, **146**, 181–191.

Aebischer, N.J. and Ewald, J.A. (2010). Grey partridge *Perdix perdix* in the UK: recovery status, set-aside and shooting. *Ibis*, **152**, 530–542.

Aebischer, N.J., Robertson, P.A., and Kenward, R.E. (1993). Compositional analysis of habitat use from animal radio-tracking data. *Ecology*, **74**, 1313–1325.

Anttila, I., Putaala, A., and Hissa, R. (1995). Tarhattujen ja villien peltopyyn poikasten kayttaytymisesta. *Suomen Riista*, **41**, 53–65.

Baillie, S.R., Marchant, J.H., Leech, D.I., et al. (2013). *Bird-Trends 2012: trends in numbers, breeding success and survival for UK breeding birds*. BTO Research Report No. 644. BTO, Thetford.

Baker, H., Stroud, D.A., Aebischer, N.J., Cranswick, P.A., Gregory, R.D., McSorley, C.A., Noble, D.G., and Rehfisch, M.M. (2006). Population estimates of birds in Great Britain and the United Kingdom. *British Birds*, **99**, 25–44.

Beani, L. and Dessi-Fulgheri, F. (1998). Anti-predator behaviour of captive Grey partridges (*Perdix perdix*). *Ethology Ecology and Evolution*, **10**, 185–196.

Boissy, A. (1995). Fear and fearfulness in animals. *The Quarterly Review of Biology*, **70**, 165–191.

Browne, S.J., Buner, F., and Aebischer, N.J. (2009). A review of gray partridge restocking in the UK and its implications for the UK Biodiversity Action Plan. In: Cederbaum, S.B., Faircloth, B.C., Terhune, T.M., Thompson, J.J., and Carroll, J.P. eds., *Gamebird 2006: Quail VI and Perdix XII. 31 May—4 June 2006*: 380–390. Warnell School of Forestry and Natural Resources, Athens, GA, USA.

Brun, J.-C. and Aubineau, J. (1989). La reconstitution des populations de perdrix rouges (*Alectoris rufa*) et grises (*Perdix perdix*) à l'aide d'oiseaux d'élevage. *Gibier Faune Sauvage*, **6**, 205–223.

Buner, F. and Aebischer, N.J. (2008). *Guidelines for re-establishing grey partridges through releasing*. The Game and Wildlife Conservancy Trust, Fordingbridge. ISBN: 978–1–901369–17–5.

Buner, F.D., Browne, S.J., and Aebischer, N.J. (2011). Experimental assessment of release methods for the re-establishment of a red-listed galliform, the grey partridge (*Perdix perdix*). *Biological Conservation*, **144**, 593–601.

Buner, F. and Schaub, M. (2008). How do different releasing techniques affect the survival of reintroduced grey partridges *Perdix perdix*? *Wildlife Biology*, **14**, 26–35.

Caro, T.M. (2005). *Antipredator defenses in birds and mammals*. University of Chicago Press, Chicago.

Curio, E. (1996). Conservation needs ethology. *Trends in Ecology and Evolution*, **11**, 260–263.

Dowell, S.D. (1990). Differential behaviour and survival of hand-reared and wild grey partridge in the United Kingdom. In K.E. Church, R.E. Warner, and S.J. Brady, eds., *Perdix V: Gray Partrige and Ring-necked Pheasant Workshop*, pp. 230–241. Kansas Department of Wildlife and Parks, Emporia.

Eaton, M.A., Brown, A.F., Noble, D.G., et al. (2009) Birds of Conservation Concern 3: the population status of birds in the United Kingdom, Channel Islands and the Isle of Man. *British Birds*, **102**, 296–341.

Fischer, J. and Lindenmayer, D.B. (2000). An assessment of the published results of animal relocations. *Biological Conservation*, **96**, 1–11.

Game Advisory Booklet (1964). *Partridge Rearing*. Eley Game Advisory Service, Fordingbridge.

Gaudioso, V. R., Sanchez-Garcia, C., Perez, J. A., et al. (2011). Does early antipredator training increase the suitability of captive red-legged partridges (*Alectoris rufa*) for releasing? *Poultry Science*, **90**, 1900–1908.

Griffith, B., Scott, J.M., Carpenter, J.W., and Reed, C. (1989). Translocations as a species conservation tool: status and strategy. *Science*, **245**, 477–480.

GWCT (2012) <http://www.gwct.org.uk/game/research/species/pheasant/>and<http://www.gwct.org.uk/game/research/species/red-legged-partridge/>. 27th February 2015.

Håkansson, J. and Jensen, P. (2005). Behavioural and morphological variation between captive populations of red junglefowl (*Gallus gallus*)—Possible implications for conservation. *Biological Conservation*, **122**, 431–439.

Håkansson, J. and Jensen, P. (2008). A longitudinal study of antipredator behaviour in four successive generations of two populations of captive red junglefowl. *Applied Animal Behaviour Science*, **114**, 409–418.

Harrington, L.A., Moehrenschlager, A., Gelling, M., Atkinson, R.P.D., Hughes, J., and Macdonald, D.W. (2013). Conflicting and complementary ethics of animal welfare considerations in reintroductions *Conservation Biology*, **27**, 486–500.

Harris, S. and Woollard, T. (1990). The dispersal of mammals in agricultural habitats in Britain. In R.G.H. Bunce and D.C. Howard, eds., *Species Dispersal in Agricultural Habitats*, pp. 159–188. Belhaven Press, London.

IUCN (World Conservation Union) (1998). *Guidelines for reintroductions*. IUCN/SSC Re-introduction specialist group, IUCN, Gland and Cambridge.

Jackson, C.D. and Brown, G.E. (2011). Differences in antipredator behaviour between wild and hatchery-reared juvenile Atlantic salmon (Salmo salar) under seminatural conditions. *Canadian Journal of Fisheries and Aquatic Sciences*, **68**, 2157–2165.

Jenkins, D. (1961). Social behaviour in the Partridge Perdix perdix. *The Ibis*, **103a**, 155–191.

Kenward, R.E. (2001). *A Manual for Wildlife Radio Tagging*. Academic Press, London.

Lewis, O.T. and Thomas, C.D. (2001). Adaptations to captivity in the butterfly Pieris brassicae (L.) and the implications for ex situ conservation. *Journal of Insect Conservation* **5**, 55–63.

Macdonald, D.W. (1976). Food caching by red foxes and some other carnivores. *Zeitschrift für Tierpsychologie*, **42**, 170–185.

Macdonald, D.W. (2009). Lessons learnt and plans laid: seven awkward questions for the future of reintroductions. In M. Hayward and M. Somers, eds., *Reintroduction of Top-Order Predators*, pp. 411–448. Wiley-Blackwell, Oxford.

Macdonald, D.W. and Feber, R.E., eds. (2015). *Wildlife Conservation on Farmland. Conflict in the Countryside*. Oxford University Press, Oxford.

Martin, J. (2011). The transformation of lowland game shooting in England and Wales since the Second World War: the supply side revolution. *Rural History*, **22**, 207–226.

Mathews, F., Orros, M., McLaren, G., Gelling, M., and Foster, R. (2005). Keeping fit on the ark: assessing the suitability of captive-bred animals for release. *Biological Conservation*, **121**, 569–577.

McNamara, J.M., Houston, A.I., and Lima, S.L. (1994). Foraging routines of small birds in winter: A theoretical investigation. *Journal of Avian Biology*, **25**, 287–302.

McPhee, M.E. (2003). Generations in captivity increases behavioral variance: considerations for captive breeding and reintroduction programs. *Biological Conservation*, **115**, 71–77.

Moburg, G.P. (1991). How behavioural stress disrupts the endocrine control of reproduction in domestic animals. *Journal of Dairy Science*, **74**, 304–311.

Moehrenschlager, A., Shier, D.M., Moorhouse, T.P., and Stanley Price, M.R. (2013). Righting past wrongs and ensuring the future: challenges and opportunities for effective reintroductions amidst a biodiversity crisis. In D.W. Macdonald and K.J. Willis, eds., *Key Topics in Conservation Biology 2*, Wiley-Blackwell, John Wiley & Sons Ltd., Oxford.

Moorhouse, T.P., Gelling, M., McLaren, G.W., Mian, R., and Macdonald, D.W. (2007). Physiological consequences of captive conditions in water voles (*Arvicola terrestris*). *Journal of Zoology*, **271**, 19–26.

Morgan, K.N. and Tromborg, C.T. (2007). Sources of stress in captivity. *Applied Animal Behaviour Science*, **102**, 262–302.

Morris, A.J. and Gilroy, J.J. (2008). Close to the edge: predation risks for two declining farmland passerines. *Ibis*, **150**, 168–177.

Parish, D.M.B. and Sotherton, N.W. (2007). The fate of released captive-reared grey partridges Perdix perdix: implications for reintroduction programmes. *Wildlife Biology*, **13**, 140–149.

Potts, G. (1980). The effects of modern agriculture, nest predation and game management on the population ecology of partridges Perdix perdix and Alectoris rufa. *Advances in Ecological Research*, **11**, 2–79.

Potts, G.R. (1986). *The Partridge—Pesticides, Predation and Conservation*. Collins Professional and Technical Books, London.

Potts, G. and Aebischer, N. (1995). Population dynamics of the Grey Partridge Perdix perdix 1793–1993: monitoring, modelling and management. *Ibis*, **137**, 29–37.

Price, E.O. (1984). Behavioral aspects of animal domestication. *Quarterly Review of Biology*, **59**, 1–32.

Price, E.O. (1999). Behavioral development in animals undergoing domestication. *Applied Animal Behaviour Science*, **65**, 245–271.

Putaala, A., Turtola, A., and Hissa, R. (2001). Mortality of wild and released hand-reared grey partridges (*Perdix perdix*) in Finland. *Game and Wildlife Science*, **18**, 291–304.

Rands, M. (1985). Pesticide use on cereals and the survival of Grey Partridge chicks: a field experiment. *Journal of Applied Ecology*, **22**, 49–54.

Rands, M.R.W. and Hayward, T.P. (1987). Survival and chick production of hand-reared gray partridges in the wild. *Wildlife Society Bulletin*, **15**, 456–457.

Rantanen, E.M.I. (2009). *Behaviour of captive-bred grey partridges (Perdix perdix) and its implications for reintroduction success*. D.Phil. Thesis, University of Oxford, Oxford.

Rantanen, E.M., Buner, F., Riordan, P., Sotherton, N.W., and Macdonald, D.W. (2010a). Habitat preferences and survival in wildlife reintroductions: an ecological trap in reintroduced grey partridges. *Journal of Applied Ecology*, **47**, 1357–1364.

Rantanen, E.M.I., Buner, F., Riordan, P., Sotherton, N.W., and Macdonald, D.W. (2010b). Vigilance behaviour and time budgets in reintroduced captive-bred grey partridges Perdix perdix. *Applied Animal Behaviour Science*, **127**, 43–50.

Robinson, R.A. (2009). *BirdFacts: profiles of birds occurring in Britain and Ireland (v1.24, June 2009)*. BTO Research Report 407, BTO, Thetford.

Rymešová D, Šmilauer P, and Šálek M (2012). Sex- and age-biased mortality in wild grey partridge Perdix perdix populations. *Ibis*, **154**, 815–824.

Rymešová, D., Tomášek, O., and Šálek, M. (2013). Differences in mortality rates, dispersal distances and breeding success of commercially reared and wild grey partridges in the Czech agricultural landscape. *European Journal of Wildlife Research*, **59**, 147–158.

Shepherdson, D.J. (1998). Tracing the path of environmental enrichment in zoos. In D.J. Shepherdson, J.D. Mellen, and M. Hutchins, eds., *Second Nature: Environmental Enrichment for Captive Animals*, pp. 1–12. Smithsonian University Press, Washington, DC.

Shier, D.M. and Owings, D.H. (2007). Effects of social learning on predator training and postrelease survival in juvenile black-tailed prairie dogs, *Cynomys ludovicianus*. *Animal Behaviour*, **73**, 567–577.

Snyder, N.F.R., Derrickson, S.R., Beissinger, S.R., et al. (1996). Limitations of captive breeding in endangered species recovery. *Conservation Biology*, **10**, 338–348.

Tapper, S.C. (1992). *Game Heritage: an ecological review from shooting and gamekeeping records*. The Game Conservancy Ltd, Fordingbridge, Hampshire.

Tapper, S.C., Potts, G.R., and Brockless, M.H. (1996). The effect of an experimental reduction in predation pressure on the breeding success and population density of grey partridges *Perdix perdix*. *Journal of Applied Ecology*, **33**, 965–978.

Thirgood, S.J., Redpath, S.M., Hudson, P.J., and Donnelly, E. (1998). Estimating the cause and rate of mortality in red grouse *Lagopus lagopus scoticus*. *Wildlife Biology*, **4**, 65–71.

Tillmann, J.E. (2009). Fear of the dark: night-time roosting and anti-predator behaviour in the grey partridge (*Perdix perdix* L.). *Behaviour*, **146**, 999–1023.

Vickery, J.A., Bradbury, R.B., Henderson, I.G., Eaton, M.A., and Grice, P.V. (2004). The role of agri-environment schemes and farm management practices in reversing the decline of farmland birds in England. *Biological Conservation*, **119**, 19–39.

Wallace, M.P. (2000). Retaining natural behaviour in captivity for re-introduction programmes. In L.M. Gosling and W.J. Sutherland, eds., *Behaviour and Conservation*, pp. 300–314. Cambridge University Press, Cambridge.

Watson, M., Aebischer, N.J., and Cresswell, W. (2007). Vigilance and fitness in grey partridges *Perdix perdix*: the effects of group size and foraging-vigilance trade-offs on predation mortality. *Journal of Animal Ecology*, **76**, 211–221.

Wolf, C.M., Griffith, B., Reed, C., and Temple, S.A. (1996). Avian and mammalian translocations: Update and reanalysis of 1987 survey data. *Conservation Biology*, **10**, 1142–1154.

Zidon, R., Saltz, D., Shore, L.S., and Motro, U. (2009). Behavioral changes, stress, and survival following reintroduction of Persian fallow deer from two breeding facilities. *Conservation Biology*, **23**, 1026–1035.

CHAPTER 14

Water vole restoration in the Upper Thames

Tom P. Moorhouse, Merryl Gelling, and David W. Macdonald

If something's broken, fix it. If you don't know how, learn.

Anon

14.1 Restoration and reintroductions

Restoration is a conservation action intended to reverse a process of deterioration and loss, or to reinstate a species or ecosystem process that has been lost (Macdonald et al. 2002; Moehrenschlager et al. 2013; Sandom et al. 2013). The concept of restoration is a subset of that of conservation, differing in that the goal of conservation is to prevent deterioration, while that of restoration is to reverse it. Obviously, it is restoration if the aim is to repair, or even recreate, the habitat that a particular species or community requires but, less obviously, restoration may involve not only the fostering of beneficial factors but the removal of inimical ones. Restoration may remedy problems arising from any or all of direct exploitation, introduction of exotic predators, competitors or disease, and habitat loss or fragmentation (Diamond 1989), and may take all or some of various forms, including reintroduction of the species, removal of predators, competitors or disease, and restoration of the habitat (Macdonald et al. 2002).

Where an animal species has been extirpated, or its numbers have fallen below the level from which they can recover, the available restoration options for the species are reintroduction or supplementation. Reintroduction projects attempt to re-establish species within their historical ranges through the release of wild or captive-bred individuals following local extirpation or extinction in the wild (IUCN 1998). Reintroductions have become an important tool in conservation biology and wildlife management (Griffith et al. 1989; Seddon 1999; Fischer and Lindenmayer 2000; Seddon et al. 2007). However, early attempts at reintroductions frequently failed to fulfil their potential (Beck et al.

1994; Wolf et al. 1996; Fischer and Lindenmayer 2000). Of 116 published reintroduction studies reviewed by Fischer and Lindenmayer in 2000, only 30 (26%) were classified as 'successful'. Such observations led to increased interest in the factors that influence reintroduction 'success' (Fischer and Lindenmayer 2000; Seddon et al. 2007). Teixeira et al. (2007) consider the three main objectives of a reintroduction to be: (1) survival of the animals after release, (2) settlement of the animals into the release area, and (3) successful reproduction in the release area. Seddon (1999) and Armstrong and Seddon (2008) additionally require persistence of the established population for at least a given quantity of time. It is, however, difficult to specify how long a population must persist following the onset of breeding for the reintroduction to be deemed a 'success'.

Reintroduction biology is a rapidly growing but relatively young discipline (Seddon et al. 2007). Seddon et al. (2007) and Armstrong and Seddon (2008) emphasize the value of experimental approaches to test hypotheses on reintroduction success, as opposed to inferring patterns from *post hoc* interpretation of monitoring results. For these purposes, therefore, an ideal reintroduction project is one in which the practical conservation actions (namely the establishment of new populations of a previously extirpated species within its historical range) are accompanied by an experimental design which permits the testing of the factors that may influence reintroduction success. Obviously, it is not appropriate to consider reintroduction if the causes of the original decline have not been understood and remedied (Sarrazin and Barbault 1996). The most significant of these issues, but far from the only one, is habitat quality. The most likely cause of failure for a

Wildlife Conservation on Farmland. Managing for Nature on Lowland Farms. Edited by David W. Macdonald and Ruth E. Feber.
© Oxford University Press 2015. Published 2015 by Oxford University Press.

reintroduction is the release of a species into unsuitable habitat (Griffith et al. 1989; Balmford et al. 1996). The quality of the habitat into which animals are released has the potential to influence a large number of population parameters, such as mortality rates, breeding rates, range sizes, and population densities.

A further issue is that the conditions in which individuals are maintained prior to release can adversely affect their health and well-being (Gelling et al. 2010, see Box 14:1). To optimize success in reintroductions, animals should be released in the best condition possible (Seddon et al. 2007). The maintenance of animals in captivity as part of a larger programme to ensure a particular species' survival is becoming increasingly commonplace (Teixeira et al. 2007), and within the realms of species reintroduction and translocation, individuals or groups of some species are routinely bred and/or held in captivity prior to being released into the wild. Animals are often housed under conditions which may not be representative of their preferred social structure, and this is particularly likely during transportation to the release site (Morgan and Tromborg 2007). Indeed, in many cases, housing conditions are defined by convenience for the establishment, rather than animal welfare considerations (Olsson and Westlund 2007). Although much research has been conducted into social group size and optimal housing conditions in, for example, zoos (Price and Stoinski 2007) and for some species groups, little attention has focused upon those animals being housed or transported for the purposes of reintroduction or translocation.

14.2 The aims of the study

The purpose of the Upper Thames water vole restoration project was to reverse the threat status of an endangered small mammal—the water vole *Arvicola amphibius* (formerly *A. terrestris*)—in the Upper Thames catchment, and in so doing also to learn about the process of reintroduction and derive guidelines to inform future conservation actions. Our aims were two-fold. First, from a species-specific perspective, we aimed to re-establish populations of water voles along rivers in the Upper Thames, and to test how riparian habitat enhancements resulting from management practices such as creating riparian buffer strips, livestock fencing, and scrub clearance (Strachan et al. 2011; Chapter 15, this volume) might affect the success of water vole reintroductions. Second, we wished to derive generally applicable lessons for reintroduction ecology as a whole. We accordingly devised a series of

experimental water vole releases to answer the following questions: What effect does habitat quality have on the success of re-introductions? Are there any impacts of habitat quality on population density and growth rates in the new populations? Could conditions in captivity pre-release influence the outcome of reintroductions? How soon do 'clean' captive-bred individuals acquire parasites and pathogens when in the wild?

14.3 Why water voles? Ecology, conservation, and suitability for reintroduction

Water voles (Fig. 14.1) are Arvicoline rodents (the subfamily of Rodentia that contains the voles, lemmings, and muskrats), and were once commonplace throughout the British Isles.

The species has declined more rapidly than any British mammal in the last century (Telfer et al. 2003), and is now on the official endangered list for the United Kingdom (see Chapter 15, this volume; Jefferies et al. 1989; Strachan and Jefferies 1993; Macdonald and Strachan 1999; Strachan et al. 2000). This decline results from a combination of habitat loss, and predation by the invasive American mink *Neovision vison* (Barreto et al. 1998a, b; Woodroffe et al. 1990; Lawton and Woodroffe 1991; Barreto and Macdonald 2000; Macdonald and Feber 2015: Chapter 6).

In many areas the degree of fragmentation of the national population of water voles has become so great that, although good quality water vole habitat may exist, it is likely to remain unoccupied, simply because it is too far from another existing population to be recolonized by natural dispersal (Telfer et al. 2001; Bonesi et al. 2002). Indeed there is evidence that water voles may have become completely extinct in some counties in the UK. Restoring water voles to the wider countryside in these cases is likely to be accomplished most effectively by the reintroduction of individuals from captive bred populations (Strachan et al. 2011).

Water voles live in a variety of UK habitats, from wetlands to estuaries, lakes, rivers, ditches, and streams (Strachan et al. 2011; Chapter 15, this volume). A large proportion of these habitats are located on farmland, and have been degraded or destroyed by agricultural intensification, for example through the drainage of farm wetlands, or increased stocking densities of livestock leading to poaching of waterway banks (Strachan et al. 2011). Water voles require large quantities of riparian vegetation, which serves as both food and shelter from predation (Zejda and Zapletal 1969;

Figure 14.1 A water vole perched on the lip of a potato crisp tube. These tubes are an effective way of handling water voles as they will run into the tube freely, and can then be weighed and measured and easily released back into the pen/wild as appropriate. Photograph ©A. Griffiths.

Woodall 1977; Barreto et al. 1998a, b; Telfer et al. 2001; Strachan et al. 2011); this vegetation needs to grow from earth or silt-shored banks of slow-flowing water courses into which they are able to burrow (Zejda and Zapletal 1969; Woodall 1993; Telfer et al. 2001; Strachan et al. 2011). In most suitable habitats in the UK, water vole ranges comprise a narrow (1–2 m) width of habitat over a much larger length; typically 30–300 m (Stoddart 1970; Lawton and Woodroffe 1991; Barreto et al. 1998a; Macdonald and Strachan 1999). Measurements of water vole ranges are therefore usually considered, not in terms of *area*, but in terms of *length* of occupied habitat (Stoddart 1970).

Water voles are an ideal model species for testing the effects of habitat quality upon reintroduction success, for a number of reasons. First, the causes of the species' original decline are both well understood and able to be remedied. Second, the linear nature of the water voles' habitat in our farmland study area meant

that we were able to locate replicate lengths of river that varied in habitat quality. Variations in amounts of riparian vegetation affect predation risk and food abundance for water voles (Barreto et al. 1998a), and thereby affect survival rates (Efford 1985). Selecting stretches of river of the same length, but with different widths of suitable vegetation bordering the water, allowed us to test how survival rates in reintroductions are affected by vegetation abundance, by creating a number of experimental replicates, each representing a different abundance of suitable vegetation per unit range length. Third, water voles are short-lived (a maximum of two winters in the wild), have short gestation times (20–23 days), and large litter sizes (3–6 pups), and newborn animals can achieve breeding condition within a month of birth (Efford 1985). They breed well in captivity, can be released in large numbers, and are solitary, and so captive-bred individuals are likely to have lost relatively few behavioural traits necessary for survival, compared with social species (e.g. Gusset et al. 2006; Shier 2006). These traits mean that the success or failure of a reintroduction can be assessed within a relatively short time period, enabling conclusions to be drawn from short-term studies.

14.4 Design and preparation of project

The project comprised a series of experimental water vole releases into 12 sites in the Upper Thames over a three year period from 2005 to 2007. Releases were made into 12 sites in all: five sites in 2005, five in 2006, and a further two sites in 2007. All sites were independent because, with the exception of two locations on the lower Windrush, which were situated on separate arms of the river and not in direct contact (there was no recorded over-land dispersal between the sites), sites were separated by a minimum of 5 km linear distance of waterway. Each site comprised an unbroken stretch of suitable riparian habitat approximately 800 m in length, along which voles were released into a series of pens. Each 800 m (± 40 m) release site was bounded at each end by unsuitable habitat in a location at which water voles had been previously recorded in the past 20 years by the local wildlife non-governmental organization (the Bedfordshire, Buckinghamshire and Oxfordshire Wildlife Trust, BBOWT), but from which they had since been eradicated by mink predation.

Six of the twelve sites had an overall width (perpendicular to the water course) of suitable habitat of 3 m or less. The remaining six sites had suitable widths of between 3–6 m. Figure 14.2 shows a site supporting an approximately 3 m margin. These widths represent the typical range found in wild water vole habitats (e.g. Moorhouse and Macdonald 2008), and, while sites in practice represented a continuum of habitat widths, selecting within these categories ensured an even spread across an appropriate range of vegetated widths. To measure the actual vegetation abundance at each site, we employed quadrat techniques (e.g. Zejda and Zapletal 1969; Lawton and Woodroffe 1991; Telfer et al. 2001). The quadrats each comprised a strip 1 m in length along the bank, with the same width (perpendicular to the water course) as the habitat present. We measured (to the nearest 5 cm): (1) the width of the riparian bankside vegetation (suitable riparian vegetation—e.g. reed canary grass *Phalaris arundinacea*, or reed sweet grass *Glyceria maxima*—growing from the bank near the water's edge, as distinct from other vegetation types located away from the water's edge, usually dominated by nettles *Urtica dioica*, and dry grassland species; see Fig. 14.2); and (2) the width of emergent vegetation (riparian or aquatic vegetation growing from the water—e.g. water cress *Rorippa nasturtium-aquaticum* or reed sweet grass). We also measured the percentage cover of both vegetation types. For each quadrat, vegetation abundance was calculated as [(width of emergent × percentage cover) + (width of riparian bankside × percentage cover)]. Mean vegetation abundance for each site in each session was the mean of these values across all quadrats.

To determine how many released water voles each 800 m length could potentially support, we undertook population modelling (using Vortex 9) based on parameters derived from previous capture-mark-recapture studies (Moorhouse 2003; Moorhouse and Macdonald 2005). The models suggested that each 800 m length of habitat, with the minimal habitat width, would be sufficient to support a release population of 44 water voles with a 75% probability of population survival over 50 years. Moreover, although each 800 m replicate length was bounded at each end by unsuitable habitat, additional suitable habitat was available on each water course within 1 km linear distance up or downstream, and therefore within potential dispersal distance (see Telfer et al. 2001, 2003).

Crucial to the experiment was the removal of American mink from the release area. Mink predate heavily upon water voles (Barreto et al. 1998a, b), and, even in sites with good quality habitat and large numbers of water voles, can eradicate a population (Strachan et al. 2011; Chapter 15, this volume). A prerequisite of any water vole reintroduction is therefore the absence of mink (Strachan et al. 2011). Mink trapping began in advance of vole releases, with the aim of ensuring no contact between the released animals and mink throughout the study. Mink were captured in Game and Wildlife Conservation Trust (formerly the Game Conservancy Trust) mink rafts, each comprising a floating platform supporting a tunnel and clay tracking plate to record footprints of any mink investigating the raft (see Reynolds et al. 2007). On each tributary of the Thames hosting one of the 12 sites, mink rafts were deployed at 1 km intervals for a minimum of 4 km upstream and downstream of each 800 m release site. Rafts were monitored once a week, and, if mink prints were discovered, the tracking plate was replaced with a live-capture trap, checked daily. Captured mink were euthanized by the trapper. Mink trapping at the release sites was conducted partly by a concurrent mink control project (see Harrington et al. 2008; Macdonald and Feber 2015: Chapter 6), and partly by land-owners, coordinated in partnership with BBOWT.

Figure 14.2 A stretch of river from one of the release sites, showing a release pen (dug in prior to release) and vegetation suitable for water voles (in this case the plants growing out of the water, from flat beds just above the water, and up the bank to the front of the pen) which represents the approximate mean of the range of widths used across sites in the study. Photograph ©T. Moorhouse.

We followed the definition of reintroduction 'success' supplied by Teixeira et al. (2007)—survival of the animals after release; settlement of the animals into the release area, and successful reproduction in the release area—because it includes a measurable end-point (it requires successful reproduction in the release area). However, this definition relates only to the population establishment process (*sensu* Armstrong and Seddon 2008). We also measured short-term post-establishment persistence after the non-breeding, overwinter period (October to March). During this period, wild water vole populations typically decline by 70% (Strachan et al. 2011). This second measure provides an indication of short-term, but not long-term persistence.

14.5 Release methods

Voles were released during the first two weeks of May in each of the three years. Water voles typically breed from April through to October, so this timing allowed the maximum time for breeding and recruitment, whilst ensuring that seasonal on-site vegetation was fully grown before release. All founder animals were captive-born adults, born late in the previous breeding season, and housed in sibling groups until translocation.

In total, 532 water voles were released, in cohorts of 44 or 45 individuals per site. Water voles were sourced from two captive-breeding establishments, and housed in laboratory cages until release into pre-designated 'soft-release' pens on site. 'Hard-release' strategies (in which individuals are directly released into the environment, as opposed to soft-release where they first can acclimatize in predator-proof pens) have been shown to be ineffective in earlier projects (Strachan et al. 2011). Between 20 and 22 open-floored 'release pens' were dug into river banks at each site, close to the water in tall riparian vegetation (see Moorhouse et al. 2009 for full details; Fig. 14.2). The pens comprised a predator-proof box containing a straw bale, food, and water. The open bottoms of the pens prevented immediate escape but allowed the water voles to construct a burrow system which eventually emerged on the outside. Then as the released animals explored the site, they were able to return to a pre-constructed, familiar burrow system. The release pens at each site were spaced at 40 m intervals (each covering approximately one female territory at moderate densities: Stoddart 1970; Moorhouse 2003; Moorhouse and Macdonald 2005), and were constructed three weeks prior to the release, to allow time for vegetation recovery. Most individuals burrowed out within two or three days, and at least half of the food supplied was left untouched in each pen.

14.6 Fates of released cohorts

Of the 12 released populations, seven established successfully, with the established individuals reproducing, and the populations surviving at least their first overwinter period. These reintroductions were therefore considered a success. At the time of writing, in 2012, six of these populations on two separate rivers have expanded to the extent that they have now joined to form two continuous populations. Of the five unsuccessful reintroductions, three failed to establish and two established, and started breeding on site, but were later extirpated by American mink (Table 14.1).

Table 14.1 Summary of the results of the reintroductions and causes of failure. Numbers in brackets represent the number of populations that were lost for a particular reason during the period. These releases took place over three years, but year is omitted for clarity. From Moorhouse et al. (2009). Reproduced with permission from Elsevier.

Stage of reintroduction	Month (end)	Number of extant populations	Reason for loss of population(s)
Initial release	May	12	/
Establishment	June	9	Flooding post release (1) Failure of mink control (2)
Survival to end of breeding season	October	8	Failure of mink control (1)
Successful overwintering of population	April	7	Failure of mink control (1)

Of the three populations that failed to establish, two were destroyed by incoming American mink so rapidly that, despite mink trapping, no individuals remained alive by the third month after release. The third population was damaged by a flooding event (due to atypically severe rainfall in Britain during 2007) two weeks following release, which killed or displaced all but six individuals (Table 14.1). One of the two release cohorts that established and bred, but were later extirpated by mink, was invaded by mink during the third month after release, and the other was lost during the overwinter period.

14.7 Impact of vegetation abundance on establishment, survival rates, and population densities

At the nine sites at which populations initially established, 48% of released individuals were subsequently recaptured during post-release monitoring. Females had a 57% probability of recapture, whereas males had a 40% probability of recapture post release. Figure 14.3 presents mean survival rates, averaged across all sites for both sexes for each monitoring period post release. Male survival rates for the initial post-release period (the first month between the release and the first monitoring session, during which individuals establish home ranges in the wild) were approximately half those of the females (Fig. 14.3), after which survival rates of both sexes were approximately equal for all periods (Moorhouse et al. 2009; Fig. 14.3). The probability of recapturing water voles (indicative of survival rates and / or the likelihood of the voles dispersing away from the habitat) following the initial post-release period was correlated with mean vegetation abundance on site. The probability of recapture varied between 0.52 and 0.69 for females, and 0.34 and 0.53 for males, over the range of vegetation abundances: 71 cm^2 of vegetation per metre of bank to 361 cm^2 per metre of bank.

Post-establishment survival rates and population densities were also positively correlated with vegetation abundance (Fig. 14.4; Fig. 14.5). At all sites where water voles initially established, independent young were captured two months after release. Recruitment rates (measured as new offspring per capita of the population) were not correlated with vegetation abundance, indicating that increased quantities of vegetation influenced population densities through offering increased protection from predation; as opposed to increasing food resources for breeding (Moorhouse et al. 2008). Our results imply that thick and extensive

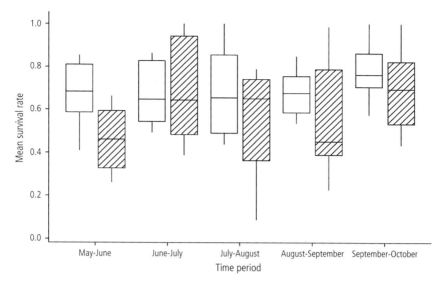

Figure 14.3 Boxplot of mean survival rates averaged across all successful sites for male and female water voles for each trapping period. Females are represented by open boxes, males by hatched boxes. Boxes show median, interquartile range, and outlier whiskers. From Moorhouse et al. (2009). Reproduced with permission from Elsevier.

riparian vegetation promotes a higher likelihood of individuals establishing on site, higher survival rates for established individuals, and higher population densities (Fig. 14.4; Fig. 14.5; Moorhouse et al. 2009).

The difference in initial recapture rates of each sex apparently results from their reaction to the process of establishment, as opposed to low male survival being

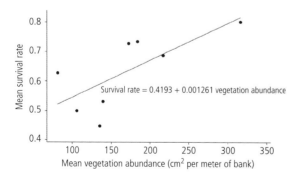

Figure 14.4 Regression plot demonstrating the relationship between mean survival rate and mean vegetation abundance at each site where water voles established. Survival rates are represented as mean survival across all capture sessions for each site, averaged from the June–July period onwards. Vegetation abundance is represented as mean abundance across all capture sessions from June onwards. From Moorhouse et al. (2009). Reproduced with permission from Elsevier.

a general trait in water voles. After the initial establishment period, male survival rates were equal to those of the females (Fig. 14.3). Male water voles have larger range-sizes than do females in wild populations (Moorhouse and Macdonald 2005, 2008), and at all release sites the nearest capture location of males from their release pen was double that of females on average (103 m as opposed to 51 m). Hence, this initial discrepancy in recapture rates is probably the result of increased mortality of males ranging further than females following release. At sites with low initial recapture rates, both sexes ranged significantly further from their point of release than at sites with high initial recapture rates. This suggests that water voles released into poor quality habitat ranged further from their point of release—conceivably to locate a better quality home range—and were more likely to suffer early mortality than those in better habitats, which stayed closer to their release point.

While it is not possible to establish a causal relationship between habitat quality, post-release ranging, and survival from the above data, Gelling et al. (2012a) showed that faecal cortisteroid or corticosterone levels (one measure of the stress response) in the reintroduced water voles were inversely correlated with site quality. For example, at sites with high vegetation abundance, the water voles had low faecal cortisol levels, suggesting that these voles experienced less physiological

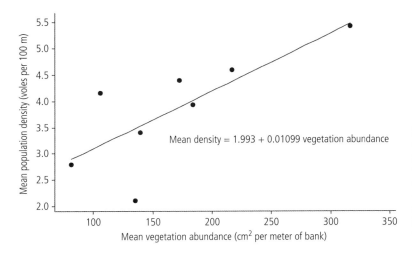

Figure 14.5 Regression plot demonstrating the relationship between mean population density and mean vegetation abundance at each site where water voles established. Density and vegetation abundance are averaged across all trapping sessions from June onward. From Moorhouse et al. (2009). Reproduced with permission from Elsevier.

stress than at sites with lower vegetation abundance. This finding adds greater weight to the suggestion that habitat quality may have influenced the water voles' ranging behaviour, and also emphasizes the importance of ensuring that reintroductions should be made only into the best possible habitat.

14.8 Effect of habitat quality on water vole growth rates

Our experimental reintroductions provided an ideal platform from which to examine the relationship between habitat quality, population density, and growth rates. Our results indicated that greater abundance of riparian vegetation increased individual survival rates and population densities in the established populations. By repeatedly sampling the body weights of established individuals, and those of the young born on site, we were also able to compare individual growth rates between sites, and calculate the mean time for young to reach sexual maturity under a given treatment of habitat width (water voles reach breeding condition at a given weight, not at a given age, and so growth rates may affect the age of sexual maturity; Moorhouse et al. 2008).

At the reintroduction sites, the significant between-population variation in growth rates caused differences between populations in the time taken for young to reach sexual maturity. Individuals' growth rates positively correlated with the length of their range; in turn, range lengths of individual voles were negatively correlated with population density (Moorhouse et al. 2008). That larger population densities are associated

with smaller range lengths in both male and female water voles is well known (Moorhouse et al. 2008), and so it is not unexpected that water voles living at release sites with more vegetation, and therefore higher population densities, should have smaller range sizes.

Water voles typically feed only in the first 1 m width of riparian habitat—approximately 99% (837 of 848) of the feeding sign surveyed was within 1 m of the water's edge (Moorhouse et al. 2008), and so the mean amount of forage available per individual at a site increased with mean range length (Moorhouse et al. 2008). From these results we infer that the time taken for young to reach maturity was greater in higher density populations, because the decreased range lengths in these populations reduced the forage available per individual, resulting in slower individual growth rates.

The above observations comprise the first evidence from wild populations of small mammals that demography can be regulated, in part, by a feed-back mechanism derived from the effects of variation in range sizes in response to population density upon forage availability, and therefore maturation rates (Moorhouse et al. 2008).

14.9 Health and welfare in reintroduced water voles

14.9.1 Effect of housing conditions on stress

Animals destined for release spend periods in captivity under conditions which do not occur in the wild (Morgan and Tromborg 2007). In water vole populations, females are territorial and males, while not territorial, compete for access to females, and both sexes are therefore typically solitary (Moorhouse and Macdonald 2008). In our study the water voles arrived from the captive breeding facility in laboratory cages, some containing single occupants, and others with up to eight members of a family group (Gelling et al. 2010). We wished to measure how the stress of these individuals varied with group size. Measuring stress in animals presents a challenge, especially where—as in our captive water voles—behaviour is difficult to record. Two proxy measures in these circumstances are body weight (McLaren et al. 2004), and leukocyte coping capacity (LCC) (McLaren et al. 2003). Measurements of LCC are made by chemically simulating an immune challenge in a small amount of whole blood *in vitro* to produce a quantifiable measure of an individual's capacity to mount an immune response. These results are then compared to the same animal's basal (unchallenged) immune system response (McLaren et al. 2003). LCC correlates well with physiological stress levels (McLaren et al. 2003; Montes et al. 2004; McLaren et al. 2004); lowered LCC scores suggest immuno-suppression, and therefore physiological stress in mammals, and have previously been shown to be affected by trapping and handling in wild small mammals (Gelling et al. 2009). Body weight is also affected by stress (Zhou et al. 1999).

Box 14.1 Animal welfare in reintroductions

Animal welfare typically focuses on the needs of individual animals (usually those held in captivity, in laboratories, zoos, and on farms, e.g. Sainsbury et al. 1995), whereas animal conservation, and therefore reintroduction, primarily addresses the viability of wild populations. Reintroductions certainly pose moral concerns (see e.g. Bekoff 2001)—losses of individual animals, particularly during the early stages can be high, and, even under the best conditions, reintroductions can have significant welfare costs to the individual animals being reintroduced (Swaisgood 2010), and some level of stress is inevitable (Dickens et al. 2010)—but some perceive the two fields as being incompatible (e.g. Callicott 1980; but see Callicott 2011). Given that the well-being of a population cannot be entirely separated from that of its component individuals (Rolston 1988; McLaren et al. 2007), we argue in Harrington et al. (2013) that increasing consideration of individual animal welfare[1] at all stages of a reintroduction will improve the efficiency, effectiveness, humaneness, and credibility of such projects.

Using keywords as an indicator of the aspects of a reintroduction that authors considered most important, Harrington et al. (2013), from published reports of 199 reintroduction projects carried out worldwide since 1990, found that only 6% referred explicitly to animal 'welfare'. In contrast, 84% referred to 'success', and 36% to 'cost' (meaning time, resource, or financial cost). Few projects reported actually measuring animal welfare of released individuals directly—2% of projects reported monitoring animal stress levels, 18% monitored body condition (usually only body weight).

We found that most (79%) projects already incorporated some kind of supportive measure: usually provision of food or shelter post release, or the use of pre-release pens to allow animals to acclimatize to the release site, but also veterinary treatment and pre-release training for captive-bred animals, and attempts to reduce stress during transport. Such measures are not necessarily used by practitioners with animal welfare in mind, but nonetheless have the potential to increase animals' well-being. However, the nature and extent of such support varied greatly across projects, and few tested experimentally the effectiveness of these measures or compared alternative approaches. Perhaps most importantly, although many of the reintroduction projects reviewed would have been required to consider potential animal welfare issues within the ethics committees of their own organizations[2], the lack of reporting of such considerations, and how they might have been addressed, means that projects cannot easily learn from others.

Harrington et al. (2013) recommend greater systematic and transparent evaluation of animal welfare issues[3], and highlight a number of research themes that warrant further research, including (amongst others): the health risks associated with reintroduction, post-release stress, and the effectiveness of supportive measures.

---

[1] The authors consider positive welfare conditions met if animals are able to access suitable shelter and food and behave normally, and are as able as their wild counterparts to avoid pain, injury, and disease.

[2] For example, the water vole reintroductions described in this book were subject to local ethical review within Oxford University.

[3] A decision tree is provided in Harrington et al. (2013) to guide practitioners through the relevant welfare considerations, and to illustrate the type of decision-making process required to improve animal welfare in reintroduction projects.

We found measurable differences in the LCC scores of individuals, which correlated negatively with the number of water voles in each of 37 laboratory cages. In other words, water voles housed in larger groups were more immunosuppressed, which we attribute to increased (social) stress, than those housed in cages containing fewer animals (Gelling et al. 2010). Body weights, however, were not affected by the group size of water voles in each cage, despite a previous study demonstrating that changes in housing conditions (from external enclosures into indoor, singly housed lab cages) correlated with weight (Moorhouse et al. 2007). A key difference between the water voles sampled between these studies is that animals in the pre-release study were born in late autumn, and would not be expected to have reached full adult weight at the time of sampling (May), whereas those in the previous study were sampled later in the year (July/August) when the animals were full adults. In this study we were unable quantifiably to distinguish between the effects of stress on growth rate and upon recorded body weight per se, because only one sample was taken per individual.

The above results suggested that the individuals housed in large group sizes may have been physiologically stressed compared with less densely housed individuals. Future recommendations for husbandry practice should encourage breeding establishments, and consultants translocating animals, to house captive water voles in smaller groups, or preferably individually, even if they are only being contained for short periods of time. Housing for future reintroduction programmes can be improved quite simply, and it should be easier to reduce the additional stress of overcrowding than the many other stressors which might be unavoidable.

Regardless of the mitigable nature of some stresses arising from captive conditions, captivity itself may still represent a source of stress. Gelling et al. (2012a) showed that the reintroduced voles' body weights, and levels of hydration increased significantly post release when compared with the captive values, whereas corticosterone levels (one measure of stress) decreased, suggesting that the voles were less physiologically stressed post release. These findings suggest that time in captivity should be minimized wherever possible prior to a reintroduction.

14.9.2 Leptospirosis in reintroduced voles

Health status of reintroduced animals is a potentially important factor determining reintroduction success,

but is rarely monitored in reintroduction programmes (Mathews et al. 2006). We measured various health indices of our individual water voles to investigate the possible effects of common environmental pathogens on reintroduction success. Pathogens and disease have the potential to regulate population dynamics for their hosts, even pathogens that do not stimulate clinical signs. They may delay attainment of sexual maturity, or reduce growth or survival rates (Scott 1988; Telfer et al. 2005), which may ultimately reduce the likelihood of reintroduction success. We investigated the uptake of pathogens by comparing the prevalence of *Leptospira* spp. (long suspected to infect British water voles: Strachan et al. 2011) in the 'clean' captive-bred voles, reintroduced to the experimental reintroduction sites in 2007, with levels found in wild water voles live-captured from extant populations throughout the UK (Gelling et al. 2012b).

Water voles were anaesthetized in the field (see Mathews et al. 2002), and urine and blood samples collected during each recapture session post release, to investigate the presence of Leptospire antibodies in serum. Urine samples were cultured for *Leptospira* spp., and blood samples tested by the Leptospirosis Reference Unit though a micro-agglutination test (MAT; Palmer et al. 1987).

Forty-four water voles were screened for *Leptospira* spp. via either MAT or urine culture pre-release; 16 during the first recapture, 19 in the second, 21 in the third, and 28 in the fourth recapture (therefore over a four month period). Of the 141 water voles sampled across all time-points, 21 voles were sampled on two occasions, four voles on three occasions, and five voles on each of four separate occasions, totalling 97 individual reintroduced or wild-born voles sampled. Of the three voles sampled on more than one occasion and found to be positive, none later gave negative results. No captive-bred water voles were positive for *Leptospira* spp. before release to the wild. Urine culture indicated that no individual began excreting leptospires until four months post release. However the serology results indicated that 1/14 (7.1%) of individuals sampled in the second month post release had been exposed to *Leptospira* spp. By three months post release this increased to 5/14 (35.7%), and 10/25 (40%) by four months post release. By month four 12/28 (42.9%) of individuals screened by either MAT or urine culture were positive.

The overall prevalence rate of nearly 43% in reintroduced voles was significantly higher than the 6.2% found in extant populations of wild voles, suggesting that the reintroduced voles may be immuno-compromised, and thus more susceptible to acquiring leptospires. However, water vole population density may also have contributed to the difference in disease prevalence between extant and reintroduced populations. Low density extant populations may have been too depleted to maintain leptospire infection, and conversely the reintroduced water voles were released at comparatively high densities (although within normal limits for water vole populations: Moorhouse and Macdonald 2008). While a large founder population is likely to predict population persistence (Griffith et al. 1989), our results raise the possibility that this strategy may also have negative effects if, during population establishment when animals range widely in search of territories, the increase in interactions with other voles leads to greater competition for resources, stress, and transmission of infectious disease (Bar-David et al. 2006).

14.10 Conclusions

Of the 12 released populations, seven reintroductions were successful. Our project therefore directly contributed to the practical conservation of water voles in the Upper Thames, and resulted in the first reversal of the population decline in this area. Just as importantly, however, the experimental reintroductions were conducted so that we were able to learn lessons for future water vole reintroductions, and for reintroduction ecology as a whole.

The principal reason for failed reintroductions in this project was ineffective mink control (see also Macdonald and Feber 2015: Chapter 6). We demonstrated that variations in habitat suitability between sites at which reintroductions were 'successful' resulted in large variations in the percentage of the released animals that established (43% to 61%), in post-establishment survival rates (0.45–0.80), and in the number and density of individuals (range 2.1–5.4 voles per 100 m length of habitat) that the habitat supported. Populations released into sites with more abundant vegetation potentially had a better chance of long-term survival, and the opportunity to colonize further areas. Given the limited resources available for most conservation reintroductions, this study emphasizes the need to ensure that any habitat selected for a reintroduction is the best obtainable. Similarly, our results highlight the importance of simple, practical measures for water vole conservation on farmland. If fencing of livestock (which permits the regrowth of thick swathes of riparian vegetation in intensively stocked areas; Chapter 15, this volume), creation of riparian buffer zones (both of which are eligible for AES funding; Chapter 11, this volume), and,

importantly, mink control, were employed widely, the national water vole population would be expected to derive substantial benefits (Chapter 15, this volume).

Our LCC results have clear implications for pre-release housing, specifically for water voles, but potentially also for other species being translocated or reintroduced. Water voles destined for conservation restoration programmes are bred in large outdoor pens. It is logistically unavoidable that transfer to the reintroduction site requires housing in laboratory cages, or equivalent, for an intermediate period. The data from this project (Gelling et al. 2010; 2012a), and that of Moorhouse et al. (2007) indicate that this time should, however, be minimized as far as possible. While it is expedient to house water voles in groups in laboratory cages for ease of both transport and release, the need to ensure that individuals are in the best physical condition possible for release, and the associated ethical considerations of ensuring good standards of animal husbandry and welfare, suggest that housing water voles in single units should be a standard practice.

The potential risk to humans of contracting Leptospirosis from infected water voles should be considered when developing trapping and handling protocols, but not disproportionately, since our confirmation that water voles are a host for *Leptospira* spp. does not worsen the long-established appreciation that people working near water are at risk of Weil's disease (Faine 1998; Macdonald and Feber 2015: Chapter 11).

This project was successful in terms of providing tangible conservation benefits, in establishing seven new populations of water voles, and reversing the local decline in their numbers in the Upper Thames for the first time, but also in providing a rigorous investigation of the fundamental processes that underpin the success or failure of a reintroduction, and indeed the fundamental ecology of the species. While the findings of the component studies are necessarily species specific, they have wider applicability, demonstrating that variations in the quality and quantity of habitat, and in the conditions and duration of captive housing, have real-world consequences for the numbers of individuals that establish, the resultant population densities and survival rates, and for the physiological condition of the released individuals. Above all, however, our findings serve to reemphasize the need in all wildlife reintroductions to ensure that the habitat requirements of the species are well understood, enabling sites selected for release to be the very best available, and that the original causes of the decline have been irrevocably removed.

Acknowledgements

This study was funded by the Holly Hill Trust, Environment Agency, and the People's Trust for Endangered Species. We thank BBOWT for their support. We thank English Nature for permission to access the Bure Marshes and the many landowners and volunteers who helped with the fieldwork during this study, especially Ian Ellis, Rebecca Dean, Gillian McCoy, Ruth Dalton, and Phil Davies. We are grateful to Paul Johnson for helpful conversations regarding data analysis.

References

Armstrong, D.P. and Seddon, P.J. (2008). Directions in reintroduction biology. *Trends in Ecology and Evolution*, **23**, 20–25.

Balmford, A., Mace, G.M., and Leader-Williams, N. (1996). Designing the Ark, setting priorities for captive breeding. *Conservation Biology*, **10**, 719–727.

Bar-David, S., Lloyd-Smith, J.O., and Getz, W.M. (2006). Dynamics and management of infectious disease in colonizing populations. *Ecology*, **87**, 1215–1224.

Barreto, G.R. and Macdonald, D.W. (2000). The decline and local extinction of a population of water voles, *Arvicola terrestris*, in southern England, *Zeitschrift für Saugetierkund*, **65**, 110–120.

Barreto, G.R., Macdonald, D.W., and Strachan, R. (1998a). The tightrope hypothesis, an explanation for plummeting water vole numbers in the Thames catchment. In R.G. Bailey, P.V. Joseand, B.R. Sherwood, eds., *United Kingdom Floodplains*, pp. 311–327. Westbury Academic and Scientific Publishing, Otley.

Barreto, G.R., Rushton, S.P., Strachan, R., and Macdonald, D.W. (1998b). The role of habitat and mink predation in determining the status of water voles in England. *Animal Conservation*, **2**, 53–61.

Beck, B.B., Rapaport, L.G., Stanley-Price, M.R., and Wilson, A.C. (1994). Reintroduction of captive-born animals. In P.J.S. Olney, G.M. Mace, and A.T.C. Feistner, eds., *Creative Conservation, Interactive Management of Wild and Captive Animals*, pp. 265–286. Chapman and Hall, London.

Bekoff, M. (2001). *Human-carnivore interactions: adopting proactive strategies for complex problems*. In J.L. Gittleman, S.M. Funk, D. Macdonald, and R.K. Wayne, eds., *Carnivore Conservation*, 179–195. Cambridge University Press, Cambridge, UK.

Bonesi, L., Rushton, S., and Macdonald, D. (2002). The combined effect of environmental factors and neighbouring populations on the distribution and abundance of *Arvicola terrestris*. An approach using rule-based models. *Oikos*, **99**, 220–230.

Callicott, J.B. (1980). Animal liberation: a triangular affair. *Environmental Ethics*, **2**, 311–328.

Callicott, J.B. (2011). *An introductory palinode*. In P. Galvão, ed., *Do animals Have Rights?* 121–131. Dinalivros,

Lisbon [in Portuguese] English translation available from <http://jbcallicott.weebly.com/introductory-palinode.html>. Accessed May 2012.

Diamond, J.M. (1989). The present, past and future of human-caused extinctions. *Philosophical Transactions of the Royal Society of London Series B, Biological Sciences*, **325**, 469–477.

Dickens, M.J., Delehanty, D.J., and Romero, L.M. (2010). Stress: an inevitable component of animal translocation. *Biological Conservation*, **143**, 1329–1341.

Efford, M.G. (1985). *The structure and dynamics of water vole populations*. DPhil Thesis, University of Oxford.

Faine, S. (1998). Leptospirosis. In A.S. Evans and P.S. Brachman, eds., *Bacterial Infections of People, Epidemiology and Control*, pp. 395–420. Springer.

Fischer, J. and Lindenmayer, D.B. (2000). An assessment of the published results of animal relocations. *Biological Conservation*, **96**, 1–11.

Gelling, M., Johnson, P., Moorhouse, T.P., and Macdonald, D.W. (2012a). Measuring animal welfare within a reintroduction: an assessment of different indices of stress in water voles *Arvicola amphibius*. PLOS ONE, **7** (7).

Gelling, M., Macdonald, D.W., Telfer, S., Birtles, R., Jones, T., and Mathews, F. (2012b). Parasites and pathogens in wild populations of water voles (*Arvicola amphibius*) in the UK. *European Journal of Wildlife Research*, **58**, 615–619.

Gelling, M., McLaren, G.W., Mathews, F., Mian, R., and Macdonald, D.W. (2009). Impact of trapping and handling on Leukocyte Coping Capacity in bank voles (*Clethrionomys glareolus*) and wood mice (*Apodemus sylvaticus*). *Animal Welfare*, **18**, 1–7.

Gelling, M., Montes, I., Moorhouse, T.P., and Macdonald, D.W. (2010). Captive housing during water vole (*Arvicola terrestris*) reintroduction: does short-term social stress impact on animal welfare? PLOS ONE, **5**, e9791.

Griffith, B., Scott, J.M., Carpenter, J.W., and Reed, C. (1989). Translocation as a species conservation tool—status and strategy. *Science*, **245**, 477–480.

Gusset, M., Slotow, R., and Somers, M.J. (2006). Divided we fail, the importance of social integration for the reintroduction of endangered African wild dogs (*Lycaonpictus*). *Journal of Zoology*, **270**, 502–511.

Harrington, L.A., Harrington, A.L., and Macdonald, D.W. (2008). Estimating the relative abundance of American mink *Mustela vison* on lowland rivers, evaluation and comparison of two techniques. *European Journal of Wildlife Research*, **54**, 79–87.

Harrington, L.A., Moehrenschlager, A., Gelling, M., Atkinson, R.P.D., Hughes, J., and Macdonald, D.W. (2013). Conflicting and complementary ethics of animal welfare considerations in reintroductions. *Conservation Biology*, **27**, 486–500.

IUCN (1998). *IUCN/SSC Guidelines for Reintroductions*. IUCN/SSC Reintroduction specialist group, IUCN. Gland, Switzerland.

Jefferies, D.J., Morris, P.A., and Mulleneux, J.E. (1989). An enquiry into the changing status of the water vole *Arvicola terrestris* in Britain. *Mammal Review*, **19**, 111–131.

Lawton, J.H. and Woodroffe, G.L. (1991). Habitat and distribution of water voles, why are there gaps in a species' range? *Journal of Animal Ecology*, **60**, 79–91.

Macdonald, D.W. and Feber, R.E., eds. (2015). *Wildlife Conservation on Farmland. Conflict in the Countryside.* Oxford University Press, Oxford.

Macdonald, D.W., Moorhouse, T.P., and Enck, J.W. (2002). The ecological context, a species population perspective. In M.R. Perrow and A.J. Davy, eds., *Handbook of Ecological Restoration. Volume 1, Principles of Restoration*, pp. 47–65. Cambridge University Press, Cambridge.

Macdonald, D.W. and Strachan, R. (1999). *The mink and the water vole, analyses for conservation.* Wildlife Conservation Research Unit, University of Oxford, Oxford.

Mathews, F., Honess, P., and Wolfensohn, S. (2002). Use of inhalation anaesthesia for wild mammals in the field. *Veterinary Record*, **150**, 785–787.

Mathews, F., Moro, D., Strachan, R., Gelling, M., and Buller, N. (2006). Health surveillance in wildlife reintroductions. *Biological Conservation*, **131**, 338–347.

McLaren, G., Bonacic, C., and Rowan, A. (2007). *Animal welfare and conservation: measuring stress in the wild.* In D.W. Macdonald and K. Service, eds., *Key Topics in Conservation Biology*, 120–133. Blackwell Publishing, Oxford, UK.

McLaren, G.W., Macdonald, D.W., Georgiou, C., Mathews, F., Newman, C., and Mian, R. (2003). Leukocyte coping capacity, a novel technique for measuring the stress response in vertebrates. *Experimental Physiology*, **88**, 541–546.

McLaren, G.W., Mathews, F., Fell, R., Gelling, M., and Macdonald, D.W. (2004). Body weight changes as a measure of stress, a practical test. *Animal Welfare*, **13**, 337–341.

Moehrenschlager, A., Shier, D.M., Moorhouse, T.P., and Stanley Price, M.R. (2013). Righting past wrongs and ensuring the future: challenges and opportunities for effective reintroductions amidst a biodiversity crisis. In D.W. Macdonald and K. Willis, eds., *Key Topics in Conservation Biology II.* Wiley-Blackwell, John Wiley & Sons Ltd., Oxford.

Montes, I., McLaren, G.W., Macdonald, D.W., and Mian, R. (2004). The effect of transport stress on neutrophil activation in wild badgers (*Meles meles*). *Animal Welfare*, **13**, 355–359.

Moorhouse, T.P. (2003). *Demography and social structure of water vole populations, implications for restoration.* DPhil Thesis, University of Oxford.

Moorhouse, T.P., Gelling, M., and Macdonald, D.W. (2008). Effects of forage availability on growth and maturation rates in water voles. *Journal of Animal Ecology*, **77**, 1288–1295.

Moorhouse, T.P., Gelling, M., and Macdonald, D.W. (2009). Effects of habitat quality upon reintroduction success in water voles: Evidence from a replicated experiment. *Biological Conservation*, **142**, 53–60.

Moorhouse, T.P., Gelling, M., McLaren, G.W., Mian, R., and Macdonald, D.W. (2007). Physiological consequences of captive conditions in water voles (*Arvicola terrestris*). *Journal of Zoology*, **271**, 19–26.

Moorhouse, T.P. and Macdonald, D.W. (2005). Indirect negative impacts of radio-collaring, sex ratio variation in water voles. *Journal of Applied Ecology*, **42**, 91–98.

Moorhouse, T.P. and Macdonald, D.W. (2008). What limits male range sizes at different population densities? Evidence from three populations of water voles. *Journal of Zoology*, **274**, 395–402.

Morgan, K.N. and Tromborg, C.T. (2007). Sources of stress in captivity. *Applied Animal Behaviour Science*, **102**, 262–302.

Olsson, I.A.S. and Westlund, K. (2007). More than numbers matter: The effect of social factors on behaviour and welfare of laboratory rodents and non-human primates. *Applied Animal Behaviour Science*, **103**, 229–254.

Palmer, M.F., Waitkins, S.A., and Wanyangu, S.W. (1987). A comparison of live and formalised leptospiral microscopic agglutination test. *Zentralblatt fur Bakteriologie, Mikrobiologie und Hygiene A*, **265**, 151–159.

Price, E.E. and Stoinski, T.S. (2007). Group size: Determinants in the wild and implications for the captive housing of wild mammals in zoos. *Applied Animal Behaviour Science*, **103**, 255–264.

Reynolds, J.D., Short, M., Porteus, T., Rodgers, B., and Swan, M. (2007). *The GCT mink raft.* <http://www.gct.org.uk/>, (04 August 2008).

Rolston, H III. (1988). *Environmental Ethics. Duties To and Values In the Natural World.* Temple University Press, Philadelphia, USA.

Sainsbury, A.W., Bennett, P.M., and Kirkwood, J.K. (1995). The welfare of free-living animals in Europe: harm caused by human activities. *Animal Welfare*, **4**, 183–206.

Sandom, C., Donlan, J., Svenning, J., and Hansen, D. (2013). Rewilding. In D.W. Macdonald and K. Willis, eds., *Key Topics in Conservation Biology II.* Wiley-Blackwell, John Wiley & Sons Ltd., Oxford.

Sarrazin, F. and Barbault, R. (1996). Reintroductions, challenges and lessons for basic ecology. *Trends in Ecology and Evolution*, **11**, 474–478.

Scott, M.E. (1988). The impact of infection and disease on animal populations: implications for conservation biology. *Conservation Biology*, **2**, 40–56.

Seddon, P.J. (1999). Persistence without intervention, assessing success in wildlife reintroductions. *Trends in Ecology and Evolution*, **14**, 503.

Seddon, P.J., Armstrong, D.P., and Maloney, R.F. (2007). Developing the science of reintroduction biology. *Conservation Biology*, **21**, 303–312.

Shier, D.M. (2006). Effect of family support on the success of translocated black-tailed prairie dogs. *Conservation Biology*, **20**, 1780–1790.

Stoddart, D.M. (1970). Individual range, dispersion and dispersal in a population of water voles (*Arvicola terrestris* (L.)). *Journal of Animal Ecology*, **39**, 403–425.

Strachan, R. and Jefferies, D.J. (1993). *The water vole* Arvicola terrestris *in Britain 1989–1990, its distribution and changing status.* The Vincent Wildlife Trust, London.

Strachan, S., Moorhouse, T.P., and Gelling, M.G. (2011). *Water Vole Conservation Handbook, third edition.* Wildlife Conservation Research Unit, University of Oxford, Oxford.

Strachan, C., Strachan, R., and Jefferies, D. J. (2000). *Preliminary report on the changes in the water vole population of Britain as shown by the national surveys of 1989–1990 and 1996–1998.* The Vincent Wildlife Trust, London.

Swaisgood, R.R. (2010). The conservation-welfare nexus in reintroduction programmes: a role for sensory ecology. *Animal Welfare*, **19**, 125–138.

Teixeira, C.P., Schetini de Azevedo, C., Mendl, M., Cipreste, C.F., and Young, R.J. (2007). Revisiting translocation and reintroduction programmes, the importance of considering stress. *Animal Behaviour*, **73**, 1–13.

Telfer, S., Bennet, M., Bown, K., et al. (2005). Infection with cowpox virus decreases female maturation rates in wild populations of woodland rodents. *Oikos*, **109**, 317–322.

Telfer, S., Holt, A., Donaldson, R., and Lambin, X. (2001). Metapopulation processes and persistence in remnant water vole populations. *Oikos*, **95**, 31–42.

Telfer, S., Piertney, B., Dallas, J.F., et al. (2003). Parentage assignment detects frequent and large-scale dispersal in water voles. *Molecular Ecology*, **12**, 1939–1949.

Wolf, C.M., Griffith, B., Reed, C., and Temple, S.A. (1996). Avian and mammalian translocations, update and reanalysis of 1987 survey data. *Conservation Biology*, **10**, 1142–1154.

Woodall, P.F. (1977). *Aspects of the ecology and nutrition of the water vole.* DPhil Thesis, University of Oxford.

Woodall, P.F. (1993). Dispersion and habitat preferences of the water vole (*Arvicola terrestris*) on the River Thames. *Zeitschrift für Saügetierkunde*, **58**, 160–171.

Woodroffe, G.L., Lawton, J.H., and Davidson, W.L. (1990). Patterns in the production of latrines by water voles (*Arvicola terrestris*) and their use as indices of abundance in population surveys. *Journal of Zoology (London)*, **220**, 439–445.

Zejda, J. and Zapletal, M. (1969). Habitat requirements of the water vole (*Arvicola terrestris* Linn.) along water streams. *Zoologické Listy*, **18**, 225–238.

Zhou, J., Yan, X., Ryan, D.H., and Harris, R.B.S. (1999). Sustained effects of repeated restraint stress in muscles and adipocyte metabolism in high-fed rats. *American Journal of Physiology*, **277**, 757–766.

CHAPTER 15

What does conservation research do, when should it stop, and what do we do then? Questions answered with water voles

Tom P. Moorhouse, David W. Macdonald, Rob Strachan, and Xavier Lambin

Science is facts; just as houses are made of stones, so is science made of facts; but a pile of stones is not a house and a collection of facts is not necessarily science.

Henri Poincaré

Vision without action is a daydream. Action without vision is a nightmare.

Japanese proverb

15.1 Introduction

The general public, at least in Great Britain, might imagine species conservation research along the lines of this narrative: a species has been declining as a result of a number of factors, and now scientists (in this case, ecologists) are embarking on a programme designed to remove the inimical factor(s) and restore the population. This narrative is reflected in such BBC news headlines as 'Dormice reintroduced in Warwickshire wood', 'Ospreys get a 'super-nest', or 'Kent mink eradication will help water vole return'. Predictably, details of the research process are sacrificed at the altar of sound bites. While the quest for distillation is understandable and sometimes clarifying, it often distorts beyond recognition (and comprehension) the reality of the journey from science to practice. As inconvenient as it may be, complexity, incommensurables, and judgement are not obfuscating distractions on that journey, but the daily reality of conservation research projects. Enacting wide-scale resolution of a conservation issue is likely to be substantially more complex and costly, and demand far harder decisions to be made, than is implied by catchy headlines.

There are lessons to be learnt from taking a broad perspective on the entire conservation journey from the initial discovery that a species is declining, through diagnostic research into the causes, then development and testing of solutions, to the production and implementation of a strategy for the species' recovery. This is our purpose, reflecting on the case of the water vole *Arvicola amphibius* (formerly *A. terrestris*; see Chapter 14, this volume) in the UK. Water voles provide a vivid example, for which almost the entire journey has been concentrated into little over 20 years of intensive research effort with which, along with many others, we have been privileged to engage since almost the beginning. Our perspective (born of a shared journey which the WildCRU travelled largely in the English lowlands while XL's perspective was shaped largely in the Scottish highlands) draws particularly on our own early research on diagnosing the problem, and extends to work on the wider community of waterside predators (Macdonald and Feber 2015: Chapters 6 and 7) and to solution testing (Chapter 14, this volume). We conclude with a discussion of the current state of water vole conservation, flushing out the questions that now face society as it decides the voles' future in the UK.

We address questions from two broad phases of conservation. Initially we ask: *How did it come to light that the species was a conservation concern? What evidence was required, and how was that evidence accumulated? How were the causes of the decline diagnosed? How were the remedies devised and tested? How were the remedies*

Wildlife Conservation on Farmland. Managing for Nature on Lowland Farms. Edited by David W. Macdonald and Ruth E. Feber.
© Oxford University Press 2015. Published 2015 by Oxford University Press.

implemented? We then address the questions: *When do we know enough that research is no longer necessary? Is society prepared to foot the bill, both financially and ethically, to enact the conservation actions required?* These questions are distributed through what follows, and the essay comprises our answers.

15.2 The stages of conservation

At the heart of conservation lies a dichotomy. While the problems facing, and solutions to, species conservation distil for each taxon to starkly few, often rather predictable, generalities (Macdonald and Sillero-Zubiri 2004), when it comes to practicalities, each species, indeed each situation, necessitates highly customised understanding and action. The same principles of habitat loss, degradation, and fragmentation may reverberate through the problems faced by big cats from Brazil to Borneo (Macdonald et al. 2010), the primates living alongside those cats (Macdonald et al. 2012), moths traversing UK forest blocks (Chapter 8, this volume), and damselflies searching for a diminishing supply of ponds (Chapter 10, this volume). But despite this unifying framework, and inevitable convergence in principle of the solutions, the details of each case are totally different. Likewise, the conservation process itself remains almost identical across all species, in terms of the broad stages involved in learning about the species and the types of research required for understanding to advance to the outcomes which, in detail, can be extraordinarily varied. The idealized trajectory of a conservation research programme follows a series of steps (Fig. 15.1) in two phases: research and implementation.

15.2.1 Research

1. *Revealing the decline.* Naturalistic insight and anecdote raise concerns that prompt (preferably systematic) surveys. A problem is thereby revealed and its magnitude evaluated, in comparison to the expected status quo (the absence of a reliable baseline is often a serious stumbling block).
2. *Diagnostic research.* A problem, once identified, needs to be understood before it can be solved, so hypotheses are formulated as to the likely causes and mechanisms. This diagnostic research typically involves exploration of associations, for example matching the circumstances of locations, where the species is either scarce or abundant, to a suite of candidate explanatory variables (e.g. habitat characteristics, the abundance of known predators or prey, etc.).

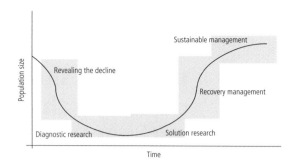

Figure 15.1 Theoretical species recovery curve for the water vole (redrawn from Strachan et al. 2005). This curve, identified for water voles, can be generalized to any species that is endangered and for which research is required to identify the causes and solutions before appropriate management is put in place.

3. *Solution research.* Once the causes have been diagnosed the cures must be developed, requiring research into candidate solutions, including experimental trials, to confirm the benefits of potential mitigation programmes.

15.2.2 Implementation

1. *Recovery management.* Once the solutions have been demonstrated then management at a large scale is required to implement them and foster the species' recovery. Implementation still incorporates a requirement for research, however: having demonstrated a solution in principle, implementing it over meaningful spatial scales requires adaptive management (that is 'learning by doing by learning') to improve its effectiveness. This approach incorporates both ecological and social elements, especially when the remedy is applied over long time periods, to overcome the fickleness of funding, policy directions, and public involvement.
2. *Sustainable management.* Once the population has recovered, the need for management is likely to continue, perhaps in perpetuity.

Now we will follow each step of this trajectory, as it unfolded through research on water voles.

15.3 Revealing the decline

How did it come to light that the species was a conservation concern? What evidence was required, and how was that evidence accumulated?

To discern a decline requires a baseline against which to compare a species' numbers and distribution.

Counting mammals is often technically difficult and, in the UK, the absence of systematic, national monitoring schemes (such as the MaMoNet—Mammal Monitoring Network, proposed by Macdonald et al. 1998, 2000) has been a long-standing problem. As recently as 1950, records of most UK mammals were sparse, with the exception of quarry animals (e.g. hares and otters) recorded in game bags (Jefferies et al. 1989). Increasing interest in mammals in the 1950s was reflected in the formation of The Mammal Society in 1954 (The Mammal Society 2013), the Society's first national mammal recording programme in 1965, and the publishing of the resultant British Mammal Atlas in 1970 (Corbet 1971) and its 1984 update (Arnold 1984), which included the first distribution maps for each British mammal species.

Two reports by Jefferies and Arnold (1980, 1982), produced for the Huntingdonshire Fauna and Flora Society (located in South East England), provided the first evidence that water voles had been lost during the 1970s from some habitats in which they had long histories of occupation. These were sufficiently worrying to stimulate a desk-study, commissioned by the Nature Conservancy Council (the forerunner of a succession of statutory bodies in the UK, culminating in the evolution of Natural England). The study (1985–1986; Jefferies et al. 1989) analysed published literature from County Mammal Reports and local natural history societies from the whole of the British Isles, but excluding Ireland, going back to the early 1900s, and compared these results with questionnaire and observational data from a contemporary Waterways Bird Survey, conducted by the British Trust for Ornithology (BTO). Ironically, the number of reports in the literature of locations occupied by water voles peaked during the 1970s (Fig. 15.2), but this was an artefact of increasing observer effort (Jefferies et al. 1989).

To avoid such biases, the desk study inferred trends from the *types* of references made to water voles in the literature: from 1900, water voles were increasingly less frequently described as 'common' (Fig. 15.3), and increasingly as 'declining' (in 0–4% of reports prior to 1930 and 10–19% thereafter). Of 57 BTO questionnaire respondents who routinely visited sites occupied by water voles, 42% (24 sites) considered water vole abundance to be unchanged, 9% (5 sites) thought they were more numerous, and 49% (28 sites) thought they had declined (Jefferies et al. 1989).

From 184 literature studies that attributed a cause to changes in water vole numbers, 68% (126 cases) listed predators and 12% (22 cases) habitat destruction, with human disturbance (5%; 9 cases), pollution (5%;

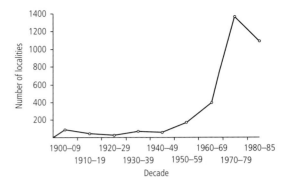

Figure 15.2 Total number of water vole localities reported in the literature for each decade. The 1980–85 half-decade figure has been doubled to be consistent with figures for earlier decades. From Jefferies et al. 1989. Reproduced with permission from John Wiley & Sons.

10 cases), and climatic factors (5%; 17 cases) also mentioned. Of the 126 studies mentioning predators, 13% (16) listed American mink (*Neovison vison*, a non-native predator; Macdonald and Feber 2015: Chapter 6), and 13% (16) listed stoats and weasels, while 24% (30) listed owls (Jefferies et al. 1989). Twenty nine per cent (51 cases) of BTO respondents also listed predators as a cause—of which mink were the top with 14 citations—with habitat destruction (26%; 46 cases), human disturbance (19%; 33 cases), pollution (12%; 21 cases), and climate factors (5%, 9 cases). The study concluded that water voles may have suffered a long-term decline in Britain since 1900, plausibly deriving from habitat destruction, American mink predation, human interference, pollution, and climate change (Jefferies et al. 1989).

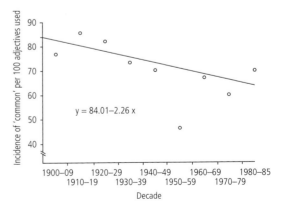

Figure 15.3 Decrease in the incidence of the word 'common' per 100 relevant adjectives used in the literature each decade (1950–59 point was not used for calculating the regression). From Jefferies et al. 1989. Reproduced with permission from John Wiley & Sons.

Such desk studies provide a quick confirmation (or rejection) of a perceived trend but are limited by their (varied, often inaccurate and patchy, with non-random spatial coverage) source material, which is typically sufficient to form only broad and speculative conclusions (Jefferies et al. 1989). Such analyses, based on the collective weight of anecdote and professional opinion, are open to errors arising from pseudoreplication, meaning that a given piece of data may fail to be independent from the others in a number of ways. If, for example, opinions are culled from 16 people, 15 of whom derive their viewpoint from a number of anecdotes shared between their group, the real sample size may be only two. And if opinions differ between those 15 and the remaining interviewee, the true ratio of independent opinion is not 15:1 but 1:1. Likewise, people may tend to make similar types of judgement when faced with similar evidence, and the majority can be wrong: the reported role of owls in the water vole's decline, for example—although indicated in 30 studies—was eventually discounted because water voles make up < 1% of their diets (Jefferies et al. 1989). Similarly, the authors admitted that the decline in the word 'common' was thin evidence for a species' decline and so a key recommendation of the report was for an objective field survey to provide a firm basis for future conservation and monitoring (Jefferies et al. 1989). The first national water vole survey (1989–1990) by a non-governmental organization (NGO), the Vincent Wildlife Trust, comprised two parallel surveys: a systematic baseline survey of 1926 sites across Britain, pre-selected from 1 x 1 km sites within a grid of 418 squares of 10 x 10 km (a 'Quin-Quox' grid; Macdonald et al. 1998; 2000) which could be re-visited to ascertain future population and distributional changes—and a survey of 1044 sites known to have been historically occupied, selected randomly from the Institute of Terrestrial Ecology Mammal Atlas database, to give an immediate indication of population changes (Strachan and Jefferies 1993). Both surveys recorded the presence and density of field signs, including the droppings and feeding remains left by water voles (see Strachan and Jefferies 1993; Strachan and Moorhouse 2006; Strachan et al. 2011 for details). At the same time, habitat criteria and the occurrence of footprints and faeces of mink were also recorded (Strachan and Jefferies 1993; Macdonald and Feber 2015: Chapter 6).

The survey results demonstrated that water vole populations had declined nationwide in England, Scotland, and Wales (they were always absent from Ireland): from all 2970 sites, 47.7% held water vole populations, but only 32.2% of the sites known to have been occupied prior to 1939 were still occupied in 1990 (Strachan and Jefferies 1993). A second national survey followed in 1996–1998, revealing the rate of decline to be alarming (Fig. 15.4), with the further loss of 67.5% of occupied sites in the seven intervening years. By 1998 the cumulative mean loss of water vole occupied sites across all regions of England, Scotland, and Wales from the 1939 baseline was 98.7%.

So, the detection of a problem for water voles followed an exemplary progression from initial, small-scale surveys that revealed local population losses, through confirmation with a desk study, and then national surveys that produced quantitative evidence and allowed the rate of the decline to be measured. Of note in this process is that, however thorough the desk study, the evidence it uncovers is insufficient to state with confidence that a decline has occurred, let alone to quantify losses or diagnose causes. A desk study cannot be used as a basis for mitigative action due to these uncertainties and so an objective and well-constructed field survey is required at a relevant geographical scale to quantify the losses and begin to establish causes. The process takes time: a decade elapsed between the first observed losses and field confirmation, during which period the national water vole population suffered a substantial further decline. It is also worth noting that these surveys would probably not have occurred had a millionaire son of a Glaswegian ship-building family not eccentrically insisted on funding the Vincent Wildlife Trust (VWT) surveys (and paid one of the authors, RS, to do them). This raises key points, to which we will refer again in this essay, concerning the funding of conservation, society's valuation of wildlife, and the importance of evidence. At the time of the surveys, little alternative funding was available for research on water voles and without The Rt Hon. Vincent Weir's contribution, British water vole conservation may well have foundered. The fact that it did not could be viewed as a measure of wider society's regard for the water vole, even in the absence of governmental funding. By producing quantitative evidence, based on a robust design, the hand of government was forced to do something about water voles—the alternative was to be shamed by inaction in the face of a decline of greater magnitude than that of the Sumatran tiger, the mountain gorilla, or the white rhinoceros (an uncomfortable prospect for a country that aspires to be a leader in conservation, and offers other nations advice about their own backyards).

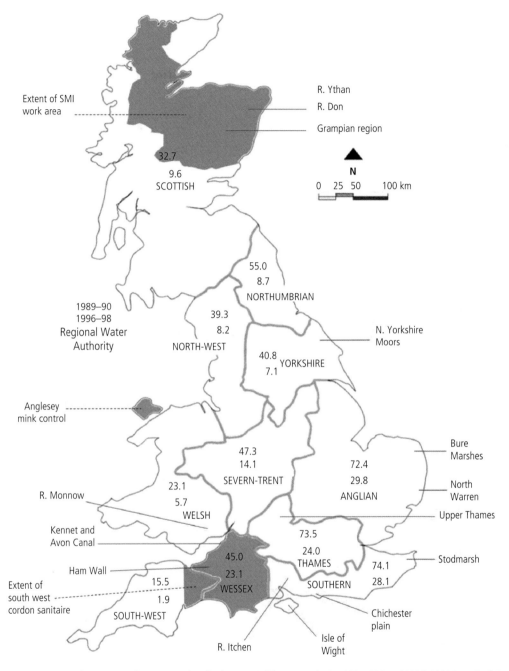

Figure 15.4 Comparison of percentage of survey sites found to be occupied by water voles in 1989–1990 and 1996–1998 in each of the main river catchments administered by the pre-1989 Regional Water Authority. Percentages include both baseline and historical series of survey sites. There has been a 67.5% total loss of occupied sites in between surveys (modified from Strachan et al. 2000). Ireland is not shown as it does not have water voles. Sites mentioned in this chapter are highlighted.

15.4 Diagnostic research

How were the causes of the decline diagnosed?

Having identified a problem, the next step is to diagnose its cause. The 1977 Handbook of British Mammals (prior to reports of the water vole decline) noted that water voles were vulnerable to American mink predation (Stoddart 1977), and the initial desk survey (Jefferies et al. 1989) narrowed the decline's causes to habitat degradation, predation by American mink, water pollution, and human disturbance. By resurveying sites known to have been historically occupied, the first national survey indicated that water voles had been declining since the decade 1930–1940: only 32.3% of sites occupied in that decade were still occupied in the 1989–1990 survey, indicating a decline of 67.7% over 50 years (Strachan and Jeffereies 1993). Prior to 1930 there was no substantial evidence of a decline (the same percentage—32.3%—of sites occupied in 1900–1910 were still occupied in 1990). It may be tempting to imagine that the loss of water vole populations occurred linearly throughout the period 1930–1990, but rates of decline are rarely linear. Table 15.1, shows the percentage of sites occupied in a given decade that were still occupied in the 1989–1990 survey, versus what would be expected if the rate of decline had been linear. These data indicate that the rate of decline during the period 1980–1990 was considerably higher than in previous decades. It would be expected that 95% of sites occupied in 1985 would still be occupied four to five years later at the time of the survey, but only 79% were still occupied, indicating a 21% decline in site occupation in under a decade.

The first period in which water vole losses were evident in the literature (1940–1950) coincides with the start of post-war agricultural intensification, which linearized water courses, increased stocking densities, and drained farm wetlands (Barreto et al. 1998a), resulting in loss and degradation of water voles' habitat. The severe losses in the 1980s coincide with a separate phenomenon: the spread of the non-native American mink (see Chapter 14, this volume; Macdonald and Feber 2015: Chapter 6). The VWT national survey revealed a negative association between the presence of mink sign and the water vole sign (Fig. 15.5). This association was country-wide and could not be discounted as mink and water voles having different habitat preferences (Strachan and Jefferies 1993). For example, overall mean latrine counts per 100 m (a coarse measure of water vole density; Woodroffe and Lawton 1990) were 9.2 per 100 m, in a region where 21% of sites were occupied by mink, and 3.6 per 100 m when mink occupied 66% of sites (Fig. 15.5). The survey did not uncover any consistent evidence for an effect of water pollution or human disturbance (which were identified as potential causes in the original desk studies, possibly because especially pollution—with organochlorines—had previously been implicated in the decline of another semi-aquatic mammal, the otter; Macdonald and Feber 2015: Chapter 7).

Two research papers, separate from and prior to the first national survey report, were pivotal in revealing that mink could be responsible for losses of water vole populations (Woodroffe et al. 1990; Lawton and Woodroffe 1991). The authors, working on selected 100–300 m lengths of rivers in the North Yorkshire Moors (Fig. 15.4), found three reasons why water voles could be absent from a location: approximately 45% of sites held habitat that was simply unsuitable for water voles; of the remaining sites 30% were unoccupied

Table 15.1 The actual versus expected percentage of sites occupied in a given year that were still occupied in the 1989–1990 national water vole survey. For example, 79.3% of sites occupied in 1970–1980 were still occupied by the time of the survey (a 20.7% decline since that decade). Had the rate of loss been consistent across decades, we would have expected 82.5% of sites to still be occupied (a 17.5% decline). The difference between observed and expected therefore shows that the decline since 1970 has been faster than expected. (Reproduced from text and data in Fig. 31 from Strachan and Jefferies 1993).

Date	Percentage of occupied sites that were still occupied in 1989–1990	Expected percentage of sites still occupied in 1989–1990 if rate of decline were consistent	Difference from expected
1930–40	32.3	32.3	0
1940–50	44.0	44.9	−0.9
1950–60	55.9	57.4	−1.5
1960–70	73.4	69.9	+3.5
1970–80	79.3	82.5	−3.2
1980–90	79.2	95.0	−15.8

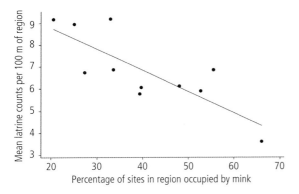

Figure 15.5 The relationship between regional site occupation by American mink and mean latrine counts of water voles per 100 m for water courses in that region in the 1989–1990 survey. (Redrawn from Strachan and Jefferies 1993).

either because they were too isolated or because the voles had been extirpated by American mink[1]. Although the authors were unable to disentangle the relative contributions of isolation and mink predation, they witnessed the extermination of water voles from high density sites between 1987 and 1989 by incoming mink (Lawton and Woodroffe 1991). Woodroffe et al. (1990) concluded, 'At the present time, water voles and mink coexist . . . in the North Yorkshire Moors . . . Voles may persist providing mink are not equally abundant throughout a water course . . . But by reducing population size and fragmenting water vole colonies mink pose a serious long-term threat to the survival of water voles on British rivers'.

With hindsight, therefore, by 1993 the cause of the water vole's decline was clearly attributed to the joint forces of habitat loss (Strachan and Jefferies 1993) and predation by American mink. Indeed, in their conclusion to the first national survey, Strachan and Jefferies (1993) asked why there had been such a 'long debating period' concerning the involvement of mink. Why, then, did the conservation community not immediately instigate a mink control programme? Such a programme could perhaps have prevented, or at least ameliorated, what became the loss of a further 68% of water vole populations by 1998. The hesitancy stemmed at least in part from uncertainty (and

optimism) based on early research on mink where water voles did not feature in their diet (i.e. in western Britain along rocky rivers that did not support water voles, Birks and Linn 1982; or at coastal sites, Birks and Dunstone 1984) and on observations suggesting that mink and water voles might co-exist in habitats with extensive areas of riparian vegetation, particularly wetlands (e.g. Stodmarsh in Kent, where mink and water voles had already co-existed for over 20 years), and some contemporaneous reports suggesting that the mink population decreased in areas to which otters had returned (after themselves suffering a population decline in the 1950s, Macdonald and Feber 2015: Chapter 6) (Strachan and Jefferies 1993). In essence, it was not certain whether water voles were vulnerable to mink only in areas where habitat isolation and degradation were also contributing factors (e.g. Lawton and Woodroffe 1991), and the hope existed that the returning otter population in England would spare everyone the expensive, logistically difficult, and morally tricky necessity of killing mink. (As it turns out, the return of the otter has, fascinatingly, not led to the substantial reduction in mink numbers that at first appeared likely (Macdonald and Feber 2015: Chapter 6) and it is noteworthy that the Scottish otter population remained thriving but water vole declines occurred there regardless.) Uncertainties such as these lead to hesitation, in this case by weakening the mandate for large-scale mink control, which can permit a problem to worsen. For this reason firmly, and unequivocally, establishing an early consensus on causes is crucial to forming a basis for conservation action.

Understanding which conservation actions were likely to be beneficial for water voles hinged on untangling the effects of mink predation and habitat loss, a situation encapsulated in the 'tightrope hypothesis'. The tightrope hypothesis proposed that habitat loss resulted in water vole populations becoming restricted to thin, linear 'tightropes' of habitat that made them vulnerable to predation by mink, because the mink now had only to hunt a narrow band of habitat, rather than a matrix of wetlands, rivers, ditches, and reedbeds. The idea, then, was not so much that the lost habitat was critical, but rather that the remaining habitat was configured in a way that made it fatally easy for the mink to find the voles. The hypothesis arose from our study in 1995 in the Thames catchment in which Barreto et al. (1998b) revisited 161 sites from Strachan and Jefferies' original surveys and statistically associated the distribution of water voles with 32 habitat characteristics pertaining to physical river attributes

---

[1] This conclusion overlooks the fact that patches can be empty even in a stable water vole population—because the concept of the 'metapopulation' on which our understanding is now based, and which we describe later, was in its infancy at the time; an example of how developments in fundamental ecology can have applied relevance.

(e.g. riffles, pools, channel substrate), physical bank attributes (e.g. bank height, profile, and material), river vegetation (e.g. vegetated bars and emergent broad-leaved vegetation) and bank vegetation (e.g. surrounding land use, the presence of trees, vegetation, structure), as well as with the presence of mink signs (footprints, scats, and sightings). The statistical model that best explained the presence of water voles for the fewest explanatory variables (i.e. used only the most influential factors, winnowed from the original 36) contained effects of water quality, wetlands, submerged vegetation, and mink; and the effect of the presence of mink was larger than any of the habitat variables.

While these correlative results broadly accord with Strachan and Jefferies' (1993) conclusions, they differ in that submerged vegetation and water quality had not been important in the earlier study. The probable explanation for this change is that, in 1990, water voles were common and mink had colonized only 24.6% of sites, whereas, five years later, mink in our Thames study occupied 46.2% of sites (Barreto et al. 1998a; Fig. 15.6). The increased importance of submerged vegetation plausibly reflected its role in providing areas where mink found it difficult to hunt (which requires submerged plants to hinder mink sufficiently to allow water voles to escape), and that the association with water quality was explicable because the best water quality was to be found in headwaters—where the mink may not have yet reached—and in fisheries, where the land managers pursued intense mink control programmes (Barreto et al. 1998b; Macdonald and Strachan 1999). The conclusion from both the 1995 River Thames survey and a 1994 survey covering 22 km of the River Soar in Leicestershire, was that the arrival of mink in both catchments was associated with catastrophic declines in water vole numbers (Fig. 15.6; Strachan et al. 1998). On the Soar, water voles' numbers were severely reduced in just one breeding season following the arrival of mink, and they were eradicated within two years. Mink were heavily implicated—more than a third of the volume of 863 mink scats collected along the Soar comprised mammalian prey, of which water voles were the predominant species (13.2% of remains)—but hope persisted that large areas of riparian habitat could provide conditions under which mink and water voles could co-exist (Strachan et al. 1998).

A key feature of the tightrope hypothesis was that it suggested that, although mink predation had clearly aggravated water vole declines, the ultimate cause was habitat degradation, fragmentation, and linearization,

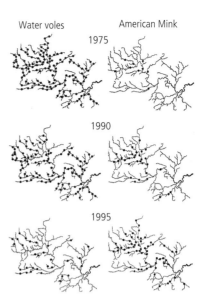

Figure 15.6 The simultaneous increase in the distribution of mink and decrease in the distribution of water voles in the Upper Thames catchment area from 1975 to 1995. (From Macdonald and Strachan, 1999). By 1995, sites that had been recorded as 'water vole only' in 1975 were found to be replaced by 'mink only' evidence. Water voles were largely restricted to the upper most reaches of the Thames tributaries.

and so mink control might be less necessary if suitable (less linear) habitat were available and appropriately managed: i.e. water voles would not decline in the presence of mink if the habitat were pristine. Evidence from Stodmarsh in Kent, where water voles and mink had co-existed for nearly 30 years, as well as from reedbed populations in North Warren, Suffolk, and Ham Wall, Somerset (Fig. 15.4), showed that although mink cause high levels of mortality on the edges of reedbeds near main channels, predation rates decline sharply with distance, such that water voles 150 m from a main channel had half the predation risk of those 10 m from the channel (Fig. 15.7; Carter and Bright 2003). Large marshes were therefore potential havens from predation by mink, and may support populations in the wider landscape by acting as a source of dispersing individuals for surrounding areas (Carter and Bright 2003). However the three study sites covered 80, 40, and 25 ha respectively, and the estimated mink predation rate approximated zero only at 450 m from the edge (Fig. 15.7; Carter and Bright 2003), which leaves the possibility that the population may still decline to extinction, but over a longer period than in linear waterways.

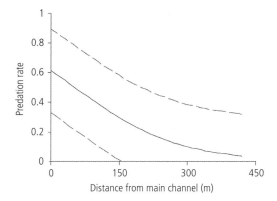

Figure 15.7 The relationship between predation rates of water voles by mink and distance from a main water channel (>10 m wide) of a water vole's centre of activity. The solid line shows the main relationship and the broken lines show the 95% confidence limits. Figure reproduced from Carter and Bright (2003) with permission from Elsevier.

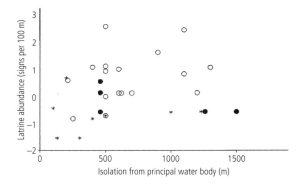

Figure 15.8 The effect of isolation from principal watercourses on water vole latrine abundance at large sites (filled circles), medium sites (stars), and small sites (open circles) in Belarus. From Macdonald et al. (2002). Reproduced with permission from John Wiley & Sons.

Opportunities to create such large swathes of habitat in Britain's agricultural landscape are severely limited, and so it was still unknown whether restoring the matrix of river habitats, small wetlands, ponds, and reedbeds that existed in much of lowland UK prior to the 1950s would preserve water voles from mink predation. Macdonald et al. (2002) approached this question—since the relevant habitats no longer existed in the UK—by locating field sites in Belarus where American mink had recently (early 1990s; Sidorovich 1992) invaded and where this matrix of habitats, with incumbent water voles, still existed. They found that water voles were exterminated from at least some of all of the different habitats studied, regardless of their size, and that small patches (< 15 m²) isolated from a main waterway by 500–1500 m were four times more likely to contain water voles than large (> 100 m²) patches connected to waterways (Fig. 15.8). These patches were not of the multiple-hectare scale of the reedbeds in Carter and Bright's (2003) study and so the impacts of American mink were not reduced by patch size, but by the difficulties of hunting in sparsely distributed small patches. This interacted with the water voles' dispersal ability to create a water vole population that was, while severely impacted (the population was more than decimated), stable across the landscape as a whole (Macdonald et al. 2002).

The above studies demonstrate two points. First, the tightrope hypothesis was partly correct in that while good habitat can mitigate, or delay, the impact of mink under some circumstances, the decimation of

water vole numbers in Belarus (Macdonald et al. 2002) and Scotland (as we will describe below, Aars et al. 2001) suggested that any UK habitat restoration programme would require an associated programme of mink control (Macdonald and Strachan 1999). It was therefore wrong in the sense that good quality, non-linear water vole habitat outside of large wetlands in the UK could not be expected to be sufficient protection from mink predation to prevent population extinction. Second, not only the size but also the spatial juxtaposition of habitats, and the ability of water voles and mink to disperse between them, are important factors in determining the stability of water voles across a landscape.

The thinking behind the tightrope hypothesis developed in lowland habitat, and there a key prediction was that where habitat had not been modified and linearized, the magnitude of water vole decline in the presence of mink would have been less. This was broadly supported in lowland wetland sites and in our study in Belarus. However, it was not supported at all by the University of Aberdeen's work in the Highlands of Scotland. Indeed, work in both upland (Aars et al. 2001) and lowland (Telfer et al. 2001) Scottish water vole populations (and described in two even earlier reports: Lambin et al. 1996; Lambin et al. 1998) also demonstrated that mink could eradicate water voles from pristine habitats (Aars et al. 2001—see below) and that water vole populations there typically comprise a series of small populations (each perhaps only a few individuals), separated by unsuitable habitats, any one of which has a high probability of going extinct in a given year, but also of being recolonized by dispersal from another population.

Water voles are extremely well adapted to living in such 'metapopulations' (e.g. Lambin et al. 2012), which can exist in the UK in any habitats in which water voles are patchily distributed: increasingly, therefore, anywhere outside of large, protected wetlands and reedbeds. For example, working on the River Itchen and across the Isle of Wight, both with thriving populations of water voles located in mink-free areas, our team found that habitats that were occupied, whether suboptimal or optimal, had between 3–4 neighbouring occupied patches within a 1 km radius, whereas unoccupied habitats had 0–2 occupied neighbours; and even the best habitats could be unoccupied if too isolated (Fig. 15.9; Bonesi et al. 2002). Indeed, studies of Scottish water vole metapopulations have estimated the mean effective dispersal distance of water voles (the distance beyond which recolonization rates are so small as to be practically negligible) as being 1.39–1.96 km from their nearest colonies (Telfer et al. 2003; Lambin et al. 2004). The implications of these findings for conservation are that some suitable patches of habitat may be unoccupied by chance in a given year, but recolonized at a later date; therefore, destruction of habitat that is unoccupied may still threaten the metapopulation as a whole, and any patch of unoccupied habitat greater than 2 km from an existing population is likely to remain unoccupied simply because it is too far away to be recolonized. Similarly, one effect of the post-war loss of farm wetlands and ponds and the canalization and clearing of drainage ditches was to limit the number of neighbours of river water vole populations; hence sources of potential dispersers were reduced to just two, upstream and downstream, which concomitantly increased the distance between them and reduced the ability of water vole populations to exchange individuals. For example MacPherson and Bright (2009) modelled the persistence of ten individual populations of water voles in upland Wales and showed that removing three habitat patches resulted in long-term persistence, but with large individual fluctuations in extinction risk in a given patch, while removing four patches resulted in the metapopulation tending to extinction.

As habitats become progressively more fragmented, the importance of individual water voles' ability to disperse between patches increases until patches become too widely separated (Rushton et al. 2000). In a computer-simulated landscape, Rushton et al. (2000) found that dispersal ability increased in importance in determining metapopulation viability up to fragmentation levels of 50%, but where the suitable habitat was reduced to < 20%, dispersal was less important than adult and juvenile mortality rates. Rushton et al. (2000) also modelled the effects of habitat fragmentation and mink on the water vole and correctly predicted the extermination of water voles on the River Windrush, Oxfordshire (Rushton et al. 2000; Macdonald and Rushton 2003); this extermination later provided the opportunity for a reintroduction campaign onto this river (Chapter 15, this volume).

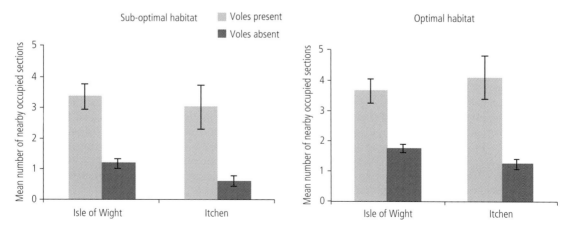

Figure 15.9 Mean number of nearby occupied sections surrounding: (left) a section with or without water voles in sub-optimal habitat (Isle of Wight N1 = 20, N2 = 148; River Itchen: N1 = 4, N2 = 30); (right) a section with or without water voles in optimal habitat (Isle of Wight: N1 = 65, N2 = 101; River Itchen: N1 = 16, N2 = 66). Bars represent standard errors. From Bonesi et al. (2002). Reproduced with permission from John Wiley & Sons.

In the absence of mink, the extinction of individual colonies in metapopulations may be caused by native predators or chance events. The presence of mink tips the balance between extinction and recolonization towards the former. Additionally, because a mink's range can encompass multiple water vole colonies, they can cause the simultaneous extinction of adjacent colonies (spatial synchrony in extinction), which severely hinders the natural process of dispersers from neighbouring patches recolonizing the empty habitat. A study of water voles in a series of lowland Scottish burns in an intensive arable farmland area from 1996 to 1999, concluded that the ingress of mink into this system—which began in 1998 with the local extinction of one of the largest local water vole populations in only one month—would increase the rate at which local extinctions accumulated, and therefore the effective degree of fragmentation and isolation of the populations, until the whole region became unoccupied (Telfer et al. 2001). Similarly, Aars et al. (2001) document the destruction by mink, and in just four months, of 11 adjacent colonies of a water vole metapopulation in 25 km² of Scottish upland. Conversely, it is possible, as in the Belarus example, that water vole metapopulations might be protected from mink predation if they are sufficiently isolated that mink are unlikely to disperse that far (juvenile dispersing mink can move in October an average distance of 30 km; Bonesi et al. 2007; Macdonald and Feber 2015: Chapter 6). Also, areas characterized by scattered, small water vole populations may not support sufficient prey for a mink to be able to exist there permanently i.e. the water vole patches are 'low productivity refuges' where the voles live at densities too low for mink to survive in the absence of other prey sources. In such areas, a mink may make incursions and predate on water voles, ground nesting birds, and post spawning salmon, but may be unable to persist, having to eventually return to more profitable locations.

The relationship between patch quality, isolation, and mink predation is complicated by the involvement of other species. A study in Scotland's Grampian region (Oliver et al. 2009), where mink were present, examined 310 patches suitable for water voles in moorland habitat and 3122 in habitat suitable for rabbits and found that, if water vole patches were highly connected to rabbit habitats (as measured by distance to, and size of, rabbit habitats), occupation of water vole patches tended towards zero; if, conversely, water vole patches were relatively unconnected to rabbit habitats, then occupancy approximated 30% (Oliver et al. 2009). In a separate location (comprising 88 water vole patches and 1275 rabbit patches) in which mink were

absent, water vole occupation was weakly positively correlated with connectivity to rabbit habitats (which is reasonable given the expected correlation between the pasture habits preferred by rabbits and the vegetated banksides required by water voles) (Oliver et al. 2009). The implication of these findings is that rabbits (also an invasive/naturalized species and a preferred prey item for mink) facilitated the American mink's invasion to the detriment of water vole populations that may otherwise have been protected by their isolation. This finding accords with speculation from an earlier study on the upper Thames, in which we suggested that since mink sometimes den away from watercourses, the presence of blocks of woodland and hedgerows that contain rabbits may play a role in determining a river's carrying capacity for, and the distribution of, mink (Halliwell and Macdonald 1996; Yamaguchi et al. 2003). In the Scottish study, the presence of rabbit patches had a strong effect on water vole patch occupation as much as 10 km away (Oliver et al. 2009). Thus, only the sparsest water vole populations might be safe from the impact of mink.

In the above case, there is hope that research efforts can lead to action to prevent the spread of mink. Recent findings from a study examining the genetic structure and connectivity of Scottish mink populations suggest that the presence of mountains restricts dispersal while valleys facilitate movement between populations, but that mink disperse and exchange genes freely over large distances over much of the lowlands of Scotland, and by analogy, over most of England (Zalewski et al. 2009; Fraser et al. 2013). Information like this can help to define management units and establish priority areas for targeting control efforts. In particular, the study identified the mink populations that were contributing to the species' range expansion (Fraser et al. 2013), information that allows targeted mink control to prevent encroachment into the as yet un-invaded area of northwest Scotland, which is a stronghold for water voles (Aars et al. 2001). Unfortunately, these findings also imply that defining contained management areas in lowland England, outside of large wetlands, will be challenging because of unrestricted movement/ingress of mink from other locations.

In conclusion, by the early 2000s, it was clear that water voles had declined due to the sequential impacts of habitat loss and linearization (in lowland England), and predation by American mink, exacerbated by additional population extinction arising from the resultant fragmentation of water voles' populations. It was also clear that restoring swathes of riparian habitat would not be an effective solution without first controlling

American mink, and without ensuring that any restored habitats were sufficiently close to an existing source of potential dispersers to be recolonized. Whatever solution was proposed, therefore, would have to: (a) incorporate prior and ongoing mink control, and; (b) account for the spatial distribution of water vole populations and, perhaps, consider their reintroduction where no suitable population remained within the locality. The crucial, general point arising from the synthesis of this diagnostic research is again that the correct conservation actions remained unimplemented while uncertainties existed as to the relative degree of involvement and importance of the contributing factors. Conservation actions therefore hinge on establishing solid, unequivocal evidence for the involvement of inimical factors.

15.5 Solution research

How were the remedies devised and tested?

The diagnosis of the causes of a species' decline automatically indicates the direction of travel towards the cure (it was hopefully apparent as the previous section progressed that the research questions were increasingly strongly linked to practical action), but practicalities may be daunting. Simply stating that mink control, habitat restoration and, perhaps, reintroductions, are required, for example, ignores the practical questions of how intense and geographically wide-ranging mink control must be, how much habitat to restore and in what configuration, and what determines whether population translocation and/or reintroduction are required in a given case. These practical questions are answered by mitigation research.

One of our approaches to mitigation research was the Chichester coastal plain sustainable farming partnership project, created in 2000 (Macdonald and Feber 2015: Chapter 14). The project recognized that most British waterways are located within farmland (as opposed to nature reserves), and that because water vole habitats had been lost through mink incursion and agricultural intensification, then mitigation should adopt a landscape-scale approach to removing mink and improving the connectivity of rivers to their farmland backwaters, ponds, wet grasslands, marshes, reedbeds, and ditches. The aim was to create a 'demonstration area of best practice for sustainable farming and biodiversity enhancement, where land managers, farmers, and practitioners could experience sympathetic management first hand' (Strachan and Holmes-Ling 2003).

On 8400 ha of the Chichester Plain we targeted 42 farms, working with a number of partners, including the Farming and Wildlife Advisory Group, the Environment Agency, English Nature, and Sussex Wildlife Trust. We offered all farmers in the study area a free biodiversity audit and a Whole Farm Conservation Plan (WFCP) of their holdings through FWAG's Landwise package. The WFCP incorporated advice on best practice management of watercourses, ditches, wetlands, and the targeting of waterside land. We presented farmers with this funded service to encourage landowners to apply for agri-environment scheme (AES) payments and much of the enhancement work for improving water vole relevant habitats was met by acceptance into the Countryside Stewardship Scheme (the AES available at the time), for which we guided farmers through the funding application process. Initial water vole surveys in 2000 provided a population estimate of only 103 individuals across all farms (Fig. 15.10a; Strachan and Holmes-Ling 2003). A full-time mink trapper was employed over three consecutive winters (2000–2003) and the project installed field margins and waterside buffer strips to provide thick stands of riparian vegetation, and livestock management (fencing) alongside watercourses. A 2002 survey showed that water voles had become strongly associated with newly fenced banks that supported dense bankside vegetation—88% of all occupied stretches were protected from cattle grazing and trampling by fences and only 16% of unfenced sites had evidence of water vole presence (Fig. 15.11)—as well as sites with steep bank profiles, nearby suitable ditches, and appropriate soil for burrowing soil (Strachan and Holmes-Ling 2003). The project also restored wetlands, ponds, and ditches, with accompanying installation of check-weirs into ditches to maintain summer water levels. These weirs and junction ponds were quickly occupied by water voles, which expanded their range and increased population size (Fig. 15.10b; Strachan and Holmes-Ling 2003). In 2002, ten pairs of captively bred water voles were released into a 1 km section of one watercourse. The first wild-born juveniles were captured two months after release, and by week 12, a total of 42 animals were estimated over the watercourse. By week 20 (October 2002), 57–63 total individuals were estimated.

By the end of the project in 2003, 4556 ha of the 8400 ha area had entered the Countryside Stewardship Scheme and a further 1328 ha had improved amendments. Twenty seven farm ponds and 28 ditch ponds had been created and summer water levels were raised over 42 km of ditches; 7200 m of fencing were erected to

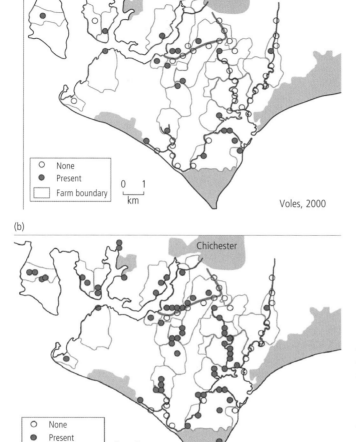

Figure 15.10 Relative distribution of water vole colonies (none/present) across the Chichester Coastal Plain farms in (a) 2000–2001 and (b) following habitat restoration in 2002–2003. The most easterly watercourse was a control area receiving little/no proactive management for water voles apart from mink removal. From Strachan and Holmes-Ling (2003) and Dutton et al. (2008). Reproduced with permission from John Wiley & Sons.

protect watercourses from livestock, and the Environment Agency revised its Flood Defence maintenance work to make it more sensitive towards water voles (and other species of conservation concern). Water vole numbers responded by more than tripling in three years to an estimated 348 individuals (Fig. 15.10b) before the breeding season (i.e. after winter and before the population expands; Chapter 14, this volume).

It would have been entirely irresponsible to set up such a project and not ensure the tenure of the restored water vole populations. The Environment Agency and the Manhood Wildlife and Heritage Group, both members of the original project steering group, currently continue to map and monitor the water voles, and mink control continues to be carried out by the original trapper. The Manhood Wildlife and Heritage Group (<http://mwhg.org.uk/water-voles/>) report that water voles remain widespread through the area, have consolidated their range at core sites, and have expanded along the Pagham rife, where they were absent in 2003. The Chichester Coastal Plain remains one of the few extant populations of water voles in Sussex.

The two elements of the Chichester project that were not fundable through agri-environment schemes were the mink control and water vole captive releases, both of which activities are likely to be necessary steps for the restoration of the national water vole population (e.g. Jefferies 2003). Large-scale mink control became considerably easier following the invention in 2004 of the Game & Wildlife Conservation Trust (GWCT)

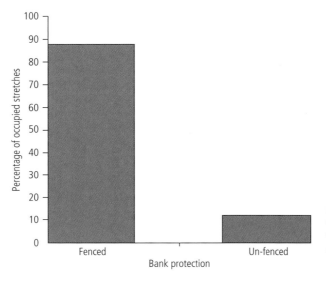

Figure 15.11 Response of water voles to banks fenced or unfenced on the Chichester Coastal Plain project experiments. From Strachan and Holmes-Ling (2003).

mink raft (Reynolds et al. 1994, 2007; see Chapter 14, this volume, and Macdonald and Feber 2015: Chapter 6 for a full description of its use), and research described elsewhere in this volume (Chapter 14) and in Macdonald and Feber (2015: Chapter 6) was dedicated to ensuring that mink control could be conducted efficiently and to a high standard of animal welfare. Similarly, reintroductions had clear potential to benefit water vole conservation but a number of unknowns existed. A pre-requisite for any reintroduction is that the original causes of the species' decline must have been understood and remedied—and in general all reintroductions should conform to the IUCN guidelines (IUCN 1998, 2013)—which in the case of water voles requires enough habitat for a sufficiently large water vole population to establish (Chapter 14, this volume) and adequate mink control over a large enough area. How large this area needs to be is unknown and depends on mink immigration rates, which in turn relate to the number and efficacy of nearby mink control projects. Since water voles reproduce quickly—Chapter 14, this volume—it may be unnecessary to ensure zero impacts from mink as long as the impacts occur when the vole population is established, and the mink are captured before they have kits.

Many water vole reintroductions transgress these rules, in part because of natural enthusiasm and the publicity opportunities associated with reintroductions, and also, perversely, due to the legal framework associated with the mitigation of developments that affect water voles (Strachan et al. 2011). Such reintroductions commonly fail. Indeed a failure of mink control,

related to waning landowner participation, quickly resulted in the failure of one of our own research reintroductions (Chapter 14, this volume). Similarly, controlling mink over too small an area may result in simply cropping an endless supply of dispersing individuals (mink disperse over 15–30 km; Macdonald and Feber 2015: Chapter 6). While we authors flagellate ourselves in this essay, one of us observes that several years spent killing mink within first 30 km^2 and then 100 km^2 of treasured water vole colonies was little more than a mink harvest until efforts were scaled up to an entire river catchment (1000 km^2), and to multiple river catchments (10–20 000 km^2) (Bryce et al. 2011). These personal experiences are salutary lessons in the importance of ensuring that the detailed practicalities of the broader concepts are firmly researched. In the case of water vole reintroductions (Chapter 14, this volume) these details included: (i) on what length (500 m, 1 km, 2 km?) of water course habitats to reintroduce, (ii) of what width (1 m, 3 m, 6 m?) of bordering vegetation, (iii) how many males, females, and juveniles of what ages to release, (iv) by what methods (immediate 'hard' release, or 'soft' release in provisioned pens to permit acclimatization?), (v) what conditions to house them under prior-release (singly, in pairs, in family groups?) (vi) how far to extend mink control from the release site (2 km, 6 km, 20 km, further?) to give the best hope of establishing a population, and (vii) what are the disease and welfare implications for the released animals? The key point is that any of the myriad practical details of any aspect of any conservation project might understandably be overlooked by conservation practitioners, but nevertheless could severely adversely affect the outcome of the project.

A case study highlighting the importance of these details is that of the use of radio-tracking. Radio-tracking is used ubiquitously for studying a vast array of wildlife species, and key assumptions are that radio-tags will not alter the behaviour of individuals nor compromise their welfare. In a 2005 study, however, we discovered that the radio-collaring of 38 water voles (20 male and 18 female) in an extant wild population in the Bure Marshes, Norfolk, correlated with a decline in the number of females in that population (Fig. 15.12; Moorhouse and Macdonald 2005a). Our study compared the population estimates for males and females in the Bure Marshes before and after radio-tracking, as well as estimates for a separate control population on the Kennet and Avon canal, in which water voles were not radio-tracked and hence had never been radio-collared (Fig. 15.12).

Our population estimates revealed a decline in female, but not male, numbers at the Bure Marshes only during 2002 (when radio-tracking occurred). No such female decline was evident for the control Kennet and Avon population (where tracking did not take place; Fig. 15.12). The decline of females did not result from increased female mortality, as survival rates for females throughout were greater than for males, but rather through a 48% decrease in numbers of females being recruited into the population. This decrease was similar to the proportion (0.49) of the female

population that had been radio-collared. We concluded that the observed decline in female numbers resulted from a shift from the expected birth sex ratio of 1:1 to a ratio of 43:13 in favour of males associated with the attachment of radio-collars to female water voles, potentially arising from an evolutionary mechanism to mitigate impacts of suboptimal habitats (Moorhouse and Macdonald 2005a). The explanation for this mechanism, and its relevance to radio-collaring, derives from the difference between male and female water vole life strategies. Female water voles defend territories which, although their positions are not static over time, and drift geographically over weeks and months (Moorhouse and Macdonald 2005b), exclude other females and so become smaller as population densities increase (e.g. 150 m at low densities and 30 m at high densities; Moorhouse and Macdonald 2008; Strachan et al. 2011). Males are non-territorial, occupying ranges that overlap with those of multiple other males and females, and remain so irrespective of population density (Moorhouse and Macdonald 2008). Female water voles are likely to pass their territories to their female offspring, but if their territory becomes sufficiently suboptimal (e.g. if severely reduced in size at high densities due to increasing pressure from competing females), evolutionarily it may be better to raise male offspring which leave the territory and compete for (now abundant) females; such sex ratio selection may occur if the mother becomes stressed by, for example, food shortages, increased aggressive encounters with

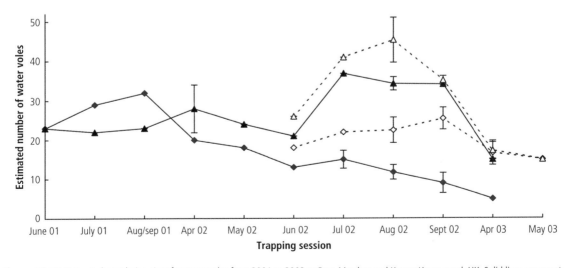

Figure 15.12 Estimated population sizes for water voles from 2001 to 2003 at Bure Marshes and Kennet/Avon canal, UK. Solid lines represent Bure Marshes populations in 2001 (when no radio-tracking occurred) and 2002 (radio-tracking from April onwards), dotted lines represent Kennet and Avon populations (had never been radio-tracked). Triangular points represent males, rhomboid points represent females. Error bars represent SE of population estimate. From Moorhouse and Macdonald (2005a). Reproduced with permission from John Wiley & Sons.

neighbouring females, or in this case, by the attachment of a radio-collar (Moorhouse and Macdonald 2005a).

We found evidence for radio-collars causing short-term stress to water voles by measuring their mean leucocyte coping capacity (LCC: a measure of their ability to mount an immune response, which is known to be affected by stress) prior to and after collaring in captivity. Mean LCC decreased from 308 to 240 in the week post collaring (Moorhouse et al. 2007). Our study also demonstrated that water voles raised in large outdoor enclosures were heavier (240 g as opposed to 203 g), more hydrated, and had greater LCC (262 units as opposed to 151 units) than water voles maintained in laboratory cages. The mean body weight of laboratory cage voles had been 254 g before being transferred into the cages and fell to 209 g within two weeks, suggesting that they lost body condition due to this type of housing (Moorhouse et al. 2007). This is only one example of how details, such as how captive water voles were housed and the method used to monitor them post release, could have led to reintroduced water voles that were stressed, underweight, immuno-compromised, and likely to produce male-skewed litters, all of which have welfare implications and could potentially result in a failure of the population to establish successfully (Chapter 14, this volume).

Mitigation research offers a testing ground to prevent the details of how conservation is conducted from jeopardizing the intended outcome, and to ensure that the money and effort expended do not contribute to ineffective programmes or compromise the welfare either of the species to be conserved or of any inimical species that may need to be controlled.

15.6 Recovery management

How are the remedies implemented?

Management activities inevitably require the involvement of a large and diverse body of people. In the case of water voles this list includes farmers, reserve managers, wildlife trusts and interest groups, housing developers, water bailiffs, game keepers, fishermen, statutory agencies, and the general public, all of whom need, at least to some extent, to appreciate the research results. It is also unavoidable that initial attempts at implementation are more costly than their successors, which benefit from the experience gained. Thus not only should information be widely disseminated but, in addition, also continually updated and improved, so as to make implementation more cost effective.

Studies published in the academic literature, however, often remain inaccessible to the people who provide practical conservation management. Recognizing this problem, WildCRU released a series of 'grey-literature' publications, aimed at communicating the findings of academic research more widely. The Water Vole Conservation Handbook (Strachan 1998) and its two revisions (Strachan and Moorhouse 2006; Strachan et al. 2011) drew together the current research to offer best-practice management advice for habitat managers, conservationists, and developers, and The Mink and the Water Vole (Macdonald and Strachan 1999) provided a breakdown of the academic analyses underpinning that advice. The Chichester report (Strachan and Holmes-Ling 2003), exemplifies the joint requirements of communicating research to stakeholders and trialling mitigation techniques.

The findings of research, when mixed with judgement and, often, tough choices, underpin policy and thus management (Macdonald et al. 2006). The report on the first national water vole survey (Strachan and Jefferies 1993) coincided with the Convention on Biological Diversity, with its doctrine of thinking globally but acting locally to halt biodiversity loss worldwide. In response, the UK Government published its first tranche of Biodiversity Action Plans in 1994, heralding the water vole as a priority species and stimulating the creation of the UK Water Vole Steering Group. The Water Vole Species Action Plan, sought for the water vole to be granted legal protection through its inclusion on Schedule 5 of the Wildlife and Countryside Act 1981 (as amended 1998, although only partial protection was given initially, so that the habitat rather than the voles themselves were protected; Strachan et al. 2000), a step which permitted the later provision of full legal protection for water voles in England under Section 9 of Provision 5, in 2008. Similarly, the designation of 15 wetlands as National Key Sites for water voles in England and Wales in 1997 stemmed from the earlier work by Carter and Bright (2003) showing that large wetlands were potential havens for water voles from mink predation (Strachan et al. 2011). In terms of oversight of water vole conservation, the UK Water Vole Steering Group, chaired by the Environment Agency, coordinates water vole conservation attempts, and works with the Wildlife Trusts to provide mapping of water vole distributions. These distribution maps are used to assess the feasibility of survival of the water voles in different localities and to prioritize habitat restoration and mink control projects. Unfortunately, resources are not available to repeat the VWT national surveys and so no directly comparable quantitative data are available nationally to describe whether the national water vole decline is slowing or being reversed. This raises the crucial point that management action should always include thorough monitoring and reporting of the consequences (Strachan et al. 2011; Moehrenschlager et al. 2013) so that the collective effort can learn from successes and failures. Monitoring, though, is often overlooked when resources are allocated to management action, leading in our case to a distinct risk that perceptions on the fate of water voles may become dominated by well advertised, localized good news, rather than being balanced against the (expected, and less frequently reported) failures and the ongoing loss of populations elsewhere.

Mink control is a key requirement of water vole conservation and a number of such projects are ongoing. In the following descriptions, although the figures for mink captured and project costs are available in the public domain, we deliberately avoid using number of mink captured per unit cost as a measure of a mink control project's 'success', because 'cost per head of mink' is a poor measure. This is, first, because the in-kind contribution of the countless volunteers (in both time and expertise) cannot be represented in estimated costs and, second, because the true 'success' of any such project is measured in the number of kilometres of mink-free water course which are able to be ensured in perpetuity. This latter point is crucial: capture rates of mink depend to a large extent on their density, which will, ironically, decrease as the project becomes more successful at maintaining mink-free areas. As with many issues in conservation, at the outset, remedial action seems positive and exciting (a good news story) and captures the public's imagination ('100 mink captured in a bid to save the water vole'). The greatest gains are inevitably made, however, years after the headlines have ceased and when public interest in the project has dissipated, or indeed, when attempting to safeguard unaffected areas. As an example, in an attempt to keep the Isle of Anglesey a mink-free island, 105 mink rafts have been deployed and are monitored across the island in the knowledge that only one or two mink may be caught each year as they arrive on the island. An ideal project, therefore, may eventually capture only a few mink per year but effectively safeguard many hundreds of kilometres of water course[2].

In addition to the mink trapping outlined in Macdonald and Feber (2015: Chapter 6) and the water vole reintroductions outlined elsewhere in this book (Chapter 14), in spring 2003 the Environment Agency (EA) initiated a *cordon sanitaire* strategy for creating a mink-free area southwest of the line from the Bristol Channel to the Dorset coast. The project was part funded by the EA and coordinated by the British Association for Shooting and Conservation, working with local Wildlife Trusts, local farmers and landowners, members of Natural England, and the Royal Society for the Protection of Birds. Starting in autumn 2003, the team of 70 trappers began trapping mink. In response, water voles have recolonized many local areas. Similarly, Reynolds et al. (2013) undertook a four year (2006–2010) mink eradication programme on the River Monnow Catchment in western Britain, using mink rafts, with a concurrent reintroduction of water voles over two years (2006–2008). The project resulted in a persistent population of water voles distributed across 13.3 km of the river, for the first time since they were originally eradicated in the 1980s. Working in Scotland over three years (2006–2009), Bryce et al. (2011) removed mink from 10 570 km² of the Cairngorms National Park with the involvement of 186 volunteers, resulting in the 'largest mainland invasive species eradication effort worldwide'. However, although there was some evidence of localized water vole expansions from their upland strongholds, they reported an 'inherent lag to the recolonization process which we expect to witness during the future phases of the project without recourse to the release of captive bred individuals'.

Building on the success of this project, the Scottish Mink Initiative was created in May 2011 with the aim of removing breeding American mink from north Scotland and covering an area of 20 000 km² from northern Tayside across Aberdeenshire, Moray, and the Cairngorms National Park, to the north and east Highlands. The initiative currently incorporates 212 landowners with 972 mink rafts operated by 519 volunteers (correct as of March 2013). The key features of this project are, first, that it is not a restoration but a safeguarding project, designed to protect large, but slowly declining, water vole populations in the Cairngorms that were little affected by mink, and second, that water vole populations existed that could recolonize those areas at the periphery of the region that had been affected by mink. For example, the upper Don was empty of water voles when surveyed in 1996 and now supports a thriving population. The project is now managed by a local Rivers and Fisheries Trust, who act as custodians of the mink-free status delivered by the partnership.

---

[2] As another example, the per capita cost of removing the animals toward the end of eradication campaigns commonly escalates: the majority of the 79 569 goats removed from Pinta island in the Galapagos cost between US $10–100 per goat to remove. In contrast, the cost of removing the final goats was over $10,000 each (Carrion et al. 2011).

Given the above projects, it is clear that our original theoretical species recovery curve (Fig. 15.1) for the water vole has passed firmly into the recovery management stage. The final questions we posed in the Introduction, therefore, need addressing: *When do we know enough that research is no longer necessary*? And *Is society prepared to foot the bill, both financially and ethically, to enact the conservation actions required*?

15.7 Conclusion: management, policy, and societal judgement

We set out at the beginning of this chapter to provide a top-to-tail account of a single species conservation research effort—a story that parallels numerous other such projects on a range of other species. After 20 years, the water vole conservation story is now in the early stages of recovery management and there is cause for cautious optimism, given that the conservation actions required to secure the water vole's future are known—and supported by a wealth of conservation research—, the actions are certainly feasible, and can be implemented to positive effect over substantial areas. The cautionary note to the optimism, however, is decisive: it is that all conservation activities require funding for people and equipment. People are often willing to give their time and expertise *gratis* to conservation projects, but even those projects that rely heavily on a volunteer workforce require substantial funding for coordinating roles (Macdonald and Feber 2015: Chapter 15), volunteers typically only give a finite commitment, and the determinants of volunteer retention are many and varied (for example, in the case of the Scottish Mink Initiative, volunteers were retained longer if they had detected a mink within the previous six months; Beirne and Lambin 2013).

In essence, Britain now faces a choice between two future water vole conservation scenarios: The first is *American mink are fully eradicated from the countryside and water vole populations can be maintained in perpetuity with targeted habitat restoration and management*. With a national programme of mink control, coupled with limited targeted habitat restoration and reintroductions, the water vole population could almost certainly be restored if not to its pre-1930s level, then at least to its pre-1970s level, representing a substantial conservation coup and benefiting a range of other species that use these habitats (e.g. odonates and freshwater macroinvertebrates/plants, Chapters 10 and 11, this volume). The second is *American mink are not nationally eradicated and mink control and habitat restoration activities continue to be limited to given*

geographical locations, and are therefore reliant on the vagaries of ongoing funding. Under this scenario, the risk is that funding for conservation in a given location may cease, resulting (in many areas) in a swift return of American mink and the loss of local water vole populations; as well as all of the years of preceding effort and the good will of everyone who gave time and expertise to the project. Ultimately, if all funding for water vole conservation were to be withdrawn, the likelihood is that Britain's water vole population would eventually become restricted to a very few large wetland and mountain-top sites (therefore protected from mink by their habitats).

In the middle ground between these two scenarios lies a strategy of organically spreading mink control. Such projects begin in defensible strongholds defined by landscape (whether reedbeds or mountains), and spread, buoyed by early success and iteratively refining the practicalities, so that areas can be gradually extended by subsuming contiguous localities, again prioritized by topography and size, to ensure a minimal risk of reinvasion by mink. The approach here is to refine mink control in 'low-risk' locations and to spread outward to avoid costly (financially, morally, and reputationally) failures. As the successes accumulate they may be more likely to attract funding and so permit the creation of large-scale mink-free areas. But the key point remains that, unless the projects eventually grow large enough to eradicate mink on a national basis, then the fate of the water vole cannot be ensured in perpetuity: otherwise the projects will still rely upon continuity of funding to preserve the areas they have so painstakingly safeguarded.

Macdonald and Strachan (1999) noted that the then estimates for the national eradication of mink were around £30 million. While this figure predated the invention of the mink raft, and future refinements that may lower the cost per km of mink control, the present-day costs are still likely to be substantial. So the question is still relevant of whether spending such funds on mink control—as opposed to any other activity—is a priority for British conservation (Macdonald and Willis 2013). However, such a figure, applied over the short term and then able to be withdrawn, could well be cheaper in the long run than a perpetual commitment to maintaining the status quo of mink control.

When do we know enough that further research is no longer necessary? The wider point that the above scenarios demonstrates is that the role of conservation research is necessarily limited, not least because academic funding for a given project is typically of a short duration, but also because if performed correctly, the research delivers the rigorously tested methods and techniques required for a species' restoration that then need to be enacted on a large scale. The water vole conservation scenarios above no longer reflect a need for 'more research' as much as a need for society to step forwards and make a decision concerning what the water vole's future will be. The transition from 'mitigation research' to 'recovery management' represents a passing of responsibility from the research community to society as a whole, but still with a key role for researchers working in partnership with managers. Until this point, it could be claimed, with some justification, that if key information were missing the precautionary principle would imply that undertaking a national conservation effort would be unwise. Conversely, however, once the bulk of the key information is known—as it clearly is in the case of the water vole—then any further delay in restoring a species increasingly becomes a judgement on how it is valued by society. While not forgetting that the money that research ecologists receive to pursue their work is itself a reflection of how society values these species, the danger remains that beyond a certain point the relatively cheap research activities increasingly become used as an excuse for prevarication; in particular, those research papers that purport to be about conservation but contribute little of practical use. Although research for its own sake is certainly both valid and interesting, with respect to species conservation research it must also be useful and not distract from the hard decisions. A noticeable feature of the water vole's story (albeit observed with the clarity of hindsight) is that at each stage the expensive and ethically difficult option of controlling mink was delayed by the promise of a cheaper and less ethically demanding alternative that required researching. Arguably, calls for further research now, unless performed as an integrated part of a national (incorporating England, Wales, and Scotland) action programme, could simply delay the process of society deciding what it wants, by acting as a diversion.

This argument raises the uncomfortable thought that perhaps we, the authors, reached this point some time ago but were too uncritical, or even cowardly, to recognize it. For example, were we naively eager to clutch the idea that recovering native otters might oust the inconveniently alien mink; or did we hope too hard that steps to de-linearize water vole habitat, potentially achievable through relatively easily funded agri-environment scheme payments, would similarly avoid more difficult mink control and restoration strategies? When controlling mink or reintroducing water voles did we opt for schemes that took place over too

small a geographical area because such schemes are more easily manageable, rather than beginning more difficult schemes and studies at a larger spatial scale that were more likely to provide benefits? Actually, although in writing this chapter we have challenged ourselves to face this possibility with potentially painful honesty, and certainly, we acknowledge that many things we assumed, hypothesised, and predicted were later proven to be wrong, it is only with hindsight that it has become starkly clear that the only operationally relevant truth, however inconvenient, is that conserving water voles requires killing mink.

The details of any collective approach to mink removal are certainly relevant, and it is here that conservation research would still expect to have an impact. By using knowledge of ecological theory to design programmes and prioritize locations, and in following an adaptive management approach to mink control—monitoring outcomes of programmes in various locations, exploring volunteers' motivations, and feeding back the information in such a way that subsequent strategies are improved and made more cost effective—, ecological research can refine ongoing programmes for maximum effectiveness. However, choosing between our two future scenarios ultimately remains a policy judgement. Macdonald and Willis (2013) point out that how many of a species we *need* sets a target level of biodiversity below which it would be functionally precarious to dip (because, for example, it may jeopardize the provision of ecosystem services on which our well-being depends), whereas the question of how much biodiversity we *want* largely concerns policy, and recognizes the reality that accepting one option generally involves rejecting others; thereby decisions about biodiversity become, instantly, political and require many trade-offs (Macdonald et al. 2006; Barrett et al. 2013). How many water voles we *need* from a human well-being or utilitarian viewpoint may be close to zero, except insofar as their loss may alter riparian plant communities, with potential for the loss of biodiversity (Bryce et al. 2013), and that the sight of them may gladden people walking beside rivers, with possible contributions to human physical and mental health (Hughes et al. 2013; Macdonald and Feber 2015: Chapter 1). Nevertheless, we may collectively feel a sense of loss if a once common species were allowed to go extinct, or become severely restricted only to a few habitats. How many water voles UK society *wants*, therefore, may be considerably more than a strictly utilitarian baseline, and this is a judgement that now needs to be made; especially, if Britain is to retain moral credibility when enticing developing countries, with major poverty alleviation needs, to divert resources and land toward protecting their own species.

Is society prepared to foot the bill, both financially and ethically, to enact the conservation actions required? We have reached the point where undertaking more research is no longer a sensible precaution for water voles. The causes of, and solutions to, the water vole's decline are both understood and mitigable, and safeguarding its future will not require a trade-off with human well-being. Given that the return of the water vole has vanishingly few negative trade-offs and, as various volunteer programmes have shown, there is also no shortage of public support either for habitat restoration or the ethically more thorny mink eradication (e.g. Bryce et al. 2011)—and the position of the UK Water Vole Steering Group likewise embodies the policy decision that to have water voles, one must remove mink—the only substantive barrier to the water vole's restoration remains financial. In concluding this chapter we acknowledge that the conservation community has not 'presented a bill' in the sense that, to our knowledge, no fully costed proposal has ever been created for the national eradication of mink or the accompanying restoration of selected habitats for water vole conservation. The knowledge required to produce such a proposal is certainly available, however: if every conservation professional involved with mink control or water vole conservation were to meet, and present an estimated cost per kilometre for mink control over a reasonable time period in their area, a total cost could presumably be calculated with a few days' work. The bill would doubtless be extremely large—but certainly less than the cost of a single jet fighter, let alone a pair of aircraft carriers—and in this sense, the only remaining question concerning water vole conservation that needs answering is this: *Does society value the water vole sufficiently to fund the needed course of action to the end?*

Acknowledgements

The work we describe above is the product of a research community comprising a huge number of researchers, practitioners, and funders, all of whose contributions and dedication we gratefully acknowledge. While it would normally be improper in this context to highlight the efforts of one individual, we wish to take this opportunity to commemorate Rob Strachan, an author on this chapter, who passed away in May 2014. Rob was a stunning natural historian, much of whose professional and personal life was dedicated to the wildlife he loved, and who played a pivotal role in uncovering, and working to combat, the decline in UK water voles. This book chapter forms a small but, we hope, lasting part of his legacy. He is much missed.

References

Aars, J., Lambin, X., Denny, R., and Griffin, C. (2001). Water vole in the Scottish uplands: Distribution patterns of disturbed and pristine populations ahead and behind the American mink invasion front. *Animal Conservation*, **4**, 187–194.

Arnold, H.R. (1984). *Distribution maps of mammals of the British Isles*. Institute of Terrestrial Ecology, Abbots Ripton, Huntingdon.

Barreto, G.R., Macdonald, D.W., and Strachan, R. (1998a). The tightrope hypothesis: an explanation for plummeting water vole numbers in the Thames catchment. In R.G. Bailey and P.V. Jose, eds., *UK Flood Plains*, pp. 311–327. Sherwood Westbury Academic & Scientific Publications, Otley, UK.

Barreto, G.R., Rushton, S.P., Strachan, R., and Macdonald, D.W. (1998b). The role of habitat and mink predation in determining the status and distribution of water voles in England. *Animal Conservation*, **1**, 129–137.

Barrett, C., Bulte, E., Ferraro, P., and Wunder, S. (2013). Economic instruments for nature conservation. In D.W. Macdonald and K.J. Willis, eds., *Key Topics in Conservation Biology 2*, pp. 59–73. Wiley-Blackwell, John Wiley & Sons Ltd., Oxford.

Beirne, C. and Lambin, X. (2013). Understanding the determinants of volunteer retention through capture-recapture analysis: answering social science questions using a wildlife ecology toolkit. *Conservation Letters*, **6**, 391–401.

Birks, J.D.S. and Dunstone, N. (1984). *A Note on Prey Remains Collected from the Dens of Feral Mink (Mustela-Vison) in a Coastal Habitat. Journal of Zoology*, **203**, 279–281.

Birks, J.D.S. and Linn, I.J. (1982). *Studies of home range of the feral mink, Mustela vison. Symposia of the Zoological Society of London*, **49**, 231–257.

Bonesi, L., Rushton, S., and Macdonald, D. (2002). The combined effect of environmental factors and neighbouring populations on the distribution and abundance of Arvicola terrestris. An approach using rule-based models. *Oikos*, **99**, 220–230.

Bonesi, L., Rushton, S.P., and Macdonald, D.W. (2007). Trapping for mink control and water vole survival: identifying key criteria using a spatially explicit individual based model. *Biological Conservation*, **136**, 636–650.

Bryce, R., Oliver, M.K., Davies, L., Gray, H. Urquhart, J., and Lambin, X. (2011). Turning back the tide of American mink invasion at an unprecedented scale through community participation and adaptive management. *Biological Conservation*, **144**, 575–583.

Bryce, R., Wal, R., Mitchell, R., and Lambin, X. (2013). Metapopulation dynamics of a burrowing herbivore drive spatio-temporal dynamics of riparian plant communities. *Ecosystems*, 1–13. doi:10.1007/s10021-013-9677-9.

Carrion, V., Donlan, C.J., Campbell, K.J., Lavoie, C., and Cruz, F. (2011). Archipelago-wide island restoration in the Galápagos Islands: reducing costs of invasive mammal eradication programs and reinvasion risk. *PLOS ONE*, **6**, e18835.

Carter, S.P. and Bright, P.W. (2003). Reedbeds as refuges for water voles (*Arvicola terrestris*) from predation by introduced mink (*Mustela vison*). *Biological Conservation*, **111**, 371–376.

Corbet, G.B. (1971). Provisional distribution maps of British mammals. *Mammal Review*, **1**, 95–142.

Dutton, A., Edwards-Jones, G., Strachan, R., and Macdonald, D.W. (2008). Ecological and social challenges to biodiversity conservation on farmland: reconnecting habitats on a landscape scale. *Mammal Review*, **38**, 205–219.

Fraser, E.J., Macdonald, D.W., Oliver, M.K., Piertney, S.B., and Lambin, X. (2013). Using population genetic structure of an invasive mammal to target control efforts—an example of the American mink in Scotland. *Biological Conservation*, **167**, 35–42.

Halliwell, E.C and Macdonald, D.W. (1996). American mink *Mustela vison* in the upper Thames catchment: relationship with selected prey species and den availability. *Biological Conservation*, **76**, 51–56.

Hughes, J., Pretty, J., and Macdonald, D.W. (2013). Nature as a source of health and well-being: is this an ecosystem service that could pay for conserving biodiversity? In D.W. Macdonald and K.J. Willis, eds., *Key Topics in Conservation Biology 2*, pp. 143–160. Wiley-Blackwell, John Wiley & Sons Ltd., Oxford.

IUCN (1998). IUCN Guidelines for Reintroduction.

IUCN/SSC (2013). *Guidelines for Reintroductions and Other Conservation Translocations. Version 1.0.* Gland, Switzerland: IUCN Species Survival Commission, viiii + 57 pp.

Jefferies, D.J. (2003). *The water vole and mink survey of Britain 1996–1998 with a history of the long-term changes in the status of both species and their causes.* The Vincent Wildlife Trust.

Jefferies, D.J. and Arnold, H.R. (1980). *Mammal report for 1979. Report for the Huntingdonshire Fauna and Flora Society*, **32**, 32–38.

Jefferies, D.J. and Arnold, H.R. (1982). *Mammal report for 1981. Report for the Huntingdonshire Fauna and Flora Society*, **34**, 39–46.

Jefferies, D.J., Morris, P.A., and Mulleneux, J.E. (1989). An inquiry into the changing status of the water vole *Arvicola terrestris* in Britain. *Mammal Review*, **19**, 111–131.

Lambin, X., Aars, J., Piertney, S.B., and Telfer, S. (2004). Inferring pattern and process in small mammal metapopulations: insights from ecological and genetic data. In I. Hanski and O.E. Gaggiotti, eds., *Ecology, Genetics, and Evolution of Metapopulations*, pp. 515–540. Elsevier Academic Press, Burlington.

Lambin, X., Fazey, I., Sansom, J., Dallas, J., Stewart, W., Piertney, S., Palmer, S.C.F., Bacon, P.J., and Webb, A. (1998). *Aberdeenshire Water vole survey: The distribution of isolated water vole populations in the upper catchments of the rivers Dee and Don.* SNH contract report.

Lambin, X., Le Bouille, D., Oliver, M.K., Sutherland, C., Tedesco, E., and Douglas, A. (2012). High connectivity despite high fragmentation: smart iterated dispersal in a vertebrate metapopulation. *Informed Dispersal and Spatial Evolutionary Ecology* eds., J. Clobert, M. Baguette, T.G. Benton, and J. Bullock), pp. 405–412. Oxford University Press, Oxford.

Lambin, X., Telfer, S., Cosgrove, P., and Alexander, G. (1996). *Survey of water voles and mink on the rivers Don and Ythan.* A report to SNH.

Lawton, J.H. and Woodroffe, G.L. (1991). Habitat and the distribution of water voles—why are there gaps in a species range. *Journal of Animal Ecology*, **60**, 79–91.

Macdonald, D. W., Burnham, D., Hinks, A. E., and Wrangham, R. (2012). A problem shared is a problem reduced: seeking efficiency in the conservation of Felids and Primates. *Folia Primatologica*, **83**, 171–215.

Macdonald, D.W., Collins, N.M., and Wrangham, R. (2006). Principles, practice and priorities: the quest for 'alignment'. In D. Macdonald and K. Service, eds., *Key Topics in Conservation Biology*, pp. 273–292. Blackwell Publishing, Oxford.

Macdonald, D.W. and Feber, R.E., eds. (2015). *Wildlife Conservation on Farmland. Conflict in the Countryside.* Oxford University Press, Oxford.

Macdonald, D.W., Loveridge, A.J., and Rabinowitz, A. (2010). Felid futures: crossing disciplines, borders, and generations. In D.W. Macdonald and A. Loveridge, eds., *Biology and Conservation of Wild Felids*, pp. 599–649. Oxford University Press, Oxford.

Macdonald, D.W., Mace, G.M., and Rushton, S. (1998). *Proposals for the future monitoring of British mammals.* A report produced for DETR and JNCC by the Wildlife Conservation Research Unit, University of Oxford.

Macdonald, D.W., Mace, G.M., and Rushton, S. (2000). British mammals: is there a radical future? In A. Entwistle and N. Dunstone, eds., *Priorities for the Conservation of Mammalian Diversity: Has the Panda Had its Day?*, pp. 175–205. Cambridge University Press, Cambridge.

Macdonald, D.W. and Rushton, S. (2003). Modelling space use and dispersal of mammals in real landscapes: a tool for conservation. *Journal of Biogeography*, **30**, 607–620.

Macdonald, D.W., Sidorovich, V.E., Anisomova, E.I., Sidorovich, N.V., and Johnson, P.J. (2002). The impact of American mink Mustela vison and European mink Mustela lutreola on water voles Arvicola terrestris in Belarus. *Ecography*, **25**, 295–302.

Macdonald, D.W. and Sillero-Zubiri, C. (2004). Conservation. From theory to practice, without bluster. In D.W. Macdonald and C Sillero-Zubiri, eds., *The Biology and Conservation of Wild Canids*, pp. 353–372. Oxford University Press, Oxford.

Macdonald, D. and Strachan, R. (1999). *The Mink and the Water Vole. Analyses for Conservation.* EA/Wildlife Conservation Research Unit, Oxford.

Macdonald, D.W. and Willis, K.J. (2013). Elephants in the room: tough choices for a maturing discipline. In

D.W. Macdonald and K.J. Willis, eds., *Key Topics in Conservation Biology 2*, pp. 469–494. Wiley-Blackwell, John Wiley & Sons Ltd., Oxford.

MacPherson, J. and Bright, P. (2009). *Reversing the water vole decline III: final report on the National Key Site project*. A report by Royal Holloway College to PTES, NE, CCW & EA.

Mammal Society, The (2013). <http://www.mammal.org.uk/history_achievements>, accessed May 2013.

Moehrenschlager, A., Shier, D.M., Moorhouse, T.P., and Stanley Price, M.R. (2013). Righting past wrongs and ensuring the future. *Key Topics in Conservation Biology 2* (D.W. Macdonald and K.J. Willis, eds.) pp. 405–429. Wiley-Blackwell, John Wiley & Sons Ltd., Oxford.

Moorhouse, T.P., Gelling, M., McLaren, G.W., Mian, R., and Macdonald, D.W. (2007). Physiological consequences of captive conditions in water voles (*Arvicola terrestris*). *Journal of Zoology*, **271**, 19–26.

Moorhouse, T.P. and Macdonald, D.W. (2005a). Indirect negative impacts of radio-collaring: sex ratio variation in water voles. *Journal of Applied Ecology*, **42**, 91–98.

Moorhouse, T.P. and Macdonald, D.W. (2005b). Temporal patterns of range use in water voles: Do females' territories drift? *Journal of Mammalogy*, **86**, 655–661.

Moorhouse, T.P. and Macdonald, D.W. (2008). What limits male range sizes at different population densities? Evidence from three populations of water voles. *Journal of Zoology*, **274**, 395–402.

Oliver, M., Luque-Larena, J.J., and Lambin, X. (2009). Do rabbits eat voles? Apparent competition, habitat heterogeneity and large-scale coexistence under mink predation. *Ecology Letters*, **12**, 1201–1209.

Reynolds, J.C., Richardson, S.M., Rodgers, B.J.E., and Rodgers, O.R.K. (2013). Effective control of non-native American mink by strategic trapping in a river catchment in mainland Britain. *The Journal of Wildlife Management*, **77**, 545–554.

Reynolds, J.C., Short, M.J., and Leigh, R.J. (1994). *Development of population control strategies for mink Mustela vison, using floating rafts as monitors and trap sites. Biological Conservation*, **120**, 533–543.

Reynolds, J.D., Short, M., Porteus, T., Rodgers, B., and Swan, M. (2007). *The GCT mink raft*. <http://www.gct.org.uk/uploads/minkraftleaflet.pdf> (04 August 2008).

Rushton, S.P., Barreto, G.W., Cormack, R.M., Macdonald, D.W., and Fuller, R. (2000). Modelling the effects of mink and habitat fragmentation on the water vole. *Journal of Applied Ecology*, **37**, 475–490.

Sidorovich, V.E. (1992). Gegenwartige Situation des Europaeischen Nerzes (*Mustela lutreola*) in Belorusland. Hypothese seines Verschwindens.—In R. Schroepfer, M. Stubbe, and D. Heidecke, eds., *Semiaquatische Saeugetiere*, pp. 316–328 Wissenschaftliche Beitraege/Martin-Luther, Univ. Halle Wittenberg.

Stoddart, D.M. (1977). Water vole. In G.B. Corbet and H.N. Southern, eds., *Handbook of British Mammals*. pp. 196–204. Blackwell Scientific Publications, Oxford.

Strachan, R. (1998). *Water Vole Conservation Handbook*. EN/EA/Wildlife Conservation Research Unit, Oxford.

Strachan, R. and Holmes-Ling, P. (2003). *Restoring water voles and other biodiversity to the wider countryside. A report on the Chichester coastal plain sustainable farming partnership*. Wildlife Conservation Research Unit, Oxford.

Strachan, R. and Jefferies, D.J. (1993). *The water vole Arvicola terrestris in Britain 1989–1990: its distribution and changing status*. The Vincent Wildlife Trust, London.

Strachan, C., Jefferies, D.J., Barretot, G.R., Macdonald, D.W., and Strachan, R. (1998). The rapid impact of resident American mink on water voles: case studies in lowland England. *Symposium of the Zoological Society of London*, **71**, 339–357.

Strachan, R. and Moorhouse, T. (2006). *Water Vole Conservation Handbook*, 2nd ed. Wildlife Conservation Research Unit, Oxford.

Strachan, R., Moorhouse, T.P., Bonesi, L., and Harrington, L. (2005). Restoring and reintroducing water voles. In D.W. Macdonald, ed. The second WildCRU review, Wildlife Conservation Research Unit, pp. 121–127.

Strachan, R., Moorhouse, T., and Gelling, M. (2011). *Water Vole Conservation Handbook*, 3rd ed. Wildlife Conservation Research Unit, Oxford.

Strachan, C., Strachan, R., and Jefferies, D.J. (2000). *Preliminary report on the changes in water vole population of Britain as shown by the National Surveys of 1989–1990 and 1996–1998*. The Vincent Wildlife Trust, London.

Telfer, S., Holt, A., Donaldson, R., and Lambin, X. (2001). Metapopulation processes and persistence in remnant water vole populations. *Oikos*, **95**, 31–42.

Telfer, S., Piertney, B., Dallas, J.F., Stewart, W.A., Marshall, F., Gow, J.L., and Lambin, X. (2003). Parentage assignment detects frequent and large-scale dispersal in water voles. *Molecular Ecology*, **12**, 1939–1949.

Woodroffe, G.L. and Lawton, J.H. (1990). Patterns in the production of latrines by water voles (*Arvicola terrestris*) and their use as indexes of abundance in population surveys. *Journal of Zoology*, **220**, 439–445.

Woodroffe, G.L., Lawton, J.H., and Davidson, W.L. (1990). The impact of feral mink *Mustela vison* on water voles *Arvicola terrestris* in the North Yorkshire Moors National Park. *Biological Conservation*, **51**, 49–62.

Yamaguchi, N., Rushton, S., and Macdonald, D.W. (2003). Habitat preferences of feral American mink in the Upper Thames. *Journal of Mammalogy*, **84**, 1356–1373.

Zalewski, A., Piertney, S.B., Zalewska, H., and Lambin, X. (2009). Landscape barriers reduce gene flow in an invasive carnivore: geographical and local genetic structure of American mink in Scotland. *Molecular Ecology*, **18**, 1601–1615.

CHAPTER 16

What next? Rewilding as a radical future for the British countryside

Christopher J. Sandom and David W. Macdonald

Feral is, in part, a counter-factual: it imagines the lives we no longer lead but might, the species that no longer exist but could, and the faculties we no longer engage but should.

George Monbiot

16.1 Introduction

Many of the chapters in this book have been about understanding the interface between agriculture in the British landscape, and how to conserve Nature. Indeed, the extent to which that understanding has advanced since Britain became part of the Common Agricultural Policy in 1973, or since we first asked farmers in 1981 what they thought about wildlife on their farms (Macdonald and Johnson 2000; Macdonald and Feber 2015: Chapter 14), is extraordinary. The WildCRU is proud to have played its part, along with many others, in that journey, as described in this book. That said, to borrow from the title of Chris Patten's (2009) book about the future of nation states: 'What next?'. Many of the countryside's ailments remain the same, so certainly there is great value in prescribing more of the same: better evidence-based solutions, packaged within creative policies to hold the line for Nature. However, Patten's answer for the future of nation states (which, after all, are directly relevant, being the operational units that formulate and deliver the environmental policy that will shape both food production and nature conservation in the countryside) was to look for realistically radical leaps to a better future. That is what we aim to do for Nature in this chapter.

A good starting point is to remember, as we describe in Chapter 1, that everything is connected to everything else, that when it comes to conservation, biology is necessary but not sufficient, and what is needed is extreme interdisciplinarity that weds the natural and social sciences and the environmental and human dimensions of the countryside. While many attributes of

Nature defy monetization, it is simultaneously the case that valuing Nature is an irresistible common denominator in audits of policy options (TEEB 2010). An insightful analogy is that of the circular economy, where waste is considered as a valuable resource rather than an expensive burden (Benyus 2009; Macarthur 2013). In circular economies, and in functioning ecosystems, energy and nutrients cycle.

Turning to that radical future, we think back with legitimate nostalgia to wonderful elements of the long-lost wild past. Large mammals have a special place in many human value systems (Macdonald et al. 2013), so think of the wolves *Canis lupus*, lynx *Lynx lynx*, and even bear *Ursus arctos* that once thrived here beside our recent ancestors, along with wild boar *Sus scrofa* (Fig 16.1), beaver *Castor fiber*, and elk *Alces alces*. There is nothing irrational or foolishly sentimental in seeking inspiration from these animals for a radical future, inspiration with a hard-nosed, science-based, and policy-relevant character. Our goal is not to recreate the past, but rather to build a present informed by the past and fit for the future (Macdonald 2009). Furthermore, thinking about the future is not an undisciplined self-indulgence, but a responsible (indeed essential) activity, itself shaped by carefully devised rules and procedures that takes an informed look at how things might be, say, 40 years from now, and asks how each plausible scenario might affect society's planning for the future of the countryside (Kass et al. 2011). It is in this spirit that we turn to the question: What next?

First, let us define our scope and terms. In our radical future we have identified three aims for the national ecosystem: (1) it supports native biodiversity,

Wildlife Conservation on Farmland. Managing for Nature on Lowland Farms. Edited by David W. Macdonald and Ruth E. Feber.

Figure 16.1 Alladale Wilderness Reserve with wild boar *Sus scrofa*. © C. Sandom.

(2) it is a bountiful resource of ecosystem services, and (3) it is as self-sustaining as possible. These goals are well aligned with the government's ambition to be the first generation to leave the natural environment in a better state than when it was inherited (Defra 2011). However, none of our stated goals are currently being met. The State of Nature report highlighted that, of 3148 species investigated, 60% have declined in the last 50 years (Burns et al. 2013), and the National Ecosystem Assessment reported 30% of ecosystem services are deteriorating in Britain (National Ecosystem Service Assessment 2011a), despite current conservation efforts. To achieve our goals, ecological restoration is required and to achieve this we turn to rewilding as one radical answer. Ecological restoration is 'the process of assisting the recovery of an ecosystem that has been degraded, damaged, or destroyed' (SER 2004). Unfortunately, clear and broadly applicable guidelines for delivering such assistance have not been forthcoming, reflecting the difficulty ecologists have had in developing community assembly rules (Keddy and Weiher 2001). As a result, restoration ecologists have largely left internal ecosystem dynamics as black boxes and focused on restoring appropriate ecological filters,

particularly environmental conditions (Belyea 2004), in the community assembly chain (see Fig. 16.2). Understanding these internal ecosystem dynamics as best we can will help practitioners design more effective and efficient restoration projects, and it is in this context that rewilding has become a key discussion point.

Rewilding received its first formal description by Soulé and Noss (1998) as a mechanism for conserving biodiversity using (1) large core protected areas, (2) connectivity between them, and (3) the reintroduction of keystone species, particularly large predators. The conservation value of well-connected large protected areas was illuminated by the theory of island biogeography (MacArthur and Wilson 1967), and large predators received special attention because of their importance in driving trophic cascades and their need of large and connected reserves (Terborgh and Estes 2010). More recently, the keystone species element emerges as particularly important and rewilding has come to mean species reintroduction designed to re-establish a lost or impoverished ecological process, as opposed to being motivated solely to preserve the returned species, although the distinction is sometimes fuzzy (Sandom et al. 2013b). In essence, in rewilding, species' function is prioritized over form. A complete suite of ecological processes links the working parts of a fully functioning ecosystem, thus rewilding can be defined as community (re-) assembly to restore ecosystem function. Restoring all natural processes is perhaps rewilding in its purest sense, but we see rewilding as a continuum where almost any landscape could be rewilded to some extent. This continuum is captured within the concepts of core areas and connectivity, with core areas supporting purer rewilding where completely functional and biodiverse ecosystems are prioritized, while rewilding in the intervening productive landscapes seeks to restore natural processes in a landscape of human endeavour to balance biodiversity, ecosystem services, and self-sustainability. A good balance will allow the land to be productive while also ensuring connectivity between core areas supporting biodiversity and self-sustainability.

Rewilding challenges practitioners to understand how past, present, and future human activity has, is, and will affect ecosystem function, and to take appropriate remedial action where possible (Fig. 16.2). Because there is no single point in history that should be recreated, rewilding focuses on re-establishing naturally dynamic processes that, through an appropriate sequence of species reintroductions, attempt

to move the ecosystem towards a more appropriately biodiverse and functional state that is self-sustaining in the present climate, and that projected for the near future (Fig. 16.3). Ultimately, in core areas, rewilding practitioners should be attempting to restore community components that will restore local ecosystems to a totipotent state, which determines their own shifting form and creates a complex mosaic of conditions that supports biodiversity, while in productive landscapes processes are restored to support ecosystem services and biodiversity. As we turn to this radical future we are teetering on the cutting edge of applied ecological theory, but even so, the restoration of species communities, and with them critical processes, is already bringing successes (Terborgh and Estes 2010; Estes et al. 2011).

Rewilding Britain is a challenging prospect. As an island, densely populated for the most part, finding space to establish large core areas and connecting them is difficult (Lawton et al. 2010). Furthermore, the British ecosystem has been heavily impacted by people for millennia. This creates difficulties in determining what processes have been lost and which are functioning insufficiently or too vociferously. It also means it has been many generations since the British public lived alongside wild and functioning ecosystems with, for

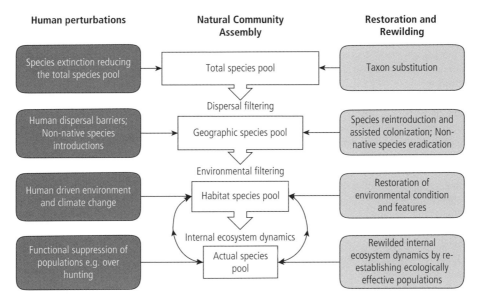

Figure 16.2 Conceptual framework of the relationship between the natural community assembly chain (centre), human drivers of ecosystem perturbation (left), and tools of restoration and rewilding to counter these perturbations (right). Traditionally, ecological restoration has targeted the restoration of appropriate environmental filtering and left the internal ecosystem dynamics as a black box. Rewilding requires the understanding and restoration of the internal dynamics; this is especially important when perturbations of the internal dynamics are a driving force for change in the ecological conditions. Community assembly chain adapted from Keddy and Weiher (2001) and Belyea (2004).

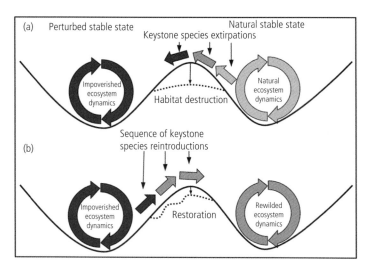

Figure 16.3 A graphical depiction of how rewilding and restoration fit within the 'alternative stable states' community assembly model. (a) A naturally functioning stable state ecosystem is moved to an alternative stable state through a combination of species extirpations and habitat destruction. (b) To achieve effective (target driven) and efficient (cost effective and long-term benefits) restoration, it is necessary to restore the internal ecosystem dynamics through species reintroduction that will naturally move the ecosystem towards a more diverse stable state. Restoration of environmental conditions can be used as short-term direct management to facilitate the reintroduction of keystone species.

example, complete communities of the large predators. Large mammalian predators, indeed large mammals in general, are at the forefront of the rewilding agenda because they have been so heavily persecuted, persecuted to the extent that they have been eradicated or functionally removed from most developed countries (Sandom et al. 2013a; Sandom et al. 2014a). Through trophic interactions, predators are intricately linked to the functioning of ecosystems (Smith et al. 2003) and so are important factors in supporting biodiversity. Because of their importance, past persecution, and present controversies we will focus on the feasibility of restoring some of the processes delivered by large mammals to Britain.

The Scottish Highlands have received greater attention than England and Wales for rewilding to date because the possibility of creating a large core area is more feasible there, due to the lower population density and land ownership structure that allow large areas to come under single land management policies. In this chapter, we will apply this experience to the whole of Britain by exploring the current state of the wild in Britain, whether a land sparing, sharing, or a combined land management structure offers the best opportunities to achieve our goals, and how three key processes, provided by large mammals, can be restored to Britain.

16.2 State of the wild in Britain

In previous interglacials similar to the Holocene, Western Europe, including Britain, is thought to have supported a wooded landscape (Svenning 2002). However, the degree to which this woodland was primarily closed, open, or heterogenous is highly contentious. The prevailing wisdom has been that it was primarily closed, although including some open patches (for a review see Svenning 2002), but this is challenged by the observations of Frans Vera at one of the world's early rewilding projects, Oostvaardersplassen (Vera 2000).

Oostvaardersplassen is a fenced area of reclaimed land in the Netherlands, not far from Amsterdam, that covers 6000 ha (including some open water). It has been allowed to go wild since the nature reserve was formed in 1968, and, importantly, part of the lost large herbivore guild, including Heck cattle and Konik horses, has been re-established and largely governs itself, bar human intervention to ease animal welfare concerns (Vera 2009). Frans Vera posited that large herbivores are a key factor in determining vegetation dynamics, creating a diverse mosaic of open, closed, and re-vegetating habitats through grazing, in combination with other natural processes (Vera 2000). It is an important hypothesis but one that, in Britain, is not strongly supported by pollen and beetle data for the early Holocene (~10 000 to 5000 years BP), prior to agriculture (Mitchell 2005; Whitehouse and Smith 2010). Evidence from beetles indicates relatively low herbivore abundances and relatively high tree cover during this period (Sandom et al. 2014b). In contrast, the Last Interglacial (Eemian, ~132 000 to 110 000 years BP) does appear to indicate greater herbivore densities and more mixed open and closed vegetation mosaics, reflecting Vera's hypothesis (Sandom et al. 2014b). No modern humans were in Britain during previous interglacials and there was a full compliment of megafauna that may have been instrumental in driving the vegetation dynamics (Sandom et al. 2014b). In this regard, despite uncertainty in past environment reconstructions, a mixed mosaic of open and closed vegetation communities may be considered as the ecological and evolutionary history of the wild British landscape that will support Britain's native biodiversity.

With the dawn of agriculture, the balance shifted back from the primarily closed woodland of the early Holocene to a greater mosaic of open, wood-pasture, and closed habitats, as land was actively cleared for agriculture and domesticated livestock were reared more extensively (Sandom et al. 2014b). Interestingly, this period may have more closely reflected interglacial vegetation structure prior to modern human arrival than the early Holocene. The negative impacts on biodiversity of the industrial revolution that allowed industrial-scale agriculture have seen the homogenization of the landscape, removing wooded areas and replacing them with open agricultural landscapes. Seventy-five per cent of Britain is now cultivated (Morton et al. 2011). Twelve per cent of the UK is woodland but only *c.* 2.3% is ancient woodland (Forestry Commission 2012). The lack of woodland, particularly mature woodland, appears to be strongly linked to the loss of biodiversity in Britain, with 40% of the species lost since 1800 associated with this habitat, considerably more than any other single habitat type (Hambler et al. 2011). Woodlands also provide a variety of ecosystem services such as improving air, soil, and water quality, and reducing the risk of flooding and, while almost all these services are currently increasing, they are doing so from a highly depressed baseline (National Ecosystem Service Assessment 2011b). Thus, there is a pressing imperative to consider woodland regeneration processes. In this chapter we will do this, by exploring the current state of processes related to woodland regeneration, and considering Britain's mammal community to see if we can identify a sequence of mammal reintroductions to restore key processes and thus lead to a more biodiverse and self sustaining British ecosystem. We do this by following the community assembly chain as outlined in Figure 16.2 to determine which functional guilds in the mammal community are either over or under represented. This is a four-step process starting with the total global mammal species pool and ending with the local internal ecosystem dynamics.

16.2.1 Total mammal species pool (global)

The Pleistocene megafauna extinction had important implications for vegetation structure and corresponding processes. In Great Britain, 13 species of large mammal[1] became either globally extinct or were continentally lost between the last interglacial (~132 000 years Before Present: BP) to 1000 years BP, with modern humans the most likely cause (Sandom et al. 2014a), abetted by climate change (Yalden and Barrett 1999). These extinctions removed incumbents of the large predator (≥10 kg, 3 species), mega-herbivore (≥1000 kg, 6 species), large herbivore (≥10 kg, 2 species), and large omnivore (2 species) functional guilds (Fig. 16.4), leaving the largest predator and herbivore guilds completely unpopulated.

16.2.2 Geographic and habitat species pools (Britain)

A further suite of extirpations removed brown bear, wolf, lynx, aurochsen *Bos primigenius*, elk, wild boar, and beaver from Great Britain during the Holocene, but they survive elsewhere in Europe (horse and reindeer were possibly lost to climate change, with small populations, respectively, returned to the New Forest and the Cairngorms). Many of these species are making their own way back across Europe (Enserink and Vogel 2006), but would need a lift to the UK, as beaver and wild boar have. Introduced non-native mammalian species have crossed the Channel with human assistance (see Fig. 16.4), often problematically (see Macdonald and Feber 2015: Chapters 6 and 8), swelling the ranks of the unpredated medium/large herbivores (Fig. 16.4), while others have become what Macdonald and Burnham (2010) term 'ecological citizens' (Macdonald and Feber 2015: Chapters 1 and 11).

Britain is now dominated by open agricultural and moorland habitats with limited native woodland communities. As a result, there is limited habitat available for woodland associated species such as lynx, wild boar, beaver, and elk, affecting their suitability to be reintroduced in most regions, and there is a continued threat to woodland dependent species (Hambler et al. 2011).

[1] Woolly rhinoceros *Coelodonta antiquitatis*[MH], spotted hyaena *Crocuta crocuta**[LP], straight-tusked elephant *Elephas antiquus**[MH], hippopotamus *Hippopotamus amphibius**[MH], Neanderthal *Homo neanderthalensis*[LO], scimitar-toothed cat *Homotherium latidens*[LP], woolly mammoth *Mammuthus primigenius*[MH], Irish elk *Megaloceros giganteus*[LH], muskox *Ovibos moschatus*[LH], cave lion *Panthera spelaea**[LP], narrow-nosed rhinoceros *Stephanorhinus hemitoechus*[MH], Merck's rhinoceros *Stephanorhinus kirchbergensis**[MH], cave bear *Ursus spelaeus*[LO]: * = reintroduction or taxon substitution consideration possible, [MH] = megaherbivore, [LP] = large predator, [LO] = large omnivore, [LH] = large herbivore.

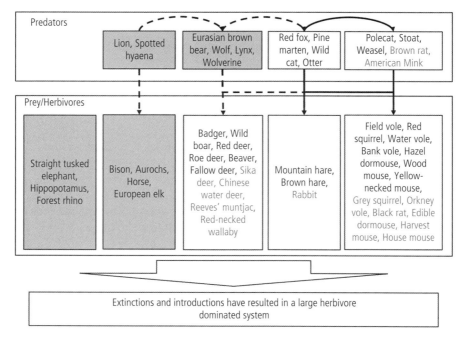

Figure 16.4 Mammalian food web structure in the UK, including all species that have been present in the UK since the Last Interglacial, except those thought to be unsuited to current climatic conditions e.g. mammoth. Grey boxes indicate extinct guilds. Grey lettering indicates introduced species. The fallow deer *Dama dama* is an example of Pleistocene rewilding as it is a species present during the Last Interglacial but reintroduced to Britain, most likely by the Normans (Yalden and Barrett 1999; Sykes 2004). Schematic is divided into two broad guilds. Boxes within these guilds group species by body size, going from large to small from left to right. Arrows indicate likely strong interactions, solid and dotted lines indicate current and lost interactions respectively. Competition within the predator guild is summarized with larger species limiting smaller species. Continentally and globally extinct species are included if it is possible they or a taxon substitute could conceivably be reintroduced, even if it is currently unworkable in reality.

16.2.3 Internal ecosystem dynamics (regional)

Humans directly impact the ecological effectiveness of particular functional guilds. Most land is under management, whether it appears wild or not. For instance, large tracts of land in the Scottish Highlands are managed for game sports (Warren 2009), red deer *Cervus elaphus* (Fig. 16.5) being a key species (but not the only species) that is suppressing natural woodland regeneration (Hobbs 2009). High densities of large deer are less prevalent in England and Wales (Macdonald and Tattersall 2001), although fallow deer *Dama dama* are widely distributed in these countries and their population is expanding (Ward 2005). Smaller cervids, such as roe deer *Capreolus capreolus* and muntjac *Muntiacus reevesi*, are rapidly expanding in distribution and possibly abundance, and may also reduce tree regeneration potential (Gill and Fuller 2007; Gill and Morgan 2010). Sheep farming on marginal land exerts strong grazing pressures in the wilder regions of England

and Wales (Fuller and Gough 1999). Wild boar and beaver offer two services that the others in their guild (Fig. 16.4) do not: rooting and dam building. Rooting is important for creating bare ground patches for new vegetation to establish (Sandom et al. in prep-a) and dam building and the felling of riparian trees can be important for creating niche space for other species (Rosell et al. 2005). The former may be important for providing germination niches and the latter for creating standing deadwood.

16.3 Rewilding Britain: wolves, wild boar, and beaver

To help achieve our stated three goals of biodiversity, ecosystem services, and self-sustainability in Britain, we identify three processes in need of restoration which are provided by three native mammals that are either already being restored to Britain (beaver and wild

Figure 16.5 Alladale Wilderness Reserve with Red Deer *Cervus elaphus.* © C. Sandom.

boar), or are considered to be possible candidates in the future (wolf). Firstly, the imbalance between large herbivores and large predators is a significant limitation for any wild land in Britain to revert to a mosaic of vegetation communities. This is particularly true of the Scottish Highlands, arguably the wildest land in Britain, where high abundances of red deer, managed at high densities, are suppressing woodland regeneration and maintaining the landscape in a biodiversity poor state. In Wales and England there are fewer large deer, but instead the 'wild' lands in these countries are equally impacted by high densities of non-native farmed sheep and potentially other, smaller deer species. On this basis we diagnose predation as the top priority for restoration in the wild regions of Scotland, and the wolf as the key species to deliver this process. While sheep farming remains the dominant land use in the wild land of England and Wales, the rewilding of predation is unlikely to be achievable. Should farming in marginal areas be abandoned, as it currently is in continental Europe (Navarro and Pereira 2012), restoration of native, wild predator–prey interactions in core areas would be a possibility, with the lynx potentially a key species (see Box 16.1). After exploring the process of restoring predation and, with it, landscape-scale

changes in woodland regeneration dynamics, we consider the important roles of rooting and dam building through restoration of wild boar and beaver populations. These species offer important processes for increasing ecosystem functionality.

16.3.1 Rewilding and the land sharing versus sparing debate

Before considering the restoration of these processes in detail, it must be decided where to focus rewilding efforts. An important and relevant debate to rewilding in Britain is consideration of the various merits of a land sharing or sparing approach to conservation[2]. Twenty-six per cent of British land is protected (IUCN and UNEP-WCMC 2011). This compares favourably with, for example, the 12% cover of protected areas in the USA (IUCN and UNEP-WCMC 2011). However, the USA's 12% includes iconic national parks, such as Yellowstone, that are far closer to a naturally

---

[2] Land sharing (or wildlife friendly farming) 'which boosts densities of wild populations on farmland but may decrease agricultural yields' compared to sparing 'which minimizes demand for farmland by increasing yield' (Green et al. 2005).

Box 16.1 Reintroducing the lynx: a simple matter of choice?

Radiocarbon-dated bones reveal that at least 12 650 years ago lynx *Lynx lynx* occurred in Britain, and persisted here as recently as 1842–1550 ^{14}C yr BP (Hetherington et al. 2006). It was once believed that the downturn in their fortunes was prompted by climate change and thus loss of forest habitat between 10 000 and 4000 years ago, but evidence of their historical persistence suggests they succumbed to persecution. There would be an argument under the EU Habitats Directive for reintroducing lynx to Britain, if people had played a hand in its demise. Furthermore, the condition that previous impediments to their survival have been reversed is certainly met in so far as nowadays, roe deer are so numerous, and still increasing, that they are commonly regarded as a pest. If lynx had the capacity to limit the current high populations of deer, that too could motivate their reintroduction. According to Hetherington (2005), neither is an especially compelling case, but recreating a large carnivore community could delight some (see Box 16.2), although it may be seen by others as foolish meddling. Hetherington et al. (2008) modelled the feasibility of such a reintroduction. They assumed lynx home range sizes characteristic of the Swiss Jura (females 45 km[2], males 74 km[2] encompassing female home ranges), and assessed the extent and connectivity of woodland in Scotland. This process identified 20 678 km[2] of potential habitat in 30 patches and two main networks of 15 000 km[2] in the Highlands, and 5330 km[2] in the Southern Uplands, with an additional contiguous 810 km[2] in England.

In Continental Europe, while lynx prefer roe deer, they will prey on ungulates up to the size of a red deer hind, taking also lagomorphs, small mammals, and ground-nesting birds (Breitenmoser et al. 2010). Hetherington et al. (2008) extrapolated from known population densities of deer in woodland to reach overall ungulate population densities in the Highlands and Southern Uplands comparable to those in the Swiss Jura, where lynx thrive. Then, using a relationship between lynx population density and log ungulate biomass/km[2], they estimated that the Highlands and the Southern

Uplands could sustain 2.63 and 0.83 lynx/100 km[-2], respectively. This equates to ~400 lynx in the Highlands and 50 in the Southern Uplands (including the Kielder Forest in England). A population viability analysis suggested such a Highland population would be viable, whereas the more southerly one might not.

What might be the pros and cons of such a reintroduction? Lynx kill sheep; for instance in Norway, where sheep are grazed in forests and roe deer density is low (Odden et al. 2006). In the Swiss Alps, 80% of farmers who lost livestock during the years 1979–1999 lost fewer than three and 95% of attacks were within 360 m of the forest edge (Stahl et al. 2001a; Stahl et al. 2002). Although the abundance of roe deer in Scotland might minimize depredation on sheep, there might also be some predation on grouse, and conservationists might fear for impacts upon capercaillies *Tetrao urogallus*, wildcats *Felis silvestris*, and pine martens *Martes martes*. Macdonald et al. (2013) conclude that this example illustrates the central importance of consumer choice in conservation. Lynx occurred in Scotland before, and conditions are now such that they could be there again. They could cause problems to some, delight to others. They would cause financial loss to some, while doubtless generating revenue for others (e.g. through ecotourism). As Macdonald et al. (2013) put it, 'it would surely not be beyond the wit of a nation such as Scotland to balance these costs and benefits with suitable economic instruments (e.g. compensation for loss of stock)'. The question is simply whether society wants lynx, and for those who think the answer should be yes, the road ahead lies in ordering their priorities and honing their advocacy. Again to quote Macdonald et al. (2013) 'The answer has some significance further afield because this is a case of a developed country considering the denizens of its own backyard; should the British public prove unenthusiastic about having lynx in its midst, this would not enhance the British conservationists' case that South Americans, Africans, and Asians should welcome, respectively, jaguars, lions, and tigers into theirs'.

---

functioning ecosystem, containing a more complete trophic structure with the presence of top carnivores (Smith et al. 2003; and see Box 16.2), than any area in Britain (Fischer et al. 2008). While the US approach is land sparing, the British approach is perhaps best described as the sharing of spared land. For instance, in England, only 6.1% of the land is primarily protected for Nature under areas designated Special Sites of Scientific Interest (SSSIs). Importantly, this land is

typically under management to sustain its biological significance and not governed by natural processes, while the large national parks and areas of outstanding natural beauty that are protected for multiple purposes cover 23.5% (Lawton et al. 2010). In spite of only 6.1% of land being spared for Nature in England, farming has become increasingly intensive elsewhere, which is linked to alarming wildlife declines over the last 50 years (Burns et al. 2013).

Box 16.2 Reintroducing swift foxes to agricultural landscapes

Can carnivores ever be reintroduced to agricultural land-scapes? The reintroduction of swift foxes *Vulpes velox* to Canada shows that indeed they can, while in Europe car-nivores are currently naturally recolonizing agricultural land (Deinet et al. 2013). Swift foxes are meso-carnivores, and ecosystem functions had been heavily disturbed at trophic levels both above and below them in the Great Plains of North America, after Europeans had 'won the Wild West'. Lessons learnt from their reintroduction to North America highlight how new ecosystem processes offer both threats and opportunities to this functional group, and highlight the importance of adaptive management under altering eco-system processes. Once, swift foxes were so abundant that over 117 000 of their pelts were traded by just one of two Canadian trapping companies in a 24-year span of the late 1800s; by 1938 they were extinct in Canada. Reintroduc-tions began in 1983, under the unpromising conditions of returning this native prairie specialist to a landscape where most of its former habitat had been ploughed, where key-stone species like prairie dogs, bison, wolves, and grizzlies had been eradicated, and where a culture of killing carni-vores was pre-eminent.

Nonetheless, lessons learned from our releases of 942 foxes over a 14-year period facilitated adaptive manage-ment (Moehrenschlager and Macdonald 2003). For example, the direct translocation of wild-caught foxes from the United States was more effective than the release of captive-bred

foxes (Moehrenschlager and Macdonald 2003). Releases ended in 1997, and nowadays the steadily increasing Canadian swift fox population consists almost entirely of wild-born individuals that comprise a genetically diverse, structured population with increasing effective population size (Cullingham and Moehrenschlager 2013), leading to the species being downlisted nationally from 'endangered' to 'threatened' in 2012.

Remnant populations of imperilled species, or the early settlers following reintroduction, are not necessarily func-tional replications of larger indigenous populations. Swift foxes were returned to a landscape in which natural prai-rie had been changed to pasture, and the predatory guild up-ended. Human settlement fostered the invasion of red foxes, the slaughter of wolves released pressure on coyotes *Canis latrans*, and both of these beneficiaries of agricul-ture were morbidly inimical to swift foxes. However, this new terrain suited American badgers *Taxidea taxus*, whose bur-rows provided swift foxes with bolt holes to evade coyotes. Once swift foxes gained a foothold, their rapid reproduct-ive rate allowed them to increase (Moehrenschlager et al. 2007). In a heavily degraded agricultural landscape, devoid of protected areas, the return of the swift fox to Canada could claim to be the most successful national-scale in-stance of carnivore reintroduction—and offers hope for rewilding and land sharing in the also entirely modified UK environment.

Considering this debate in the context of rewild-ing highlights the need to combine sparing and shar-ing to create a coordinated ecosystem-level approach to achieving our stated goals. Creating core areas is consistent with land sparing, while maintaining con-nectivity will require process-targeted rewilding in productive landscapes to achieve greater sharing with nature. Yet this is a simplistic representation, the real challenge will be striking the balance between nature and pressing human needs, primarily food production. Functioning nature is essential for delivering produc-tion, especially with regard to maintaining produc-tion capacity for future generations, but agricultural intensification has allowed increases in agricultural production that feed the growing human population and its demand for a better quality of life, which is currently synonymous with increased consumption. Careful consideration is required to balance current

and future production requirements with responsible consumption, to ensure sufficient land sparing and sharing is possible to allow nature to function for the benefit of ourselves and biodiversity. It is not the pur-pose of this chapter to resolve this debate but, instead, to explore the potential to rewild both core areas and productive landscapes in Britain, and to set out what the potential opportunities and costs of doing so are in relation to our stated goals.

Using our three case study processes of predation, rooting, and dam building, delivered by three large mammals, wolves, wild boar, and beavers, we explore how they could rewild core areas and productive land-scapes in Britain. We begin with predation and our wolf case study and the opportunities to create core areas in Britain. In our second and third case studies we explore wild boar and beavers returning to Britain in the context of land sharing.

16.3.2 Case study 1. Restoring predation through wolf reintroduction

The wolf was formerly, globally, the most widely distributed non-commensal, terrestrial mammal (Mech and Boitani 2003). Persecution severely reduced its extent but, in recent years, it has enjoyed a comeback in Europe, having recently been seen in Belgium, Netherlands, and Denmark for the first time in at least a century (Deinet et al. 2013a). The notion of reintroducing wolves to Scotland has been a recurrent source of titillation amongst British conservationists for decades, most commonly with a view to limiting the numbers of red deer on the Island of Rum. But it also receives serious academic attention on the mainland, again in the context of deer control and the economics of sheep farming (Nilsen et al. 2007). Beyond the Scottish Highlands, wolves have not received serious attention in Britain. This is perhaps because, when considering the landscape scale, rewilding the Scottish Highlands is especially attractive with its low human population density, wide open spaces, and abundant deer populations.

Lessons learnt from the wolf reintroduction to Yellowstone National Park presage the cascade of good things awaiting Scotland with wolves—in Yellowstone, the red deer (known there, confusingly, as elk; but classified as *Cervus elephas* by the IUCN nonetheless) have declined and a changed landscape of fear[3] has catalysed a trophic cascade that has increased riparian woodland, and with it, beavers, stabilizing the watercourses, reducing erosion, and increasing biodiversity (Ripple and Beschta 2012). In Scotland, heavy grazing by sheep or red deer (the latter maintained at high densities for stalking), prevents woodland regeneration and thwarts the recreation of a functioning Caledonian Forest ecosystem (Hobbs 2009). To judge by the models of Nilsen et al. (2007) and Manning et al. (2009), wolves would reduce red deer abundance by 50% and re-establish the landscape of fear to prompt a reconfiguring of the ecosystem. Yet, more than a decade ago we concluded that the release of free-ranging wolves in Britain, although an alluringly radical idea, was unlikely for as long as sheep farming remained a predominant land use (Macdonald et al. 2000). Nevertheless, we thought other marginally less radical possibilities existed, and foremost amongst them was a fenced wilderness area. Since that 2000 essay, entitled 'British mammals: is there a radical future?', and

the rewilding aspirations of Paul Lister, owner of the 90 km² Alladale reserve, Scottish Highlands (Sandom et al. 2012; Fig. 16.6), we have sought to answer that question in two important ways. First, we explored whether an area of Scotland exists in which a suitably large fenced area might be feasible, biologically and geopolitically, and we modelled some consequences of varying its size in terms of the population dynamics therein of both the wolves and their likely main prey, red deer (see Fig. 16.5). Second, we addressed the question of how fencing a population of wolves might affect their capacity to exert top-down regulation on deer, thereby exposing an interesting paradox that might be relevant to all fenced reserves.

To reintroduce ecosystem function, both the species and the natural dynamics of the population must be restored (Bull et al. in prep) and, in the absence of a fence, predator killing outside unfenced protected areas can limit predator population density within. Source-sink population dynamics result from variability in habitat suitability within the meta-population. Importantly, even before population viability is threatened, the functionality of the population may be impacted (Bull et al. in prep). Wolves are commonly persecuted in most of their range in Europe (Linnell et al. 2008), and a reintroduction to Scotland without the use of a fenced reserve would almost certainly result in anthropogenic wolf mortality. It is expected that higher persecution would occur in unprotected compared to protected areas (Fritts and Carbyn 1995), although this is not always the case (e.g. Hilborn et al. 2006), which could create a source-sink population dynamic. Modelling suggests that, if 35% of dispersing wolves are lost from a protected area, wolves will possibly be prevented from achieving population densities sufficiently high to exert strong top-down forcing of the deer population (Bull et al. in prep). It should be noted that natural population dynamics may limit wolves' impact on deer density and is part of the totipotent state we seek to restore. If > 50% of dispersers are lost the whole population would be threatened. Hence, reintroducing wolves to a large fenced reserve in Scotland, as opposed to restoring nationally, could facilitate the restoration of predation as a process otherwise threatened by wolf killing.

Fences can pose a significant threat to wildlife, for instance, thousands of ungulates died in Botswana when they were cut off from essential water sources on their dry season migration, by a fence designed to prevent the spread of foot-and-mouth disease between wildlife and livestock (Hayward and Kerley 2009). Nonetheless, fencing game reserves in southern Africa

---

[3] Landscape of fear: 'altering foraging patterns and habitat use of herbivores under risk of predation' (Manning et al. 2009).

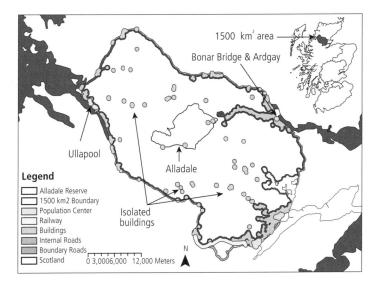

Figure 16.6 Method of calculating the maximum area of suitable land to be included within a fenced reserve. Tarmac road infrastructure and buildings 700 m from these roads ~~were~~ excluded creating a maximum boundary of 1500 km². Buildings further than 700 m from tarmac roads were deemed potentially suitable for inclusion. Map in top right corner depicts the location of the 1500 km² area in Scotland. Source Sandom et al. (2012). With kind permission from Springer Science and Business Media.

is commonplace and helps to reduce human–wildlife conflict which, Hayward and Kerley (2009) argue, can conserve large mammals more successfully than in unfenced parts of East Africa. At the time of writing, a hotly debated correspondence has arisen over whether the least worst option for conserving African lions may be the fencing of their diminishing strongholds. Essentially, Packer et al. (2013) argue that fenced reserves offer better value than unfenced reserves and preserve predation as a process, while Creel et al. (2013) contest this conclusion, highlighting that as fenced reserves are typically small, they have low actual numbers of lions, despite the lion population densities being high. A landscape-scale fenced wilderness reserve in Scotland, based on the southern African model, would raise numerous ecological, economic, social, legal, and ethical questions (Deer Commission Scotland et al. 2004). For such a reserve to be realized, all of these aspects need to be addressed and challenging issues resolved; we explored three pertinent ecological and geographical questions: (1) how much space is required for a viable wolf population?, (2) will wolves limit deer numbers within a fenced reserve?, and (3) how much space is available around Alladale for such a reserve?

Tackling the issue of how much space is required is greatly aided by the intensive study of the wolves of Isle Royale, situated in Lake Superior in North America. Wolves have been present since they introduced themselves to the 544 km² island in the late 1940s via an ice bridge (Mech 1966; Vucetich et al. 2012), illustrating that an area this size can feasibly sustain a small

wolf population for at least 60 years. We used agent-based modelling to further this empirical example and revealed that a reserve of 600 km² had an 88% chance of sustaining a wolf population after 100 years (Sandom et al. 2012), although our model did not take into account catastrophic events such as disease outbreaks, nor did we consider genetics. Disease would need to be treated and inbreeding managed, for instance through a managed metapopulation (Akçakaya et al. 2007). Isle Royale also reveals that genetic rescue of small populations should occur before the population becomes heavily inbred so as to avoid an immigrant dominating breeding opportunities and preventing improvements in population demographics. In 1997 an immigrant arrived on Isle Royale and within 2.5 generations was related to every individual in the population (Adams et al. 2011).

The scenario that recorded an 88% probability of a reserve sustaining a wolf population after 100 years also assumed that maximum pack density would not exceed one pack/200 km², a contentious assumption based on the intriguing population dynamics recorded on Isle Royale. Between 1959 and 1974 the wolf population appeared limited to around 17 to 31 animals in two packs (pack density of 1/272 km²). However, between 1975 and 1980 the population increased to 50 animals in five packs, a pack density of a little over one pack/100 km² (Peterson and Page 1988). At this density, the intermediate scenario in our modelling exercise, the probability of wolves persisting after 100 years in a 600 km² reserve was just 24% (Table 16.1).

Table 16.1 Wolf extinction probability in the range of fenced reserve sizes (200, 600, and 1200 km^2) modelled using each pack density scenario. Data are the number of repetitions with remaining wolves (out of a maximum of 50) in the given year from release. Under the limited and intermediate pack density scenarios pack density was limited at one pack/200 km^2 and one pack/100 km^2 respectively. Source Sandom et al. (2012). With kind permission from Springer Science and Business Media.

Reserve area (km^2)		Unlimited pack density scenario			Intermediate pack density scenario			Limited pack density scenario		
		200	600	1200	200	600	1200	200	600	1200
No. of years after stocking	5	49	50	50	46	50	50	46	50	50
	10	35	50	50	34	50	50	27	49	50
	15	20	46	50	26	50	50	15	48	50
	20	12	43	50	14	48	50	9	48	50
	30	5	35	49	6	47	50	4	47	50
	40	3	28	46	4	39	49	0	47	50
	60	2	12	27	2	23	25	0	45	49
	80	0	6	19	0	18	13	0	45	48
	100	0	0	9	0	12	7	0	44	48

On Isle Royale, when the pack density increased, the wolf density also increased from 57 wolves/1000 km^2 to 92 wolves/1000 km^2. In our modelling scenarios we contrasted a limited pack density scenario (one pack/200 km^2), and an unlimited pack density scenario. Under the limited scenario the average maximum wolf density was just 48 wolves/ 1000 km^2 compared to the unlimited scenario of 104 wolves/1000 km^2, and compares well with Isle Royale.

The implications of these maximum densities were dramatic, with the limited scenario having a very minor impact on deer population density (> 20/km^2) compared to the unlimited scenario where deer population densities were typically reduced to ~4/km^2 (Fig. 16.7). This dramatic reduction in prey population density caused, in turn, dramatic declines in wolf survival and ultimately is the likely cause of their extinction in most modelled instances. On Isle Royale, the wolves' primary prey species is moose *Alces americana* and, while the overall impact of top-down forcing on the population is unclear, the moose population did fall during the period of high wolf population densities (Vucetich and Peterson 2004). The wolf population crashed in the early 1980s as a result of starvation, intraspecific competition, and disease (Peterson and Page 1988; Peterson et al. 1998). Our modelling may over-estimate the risk of extinction associated with prey decline, as our model only includes a single prey species. A greater variety of prey, such as sika and roe deer, wild boar, and beaver, could allow for wolves to switch prey species when red deer become scarce, increasing their chances of survival.

In summary, the empirical and modelling evidence suggests that a fenced area of at least 544 km^2 offers the possibility of a 'viable' population[4] and a top-down effect on deer density. Such a reserve would provide a unique rewilded area in Britain and would offer the opportunity for novel scientific insight, but would have to be managed to avoid undesirable extremes and inbreeding that threaten small populations. Such experiments could prove invaluable in delivering a rewilded Britain.

So, is such a landscape-scale fenced reserve conceptually possible in Scotland? A GIS exploration of the area surrounding Alladale suggests that the limited infrastructure in the region could allow for a reserve of ~1500 km^2 without affecting any tarmac roads or railway lines and only incorporate 35 isolated buildings or small clusters of buildings (Fig. 16.6). The suitability of the land within this area varies according to a number of variables, including habitat, buildings, agriculture, other leisure uses, and Sites of Special Scientific Interest (SSSI). The majority of the area indicates a neutral or positive suitability for such a reserve with areas of high ground presenting the greatest difficulties as they are especially valued by hill walkers; fences limit the ease of access, restricting people's right to roam. A further consideration is that the introduction of keystone species would have unknown impacts on SSSIs and

---

[4] Here, we define wolf viability in the context of a monitored and managed population within a fenced reserve over the medium term; this is not applicable to unfenced, long-term scenarios.

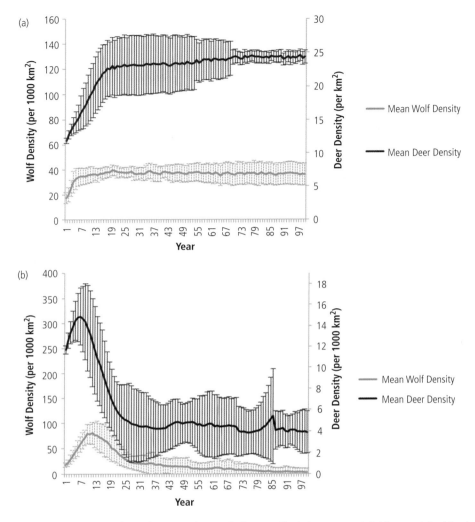

Figure 16.7 Average deer and wolf population densities in the 1200 km² reserve. Simulations were excluded from analysis of the deer population from the point wolf extinction was recorded; data are the mean and standard deviations. (a) Modelled results under the limited pack density scenario. (b) Modelled results under the unlimited pack density scenario. Source Sandom et al. (2012). With kind permission from Springer Science and Business Media.

may be considered a threat. In spite of largely encouraging projections, any such reserve would have implications for the local community, and the costs and benefits would need to be considered carefully, as well as the legal implications of housing both predator and prey (Sandom et al. 2012).

The low population density and high deer density make the Scottish Highlands an obvious area of interest for considering a fenced reserve with large predators, and England and Wales could easily be ignored and assumed to be unsuitable. This is a particularly

pertinent issue, with the reported expansion of large predators across western Europe (Deinet et al. 2013b). For instance, initial signs indicate that wolves are being welcomed back to Denmark since they arrived in 2012 (Miljominsteriet Naturstyrelsen 2013). An active reintroduction is quite different from a natural recolonization, but the question 'Is there really no room for large predators in England and Wales?' can reasonably be posed. National Parks are often important examples of a country's natural ecosystem. In England and Wales they are perhaps the finest examples of cultural rural landscapes. There are ten National Parks in England, covering 9.3% of the land area and three National Parks in Wales, covering 19.9% of the land (National Parks UK 2012). The largest National Park in England is the Lake District at 2292 km², favourable compared to our Scottish example above. However, over 40 000 people live in the National Park and, whereas deer are abundant in Scotland, there are sheep in the Lake District, which is already a cultural icon and receives in the region of 14.8 million visitors a year, spending as much as £994 million during their visits (National Parks UK 2012). In contrast, Northumberland National Park, for example, covers 1048 km², has 2500 people living in it, and is visited by 1.5 million people a year, spending a total of £190 million (National Parks UK 2012). It is primarily covered by farmed moorland and perhaps bears some similiarity to our Scottish case study, although deer densities are likely to be lower and sheep densities higher. With the decoupling of Common Agricultural Policy (CAP) subsidy payments from production and the potential for further change, there may be increasing concern that farming marginal land will become unsustainable or undesirable (Acs et al. 2010).

As a consequence, the local community may wish to be presented with alternatives for supporting the rural economy. This is a problem that is already advanced in continental Europe with what has been termed rural land abandonment (Navarro and Pereira 2012). In some instances, communities in these regions have turned to wildlife tourism as an alternative income. Charismatic large mammals are a significant draw. Staffan Widstrand, from Rewilding Europe, who has considerable experience in the wildlife tourism trade, suggests that high quality viewing opportunities of charasmatic animals can be charged at between €120–270, to see bears in Finland, and €200, to see wolves howling in Sweden (Widstrand 2013). In Sierra de la Culebra in Spain, the wolf is being seen as a new economic opportunity for the tourist industry (Richardson 2013). Further afield, the wolf reintroduction to Yellowstone National Park in 1995 was associated with a 35.5 million dollar increase in visitor spending in the local economy (Duffield et al. 2008). Rewilding a National Park in Britain is no doubt a radical and controversial proposal but, based on experiences elsewhere, one that may warrant further consideration for its financial opportunities as well as its environmental benefits. However, we want to emphasize we are proposing the exploration of the concept, an assessment of risk and opportunity, proceeding any further could only be achieved with local enthusiasm and leadership.

The predation case study has focused around the uplands where the human population density is lower. National Parks also exist in the lowlands and could also be considered as rewilding core areas. The South Downs National Park, for example, is 1624 km² and so relatively large, but with 120 000 people living in the park and the land primarily being productively farmed, a wolf reintroduction is highly unlikely. However, another large mammal, the wild boar, has been establishing itself in lowland Britain and can perhaps be accommodated, as explored in the following case study.

16.3.3 Case study 2. Restoring ground disturbance through wild boar reintroduction

Wild boar were eradicated in Britain at least 300 years ago, through habitat loss and over-hunting (Yalden 1999), a decline from perhaps as many as 950 000 animals during the Mesolithic (Maroo and Yalden 2000). Their current status is not clear, as it has been a moot point whether or not they have been reintroduced! In the late 1980s, escapees from farms took pre-emptive initiative and bypassed the smouldering debate as to whether it would be desirable to reintroduce boar to the UK.

At ~100 kg, wild boar were the largest member of the rooting guild, which also includes the badger, in the UK[5]. Rooting is an ecological trade, whose practitioners disturb ground vegetation and top soil while foraging, thereby creating ecosystem upheaval in ways that have major effects, akin to ploughing, on plants and all that live thereon. From an ecologist's perspective, the reintroduction debate has centred on this role of ground disturbance as a vital process within naturally functioning ecosystems; a process that creates niche opportunities for less robust plant species to survive in communities otherwise dominated by more competitive plants (Hobbs and Huenneke 1992). From the hunter's perspective, wild boar represent an interesting and potentially profitable quarry. Others, especially those with experience of the agricultural ravages caused by boar on continental Europe or mindful of their potential as disease vectors, view the prospect of

[5] The 'purity' of wild boar is difficult to assess. Wild boar can be crossed with domestic pig breeds, typically to increase litter size and so yields. Genetic tests are not available as standard samples for wild boar are unavailable. Certain morphological characteristics are thought to be indicative of wild boar rather than hybrids, including: straight head profile, body weight held over front legs, straight tail, and entirely black muzzle (Goulding 2014).

their reintroduction with disdain. As always, there is a question of balance—as a returning native species, wild boar would restore a missing link in Britain's temperate ecosystem and, less tangibly, would fill a void in the naturalist's sense of Nature's wholeness, but in the absence of predators and other ecosystem changes in the last 300 years they could become a pest, and potentially a bad one[6]. So where, and how, should that balance be struck?

Considering the then lively debate on invasive species, the attention being brought to bear on every detail of a proposed beaver reintroduction (see Case study 3), and the polarized views on boar in particular, Macdonald and Tattersall (2001) wryly noted the deafening silence from government (Defra) on the escaped boar. During the early years of a back-door recolonization, it might have been nipped in the bud. Thirty years later, there are established populations in Kent/Sussex (~200 animals, estimates from 2004), the Forest of Dean (probably > 100), and Dorset (< 50) (Defra 2008). Three populations of wild boar apparently flourish in Scotland (Campbell and Hartley 2010). With the passage of time, and of reports (e.g. Goulding et al. 1998), Defra has left the decision of whether or not to tolerate wild boars in the wild to individual land owners (Defra 2008).

Against this background we tackle two families of questions, remembering that wild boar are primarily a woodland species and that woodland is nowadays both limited and fragmented in Britain (see Chapter 12, this volume). First, at a landscape scale, and in the long run, can wild boar live free in the modern British countryside? Second, with rewilding in mind, how can their role in the processes of ecosystem functioning be fine-tuned to strike a balance between pest and asset of which society will approve? Boar are evidently viable in southern England and we already know from Leaper et al.'s (1999) calculations that there is sufficient suitable woodland in Scotland to support a minimum viable population. We will return to this landscape-scale question below, but first we focus at the level of the individual, as we answer questions about exactly how wild boar forage and how suitable this foraging

is to introduce naturalistic or rewilded disturbance regimes. In our example, we focus on this process in the Scottish Highlands, but our studies of wild boar behaviour have wider relevance, not only insofar as wild boar behaviour affects their impact in the ecosystem, but more immediately because such studies are a step towards answering a manifestly practical question: can heavily managed, fenced, wild boar be used cost-effectively within a conservation management system to foster desirable habitats?[7] A specific instance of this question focuses on bracken *Pteridium aquilinum*, a native to the UK, but one that is both problematic to conservationists and to stockmen, and is expanding (Robinson 2009).

So, can wild boar be employed as rewilding engineers, and specifically create niches of opportunity for less competitive species within bracken and heather dominated communities? We tackled this in a series of experiments, within a 125 ha fenced area at the Alladale Wilderness Reserve (Fig. 16.1). There, we asked: (a) at what rate do wild boar root?, (b) do wild boar use a foraging strategy that makes them effective ecosystem engineers?, and (c) can wild boar promote woodland regeneration and disturb bracken and heather communities in the Scottish Highlands?

To achieve this, eleven ~0.5 ha enclosures were established on heather moorland and stocked for at least nine weeks with varying densities of boar (between 4.5 and 14.6 boar/ha). The accumulation of rooted area was measured weekly and revealed a median per capita rooting rate of 42.4 m^2/week; higher rates later in the study period were probably due to high rainfall making rooting easier (Sandom et al. 2013d; Fig. 16.8). At natural population densities (unlikely to exceed 4/km^2) the rooting is estimated to cover less than this, at around 15 600 m^2/km^2/year (Sandom et al. 2013b).

Rooting behaviour is part of the wild boar's overall foraging strategy and targeted to exploit specific resources. To explore whether wild boar have a foraging strategy that can promote woodland expansion, beneath-canopy seedling regeneration, and/or bracken disturbance, six GPS collared wild boar were monitored intensively over a 12-month period within the 125 ha enclosure, consisting of remnant Caledonian pine woodland, broadleaved woodland, and open heather moorland (Sandom et al. 2013c).

---

[6] For example, in Luxembourg 13 276 incidences of crop damage by wild boar were reported between 1997 and 2006, costing 5.27 million euros and covering 3900 ha. The annual area affected was approximately 0.31% of agricultural land (Schley et al. 2008). Disease is also a considerable concern. For example in the 1990s, Germany recorded 424 outbreaks of classical swine fever in domestic pigs. Of the recorded primary outbreaks, 59% were associated with direct or indirect contact with infected wild boar (Fritzemeier et al. 2000).

[7] The cost-effectiveness of using naturalistic disturbance is unclear, with research indicating costs range between £184 and £1961, depending on stocking density and the rate of rooting, with only the lower end of this scale being competitive, although other business models could be more effective (Sandom et al. 2013d).

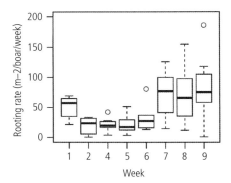

Figure 16.8 Median per capita rooting rates plotted per week (median ±IQR, n = 9). Source Sandom et al. (2013d). Reproduced with permission from John Wiley & Sons.

The ground vegetation was a mixture of bracken (7%), heather (65%), and grass communities (26%). The boar were particularly industrious during the autumn and winter, spending ~80% of their active period foraging (Fig. 16.9) and, while they were active for longer periods (14 hours) during the long summer days of northern Scotland, they spent the longest period rooting during the autumn and winter (9–9.9 hours/day). Significantly, the boar preferred a woodland canopy and favoured bracken communities while rooting in the autumn and winter, and grass communities while grazing (~28% of the active period) in the spring and summer (Sandom et al. 2013c).

At the point of impact, rooting considerably reduces vegetation biomass and species richness, and they are capable of creating relatively large patches of bare ground (within the 125 ha enclosure, 168 deep rooted patches > 1 m^2 were created after 21 months of boar activity, the average rooted area was 78 m^2 and covered a total area of 13 143 m^2, 1.05% of the enclosure, although the boar were stocked at higher than natural densities; Sandom et al. in prep-a). However, the lasting effects of rooting are dependent on the vegetation community. Rooting is typically characterized as deep (sustained rooting in a specific area that disturbed the soil layer) or shallow (surface vegetation disturbed without disturbing the soil layer); deep rooting was employed by the boar in bracken areas to exploit the rhizome network, typically found 5–15 cm below the surface.

Twenty 2 × 2 m fenced exclosures were established, half in rooted and half in unrooted bracken areas, to prevent further disturbance, and were monitored over two years. The differences between these areas were dramatic, with a 64% reduction in bracken frond density in the following spring compared with unrooted

areas. Two years post rooting, graminoids (grasses, sedges, and rushes) had 14% greater coverage in rooted areas. Forb species richness was one and two species greater in the first and second years post rooting, respectively. Bracken was, however, observed to re-establish in most, but not all, rooted areas in subsequent years after the two-year monitoring period (Sandom, personal observation). In comparison, shrubs within heather-dominated communities were greatly reduced (> 40% cover to < 15%) and slow to recover two years post-rooting (< 20%), while grasses had increased in coverage two years post rooting (~40%) compared to pre rooting (< 40%) and being reduced to ~10% cover in the first year post rooting. Rooting also stimulated the germination of seedlings, although at the expense of more established saplings that were often bark stripped or uprooted. Seedling regeneration was recorded in 40% of the rooted areas in the 125 ha enclosure, although the greatest regeneration occurred beneath broadleaved and mixed woodland stands. In a comparison of regeneration between quadrats in rooted and unrooted areas, 30% more quadrats in rooted areas had regeneration.

This series of experiments suggests that the boar could be a good tool, both to break the dominance of bracken and heather, and indeed hasten woodland regeneration beneath an established canopy, whether in a core area or a productive landscape. Insofar as boar support woodland regeneration, in a full life-cycle analysis they can claim some of the credit for associated ecosystem services such as water regulation, carbon sequestration, and recreation (National Ecosystem Service Assessment 2011b). There are, however, downsides: with their broad diet of over 400 species (Schley and Roper 2003), at high population densities the boar could damage rare species and crops. So, regarding the polarized debate as to whether to allow them back into Britain, the pros and cons of boar depend critically on the circumstances. It is certain that some individual boars will be in conflict with people and, because there is no predatory check on their numbers in the modern resource-rich landscape and they will outgrow the resources that society will feel comfortable sharing with them, it is also certain that sooner or later boar will be in conflict with many people. Indeed, on continental Europe there are reports of wild boar attacking people (Kotulski and König 2008), and where populations are present in Great Britain there are regular reports of them uprooting parks, gardens, and golf courses (Sims 2005; British Wild Boar 2015). At both scales, this is probably easily, indeed profitably, solved: the nuisance individuals, and indeed a percentage of the whole

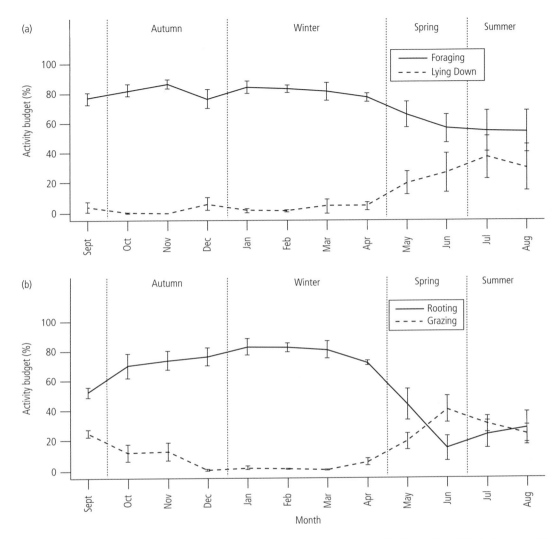

Figure 16.9 (a) Monthly variation in activity budget of lying down versus foraging (n = 4, mean, SE). (b) Monthly variation in foraging activity budget of grazing and rooting behaviour (mean ±SE, n = 4). Source Sandom et al. (2013c). Reproduced with permission from John Wiley & Sons.

population, can be shot. That boar were over-hunted to extinction before, suggests that they could be limited to acceptable numbers, if society formed a view on what numbers were acceptable.

Perhaps surprisingly, people and wild boar are already sharing productive landscapes in Britain. There are benefits associated with increasing woodland diversity and in the long term they are likely to be a key element in allowing wooded ecosystems to be self-sustaining and biodiverse. But while some enjoy their presence, whether for viewing or hunting, the potential for conflict is considerable. Thus they would

be a key component in a core area. In productive landscape predatory control, through humans, it is likely to be necessary to ensure that the balance between biodiversity and productivity is achieved.

16.3.4 Case study 3. Restoring river channel and riparian disturbance by reintroducing beaver

Beavers are ecosystem engineers in the most literal sense. They have a marked influence on their environment and on other wildlife due to their river damming and tree-felling activities (Rosell et al. 2005), qualifying them as keystone species[8] (Power et al. 1996). But the sounds of their industry have lain dormant since their demise here in Britain: by 1191, Geraldus Cambrensis[9] reported that the Teivi was the only river in Wales, or even in England, where beavers remained, and Kitchener and Conroy (1997) record their disappearance from Scotland 400 years ago. Today, however, along the shores of the lochs of Knapdale Forest[10] you can again hear the warning slaps of their paddle-like tails on the water, and see the chiselled stumps of their arboriculture as you squelch through their ponds—and it is a thrill that touches the naturalist's soul. The Knapdale beavers (currently three breeding pairs, and two single adults[11] translocated from Norway in 2009/2010 (and see Box 16.3), and at least three Scottish-born young, R. Campbell-Palmer pers. comm.) are part of an experiment[12] choreographed by Scottish Natural Heritage (government), implemented by the Scottish Wildlife Trust and the Royal Zoological Society of Scotland (non-government), in which their lives (deaths, births, health, movements, habitat use, and foraging) are monitored comprehensively (Wild-CRU is one of a number of scientific auditors[13]). Mind you, it is scarcely less thrilling to see the entirely illegal immigrant beavers of Tayside, where several escapees from captivity, or possibly renegade releases from as early as 2001, have now spawned over 100 wild-living

descendents (Campbell et al. 2012a)[14]. Whilst the Scottish Government has yet to decide on the long-term future of beavers in Scotland[15], for now beavers are back in Scotland and the first steps to rewilding Britain are underway![16] The history of their official return has been a riveting bio-political challenge, no less interesting than the biological challenge faced by the beavers in rebuilding an ecosystem (Hodder et al. 2009). Somewhere near the beginning of the saga, and over a decade before the first beavers were actually released in the wild, Macdonald et al. (1995) set out the terms of the debate in a paper entitled: *Reintroducing the European beaver to Britain: Nostalgic meddling or restoring biodiversity?* The debate involved intersecting biological and management considerations: biologically, there were questions about whether the beaver's absence had resulted in their niche being lost irretrievably; whereas in management terms, opinion was polarized over whether the habitat creation and ecological services rendered by returnee beavers would, or would not, outweigh their potential for pestilence.

What is this ecosystem service that beavers deliver, which makes them more than just another notch in the tally of species diversity? It is that they are unusual, indeed as a family unique, in their ability to dam watercourses, creating home-made habitat through tree-felling, and dam and lodge building to engineer lakes and wetlands (although Eurasian beavers are said to dam less than their Canadian cousins, this may be more to do with differences between the landscapes than between the species; Campbell et al. 2012a). These activities cause river corridors to become wider, geomorphologically more complex (Westbrook et al. 2011), biologically more diverse, and more productive (Rosell et al. 2005). Of course, this means things change; for instance, dam construction creates ponds that will alter the freshwater insect assemblage from running water species to a pond community, this may reduce local species richness (alpha diversity[17]), while increasing

---

[8] A keystone species is 'one whose impact on its community or ecosystem is large, and disproportionately large relative to its abundance.' Power et al. (1996).

[9] Geraldus Cambrensis describes how beavers, 'in order to construct their castles in the middle of rivers, make use of the animals of their own species instead of carts, who, by a wonderful mode of carnage, convey the timber from the woods to the rivers. Some of them, obeying the dictates of Nature, receive on their bellies the logs of wood cut off by their associates, which they hold tight with their feet, and thus with transverse pieces placed in their mouths, are drawn along backwards, with their cargo, by other beavers, who fasten themselves with their teeth to the raft.'

[10] Loch Coille-Bharr and its satellite, Dubh Loch (the exit of which has been beautifully dammed), Un-named (S) Loch (also known as Lily Loch), Lochan Buic, Loch Linne, Loch Fidhle, Loch na Creige Mòire (Creagmhor) and Lochan Beag all have, or had, beavers released.

[11] Four pairs, and one sub-adult believed now to be breeding and to have replaced her mother in the breeding pair; these nine animals are those that are known to have survived and remained within the release site of the 16 animals originally released (Harrington et al. 2013).

[12] The Scottish Beaver Trial—<http://www.scottishbeavers.org.uk/>.

[13] Full details, and all cited reports, available at: <http://www.snh.gov.uk>.

[14] In March 2012, the Scottish Minister for the Environment announced that beavers in Tayside would be tolerated until a decision on the future direction of beaver reintroduction in Scotland is made in 2015. Monitoring is being carried out by the Tayside Beaver Study Group, <http://taysidebeaverstudygroup.org.uk/>.

[15] A formal decision will be made by the Scottish Government in 2015, on the basis of the outcomes of the Scottish Beaver Trial at Knapdale, in conjunction with findings from the monitoring of the unlicensed Tayside beaver population.

[16] The beaver reintroduction trial at Knapdale, Argyll, is the first formal reintroduction of a mammal to Britain.

[17] Alpha diversity: biodiversity within a particular site, community, or ecosystem; usually expressed as the species richness of the site.

biodiversity at the landscape or regional scale (gamma diversity[18]) (Rosell et al. 2005). Furthermore, at a catchment scale they are responsible for filtering sediment from water that seeps through the dams, creating wetlands (Wright et al. 2002) (a habitat disappearing in many parts of Europe) that tend to be a particularly rich environment for Nature (Pollock et al. 1998), and creating a mosaic landscape which, through the eyes of some beholders, is beautiful (Rolauffs et al. 2001; Müller-Schwarze and Sun 2003).

Ecosystem services are often classified as regulatory, supporting, or cultural. In this case, beavers provide all three (Müller-Schwarze and Sun 2003; Rosell et al. 2005): (a) regulatory services: improved water quality, flooding prevention, water flow regulation, raising the water table, and the conservation of water; (b) supporting services: creating and maintaining wetland habitats with benefits for biodiversity, including benefiting some economically important fish species by raising the water temperature (this effect can however become detrimental to the fish if the temperature rises too high and reduces dissolved oxygen content); and (c) cultural services, such as tourism and aesthetics (Macdonald et al. 1995; Tattersall and Macdonald 1995; Collen and Gibson 2000; Campbell et al. 2007).

Who would choose to forego these benefits? Plenty of people, and particularly those looking at damage done by a different species, the American beaver *Castor canadensis* (Harkonen 1999; Jonker et al. 2006): not least in Tierra del Fuego where it is regarded as a toxic invasive, introduced into a landscape where many trees have not co-evolved to regenerate when felled (Skewes et al. 2006; Anderson et al. 2009). The flooding of gardens and drains, the felling of orchards and deflection of streams annoy people. Others see the beautiful beaver ponds as breeding grounds for mosquitos (Butts 2001) and giardia (Rosell et al. 2001). In the UK, salmon fishermen feared their sport would be damaged, and were unmoved by the assurances of Norwegian fisherman who tended to see beavers as an asset (Collen and Gibson 2000). The Scottish Government through Scottish Natural Heritage has established a Beaver-Salmonid Working Group to report on issues relating to the potential impacts, positive and negative, of beavers on salmonids—this will report in 2015.

Thus far, beaver-induced hydrological changes at Knapdale have been less than expected, with little damming activity taking place (Jones et al. 2013) (with the notable exception of a dam built on a drainage ditch between Loch Coille Bharr and Dubh Loch, impounding a significant amount of water and effectively increasing the size of the Dubh Loch almost 5-fold, R. Campbell-Palmer, pers. comm.). Habitat use and foraging activity (four-years post-release at the time of writing) has largely been restricted to the original release points (Harrington et al. 2013), with some localized impacts on woody vegetation (Moore et al. 2013) and aquatic vegetation (Willby et al. 2011), although it is premature to draw firm conclusions on the impacts (either positive or negative) of beavers at Knapdale.

In the more human-dominated landscape of the Tayside river catchment, Campbell et al. (2012a) similarly found beaver damming and crop foraging to be minimal (only three beaver groups of an estimated 38–39 groups, had built dams, and only two groups foraged in crops). Conflict with landowners in the Tay catchment was therefore highly localized (even after several years of beaver presence)[19], and farmers (that did not have dams on their land) and fishermen seemed unconcerned with the presence of beavers (although gillies[20] expressed some concern about fallen trees in the water snagging fishing lines; Campbell et al. 2012a). The threat to agricultural crops can be contained—or mitigated—by the establishment of a riparian buffer strip, because beavers usually forage within < 40 m of a watercourse (Campbell et al. 2012a). In Norway, Macdonald et al. (1995) reported that a Forest Owner's Association concluded that beaver damage was so minimal it was inappropriate to insure against.

Led meticulously by the UK's statutory conservation agencies, all these pros and cons, and the questions of balance between them are being exhaustively reviewed (e.g. Harrington et al. 2012; Moran and Hanley-Nickolls 2012; Moore et al. 2013/ see <http://Scottishwildbeavers.org>). In summary, there are functional traits (Luck et al. 2009) brought by beavers (dam building and tree felling), to be offset against threats that they pose to provisioning services (agriculture and forestry) and, ultimately, the balance sheet of these factors will determine the beaver's immediate future in Scotland. Should they remain, their natural expansion from their starting locations in Knapdale and the

---

[18] Gamma diversity: total species richness over a large area or region.

[19] Although Halley and Rosell (2002) note that across Europe, marginal areas that cannot sustain permanent beaver occupation tend to be those that require most beaver engineering in the form of dam construction and felling of large trees, and thus are the source of most conflict between beavers and humans—and, therefore, conflict may increase as beavers spread further across Scotland.

[20] According to the Oxford English Dictionary, a gillie is 'a man or boy who attends someone on a hunting or fishing expedition.'

Box 16.3 Beavers' response to environmental variability

To plan successful rewilding projects, it is necessary to learn as much as possible about beavers in the wild in order to determine how they might behave in the UK. To do so, we turned to Frank Rosell's (Telemark University) pioneering long-term project in Norway. There, while we were disappointed not to repeat the early observations of beaver behaviour reported by Juvenal—'*Qui se Eunuchum ipse facit, cupiens evadere damno Testiculi*'[21]—we did see a brand new beaver behaviour, tool-using (Thomsen et al. 2007), and we revealed the importance to their population dynamics of environmental variability, a factor increasingly recognized by ecologists as widely important to mammalian populations (amongst our own WildCRU work, see Campbell et al. 2012b; Campbell et al. 2013; Zhou et al. 2013).

Our study site was centred on three large rivers, the Straumen, the Gvarv, and the Sauar, in Telemark, southern Norway (Campbell et al. 2012b). Here we examined trends, from over 90 years of data, in different components of climate (including, for example, precipitation and temperature) and we used a multi-model inference procedure to analyse the effects of these components on survival and recruitment in a population of 242 Eurasian beavers over 13 recent years (Campbell et al. 2012b).

Notably, our continuous observations indicated no large-scale changes in habitat type, or other apparent causes for changing habitat productivity, which might otherwise explain variation in survival and recruitment. While mean rainfall values influenced both survival and recruitment, the effect of climate variability, in the form of variance in rainfall and temperature and seasonal amplitude in temperature, was also highly influential in a variety of predictive models for survival and recruitment rates. A higher survival rate was linked to a lower coefficient of variation (CV) of precipitation (for kits, juveniles, and dominant adults), lower residual variance of temperature (for dominant adults), and lower mean precipitation (for kits and juveniles).

Greater recruitment was linked (in order of influence) to higher seasonal amplitude of temperature, lower mean precipitation, lower residual variance in temperature, and higher precipitation CV, though the latter probably arose due to individuals from a reservoir of philopatric non-breeding adults migrating into the study area to fill the breeding vacancies created through the reduced adult survival in years of higher precipitation CV. Both climate means and variance thus proved significant to population dynamics; though, overall, components describing variance were more influential than those describing mean values (as we also report for badgers, see Macdonald and Feber (2015: Chapter 4).

Using capture-mark-recapture to monitor 198 individuals over 11 years at the Norwegian study site, we observed lower juvenile body weights after colder winters, indicating that colder winters exert a disproportionately heavy thermoregulatory burden on the smaller younger beavers (Campbell et al. 2013). Warmer spring temperatures were associated with lighter-weight adults, because the shortened spring flush of growth in warm springs reduces the time over which beavers have access to nutritious young growth.

Counter-intuitively for a herbivore, we also observed a negative association between rainfall and body weight in juveniles and adults, and also with reproductive success. Using tree cores to delve into the growth history of the alder *Alnus incana*, the principal beaver food there, we found a positive relationship with rainfall for trees growing at elevations > 0.5 m above mean water level, but a negative relationship for trees growing < 0.5 m. We deduced that, while temperature influences beavers at the landscape scale via effects on spring green-up phenology and thermoregulation, rainfall influences beavers at finer spatial scales, for example, with trees near water level prone to water-logging, producing poorer quality forage in wetter years. Unlike most other herbivores, beavers are an obligate aquatic species that utilize a restricted 'central-place' foraging range, limiting their ability to take advantage of better forage growth further from water during wetter years.

These studies reveal a multiplicity of climatic influences on beavers, marked by a reduction in population vital rates and body weights associated with trends in climate that are predicted to continue in the future. For example, weather patterns in Northern Europe are predicted to exhibit increased variability, along with warmer spring weather and wetter summer weather (IPCC 2007), all of which have been shown, in our studies, to correlate with either reduced survival, reduced reproduction, or reduced body weights, or a combination of more than one of these values. This suggests that, in the Norwegian population at least, population growth rate may be impaired in the future. All else being

continued

---

[21] 'He makes himself a eunuch, desiring to escape by the loss of his testicles', to which Cicero adds 'They ransom themselves by that part of the body, for which they are chiefly sought'. This, as Geraldus explains, is because 'When the beaver finds he cannot save himself from the pursuit of the dogs who follow him, that he may ransom his body by the sacrifice of a part, he throws away that, which by natural instinct he knows to be the object sought for, and in the sight of the hunter castrates himself, from which circumstance he has gained the name of Castor'.

Box 16.3 *Continued*

equal, those beavers returning to Britain (an island noted for its wet summers and highly changeable weather) may find their viability affected by the changing climate. On the other hand, British winters are mild (a good thing for juveniles), the rivers are empty of conspecific competitors (elsewhere, wolves play an important role in regulating beaver populations, Halley and Rosell 2002), and the vegetation is perhaps well adapted to the wetter and more variable climate, all of which may outweigh any negative effects during the initial colonization period.

Tay will be fascinating and may offer an indication of how well our riparian habitats are connected. Inevitably there are some regions that would either not be connected or take a very long time for beavers to reach.

The possibilities for a beaver re-introduction have not been confined to Scotland. A reintroduction has been also considered for Norfolk in England. Based on land cover assessments of the region, modelling was used to assess the suitablility of the habitat to support a population (South et al. 2001). It was found that Norfolk could support between 18 and 40 beaver families. Establishing other reintroduction sites such as this may help secure beavers' presence in Britain and help wider dispersal that could help further spread of their function to other suitable riparian areas in Britain.

While not all the pros and cons of living with rewilded ecosystems can or should be monetized, society's enthusiasm for rewilding will inescapably be much affected by whether the Knapdale beaver experience comes in as a profit or a loss. In a desk study that anticipated the bottom line of reintroducing beavers to Britain, Campbell et al. (2007) estimated that an individual beaver release site might generate additional local tourism worth £2 000 000 a year, whereas costs associated with beaver in Continental Europe rarely rise above €100 000 per annum. Such calculations are, however, notoriously tricky, e.g. disentangling the direct impacts of the beaver from complimentary conservation actions such as riparian woodland regeneration that also offer considerable ecosystem service benefits, such as flood mitigation (National Ecosystem Service Assessment 2011b). Even if the sums currently work against beavers locally, there are nonetheless, increasingly elegant financial mechanisms for those benefiting from regionally and nationally beneficial ecosystem services to pay those currently bearing the cost (Barrett et al. 2013). Dickman et al. (2011), focusing on predators, and thus directly relevant to our example on wolves, describe payments to encourage co-existence between people and wildlife in ways that could easily be adapted to local support for beavers.

New mechanisms, such as biodiversity offsetting and payments for ecosystem services, also offer potential new revenue streams for those providing ecosystem services with wider benefits (Jack et al. 2008; Bull et al. 2012). A notable example of such a scheme includes Water Funds, which launched in 2000 in Ecuador, and has raised $9.8 million to invest in the protection and management of the catchment area that includes the Condor Biosphere reserve. The Fund has already supported the planting of 3.5 million trees, explicitly to help provide clean water to Quito (Tallis et al. 2008).

16.4 Conclusion

If conservation is to be effective and efficient, investments in the natural world must be subject to some planning at the ecosystem level. An important message behind rewilding is that rich biodiversity with all guilds well represented, including the ones that polarize public opinion, such as large predators, are important components of ecosystem service rich and self-sustaining ecosystems, particularly in core areas. In essence, biodiversity both supports and is supported by the system. In this chapter we initially identified that woodland is an under represented habitat and resource in Britain. Wolves are not dependent on woodland but their function can help to promote tree regeneration through predation of the abundant and diverse large herbivore guild. Wild boar, in contrast, are a woodland orientated species and also provide a rooting service that can help woodland regeneration; although they are unlikely to cause sufficient ground disturbance to promote woodland expansion, they could help to develop and diversify the ecosystem in combination with wolves. Where wild boar are not currently present, wolves may initiate the trophic cascade that increases the suitability of the ecosystem to support wild boar; but where wild boar are already present wolves would aid the limitation of this species and alter their foraging strategy, again helping to support a thriving ecosystem with a well represented wooded component. The same applies to beavers, which require a wooded riparian ecosystem to support them. Where they have been reintroduced this habitat already exists, and the addition of beavers offers the opportunity of increasing the diversity of the ecosystem.

We have described a sequence of species reintroductions designed to re-assemble the mammal community that could be used to rewild three processes in Britain. Finding space for large core areas is challenging and the uplands offer the best opportunities for this land sparing approach. If an upland rural community embraced this radical future it would no doubt be a major change from business-as-usual and the benefits to the local rural economy could be significant. The most obvious benefits would be generated through tourism, but concepts such as payments for ecosystem services, carbon, and biodiversity offsetting add potential further monetary value. This might be particularly true in the context of flood alleviation through the rewilding of river catchments. To determine whether this is feasible and desirable requires a full economic, social, and legal assessment, which is beyond the scope of this chapter.

In terms of promoting increased land sharing with Nature, this book has highlighted numerous novel ways of improving society's interaction with Nature. Adding rewilding to this traditional conservation approach should encourage land managers to ask 'What ecological processes can we restore here?', as well as conserving threatened species. This is likely to include paying particular attention to processes such as pollination that are integral to maintaining food production. Finally, rewilding is an emotive concept and one that challenges society's relationship with Nature. More people now live in cities than in rural areas, creating a worrying gap between people and the environment that supports them. Wilding areas near to people, in any way that is appropriate, may also offer a miriad of health and well-being benefits (see Chapter 1, this volume), and will surely help society to maintain that connection to nature.

'Examine each question in terms of what is ethically and aesthetically right, as well as what is economically expedient.' Aldo Leopold

Acknowledgements

In addition to the editorial insights of Ruth Feber and Eva Raebel, we are grateful for help and suggestions from Ruaraidh Campbell, Kerry Kilshaw, Lauren Harrington, Axel Moehrenschlager, and Chris Newman. Andrew Kitchener provided helpful comments on an earlier draft. We gratefully acknowledge the support of Paul Lister and the European Nature Trust.

References

Acs, S., Hanley, N., Dallimer, M., Gaston, K.J., Robertson, P., Wilson, P., and Armsworth, P.R. (2010). The effect of de-coupling on marginal agricultural systems: implications for farm incomes, land use and upland ecology. *Land Use Policy*, **27**, 550–563.

Adams, J.R., Vucetich, L.M., Hedrick, P.W., Peterson, R.O., and Vucetich, J.A. (2011). Genomic sweep and potential genetic rescue during limiting environmental conditions in an isolated wolf population. *Proceedings of the Royal Society B: Biological Sciences*, **278**, 3336–3344.

Akçakaya, H.R., Mills, G., and Doncaster, C.P. (2007). The role of metapopulations in conservation. pp. 64–84 In D.W. Macdonald and K. Service, eds., Key Topics in Conservation Biology. Blackwell Publishing, Oxford.

Anderson, C.B., Pastur, G., Lencinas, M.V., Wallem, P.K., Moorman, M.C., and Rosemond, A.D. (2009). Do introduced North American beavers Castor canadensis engineer differently in southern South America? An overview with implications for restoration. *Mammal Review*, **39**, 33–52.

Barrett, C., Bulte, E.H., Ferraro, P., and Wunder, S. (2013). Economic instruments for nature conservation. pp. 59–73 In D.W. Macdonald and K.J. Willis, eds., *Key Topics in Conservation Biology 2*. Wiley.

Belyea, L.R. (2004). Beyond ecological filters: Feedback networks in the assembly and restoration of community structure. pp. 115–132 In V.M. Temperton, R.J. Hobbs, T. Nuttle, and S. Halle, eds., *Assembly Rules and Restoration Ecology: Bridging the Gap Between Theory and Practice*. Island Press, Washington.

Benyus, J.M. (2009). *Biomimicry*. HarperCollins.

Breitenmoser U, et al. (2010) The changing impact of predation as a source of conflict between hunters and re-introduced lynx in Switzerland. *Biology and Conservation of Wild Felids*, eds., Macdonald D.W. and Loveridge A.J. Oxford University Press, Oxford.

British Wild Boar (2015). <http://www.britishwildboar.org.uk/index.htm?nuisance.html>. Accessed Feb 2015.

Bull, J., Sandom, C., Ejrnæs, R., Macdonald, D.W., and Svenning, J.C. (in prep). Barriers can ensure the restoration of ecosystem function in the case of large predator reintroductions.

Bull, J. W., Suttle, K.B., Gordon, A., Singh, N.J., and Milner-Gulland, E. (2012). Biodiversity offsets in theory and practice. *Oryx*, **47** (03), 369–380.

Burns, F., Eaton, M.A., Gregory, R.D., et al. (2013). State of Nature report. The State of Nature partnership.

Butts, W.L. (2001). Beaver ponds in upstate New York as a source of anthropophilic mosquitoes. *Journal of the American Mosquito Control Association*, **17**, 85–86.

Campbell, R., Dutton, A., and Hughes, J. (2007). *Economic Impacts of the Beaver*. University of Oxford, Oxford.

Campbell, R., Harrington, A., Ross, A., and Harrington, L. (2012a). Distribution, population assessment and activities of beavers in Tayside., Scottish Natural Heritage Commissioned Report No.—540.

Campbell, S. and Hartley, G. (2010). *Wild Boar Distribution in Scotland*. Science and advice for Scottish Agriculture, Edinburgh.

Campbell, R.D., Newman, C., Macdonald, D.W., and Rosell, F. (2013). Proximate weather patterns and spring green-up phenology effect Eurasian beaver (*Castor fiber*) body mass and reproductive success: the implications of climate change and topography. *Global Change Biology*, 19(4), 1311–1324.

Campbell, R.D., Nouvellet, P., Newman, C., Macdonald, D.W., and Rosell, F. (2012b). The influence of mean climate trends and climate variance on beaver survival and recruitment dynamics. *Global Change Biology*, 18, 2730–2742.

Collen, P. and Gibson, R. (2000). The general ecology of beavers (*Castor* spp.), as related to their influence on stream ecosystems and riparian habitats, and the subsequent effects on fish, a review. *Reviews in fish biology and fisheries*, 10, 439–461.

Creel, S., Becker, M., Durant, S., et al. (2013). Conserving large populations of lions-the argument for fences has holes. *Ecology Letters.* 16(11), 1413–e3

Cullingham, C. I. and Moehrenschlager, A. (2013). Temporal analysis of genetic structure to assess population dynamics of reintroduced swift foxes. *Conservation Biology*, 27(6), 1389–1398.

Deer Commission Scotland, Forestry Commission Scotland, Scottish Natural Heritage, and Scottish Executive Environment and Rural Affairs Department (2004). Joint agency statement and guidance on deer fencing. <http://www.snh.gov.uk/docs/C249890.pdf>.

Defra (2008). Feral wild boar in England: An action plan. Department for Environment, Food and Rural Affairs, London.

Defra (2011). The natural choice: securing the value of nature. Department of Environment, Food and Rural Affairs.

Deinet, S., Ieronymidou, C., McRae, L., Burfield, I.J, Foppen, R.P., Collen, B., and Böhmm, M. (2013). *Wildlife comeback in Europe: The recovery of selected mammal and bird species*. Final report to Rewilding Europe by ZSL, Birdlife International and European Bird Census Council. London, UK:ZSL.

Dickman, A.J., Macdonald, E.A., and Macdonald, D.W. (2011). A review of financial instruments to pay for predator conservation and encourage human, carnivore coexistence. *Proceedings of the National Academy of Sciences* 108, 13937–13944.

Duffield, J.W., Neher, C.J., and Patterson, D.A. (2008). Wolf recovery in Yellowstone Park: Visitor attitudes, expenditures, and economic impacts. *Yellowstone Science* 16, 20–25.

Enserink, M. and Vogel, G. (2006). The carnivore comeback. *Science*, 314, 746–749.

Estes, J.A., Terborgh, J., Brashares, J.S., et al. (2011). Trophic downgrading of planet Earth. *Science*, 333, 301–306.

Fischer, J., Brosi, B., Daily, G.C., et al. (2008). Should agricultural policies encourage land sparing or wildlife-friendly farming? *Frontiers in Ecology and the Environment*, 6, 380–385.

Forestry Commission (2012). Forestry Statistics 2012—Environment—Ancient and semi-natural woodland.

Fritts, S.H. and Carbyn, L.N. (1995). Population viability, nature reserves, and the outlook for gray wolf conservation in North America. *Restoration Ecology*, 3, 26–38.

Fritzemeier, J., Teuffert, J., Greiser-Wilke, I., Staubach, C., Schlüter, H., and Moennig, V. (2000). Epidemiology of classical swine fever in Germany in the 1990s. *Veterinary Microbiology*, 77, 29–41.

Fuller, R. and Gough, S. (1999). Changes in sheep numbers in Britain: implications for bird populations. *Biological Conservation*, 91, 73–89.

Gill, R. and Fuller, R.J. (2007). The effects of deer browsing on woodland structure and songbirds in lowland Britain. *Ibis*, 149, 119–127.

Gill, R. and Morgan, G. (2010). The effects of varying deer density on natural regeneration in woodlands in lowland Britain. *Forestry*, 83, 53–63.

Goulding, M. (2014). <http://www.britishwildboar.org.uk>. Accessed June 2014

Goulding, M.J., Smith, G., and Baker, S. (1998). Current Status and Potential Impact of Wild Boar (*Sus scrofa*) in the English Countryside: A Risk Assessment. Conservation Management Division C, MAFF.

Green, R.E., Cornell, S.J., Scharlemann, J.r.P., and Balmford, A. (2005). Farming and the fate of wild nature. *Science*, 307, 550–555.

Halley, D. and Rosell, F. (2002). The beaver's reconquest of Eurasia: status, population development and management of a conservation success. *Mammal Review*, 32(3), 153–178.

Hambler, C., Henderson, P.A., and Speight, M.R. (2011). Extinction rates, extinction-prone habitats, and indicator groups in Britain and at larger scales. *Biological Conservation*, 144, 713–721.

Harkonen, S. (1999). Forest damage caused by the Canadian beaver (*Castor canadensis*) in South Savo, Finland. *Silva Fennica*, 33, 247–259.

Harrington, L.A., Feber, R., and Macdonald, D.W. (2012). The Scottish Beaver Trial: Ecological monitoring of the European beaver *Castor fiber* and other riparian mammals—Second Annual Report 2011. Scottish Natural Heritage Commissioned Report No. 510.

Harrington, L.A., Feber, R. and Macdonald, D.W. (2013). The Scottish Beaver Trial: Ecological monitoring of the European beaver *Castor fiber* and other riparian mammals—Third Annual Report 2012. Scottish Natural Heritage Commissioned Report No. 553.

Hayward, M.W. and Kerley, G.I.H. (2009). Fencing for conservation: Restriction of evolutionary potential or a riposte to threatening processes? *Biological Conservation*, **142**, 1–13.

Hetherington DA (2005). The feasibility of reintroducing the Eurasian lynx, Lynx lynx to Scotland. D. Phil Thesis (University of Aberdeen).

Hetherington, D.A., Lord, T.C., and Jacobi, R.M. (2006). New evidence for the occurrence of Eurasian lynx (*Lynx lynx*) in medieval Britain. *Journal of Quaternary Science*, **21**(1), 3–8.

Hetherington, D.A., Miller, D.R., Macleod, C.D., and Gorman, M.L. (2008). A potential habitat network for the Eurasian lynx *Lynx lynx* in Scotland. *Mammal Review*, **38**(4), 285–303.

Hilborn, R., Arcese, P., Borner, M., Hando, J., Hopcraft, G., Loibooki, M., Mduma, S., and Sinclair, A.R. (2006). Effective enforcement in a conservation area. *Science*, **314**, 1266–1266.

Hobbs, R. (2009). Woodland restoration in Scotland: Ecology, history, culture, economics, politics and change. *Journal of Environmental Management*, **90**, 2857–2865.

Hobbs, R.J. and Huenneke, L.F. (1992). Disturbance, Diversity, and Invasion: Implications for Conservation. *Conservation Biology*, **6**, 324–337.

Hodder, K. H., Buckland, P.C., Kirby, K.K., and Bullock, J.M. (2009). Can the mid-Holocene provide suitable models for rewilding the landscape in Britain? *British Wildlife*, **20**, 4–1.

IPCC (2007). Climate Change 2007: Working Group II: Impacts, Adaptation and Vulnerability. Cambridge University Press.

IUCN and UNEP-WCMC (2011). The World Database on Protected Areas (WDPA). In UNEP-WCMC, editor., Cambridge, UK.

Jack, B. K., Kousky, C., and Sims, K.R.E. (2008). Designing payments for ecosystem services: Lessons from previous experience with incentive-based mechanisms. *Proceedings of the National Academy of Sciences*, **105**, 9465–9470.

Jones, S., Gow, D., Lloyd Jones, A., Campbell-Palmer, R. (2013). The battle for British beavers. *British Wildlife*, **24**, 381–392.

Jonker, S.A., Muth, R.M., Organ, J.F., Zwick, R.R., and Siemer, W.F. (2006). Experiences with beaver damage and attitudes of Massachusetts residents toward beaver. *Wildlife Society Bulletin*, **34**, 1009–1021.

Kass, G., Shaw, R., Tew, T., and Macdonald, D. (2011). Securing the future of the natural environment: using scenarios to anticipate challenges to biodiversity, landscapes and public engagement with nature. *Journal of Applied Ecology*, **48**(6), 1518–1526.

Keddy, P. and Weiher, E. (2001). Introduction: The scope and goals of research on assembly rules. pp. 1–22 In E. Weiher and P. Keddy, eds., *Ecological Assembly Rules: Perspectives, Advances, Retreats*. Cambridge University Press, Cambridge.

Kitchener, A. and Conroy, J. (1997). The history of the Eurasian beaver *Castor fiber* in Scotland. *Mammal Review*, **27**, 95–108.

Kotulski, Y. and König, A. (2008). Conflicts, crises and challenges: wild boar in the Berlin City-a social empirical and statistical survey. *Natura Croatica*, **17**, 233–246.

Lawton, J.H., Brotherton, P.N.M., Brown, V.K., et al. (2010). Making Space for Nature: a review of England's wildlife sites and ecological network. Report to Defra. <http://archive.defra.gov.uk/environment/biodiversity/documents/201009space-for-nature.pdf>.

Leaper, R., Massei, G., Gorman, M.L., and Aspinall, R. (1999). The feasibility of reintroducing Wild Boar (*Sus scrofa*) to Scotland. *Mammal Review*, **29**, 239–259.

Linnell, J., Salvatori, V., and Boitani, L. (2008). Guidelines for population level management plans for large carnivores in Europe. A Large Carnivore Initiative for Europe report prepared for the European Commission.

Luck, G. W., Harrington, R., Harrison, P.A., et al. (2009). Quantifying the Contribution of Organisms to the Provision of Ecosystem Services. *Bioscience*, **59**, 223–235.

Macarthur, E. (2013). <http://www.ellenmacarthurfoundation.org>.

MacArthur, R.H. and Wilson, E.O. (1967). *The Theory of Island Biogeography*. Princeton University Press.

Macdonald, D. (2009). Lessons Learnt and Plans Laid: Seven Awkward Questions for the Future of Reintroductions. In M. Hayward and M. Somers, eds., *Reintroduction of Top-Order Predators*. Wiley-Blackwell, Hoboken, NJ, USA.

Macdonald, D.W., Boitani, L., Dinerstein, E., Fritz, H., and Wrangham, R. (2013). Conserving large mammals: are they a special case? In D.W. Macdonald and K.J. Willis, eds., *Key Topics in Conservation Biology* 2. Wiley Publications.

Macdonald, D. and Burnham, D. (2010). The state of Britain's mammals a focus on invasive species. People's trust for endangered species.

Macdonald, D.W. and Feber, R.E., eds., (2015). *Wildlife Conservation on Farmland. Conflict in the Countryside*. Oxford University Press, Oxford.

Macdonald, D. and Johnson, P. (2000). Farmers and the custody of the countryside: trends in loss and conservation of non-productive habitats 1981–1998. *Biological Conservation*, **94**, 221–234.

Macdonald, D., Mace, G., and Rushton, S. (2000). British mammals: is there a radical future? In A. Entwistle and N. Dunstone, eds., *Priorities for the Conservation of Mammalian Diversity: Has the Panda Had its Day?* Cambridge University Press.

Macdonald, D.W. and Tattersall, F. (2001). *Britain's Mammals: The Challenge for Conservation*. The People's Trust for Endangered Species, Mammals Trust UK.

Macdonald, D. W., Tattersall, F.H., Brown, E.D., and Balharry, D. (1995). Reintroducing the European Beaver to Britain: nostalgic meddling or restoring biodiversity? *Mammal Review*, **25**, 161–200.

Manning, A. D., Gordon, I.J., and Ripple, W.J. (2009). Restoring landscapes of fear with wolves in the Scottish Highlands. *Biological Conservation*, **142**, 2314–2321.

Maroo, S. and Yalden, D. (2000). The Mesolithic mammal fauna of Great Britain. *Mammal Review*, **30**, 243–248.

Mech, L. D. (1966). *The Wolves of Isle Royale*. University of Minnesota.

Mech, D. and Boitani, L. (2003). *Wolves: Behaviour, Ecology & Conservation*. The University of Chicago Press, Chicago.

Miljominsteriet Naturstyrelsen. (2013). Vildtforvaltningsrådet er klar med ulveplan.

Mitchell, F.J.G. (2005). How open were European primeval forests? Hypothesis testing using palaeoecological data. *Journal of Ecology*, **93**, 168–177.

Moehrenschlager, A., List, R., and Macdonald, D.W. (2007). Escaping intraguild predation: Mexican kit foxes survive while coyotes and golden eagles kill Canadian swift foxes. *Journal of Mammalogy*, **88**, 1029–1039.

Moehrenschlager, A. and Macdonald, D.W. (2003). Movement and survival parameters of translocated and resident swift foxes *Vulpes velox*. *Animal Conservation*, **6**, 199–206.

Moore, B.D., Sim, D.A., and Iason, G.R. (2013). The Scottish Beaver Trial: Woodland monitoring 2011. Scottish Natural Heritage Commissioned Report No. 525.

Moran, D. and Hanley-Nickolls, R. (2012). The Scottish Beaver Trial: Socio-economic monitoring—First report 2011. Scottish Natural Heritage Commissioned Report No.482.

Morton, D., Rowland, C., Wood, C., Meek, L., Marston, C., Smith, G., Wadsworth, R., and Simpson, I.C. (2011). Final report for LCM2007—the new UK Land Cover Map. Centre for Ecology & Hydrology (Natural Environmental Research Council)

Müller-Schwarze, D. and Sun, L. (2003). *The Beaver: Natural History of a Wetlands Engineer*. Cornell University Press.

National Ecosystem Service Assessment. (2011a). The UK national ecosystem service assessment. UNEP-WCMC, Cambridge.

National Ecosystem Service Assessment. (2011b). The UK national ecosystem service assessment: Woodlands. UNEP-WCMC, Cambridge.

National Parks UK. (2012). National Parks. <http://www.nationalparks.gov.uk/>

Navarro, L. and Pereira, H. (2012). Rewilding abandoned landscapes in Europe. *Ecosystems*, **15**, 900–912.

Nilsen, E.B., Milner-Gulland, E.J., Schofield, L., Mysterud, A., Stenseth, N.C., and Coulson, T. (2007). Wolf reintroduction to Scotland: public attitudes and consequences for red deer management. *Proceedings of the Royal Society B: Biological Sciences*, **274**, 995–1003.

Odden J, Linnell, J.D., and Andersen, R. (2006). Diet of Eurasian lynx, Lynx lynx, in the boreal forest of southeastern Norway: the relative importance of livestock and hares at low roe deer density. *Eur J Wildl Res* 52(4):237–244.

Packer, C., Loveridge, A. Canney, S., et al. (2013). Conserving large carnivores: dollars and fence. *Ecology Letters*, **16**, 635–641.

Patten, C. (2009). *What Next?: Surviving the Twenty-first Century*. Penguin Adult.

Peterson, R.O. and Page, R.E. (1988). The rise and fall of Isle Royale Wolves, 1975–1986. *Journal of Mammalogy*, **69**, 89–99.

Peterson, R.O., Thomas, N.J., Thurber, J.M., Vucetich, J.A., and Waite, T.A. (1998). Population limitation and the wolves of Isle Royale. *Journal of Mammalogy*, **79**, 828–841.

Pollock, M.M., Naiman, R.J., and Hanley, T.A. (1998). Plant species richness in riparian wetlands-a test of biodiversity theory. *Ecology*, **79**, 94–105.

Power, M. E., Tilman, D., Estes, J.A., Menge, B.A., Bond, W.J., Mills, L.S., Daily, G., Castilla, J.C., Lubchenco, J., and Paine, R.T. (1996). Challenges in the Quest for Keystones. *Bioscience*, **46**, 609–620.

Richardson, P. (2013). Wolf-watching in Spain. Financial Times. March 2013, <http://www.ft.com/cms/s/2/c341b948–947c43–11e2–99f0–00144feabdc0.html#axzz33UPM38By>

Ripple, W.J. and Beschta, R.L. (2012). Trophic cascades in Yellowstone: The first 15 years after wolf reintroduction. *Biological Conservation*, **145**, 205–213.

Robinson, R. (2009). Invasive and problem ferns: a European perspective. *International Urban Ecology*, **4**, 83–91.

Rolauffs, P., Hering, D., and Lohse, S. (2001). Composition, invertebrate community and productivity of a beaver dam in comparison to other stream habitat types. *Hydrobiologia*, **459**, 201–212.

Rosell, F., Bozsér, O., Collen, P., and Parker H. (2005). Ecological impact of beavers *Castor fiber* and *Castor canadensis* and their ability to modify ecosystems. *Mammal Review*, **35**, 248–276.

Rosell, F., Rosef, O., and Parker, H. (2001). Investigations of waterborne pathogens in Eurasian beaver (*Castor fiber*) from Telemark County, southeast Norway. *Acta Veterinaria Scandinavica*, **42**, 479–482.

Sandom, C.J., Bull, J., Canney, S., and Macdonald, D. (2012). Exploring the value of wolves (*Canis lupus*) in landscape-scale fenced reserves for ecological restoration in the Scottish Highlands. pp. 245–276 In M. Somers and M. Hayward, eds., *Fencing for Conservation: Restriction of Evolutionary Potential or a Riposte to Threatening Processes?* Springer, New York.

Sandom, C., Dalby, L., Fløjgaard, C., et al. (2013a). Mammal predator and prey species richness are strongly linked at macroscales. *Ecology*, **94**, 1112–1122.

Sandom, C.J., Donlan, C.J., Svenning, J.C., and Hansen, D.M. (2013b). Rewilding. In D.W. Macdonald and K.J. Willis, eds., *Key Topics in Conservation Biology 2*. Blackwell Publishing, Oxford.

Sandom, C., Ejrnæs, R., Hansen, M.D.D., and Svenning, J.-C. (2014b). High herbivore density associated with vegetation diversity in interglacial ecosystems. *Proceedings of the National Academy of Sciences*, **111**(11), 4162–4167.

Sandom, C., Faurby, S., Sandel, B., and Svenning, J.-C. (2014a) Global late Quaternary megafauna extinctions linked to humans, not climate change. *Proceedings of the Royal Society B: Biological Sciences* 281(1787).

Sandom, C.J., Hughes, J., and Macdonald, D.W. (2013c). Rewilding the Scottish Highlands: do wild boar, *Sus scrofa*, use a suitable foraging strategy to be effective ecosystem engineers? *Restoration Ecology*, **21**, 336–343.

Sandom, C. J., Hughes, J., and Macdonald, D.W. (2013d). Rooting for rewilding: quantifying wild boar's *Sus scrofa* rooting rate in the Scottish Highlands. *Restoration Ecology*, **21**, 329–335.

Sandom, C., Hughes, J., and Macdonald, D.W. (in prep-a). Rewilding with an ecosystem engineer: Can wild boar Sus scrofa promote woodland regeneration and control Bracken *Pteridium aquilinum* in the Scottish Highlands?

Schley, L., Dufrêne, M., Krier, A., and Frantz, A. (2008). Patterns of crop damage by wild boar (*Sus scrofa*) in Luxembourg over a 10-year period. *European Journal of Wildlife Research*, **54**, 589–599.

Schley, L. and Roper, T.J. (2003). Diet of wild boar *Sus scrofa* in Western Europe, with particular reference to consumption of agricultural crops. *Mammal Review*, **33**, 43–56.

SER (2004). The SER International Primer on Ecological Restoration. Society for the Ecological Restoration International.

Sims, N. (2005). The ecological impacts of wild boar rooting in East Sussex. D. Phil.Thesis University of Sussex.

Skewes, O., Gonzalez, F., Olave, R., Ávila, A., Vargas, V., Paulsen, P., and König, H.E. (2006). Abundance and distribution of American beaver, Castor canadensis (Kuhl 1820), in Tierra del Fuego and Navarino islands, Chile. *European Journal of Wildlife Research*, **52**, 292–296.

Smith, D.W., Peterson, R.O., and Houston, D.B. (2003). Yellowstone after wolves. *Bioscience*, **53**, 330–340.

Soulé, M.E. and Noss, R.F. (1998). Rewilding and biodiversity: Complementary goals for continental conservation. *Wild Earth*, 8.

South, A., Rushton, S., Macdonald, D. and Fuller, R. (2001). Reintroduction of the European beaver (*Castor fiber*) to Norfolk, UK: a preliminary modelling analysis. *Journal of Zoology*, **254**, 473–479.

Stahl, P., Vandel, J., Herrenschmidt, V., and Migot, P. (2001a). The effect of removing lynx in reducing attacks on sheep in the French Jura Mountains. *Biological Conservation*, **101**, 15–22.

Stahl, P., Vandel, J.M., Rutte, S., Coat, L., Coat, Y., and Balestra, L. (2002). Factors affecting lynx predation on sheep in the French Jura. *Journal of Applied Ecology*, **39**, 204–216.

Svenning, J.-C. (2002). A review of natural vegetation openness in north-western Europe. *Biological Conservation*, **104**, 133–148.

Sykes, N. (2004). The introduction of fallow deer to Britain: a zooarchaeological perspective. *Environmental Archaeology*, **9**, 75–83.

Tallis, H., Kareiva, P., Marvier, M., and Chang, A. (2008). An ecosystem services framework to support both practical conservation and economic development. *Proceedings of the National Academy of Sciences* **105**, 9457–9464.

Tattersall, F. and Macdonald, D. (1995). A Review of the Direct and Indirect Costs of Re-introducing the European Beaver (*Castor fiber*) to Scotland. Report, Contract No. SNH/110/95 IBB. WildCRU, Oxford.

TEEB (2010). The Economics of Ecosystems and Biodiversity: Mainstreaming the Economics of Nature: A synthesis of the approach, conclusions and recommendations of The Economics of Ecosystems and Biodiversity (TEEB).

Terborgh, J. and Estes, J.A., eds. (2010). *Trophic Cascades: Predators, Prey, and Changing Dynamics of Nature*. Island Press, Washington DC.

Thomsen, L.R., Campbell, R.D., and Rosell, F. (2007). Tool-use in a display behaviour by Eurasian beavers (*Castor fiber*). *Animal Cognition*, **10**, 477–482.

Vera, F.W.M. (2000). *Grazing Ecology and Forest History*. CABI Pub., New York.

Vera, F. (2009). Large-scale nature development—the Oostvaardersplassen. *British Wildlife*, **20**, 28–36.

Vucetich, J. A., Nelson, M.P., and Peterson, R.O. (2012). Should Isle Royale wolves be reintroduced? A case study on wilderness management in a changing world. Pages 126–147 In George Wright Forum.

Vucetich, J.A. and Peterson, R.O. (2004). The influence of top-down, bottom-up and abiotic factors on the moose (*Alces alces*) population of Isle Royale. *Proceedings of the Royal Society of London. Series B: Biological Sciences* **271**, 183–189.

Ward, A.I. (2005). Expanding ranges of wild and feral deer in Great Britain. *Mammal Review*, **35**, 165–173.

Warren, C R. (2009). *Managing Scotland's Environment*. Edinburgh University Press, Edinburgh.

Westbrook, C., Cooper, D., and Baker, B. (2011). Beaver assisted river valley formation. *River Research and Applications*, **27**, 247–256.

Whitehouse, N. J. and Smith, D. (2010). How fragmented was the British Holocene wildwood? Perspectives on the 'Vera' grazing debate from the fossil beetle record. *Quaternary Science Reviews*, **29**, 539–553.

Willby, N.J., Casas Mulet, R. and Perfect, C. (2011).The Scottish Beaver Trial: Monitoring and further baseline survey of the aquatic and semi-aquatic macrophytes of the lochs 2009. Scottish Natural Heritage Commissioned Report No. 455.

Widstrand, S. (2013). Is there really any money in wildlife watching?—a global and European perspective. In Wild10. Rewilding Europe, Salamanca.

Wright, J. P., Jones, C.G., and Flecker, A.S. (2002). An ecosystem engineer, the beaver, increases species richness at the landscape scale. *Oecologia*, **132**, 96–101.

Yalden, D. (1999). *The History of British Mammals*. Academic Press, London.

Yalden, D. and Barrett, P. (1999). *The History of British Mammals*. T & A D Poyser.

Zhou, Y., Newman, C., Chen, J., Xie, Z., and Macdonald, D.W. (2013). Anomalous, extreme weather disrupts obligate seed dispersal mutualism: snow in a subtropical forest ecosystem. *Global Change Biology*, **19**, 2867–2877.

Index

W

water habitats, *see* freshwater habitats
water quality
    in ditches 217–18
    in ponds 199–201
    in rivers 211–12
water shrew (*Neomys fodiens*) 67
water vole (*Arvicola amphibius*)
    diagnostic research 274–80
    dispersal 256, 258, 260, 276, 277, 278
    establishment 260–2
    fragmentation 255, 256, 276, 278, 279
    growth rates 262
    habitat quality 256–7, 261–2, 274–8
    health and welfare 262–4
    housing captive animals 262–4, 284
    legal protection of 284
    leptospirosis 264–5
    metapopulations 278
    mink control 258, 274–7, 279, 281–2, 285, 286, 287
    monitoring 285
    national surveys 272
    policy issues 287
    population decline 256, 270–3
    rabbit populations 279
    radio-collaring effects 283–4
    recovery management 284–6
    reintroduction 255–65, 280–4
    release methods 259–60
    solution research 280–4
    stress 262
    success of reintroduction 259
    suitability for reintroduction 257–8
    survival rates 260–2
    tightrope hypothesis 275–7
    value 12–13

vegetation abundance 256–7, 260–2, 276
way-marking 78
weasels 85
weather, *see* climate change; environmental effects
weed control 22, 32–8
well-being 13–14
white clover (*Trifolium repens*) 26, 27
wild boar (*Sus scrofa*)
    reintroduction 296, 304–7
wildflower seed mixture 22, 28, 33, 35
wildlife tourism 304
willingness to pay for conservation 14
wind, bat activity 182, 184
winter stubble 95, 96
wolf (*Canis lupus*) reintroduction 297, 300–4
wolf spiders (lycosids) 46, 121, 122, 126
woodland
    biodiversity loss 295
    connectivity 235, 237
    coppicing 232, 233
    deer-grazing effects 226–31, 296
    dispersal 224, 225, 231, 234, 237
    ecosystem services 295
    farmscape management 237
    fragmentation 224, 225, 234–7
    landscape scale 224–5, 234–7
    local scale 226–33
    management 224–37
    moths 224–5, 231–7
    Odonata populations 204
    regeneration 67, 295, 304–5
    ride widening 232
    river health 214–16

small mammals 67, 224, 225, 226–31
squirrels 225
trees as stepping stones 149, 155, 234, 235
wood mouse (*Apodemus sylvaticus*) 9, 65
    behavioural ecology 68–70
    combine harvesting 76–7
    cropped areas for 75, 78–9
    deer-grazing effects 226, 228–30
    ecology 66–7
    food selection 76, 77
    habitat use 72–3
    hedgerows 75, 86
    Heligmosomoides polygyrus infection 69, 74
    inter-specific competition 87
    molluscicide effects 78
    mowing effects 82
    population 66
    population dynamics 68–70
    population trends 72
    reproductive behaviour 70–1
    set-aside 82, 84
    way-marking 78
Wytham Woods 226–8, 230, 231, 234

Y

yellowhammer (*Emberiza citrinella*) 95, 96
yellow-necked mouse (*Apodemus flavicollis*) 67, 86
Yorkshire fog (*Holcus lanatus*) 36

Z

Zygoptera, *see* Odonata